AQUACULTURE RESEARCH TRENDS

AQUACULTURE RESEARCH TRENDS

STEPHEN H. SCHWARTZ
EDITOR

Nova Science Publishers, Inc.
New York

For permission to use material from this book please contact us:
Telephone 631-231-7269; Fax 631-231-8175
Web Site: http://www.novapublishers.com

LIBRARY OF CONGRESS CATALOGING-IN-PUBLICATION DATA

Aquaculture research trends / Stephen H. Schwartz (editor).
p. cm.
ISBN 978-1-60456-217-0 (hardcover)
1. Aquaculture--Research. I. Schwartz, Stephen H.
SH135.A78 2007
639.8--dc22

2007047831

Published by Nova Science Publishers, Inc. ✢ New York

CONTENTS

Preface vii

Expert Commentary A: Tropical Spiny Lobsters Aquaculture: How Far From Success? (Prospect for the Caribbean) 1
Erick Perera

Expert Commentary B: Diversity and Functionality of Gastrointestinal Microbiota in Farmed Fish: A Commentary on what We Have Learned and Future Directions 7
F.J. Gatesoupe

Chapter 1 Progress on Fish Disease Research in China 13
Xianle Yang and Haipeng Cao

Chapter 2 Estimating Parentage Relationships Using Molecular Markers in Aquaculture 59
Paulino Martínez and Jesús Fernández

Chapter 3 Larval Fish Nutrition and Rearing Technologies: State of the Art and Future 113
Jian G. Qin

Chapter 4 Role of Bacteria in the Chilled Storage and Cryopreservation of Sperm in Aquatic Animals: A Review 149
Subuntith Nimrat and Verapong Vuthiphandchai

Chapter 5 Growth, Gametogenic Cycle and Seasonal Changes in Biochemical Composition of the Pacific Oyster Crassostrea Gigas in Relation to the Environmental Conditions of the Bizert Lagoon, Tunisia 185
Salwa Dridi,Mohamed Salah Romdhane and M'hamed Elcafsi

Chapter 6 New Vaccination Strategies for the Improvement of DNA Vaccines Against Virus in Fish 213
Carolina Tafalla, Amparo Estepa and Julio Coll

Chapter 7	*Diplodus Puntazzo*, A New Species in Aquaculture: Biology and Culture *María Dolores Hernández, Francisco Javier Martínez, and Benjamín García García*	**237**
Chapter 8	Recent Advances in Channel Catfish Nutrition and Feeding Research *Menghe H. Li and Edwin H. Robinson*	**263**
Chapter 9	The Underwater Silviculture Approach For Reef Restoration: An Emergent Aquaculture Theme *Shai Shafir and Baruch Rinkevich*	**279**
Chapter 10	Emerging Trends of Aquaculture Research in the Arabian Gulf Region *S. M. Almatar and C.M. James*	**297**
Index		**321**

PREFACE

Aquaculture is the cultivation of aquatic organisms. Unlike fishing, aquaculture, also known as aquafarming, implies the cultivation of aquatic populations under controlled conditions. Mariculture refers to aquaculture practiced in marine environments. Particular kinds of aquaculture include algaculture (the production of kelp/seaweed and other algae); fish farming; shrimp farming, shellfish farming, and the growing of cultured pearls. This new book presents the latest research in the field.

Expert Commentary A – World fisheries landings have been declining slowly since the late 1980s, by about 0.7 millions tones per year (Pauly et al., 2002). Spiny lobsters support some of these fisheries all around the world being the most important marine resource for many countries. As for other aquatic animals, the idea of meeting growing demand for spiny lobsters through aquaculture is actually old but it is now enhanced by the fact that most fisheries are exploited at or above sustainable levels. Spiny lobsters hold excellent market price and many biological features that make them suitable candidates for aquaculture (Booth and Kittaka, 2000) but other ones so far impair a world-wide expansion of such an industry such as extended larval period and low survival rates through larval culture. After many years of research, some important progress has been made on larval culture of different species, mostly in Japan, Australia and New Zealand. However, rearing larva is still very difficult even at experimental scale for most species preventing the development of lobster aquaculture in a conventional way, at least in the nearest future. The amount of studies on larval culture presented at workshops, books and per-reviewed journals continue to increase through time indicating research efforts persist. Culture of P. argus throughout all larval stage was recently accomplished in Florida but survivals to low.

Expert Commentary B – Based on the knowledge acquired from land animals, the contribution of intestinal microbiota to fish nutrition was hypothesized, in view of the ability of some bacterial strains to produce either nutrients or enzymes, possibly active in the digestive process of the host. These early studies were reviewed by Ringø et al. (1995), for instance. However, these findings were moderated due to the relatively low level of culturable bacteria that were generally recovered from the gut of fish farmed in cold and temperate waters, in comparison with homeotherms, and with some warm water herbivorous fishes, in which intense intestinal fermentation was detected (Clements, 1997). The threshold for contributing significantly to the catabolic process of digestion was estimated to 10^7 bacterial bodies per g of digestive tract by Lésel (1993). This author wondered about possible anabolic function by releasing micronutrients, even at low population density. In most aquacultured species, the nutritional role of gut microbiota was generally considered as secondary, with few exceptions. That may change with what we have learned from new studies applying molecular methods. It is now admitted that culture-independent methods are required to account for the real complexity of gastrointestinal biota in fish (Pond et al., 2006). The proportion of culturable bacteria was estimated between 10 and 50% in human colonic biota (Zoetendal et al., 1998), and the same range was generally found in rainbow trout intestine by Huber et al. (2004), but this rate was only 2% in one fish. Such extremely low culturability was also noted in samples from Atlantic and coho salmons (Holben et al., 2002; Romero and Navarette, 2006).

Chapter 1 – China, the biggest fish farming country, is the cradle of the world's aquaculture, and the records of fish disease and its control could be tracked back to thousands of years ago in China. However, the development of fish farming industry and fish disease research in China was seriously hindered because of the social and economic limits. After People's Republic of China was founded, fish disease research in China was started and rapidly developed, and up to now, great progress was achieved in epidemiology, etiology, pathology, immunology, pharmacokinetics of fishery drugs and fish disease control. While in contrast to the current status of fish diesease in foreign developed countries, there are some problems about the detection assays and pathogenic mechanism of pathogens, fishery vaccine development and fish disease control in fish disease research in China. Therefore, some measures should be taken to promote fish disease research in China, maily reflected in the following aspects: 1. The commercialization process of rapid diagnosis means should be accelerated; 2. Study on the pathogenic mechanism centered on functional genes (or proteins) should be deepened; 3. Fish pathology and its study means should be further strengthened; 4. A simple, correct and effective evaluation method of pharmacodynamics on fishery drugs should be established; 5. New ways to effectively prevent and control fish diseases should be explored.

Chapter 2 – The inference of parentage relationships between individuals from their molecular resemblance represents a useful methodology for genetic improvement in Aquaculture. This makes possible the reorganization of broodstock in groups with low relatedness, as well as the identification of families in selection programs to avoid the harmful effects of inbreeding and to maintain the highest diversity as possible across generations. The choice of the appropriate genetic marker and the statistical methodology is essential to get the best solution for the questions considered. Depending on the availability of previous information on the genealogy, paternity or kinship analysis will be applied. The simplest approach to paternity inference, involves parent identification through the exclusion of the remaining candidates. However, this can lead to more than one solution, solvable after application of maximum likelihood procedures. The ability to identify the parents of each offspring depends on the sampling scenario (number of candidate parents and the fraction sampled), on the potential of the markers used (polymorphism, genotyping errors), and on the conformance to theoretical assumptions of the statistical models applied. A large number of paternity programs are freely available. They display usually complementary performances, and there is not the best software for all situations.

When we deal with a single group of individuals belonging to the same generation or that can not be separated into known generations, the aim is just to estimate the degree of genetic relationship between them, usually expressed as the coancestry coefficient. The basic idea is to determine how much of the molecular similarity (Identity By State) is due to the Identity

By Descent (the really important parameter). There are two groups of relationship estimators from the molecular information. One of them includes methods directed at estimating coancestry for only a pair of individuals at a time, usually relaying on the knowledge of the allele frequencies of the reference population. Estimators within this group can be further divided into those called Method of Moments Estimators (MME) and those based on Maximum Likelihood (MLE). The other group of methods uses jointly the information of all individuals to determine the more probable population/familiar structure. They perform an explicit reconstruction of the genealogy leading to the observed population (at least for one generation above). Depending on the type of estimator and the assumptions used in their development, each of them presents some advantages and/or limitations which should be taken into account when choosing the one to use. In the present study, the most relevant estimators are presented and, also, a list of free software available for their application.

Chapter 3 – Marine finfish larviculture techniques have been quickly developed in the past two decades. Although hatchery production technology still follows similar protocols worldwide, feeding regimes and culture systems have been greatly modified to improve hatchery performance and efficiency. Marine fish larvae are reared more often in green water intensive systems than in clear water systems, while extensive culture is less commonly used. The use of live food organisms, especially at first feeding, is still obligatory in most marine fish larvae mass-culture. Rotifers Brachionus spp. and brine shrimp Artemia nauplii are the main live food items for early life stages of the major finfish species, with inert feed provided at a later stage. Super-intensive rotifer culture using algal paste has improved cost-effectiveness of hatchery production, whereas Artemia nauplii have been used as biocapsules to deliver vitamins and a variety of nutrition (e.g., EPA, DAH and ARA) from different sources. Recently, the use of copepods and cladocerans as live food has been explored. However, low production of copepods is the major hurdle of mass production, but there is a growing interest in using saline water cladocerans as live food in hatchery. Until recently, it is considered impossible to feed most newly hatched marine fish species with a compound diet. Substituting a compound diet for live prey is performed several weeks after hatching, depending on the species. Ontogenetic studies suggest that most digestive enzymes are present in young larvae, but the quantity is not sufficient to digest formulated feed. Therefore, it is essential to provide live feed for the first feeding of marine finfish species. Weaning of fish larvae to inert feed at a later stage of development is easily achieved by co-feeding inert feed with live food because live food organisms consumed by the larvae assist the digestion process by donating their digestive enzymes to the gut of fish larvae. In larviculture rearing, high mortalities are likely to occur during the diet transition from endogenous to exogenous nutrition and the period of weaning, and cannibalism could contribute greatly to mortality when great size heterogeneity exists. Reduction of larval deformity is a challenge in some fish species and further understanding of the role of nutrition in fish development, especially at the molecular level will allow improving the quality of hatchery reared larvae.

Chapter 4 – Bacterial contamination is considered a major issue for the management of sperm preservation in aquatic animals. Chilled storage and cryopreservation of aquatic animal sperm have been used to meet the growing demand for seeds. Chilled storage of sperm has been effectively used in a number of fish, but in marine invertebrates has been limited to shrimp, lobster, echiura and shellfish. Successful and reproducible cryopreservation has proven to be a reliable method for preserving genetic lines of fish for aquaculture and conservation, but among invertebrates is limited to polycheates, cnidaria, crustaceans (shrimp and crab), shellfish (oysters, hard clams and abalone) and echinoderms (sea urchin and sea cucumber). Few studies have documented changes in bacterial type and number during chilled storage or cryostorage. Access of good quality chilled-stored or cryopreserved sperm is necessary for the management of risk associated with sperm preservation. The presence of bacteria in sperm of aquatic animals can result in deleterious effects on sperm quality and viability if left uncontrolled. Potential pathogens present on the preserved sperm of aquatic animals may affect the health of broodstock and offspring. This paper summarizes the current knowledge pertaining to bacterial contamination of aquatic animal sperm, including bacterial types and sources of contaminants, and gives an overview of the incidence of bacterial contamination, biosecurity and risk assessment and effects of bacterial contamination on offspring quality. These ideas, if taken into account in commercial aquaculture and conservation of aquatic animals, are almost certain to increase the value added or protection of at least some cultured species. Microbial contamination risk of chilled-stored or cryopreserved sperm after short-term or long-term storage is discussed. Future direction and new avenues of future research on chilled storage and cryopreservation of sperm in aquatic animals are proposed.

Chapter 5 - The relationship between growth, gametogenic cycle and the seasonal variations in biochemical composition in the gonad-digestive gland of the bivalve Crassostrea gigas was studied from January 2002 until January 2003. Changes in the shell growth and the body weight are associated with metabolite accumulation in the complex gonad – digestive gland, storage utilization of reserves, gametogenic events and environmental parameters, especially food availability. This study reports the first evidence of reproduction of Crassostrea gigas in the Bizert lagoon of Tunisia. Initiation of gametogenesis took place in March when water temperature reached 14 °C. Ripe oysters were observed in May at a water temperature of 20 °C. Spawning began in June and reached a peak in September when water temperature ranged between 23 °C and 27 °C. During the latter period, oysters presented gonads with ruptured follicles and residual gametes. The study found that Crassostrea gigas showed a clear cycle of energy storage and utilization in the Bizert lagoon. Total biochemical composition was about 6.67 % ± 1.34 of dry weight in lipids, 6.50 % ± 3.01 of dry weight in glycogen and 14.34 % ± 3.60 of dry weight in protein and. Lipid and glycogen concentrations in the gonad – digestive gland were inversely related: maximum concentrations of glycogen occurred between December and February and corresponded with minimum concentrations of lipids. High glycogen contents recorded in winter supported the processes of gametogenesis and then decreased progressively to attain low values at maturity and during summer corresponded to the first emission of gametes (June) and the beginning of total spawning, however lipids are accumulated in gonads during the period of maximum ripeness (spring) and decrease in summer. During autumn, the increase of total fatty acids values appeared to be related to the available food in association with an increase in the chlorophyll a concentrations. Protein contents show many variations throughout the year. High protein levels occurred in May and in July which corresponded to oocyte maturation and then decreased in late summer (beginning of total spawning). At the expense of the available food in autumn, protein concentrations increase again to assure recovery of the tissue weight of the bivalve, which induced an increase in the condition indices values. The process of growth was analyzed during the two phases of the gametogenic cycle. Correlation between the different parameters of growth models noted during sexual repose is greater than those determined during the gonadal activity, including ripeness and spawning. The gametogenic cycle influence clearly the lengthwise and weight modals of growth by affecting the bivalve with physiologic troubles traduced by win or loose of the organic matter. During the sexual repose, fatty acids are accumulated in the C. gigas tissues (total SFA, MUFA and PUFA (n-3) show the highest values (respectively 26.38 % ± 1.78; 15.84 % ± 0.56; 46.94 % ± 0.61 of the total fatty acids). The glycogen levels are also attempt the highest during the sexual repose (11,52 % ± 4,64 of the tissue dry weight). These biochemical components of the C. gigas tissues contribute in the increases of shell length, width and thickness respectively (2,04 ± 0,18 cm; 0,58 ± 0,08 cm and 0,90 ± 0,06 cm from January 02 to February 02 and 1,25 ± 0,33 cm; 0,65 ± 0,11 cm and 0,28 ± 0,07 cm from November 02 to January 03) during this phase of the gametogenic cycle. However during the gametogenic activity (March 02 – September 02) total SFA, MUFA and PUFA (n-3) show the lowest values (respectively 23.13 % ± 1.38; 12.35 % ± 0.55; 38.15 % ± 0.84 of the total fatty acids) also the glycogen levels decrease and attempt the minimum in September (2.09 % ± 1.98 of the tissue dry weight) which increase the shell size by only 1,12 ± 0,08 cm, 0,64 ± 0,80 cm and 0,57 ± 0,12 cm in length, width and thickness. Protein levels are comprised between 9,5 % ± 2,19 and 21,37 % ± 1,50 of the total tissues dry weight. At the expense of the available food in autumn, protein concentrations increase to assure recovery of the tissue weight of the bivalve. This metabolite has not any clear correlations with the shell growth but contribute in supporting the bivalve essentially during summer when lipids and glycogen reserves are depleted.

Chapter 6 – Vaccination seems like the most adequate method to control viral diseases in aquacultured fish. Up to date, very few viral vaccines are commercialised, since it has been a difficult task to obtain effective vaccines in fish that are both safe and cost-effective. In the past years, DNA vaccination has been proven as a very effective method to control some rhabdoviral infections, such as those caused by infectious hematopoietic necrosis virus (IHNV) and viral hemorrhagic septicemia virus (VHSV), responsible for great losses in aquaculture production. These vaccines consisting in a plasmid in which the viral glycoprotein gene is expressed under the control of the citomegalovirus (CMV) promoter, are able to confer protection in fish after intramuscular injection. Even though the protection conferred is high and long-lasting, these vaccines are still not being used worldwide due to safety issues concerning the viral promoter, to difficulties in the administration of the vaccine at a large scale, and to high vaccination costs. Up to date, only an IHNV DNA vaccine has been licensed for use in Canada.

Moreover, the immune mechanism through which they confer protection is still unknown, and this is a major restriction for the optimisation of DNA vaccines for other viral pathogens also of great importance in fish, such as for example nodavirus, for which DNA vaccines as formulated for rhabdovirus are not effective.

In this chapter, we discuss the trends in research dealing with DNA vaccination in fish, that will hopefully allow the improvement and finally the routine use of these vaccines in aquaculture. These will include the possible use of molecular adjuvants; the use of attenuated mutant glycoprotein genes and studies dealing with conformation versus lineal epitopes; the substitution of the CMV promoter for promoters of a non-viral origin; the use of induced plasmids or fish transposons, the study of new mass vaccination methods, or the inclusion of specific siRNA sequences within the plasmid. As well, we will review the most recent advances in studies related to the immune response towards these vaccines, some of them based on microarray technologies, that may help elucidate the immune mechanisms responsible for protection.

Chapter 7 – Diversifying the supply of cultured fish would contribute to stabilizing market prices and making aquacultural facilities profitable, while offering consumers a larger variety of cultured fish at more competitive prices than those available from extractive fishing. As a result, the quest for new fish species suitable for farming is becoming a priority in aquacultural research.

Sharpsnout seabream is a sparid fish that has recently drawn the attention of researchers and farmers alike, as it has many qualities that would make it a promising new species for culture. Pilot pre-growout and growout studies have been conducted, under both intensive and extensive culture conditions, with quite encouraging results. These results are very similar to those obtained for gilthead seabream in terms of both growth and conversion rates. In addition, sharpsnout seabream is an omnivorous species that has yielded very positive results in pilot studies designed to test the possibility of replacing fish meal for soybean meal (with replacement levels of up to 60%) as a source of dietary protein. Therefore, one may expect sharpsnout seabream commercial diets to be more economical than those for gilthead seabream, European sea bass, or other carnivorous species.

A number of commercial issues must be addressed before considering the large-scale farming of sharpsnout seabream. Consumers are not very familiar with sharpsnout seabream, as the specimens caught by fishermen tend to be locally consumed in coastal Mediterranean towns.

Chapter 8 – Channel catfish Ictalurus punctatus is the leading aquaculture species in the United States with an annual production of approximately 275,000 metric tons. Extensive research has been conducted on the nutrition and feeding of this species and as a result, its nutrient requirements and feeding characteristics are well documented. Findings from laboratory and field studies conducted over the past few years have made a significant contribution to the existing channel catfish nutrition database, which is routinely used in the formulation of cost-effective feeds. Recent research provides a better understanding of the relationship between dietary protein and digestible energy and their effects on fish growth, processed yield, and body composition. Research has also shown that traditional feed ingredients can be replaced with less expensive alternative feedstuffs to reduce feed cost. For example, marine fish meal can be reduced or replaced with other animal protein sources, such as poultry by-product meal and meat and bone/blood meal. Soybean meal can be partially replaced with a combination of distillers dried grains with solubles, cottonseed meal, and supplemental lysine. Inorganic phosphorus supplements can be completely replaced with phytase enzymes without affecting fish growth and bone mineralization. While channel catfish are more tolerant to aflatoxins and deoxynivalenol, they appear to be as sensitive to T-2 toxin as rainbow trout. Channel catfish are sensitive to ochratoxin A in their diet. Various feeding strategies have been evaluated to optimize production efficiency under various culture conditions. Feeding channel catfish in food fish production ponds once daily to apparent satiation (without adversely affecting water quality) supports maximum growth. However, depending on water temperature and water quality and on the health of the fish, it may be prudent to restrict the daily feed allowance or to feed less frequently than daily.

Chapter 9 – Coral reefs, often portrayed as "the rain forest of the sea", provide sustenance to hundreds of million of people worldwide. When severely degraded, reef restoration is expected to uphold the ecosystem biodiversity and retain, at least partially, the original biological and ecological portraits. Such active reef restoration practices are based on aquaculture of coral colonies (mariculture them in underwater nurseries) in as much as silviculture depends on rearing young trees in terrestrial nurseries. The management and cost efficiency of the nursery phase is crucial to the success of the whole restoration activities. However, contrast to terrestrial nurseries, coral nurseries are in the developing concepts and early operational regulation status. Here, we illustrate the practice of 'reef gardening' as a major reef restoration act, using common features, divergent approaches and ubiquitous rationale, as apply in terrestrial (forests) and marine (coral reefs) restoration principles. Even though basic protocols for reef gardening have not yet been thoroughly studied, several common agricultural aspects for silviculture and reef gardening are outlined. These include the use of reef restoration as mariculture theme, economic considerations and aquaculture perspectives such as dimension of nursery and its content, agricultural periods, growth rates, seedling spacing, maintenance issues, pest control, ecological goods and services and nursery construction. As restoration measures (terrestrial and marine alike) require significant amount of restoration material, simplifying coral seedling preparation and mariculturing them through low-tech technologies will ensure the success of this scientific discipline.

Chapter 10 – The Arabian Gulf Region is bordered by eight countries including Bahrain, Iran, Iraq, Kuwait, Oman, Qatar, Saudi Arabia and United Arab Emirates (UAE). All these countries have a combined coastline of about 23,000 km in the Arabian Gulf (also known as Persian Gulf) and in the Gulf of Oman, in addition to 500 km of Caspian Sea coastline bordering Iran and about 1840 km of Red Sea coast bordering Saudi Arabia. Except Oman, all other Gulf countries consume more seafood than their availability from local waters. Furthermore, wild capture fisheries in Kuwait, Iran and UAE are declining and it is anticipated that the gap between the fresh fish supply and demand will widen due to the declining natural fish stocks. To fill up the gap, several research organizations in Arabian Gulf countries are involved in intensive research programs to develop aquaculture.

Although the major focus of research is on developing marine species for aquaculture, Iran, Iraq and Saudi Arabia have also inland fishery research programs. Among Gulf countries, Kuwait was the first country to achieve captive spawning and hatchery seed production for local grouper and sobaity seabream as well as acclimation of Tilapia in seawater for commercial production. Species selection for aquaculture depends on the local demand of each of these countries and based on this research is focused on both endemic as well as exotic species. The research outcome has enabled the establishment of commercial marine fish farming using sea-cages, marine shrimp farming using earthen ponds and commercial tilapia farming as well as integrated tilapia culture along with agricultural crops.

As a research outcome, Bahrain is producing commercial quantities of fish fingerlings in their hatchery for sea-cage farming. Research efforts have enabled Iranian total aquaculture production to increase from 4,935 t in 1978 to 124,000 t in 2004. Although aquaculture production in Iraq declined from 3,622 t in 1980 to 1,600 t in 1990, it is steadily increasing and in 2005 the total production reached 12,870 t. Kuwait was the first country to establish commercial sea-cage fish farming in 1985 and its production reached up to 400 t during 2000. In Oman the total production through aquaculture was 514 t during 2004.Aquaculture is not well established in Qatar and produced 11 t during 2005. Marine shrimp farming is the major aquaculture activity in Saudi Arabia and has a targeted annual production of 13,500 t during Phase-I and 17,500 t in Phase-II. Marine fish production in UAE has increased from 570 t in 2004 to 900 t in 2006/07 period.

Each country has its own constraints and priority issues for developing aquaculture. One of the major challenges is the non-availability of fish fingerlings from hatchery for sea-cage farming, especially with reference to local prime species. Furthermore, the low domestic market price is compelling to search for export oriented production involving exotic species. In some region, shrimp farming is restricted by adverse environmental conditions. Aquaculture is expected to have a significant growth in the Arabian Gulf countries with an overall trend in species diversification as well as development of intensive culture systems.

In: Aquaculture Research Trends
Editor: Stephen H. Schwartz, pp.1-5
ISBN 978-1-60456-217-0

Expert Commentary A

TROPICAL SPINY LOBSTERS AQUACULTURE: HOW FAR FROM SUCCESS? (*PROSPECT FOR THE CARIBBEAN*)

Erick Perera
Center for Marine Research, University of Havana, Cuba

Word fisheries landings have been declining slowly since the late 1980s, by about 0.7 millions tones per year (Pauly *et al*., 2002). Spiny lobsters support some of these fisheries all around the world being the most important marine resource for many countries. As for other aquatic animals, the idea of meeting growing demand for spiny lobsters through aquaculture is actually old but it is now enhanced by the fact that most fisheries are exploited at or above sustainable levels. Spiny lobsters hold excellent market price and many biological features that make them suitable candidates for aquaculture (Booth and Kittaka, 2000) but other ones so far impair a world-wide expansion of such an industry such as extended larval period and low survival rates through larval culture. After many years of research, some important progress has been made on larval culture of different species, mostly in Japan, Australia and New Zealand. However, rearing larva is still very difficult even at experimental scale for most species preventing the development of lobster aquaculture in a conventional way, at least in the nearest future. Culture of *P. argus* throughout all larval stage was recently accomplished in Florida but survivals to low.

Given that to date there is no technology for providing sufficient spiny lobster seed from hatcheries, commercial interest is now focused on the growout of wild caught post-larvae to market size. This phenomenon is not surprising since many costal aquaculture productions still rely on wild-caught seed due to the lack of propagation techniques or because the scale of wild collection often is too superior to hatchery production (see Hair *et al*., 2002). Although land-based systems, whatever flow-through or recirculated, have been successfully used to hold spiny lobsters, less capital intensive sea cages or sea enclosures have proven to be more profitable specially in tropical areas with high sea temperature all year round. Tropical lobsters (e.g. *Panulirus ornatus* in SE Asia and *Panulirus argus* in the Caribbean) have shown the highest potential for aquaculture due to high growth rates (Jeffs and Davis, 2003).

At present, the commercial growout of tropical spiny lobsters in Vietnam with fresh food have shown to be profitable (Thuy, 2004), supporting now 100 US millions/year industry. However, there is evidence of detrimental effects on coastal environments thus proving to be an unsustainable venture based on current practices. This is partially due to management strategies that skip the many lacks in knowledge about nutritional and culture requirements of tropical lobsters. The diminishing supply and increasing cost of by-catch are also of concern. Combining fresh and shrimp pellet have produced encouraged results for some species and could be a short-term strategy to mitigate pollution by the partial replacement of fresh food.

The development of cost-effective pellets as an environmentally more sustainable alternative to fresh food has been recognised as the main obstacle for sustainability. Despite impressive advances during the last decade have been obtained (see Williams, 2006, 2007, for reviews), information on the nutritional requirements for spiny lobsters is still insufficient. Unfortunatelly, little information is available on digestibility of feed ingredients and results, in some cases, are contradictories (e. g. squid meal). Temperate lobsters historically have received more research attention and results are not always pertinent for tropical species. Intense research is now ongoing in Australia to improve feed technology for tropical *P. ornatus.* Studies in Florida and the Caribbean are providing information on culture requirements for *P. argus* at different scales including diet development. We have demonstrated that gastropod, pelecypoda and crustacean provide the most efficient use of dietary proteins (Díaz-Iglesias *et al*., 2002). None of these items are, however, available in the amounts needed for sustaining large scale growout of *P. argus*. Preliminary studies based on local fish meal dry pellets did not yield satisfactory growth, but we further obtained encouraging results when partially replace local fish meal by squid meal or high quality fish meal (Perera *et a*l. 2005). Digestive enzymes of *P. argus* are also under investigation in our laboratory.

In addition, pellets tested are much less attractive for *P. argus* than fresh food (Perera and Díaz-Iglesias, 2004) and similar results have been regularly obtained for other lobsters. The tropical *P. argus* has been an organism of choice in several neurobiological studies of the chemical sense and as a result, much is known about the sensory basis of their feeding behaviour (see Derby *et al*., 2001). The use of this information to increase feed intake is an exciting area for research in terms of both increase growth rates and reduce pollution but very little has been really done to date. A study on *P. ornatus* has suggested that including protein hydrolysates and free amino acid-rich ingredients in diet might prolong the attractiveness of pelleted feeds (Williams *et al*., 2005).

Once nutritional inconveniences are overcome for particular specie, further expansion of lobster aquaculture is being limited only by the availability of wild seed. The case of *P. argus* appears complex since it has among spiny lobsters the widest distribution range from North Carolina to Brazil as well as the Caribbean (Lellis, 1991). In Florida and the Caribbean (e.g. Cuba, Antigua, Puerto Rico, Jamaica, Mexico) high amount of post-larvae can be easily collected, but it is unlikely that this can be done in all areas throughout the distribution range of the specie. The impact of post-larvae removals on the subsequent catch was estimated to be slight in Australia (Phillips *et al*., 2003) but such an assessment is yet to be produced for *P. argus*. There is a great amount of knowledge about *P. argus* ecology including natural mortality rates, so I guess this information is coming soon.

Recruitment process is complex and number of post-larvae collected will often very markedly even at the same location. Therefore, it should be evaluated before any project comprising the removal of recruited post-larvae started. This is of a particular importance for countries where a sustained reduction in catches has been observed in last years due to either environmental factors or overexploitation. Insufficient seed supply could cause this activity to the collapse.

Although restricted gene flow among *P. argus* populations it was suggested in preliminary studies using protein isozyme electrophoresis (Menzies, 1980), more recent studies based on mitochondrial DNA support the hypothesis that *P. argus* is genetically homogeneous throughout the Caribbean (Silberman *et al.*, 1994) but distinct to Brazilian population (Sarver *et al.*, 1998). To my knowledge, there are current attempts to differentiate *P. argus* populations with more variable genetic markers but due to the extended larval stage and complex oceanic currents, I do not think the scenario will change at least for the Caribbean. Although definite studies are necessary, it seems that post-larvae could be moved out from the place of capture to pay off the shortest of seed in related areas. This is now of a minor importance given the scale of present industry but could be relevant in the future.

Two strategies have been proposed for sustainability in post-larvae lobster collection. One is the release into the wild of a percent of artificially grown lobsters for stock enhancement. The other one is the retirement of units of fishery quota to compensate for the removal of post-larvae. The reseeding option would be better for the Caribbean, since avoids the impact on fisherman incomes, especially in low productive areas. Different to clawed lobsters, restocking of spiny lobsters have received little attention. Recent results on survival of released *Jasus edwardsii* are encouraging for the potential of enhancement of wild populations (Mills *et al.*, 2005).

Diseases are not common in wild lobsters and molt death syndrome due to poor nutrition was found to be the major health problem of captive lobsters since early studies. Today, nevertheless, unsuitable culture conditions have lead to massive mortalities in Vietnam probably due to fungal or bacterial infections. A recently identified virus (Shields and Behringer 2004) also cause massive mortalities in cultured *P. argus* (Staine and Dahlgren, 2005). Research in Florida has demonstrated that the effect of this virus can be nearly eliminated by ultraviolet irradiation of sea water but it is obvious that this treatment can not be used to protect sea caged lobsters from infection. We know virtually nothing about the immune system of spiny lobsters. Our group is now providing basic information on *P. argus* immune system (Perdomo-Morales *et al.*, 2007, 2008) and studies go on towards the understanding of main defence mechanisms. The impact of environmental conditions on lobster immunity remains almost unexplored and such knowledge would be pertinent in the future if lobster aquaculture continue to expand.

It is noteworthy that most countries where *P. argus* aquaculture could be developed in a near future belong to the so called developing countries, where very few can afford lobster for dinner. This type of aquaculture has demonstrated to increase economic opportunities of costal communities in Asian countries (Hambrey *et al.*, 2001) as it is ideal for subsistence farming. Moreover, collection and sale of post-larvae to growout enterprises can provide income for sectors of the population otherwise excluded from aquaculture operations (Hair *et al.*, 2002). Looking into the future, I think that encouraging independent small-scale aquaculturist in the Caribbean, rather than large enterprises, to spiny lobster growout, could ensure costal communities to obtain the benefits of this new aquaculture venture. In any case, scaling-up to more intensive forms of production based on fresh feeding and uncontrolled removal of post-larvae can lead to deleterious effects on both environment and society.

Herein, I have provided a brief overview of recent trends and research on spiny lobster aquaculture with emphasis in tropical *P. argus*. Despite most of the above exposed is my personal view, everybody most agree that there are still many constrains to a sustainable industry and that many questions need to be answered so far. Studies are in progress and results under way.

REFERENCES

Booth, J. D. and J. Kittaka. 2000. Spiny lobster growout. Pages 556-585 in: B. F. Phillips and J. Kittaka, (eds.) Spiny Lobsters: Fisheries and Culture. Fishing News Books. Oxford, UK.

Derby, C. D., Steullet, P., Horner, A. J. and Cate, H. S., 2001. The sensory basis of feeding behaviour in the Caribbean spiny lobster, *Panulirus argus*. *Mar. Freshwater Res.*, 52, 1339-50.

Díaz-Iglesias, E., Báez-Hidalgo, M., Perera, E., Fraga, I., 2002. Respuesta metabólica de la alimentación natural y artificial en juveniles de la langosta espinosa *Panulirus argus* (Latreille, 1804). *Hidrobiológica. 12 (2): 101-112*

Hair, C., J. Bell and P. Doherty, 2002. The use of wild-caught juveniles in coastal aquaculture and its application to coral reef. In: R. R. Stickney and J. P. McVey (eds.), Responsible marine aquaculture, 17: 327-354. (CABI Publishing, U. K.).

Hambrey, J., Tuan, L. A. and Thuong, T. K., 2001. Aquaculture and poverty alleviation II. Cage culture in costal waters of Vietnam. *World Aquaculture*, Vol 32, No. 2, 34-36, 38, 66-67 pp.

Jeffs, A., and M. Davis, 2003. A assessment of the aquaculture potential of the Caribbean spiny lobster, *Panulirus argus*. *Proc. Gulf Caribbean Fish. Inst.*, 54: 413-426.

Lellis W., 1991. Spiny Lobster: a mariculture candidate for the Caribbean? *World Aquaculture* 22:60-63.

Menzies, R. A., 1980. Biochemical population genetics and the spiny lobster larval recruitment problem: an up date. *Proc. Gulf Caribb. Fish. Inst.*, 33:230-243.

Mills, D. J., Gardner, C. and Oliver, M., 2005. Survival and movement of naı¨ve juvenile spiny lobsters

Pauly, D; Christensen, V. Guenette, S.; Pitcher, T. J. Sumaila, U. R. Walters, C. J. Watson, R and Zeller, D.,2002. Towards sustainability in world fisheries. *Nature,* Vol 418, August 2002, 689-695 pp.

Perdomo-Morales, R., Montero-Alejo, V. Perera, E. Pardo-Ruiz, Z., Alonso-Jimenez, E. 2007. Phenoloxidase activity in the hemolymph of the spiny lobster *Panulirus argus*. *Fish Shellfish Immunol, 23: 1187-1195.*

Perdomo-Morales, R., Montero-Alejo, V., Perera, E., Pardo-Ruiz, Z., Alonso-Jiménez, E. (2008). Hemocyanin-derived phenoloxidase activity in the spiny lobster *Panulirus argus* (Latreille, 1804). *Biochim. Biophys. Acta* (in press), doi:10.1016/j.bbagen.2008.01.001

Perera, E. and Díaz-Iglesias (2004): Are we developing formulated diet attractive enough for spiny lobsters? *The Lobster Newsletter*, 17(1): 16-19.

Perera, E., Fraga, I., Carrillo, O., Díaz-Iglesias, E., Cruz, R., Báez, M., Galich, G., 2005. Evaluation of practical diets for the Caribbean spiny lobster *Panulirus argus* (Latreille, 1804): effects of protein sources on substrate metabolism and digestive proteases. *Aquaculture 244: 251– 262.*

Phillips, B. F., Melville-Smith, R., Yuk W. Cheng, Y. W., 2003. Estimating the effects of removing *Panulirus cygnus* pueruli on the fishery stock. *Fisheries Research* 65, 89–101.

returned to the wild. *J. Exp. Mar. Biol. Ecol.* 324 (2005) 20– 30.

Sarver, S. K., Silberman, J. D. and Walsh, P. J., 1998. Mitochondrial DNA sequence evidence supporting the recognition of two subspecies or species of the Florida spiny lobster *Panulirus argus*. *J. Crust. Biol.*, 18 (1): 177-186.

Shields J. D. and D. C. Behringer., 2004. A new pathogenic virus in the Caribbean spiny lobster *Panulirus argus* from the Florida Keys. *Diseases of Aquatic Organisms* **59**:109–118.

Silberman, J. D., Sarver, S. K., Walsh, P. J., 1994. Mitochondrial DNA variation and population structure in the spiny lobster *Panulirus argus*. *Marine Biology*, 120: 601-608.

Staine, F. and C. Dahlgren. 2005. Experimental grow-out aquaculture of the Caribbean spiny lobster in Belize.. Page 129 in *Book of Abstracts, 58th Annual Gulf and Caribbean Fisheries Institute*. San Andrés, Col.

Thuy, N. T. B., 2004. Lobsters sea-farming in the central seawaters, Vietnam. Training Workshop on Food Security 26 to 30 September, 2005, Hiroshima, Japan.

Williams, K. C., 2006. Nutritional requirements and feeds development for post/larval spiny lobster: A review. *Aquaculture, doi: 10.1016 j. aquaculture. 2006.10.019.*

Williams, K. C., Smith, D. M., Irvin, S. J., Barclay, M. C., and Tabrett, S. J., 2005. Water immersion time reduces the preference of juvenile tropical spiny lobster *Panulirus ornatus* for pelleted dry feeds and mussel flesh. *Aquac. Nutr.*, 11, 415-426.

Williams, K. C., 2007. Feeds development for post-larval spiny lobster: A review. *Bull. Fish. Res. Agen.* 20:25-7.

In: Aquaculture Research Trends
Editor: Stephen H. Schwartz, pp.7-11
ISBN 978-1-60456-217-0

Expert Commentary B

DIVERSITY AND FUNCTIONALITY OF GASTROINTESTINAL MICROBIOTA IN FARMED FISH: A COMMENTARY ON WHAT WE HAVE LEARNED AND FUTURE DIRECTIONS

F.J. Gatesoupe
INRA, Ifremer, UMR1067 Nutrition Aquaculture et Génomique,
BP 70, F-29280 Plouzané, France

Based on the knowledge acquired from land animals, the contribution of intestinal microbiota to fish nutrition was hypothesized, in view of the ability of some bacterial strains to produce either nutrients or enzymes, possibly active in the digestive process of the host. These early studies were reviewed by Ringø et al. (1995), for instance. However, these findings were moderated due to the relatively low level of culturable bacteria that were generally recovered from the gut of fish farmed in cold and temperate waters, in comparison with homeotherms, and with some warm water herbivorous fishes, in which intense intestinal fermentation was detected (Clements, 1997). The threshold for contributing significantly to the catabolic process of digestion was estimated to 10^7 bacterial bodies per g of digestive tract by Lésel (1993). This author wondered about possible anabolic function by releasing micronutrients, even at low population density. In most aquacultured species, the nutritional role of gut microbiota was generally considered as secondary, with few exceptions. That may change with what we have learned from new studies applying molecular methods. It is now admitted that culture-independent methods are required to account for the real complexity of gastrointestinal biota in fish (Pond et al., 2006). The proportion of culturable bacteria was estimated between 10 and 50% in human colonic biota (Zoetendal et al., 1998), and the same range was generally found in rainbow trout intestine by Huber et al. (2004), but this rate was only 2% in one fish. Such extremely low culturability was also noted in samples from Atlantic and coho salmons (Holben et al., 2002; Romero and Navarette, 2006).

This variability in fish gut microbiota is an important feature, which requires further investigations to understand its nature. The different characterization methods may account partly for discrepancies observed among data available in the literature (Pond et al., 2006). If culture-dependent methods are obviously biased, there is no ideal solution, and molecular methods are also subjected to PCR biases (Zoetendal et al., 2004). Verner-Jeffreys et al. (2003) noted some difficulty in amplifying genomic DNA from *Vibrio alginolyticus*, and Romero and Navarette (2006) suggested that the combination of molecular and culture-based analyses may provide a better understanding of microbiota. Beyond methodological bias, bacteria appeared to vary with sampling season (Al-Harbi and Uddin, 2004) or location (Skrodenyte-Arbaciauskiene et al., 2006), and more generally, with environmental conditions. Unlike humans, omnivorous and carnivorous fish does not seem to harbour specific and resilient intestinal biota. The frequent dominance of opportunistic proteobacteria suggests the formation of temporary consortia of strains coming from surrounding microbiota, while able to survive and to multiply in the digestive tract (Huber et al., 2004). The effects of deterministic factors from feed and environment were investigated by comparing the data collected in different experimental conditions, especially in farmed species, but stochastic factors seemed also important for the colonization of aquatic niches by bacteria (Verschuere et al., 1997). The functional niches may thus be occupied by a variety of different taxons, whereas the selection process is not fully predictable.

This unpredictability challenges the weight of characterising microbiota in particular experimental conditions, while attempting to draw conclusions of general import (Pond et al., 2006). Besides taxonomical characterization of the main components of fish gut microbiota, further research should address the relevance of general tools used to described diversity when they are applied to this particular ecosystem. Notwithstanding normal circumspection, the culture-independent approach enlarged radically the field of view on microbiota. That opened the way to abstain from tedious species naming, while overcoming the uncertainty inherent in species composition. Classical ecological indices seemed suitable to describe microbial diversity after fingerprints from partial amplicons of 16S rRNA gene, without need for sequencing (Eichner et al., 1999; Hill et al., 2003; Hewson et al., 2006), but these indices may vary with the methods. For instance, Moeseneder et al. (1999) found with Terminal-Restriction Fragment Length Polymorphism (T-RFLP) analysis more taxonomic units than when they used Denaturing Gradient Gel Electrophoresis (DGGE), which seemed thus to underestimate richness. Bias may be also expected in abundance assessment, due to different numbers of copies of the 16S rRNA gene in bacterial species (Fogel et al., 1999). Instead of ribosomal RNA genes, Giacomazzi et al. (2004) proposed to use *rpoB* gene, a single copy of which seemed always present.

Buckling et al. (2003) demonstrated that bacterial populations lose their potential to diversify, as they are adapting to a specific niche. Nelson et al. (2003) observed high diversity in bacteria from gut of wild herbivorous mammals, which are exposed to environmental changes. These authors took the example of drought -against which wildlife is more resistant than domestic animals - to hypothesize that microbial diversity could play a part in health and endurance. This aspect has not been yet explored in farmed fish, which are under process of domestication. A first instance was not in favour of this hypothesis, since Holben et al. (2002) found strong dominance of two species of *Mycoplasma* in wild salmon, whereas further richness and evenness appeared in farmed salmon. However, this single observation cannot preclude from investigating the possible impact of aquaculture on microbial diversity, and its consequences on fish health.

There has been recently a growing interest in axenic and gnotobiotic rearing of fish (Marques et al., 2006). It was demonstrated in gnotobiotic zebrafish that gut microbiota stimulated epithelial differentiation and proliferation, gut motility, protein uptake, nutrient metabolism, and innate immunity (Rawls et al., 2004; 2006; Bates et al., 2006). These functional roles are likely not limited to early ontogeny, but the information is missing for juveniles and on-growing fish. The progresses in metagenomics and metaproteomics have raised great hopes for a better understanding of metabolic activities in microbiota (Gill et al., 2006; Wilmes and Bond, 2006), but it will likely take time to apply to aquaculture, in front of the huge diversity of microbiomes. A first coarse estimate of metabolic activity may be obtained by using 16S rRNA as template for RT-PCR (Moeseneder et al., 1999), though sequences from 16S rRNA genes may correspond to highly diverse genomes (Jaspers and Overmann, 2004). The diversity indices based on the RT-PCR products from 16S rRNA cannot generally depict the actual activities expressed in the microbiome. It may be worth using in strict experimental conditions, where everything should be constant except for the tested variable. The functional response of microbiota to one environmental parameter, or to one dietary compound, may be thus described by possible changes from evenness in metabolic activity. Ward (2006) emphasized the interest to combine classical culturing and enrichment techniques to the metagenomic approach, when possible. For example, Panas et al. (2007) selected bacteria able to degrade phosphonoacetate by enrichment, then by discrimination with RT-PCR, Temperature Gradient Gel Electrophoresis (TGGE), cloning, and sequencing. Such methods may find applications for aqua feeds, like fibrous plant-based protein sources, or dietary prebiotics.

These recent advances have brought valuable ways of investigation, which can be applied to aquaculture nutrition with reasonable effort. If we cannot delineate yet the import of gut microbiota in fish nutrition, future research plans should address this issue.

References

Al-Harbi, A. H., and Uddin, M. N. (2004). Seasonal variation in the intestinal bacterial flora of hybrid tilapia (*Oreochromis niloticus*×*Oreochromis aureus*) cultured in earthen ponds in Saudi Arabia. *Aquaculture 229*, 37-44.

Bates, J. M., Mittge, E., Kuhlman, J., Baden, K. N., Cheesman, S. E., and Guillemin, K. (2006). Distinct signals from the microbiota promote different aspects of zebrafish gut differentiation. *Dev. Biol. 297*, 374-386.

Buckling, A., Wills, M. A., and Colegrave, N. (2003). Adaptation limits diversification of experimental bacterial populations. *Science 302*, 2107-2109.

Clements, K. D. (1997). Fermentation and gastrointestinal microorganisms in fishes. In R. I. Mackie, and B. A. White (Eds.), *Gastrointestinal Microbiology* (156-198). New York, NY: Chapman and Hall.

Eichner, C. A., Erb, R. W., Timmis, K. N., and Wagner-Döbler, I. (1999). Thermal Gradient Gel Electrophoresis Analysis of Bioprotection from Pollutant Shocks in the Activated Sludge Microbial Community. *Appl. Environ. Microbiol. 65*, 102-109.

Fogel, G. B., Collins, C. R., Li, J., and Brunk, C. F. (1999). Prokaryotic Genome Size and SSU rDNA Copy Number: Estimation of Microbial Relative Abundance from a Mixed Population. *Microb. Ecol. 38*, 93-113.

Giacomazzi, S., Leroi, F., L'Henaff, C., and Joffraud, J. J. (2004). *rpoB*-PCR amplified gene and temporal temperature gradient gel electrophoresis: a rapid tool to analyse bacterial strains representative of cold-smoked salmon microflora. *Lett. Appl. Microbiol. 38*, 130-134.

Gill, S. R., Pop, M., DeBoy, R. T., Eckburg, P. B., Turnbaugh, P. J., Samuel, B. S., Gordon, J. I., Relman, D. A., Fraser-Liggett, C. M., and Nelson, K. E. (2006). Metagenomic analysis of the human distal gut microbiome. *Science 312*, 1355-1359.

Hewson, I., Steele, J. A., Capone, D. G., and Fuhrman, J. A. (2006). Remarkable heterogeneity in meso- and bathypelagic bacterioplankton assemblage composition. *Limnol. Oceanogr. 51*, 1274-1283.

Hill, T. C. J., Walsh, K. A., Harris, J. A., and Moffett, B. F. (2003). Using ecological diversity measures with bacterial communities. *FEMS Microbiol. Ecol. 43*, 1-11.

Holben, W. E., Williams, P., Saarinen, M., Sarkilahti, L. K., and Apajalahti, J. H. A. (2002). Phylogenetic analysis of intestinal microflora indicates a novel *Mycoplasma* phylotype in farmed and wild salmon. *Microb. Ecol. 44*, 175-185.

Huber, I., Spanggaard, B., Appel, K. F., Rossen, L., Nielsen, T., and Gram, L. (2004). Phylogenetic analysis and in situ identification of the intestinal microbial community of rainbow trout (*Oncorhynchus mykiss*, Walbaum). *J. Appl. Microbiol. 96*, 117-132.

Jaspers, E., and Overmann, J. (2004). Ecological significance of microdiversity: Identical 16S rRNA gene sequences can be found in bacteria with highly divergent genomes and ecophysiologies. *Appl. Environ. Microbiol. 70*, 4831-4839.

Lésel, R. (1993). Does a digestive active bacterial flora exist in fish? In S. J. Kaushik, and P. Luquet (Eds.), *Fish Nutrition in Practice, IVth Int Symp Fish Nutrition and* Feeding (655-664). Paris, France: Institut National de la Recherche Agronomique (INRA).

Marques, A., Ollevier, F., Verstraete, W., Sorgeloos, P., and Bossier, P. (2006). Gnotobiotically grown aquatic animals: opportunities to investigate host-microbe interactions. *J. Appl. Microbiol. 100*, 903-918.

Moeseneder, M. M., Arrieta, J. M., Muyzer, G., Winter, C., and Herndl, G. J. (1999). Optimization of terminal-restriction fragment length polymorphism analysis for complex marine bacterioplankton communities and comparison with denaturing gradient gel electrophoresis. *Appl. Environ. Microbiol. 65*, 3518-3525.

Nelson, K. E., Zinder, S. H., Hance, I., Burr, P., Odongo, D., Wasawo, D., Odenyo, A., and Bishop, R. (2003). Phylogenetic analysis of the microbial populations in the wild herbivore gastrointestinal tract: insights into an unexplored niche. *Environ. Microbiol. 5*, 1212-1220.

Panas, P., McMullan, G., and Dooley, J. S. G. (2007). RT-TGGE as a guide for the successful isolation of phosphonoacetate degrading bacteria. *J. Appl. Microbiol. 103*, 237-244.

Pond, M. J., Stone, D. M., and Alderman, D. J. (2006). Comparison of conventional and molecular techniques to investigate the intestinal microflora of rainbow trout (*Oncorhynchus mykiss*). *Aquaculture, 261*, 194-203.

Rawls, J. F., Samuel, B. S., and Gordon, J. I. (2004). Gnotobiotic zebrafish reveal evolutionarily conserved responses to the gut microbiota. *Proc. Natl. Acad. Sci. U S A 101*, 4596-4601.

Rawls, J. F., Mahowald, M. A., Ley, R. E., and Gordon, J. I. (2006). Reciprocal gut microbiota transplants from zebrafish and mice to germ-free recipients reveal host habitat selection. *Cell 127*, 423-433.

Ringø, E., Strøm, E., and Tabachek, J. A. (1995). Intestinal microflora of salmonids: A review. *Aquacult. Res. 26*, 773-789.

Romero, J., and Navarrete, P. (2006). 16S rDNA-based analysis of dominant bacterial populations associated with early life stages of coho salmon (*Oncorhynchus kisutch*). *Microb. Ecol. 51*, 422-430.

Skrodenyte-Arbaciauskiene, V., Sruoga, A., and Butkauskas, D. (2006). Assessment of microbial diversity in the river trout *Salmo trutta fario* L. intestinal tract identified by partial 16S rRNA gene sequence analysis. *Fisheries Sci. 72*, 597-602.

Verner-Jeffreys, D. W., Shields, R. J., Bricknell, I. R., and Birkbeck, T. H. (2003). Changes in the gut-associated microflora during the development of Atlantic halibut (*Hippoglossus hippoglossus* L.) larvae in three British hatcheries. *Aquaculture 219*, 21-42.

Verschuere, L., Dhont, J., Sorgeloos, P., and Verstraete, W. (1997). Monitoring Biolog patterns and r/K-strategists in the intensive culture of Artemia juveniles. *J. Appl. Microbiol. 83*, 603-612.

Ward, N. (2006). New directions and interactions in metagenomics research. *FEMS Microbiol. Ecol. 55*, 331-338.

Wilmes, P., and Bond, P. L. (2006). Metaproteomics: studying functional gene expression in microbial ecosystems. *Trends Microbiol. 14*, 92-97.

Zoetendal, E. G., Akkermans, A. D. L., and De Vos, W. M. (1998). Temperature gradient gel electrophoresis analysis of 16S rRNA from human fecal samples reveals stable and host-specific communities of active bacteria. *Appl. Environ. Microbiol. 64*, 3854-3859.

Zoetendal, E. G., Collier, C. T., Koike, S., Mackie, R. I., and Gaskins, H. R. (2004). Molecular ecological analysis of the gastrointestinal microbiota: A review. *J. Nutr. 134*, 465-472.

In: Aquaculture Research Trends
Editor: Stephen H. Schwartz, pp.13-57

ISBN 978-1-60456-217-0

Chapter 1

PROGRESS ON FISH DISEASE RESEARCH IN CHINA[1]

Xianle Yang and Haipeng Cao
E-Institute of Shanghai Municipal Education Commission,
Aquatic Pathogen Collection Center of Ministry of Agriculture,
Shanghai Fisheries Univeristy, Shanghai 200090, China

ABSTRACT

China, the biggest fish farming country, is the cradle of the world's aquaculture, and the records of fish disease and its control could be tracked back to thousands of years ago in China. However, the development of fish farming industry and fish disease research in China was seriously hindered because of the social and economic limits. After People's Republic of China was founded, fish disease research in China was started and rapidly developed, and up to now, great progress was achieved in epidemiology, etiology, pathology, immunology, pharmacokinetics of fishery drugs and fish disease control. While in contrast to the current status of fish diesease in foreign developed countries, there are some problems about the detection assays and pathogenic mechanism of pathogens, fishery vaccine development and fish disease control in fish disease research in China. Therefore, some measures should be taken to promote fish disease research in China, maily reflected in the following aspects: 1. The commercialization process of rapid diagnosis means should be accelerated; 2. Study on the pathogenic mechanism centered on functional genes (or proteins) should be deepened; 3. Fish pathology and its study means should be further strengthened; 4. A simple, correct and effective evaluation method of pharmacodynamics on fishery drugs should be established; 5. New ways to effectively prevent and control fish diseases should be explored.

Keywords: *Fish disease research in China; History; Current status; Prospect.*

1 The first two coauthors contributed to the equal work, Email: caohaipeng19810818@yahoo.com.cn.

INTRODUCTION

As a giant country in aquiculture, China yields the most aquatic products in the world. The total volume of China's aquatic products in 2005 amounted to 51,000,000t, which took up more than 70% of the world aquatic products[1]. However, with the great development of China's aquaculture industry, some problems came into being and imposed gigantic threats on the sustainable development of China's aquaculture, such as the gradually deteriorated water quality, the frequent outbreak of fish diseases and the excessive use of the antibiotics, disinfectors and insecticide. These problems not only greatly cut down the yield of aquatic products of China, but also lead to a huge hidden trouble of food security. According to the incomplete statistical data in 2005, 207 kinds of fish diseases broke out in China and the direct loss reached 11 billion RMB, While 60%-75% of the large-scaled farming aquatic animals suffered diseases[2]. Therefore, it is very significant to reinforce the fish disease research and establish the pre-warning system of fish disease in China. In this review, the development history of fish disease research in China was recollected, the current status of fish disease research in China was summarized, the problems about the fish disease research in China were analyzed, and finally the future trend of fish disease research in China was pointed out.

1. BRIEF HISTORY ON FISH DISEASE RESEARCH IN CHINA

Fish farming has a long history in China. As early as Spring and Autumn Period and Warring States Period of China, Li Fan summarized the fish culture experience of ancient Chinese labor people and compiled *Fish Farming Seripture*, the earliest book on aquaculture in the world[3]. Additionally, the record on fish disease and its pathogen could also be tracked back to 1,300B.C.. Shi Su in Song dynasty and Guangqi Xu in Ming dynasty once recorded the fish parasite Shi and its control method in their literature, *Phase-type of Flu* and *Agricultural Governance Guidebook*, respectively. However, as a result of the limited conditions of that times, Chinese people's knowledge about fish disease still stayed in a unscientific and superficial state, and until the year1949, China made no dent in fish disease research.

Fish disease research in China began in the early 1950s, and had a history of over 50 years. Its development history could be divided into 6 different periods. (1) The first period was from 1950 to 1955, which was also the initiatory stage. During this period, early fish disease investigators such as Dashu Ni conducted the research on parasite, bacteria, fungus, protozoan, worm, *etc.*, basically solved the problems such as the species, structure, life style and harm of the grass carp parasites, found the way to control *Cryptobia branchialis*, *Trichodina* sp., *Dactylogrus* sp., Shi[4], and published some research papers such as *Fish prevention work report in 1953*[5]、*Primary study on the pathogen of red skin disease in black carp*[6], *etc.*. The achievements in this period laid a foundation for fish disease research in China. (2) The second period was from 1956 to 1966, which was also the great development stage. During this period, fish disease investigators conducted the complete studies on the epidemiology, pathogenesis, pathology and control of some serious freshwater fish diseases, formed fish disease research system and published a few important professional books such as *Fish disease (first edition)*[7], *The manual for surveys of fish disease (first edition)*[8], *etc*.. (3) The third period was from 1966 to 1977, which was also the retentive stage. During this period, fish disease investigators went on with deep research on the familiar diseases such as gill-rot disease of grass carps, white head and mouth disease of grass carps, twist disease of silver carps、*Amyloodinium ocellatum* disease, *etc.* and start to take up the studies on the preparation and application of "vaccines with indigenous methods". At the same time, some important professional books such as *The pathogeny graphy of fishes in Hubei province*[9], *Prevention manual of familiar fish disease (first edition)*[10], *etc.* were published. (4) The fourth period was from 1978 to 1989, which was also a new seedtime. During this period, fish disease investigators began the study on fish virus, published some academically valued book such as *Prevention manual of familiar fish disease (third edition)*[11], *The manual for surveys of fish disease (second edition)*[12]and *Diagnosis and prevention manual of fish disease*[13] , *etc*., held symposium of grass carp diseases, founded Chinese fish disease research institute of aquaculture and set up the courses of "pathology of cultivated animals in the sea", "fish pathology" and "fish immunology". (5) The fifth period was from 1990 to 1999, which was also the rapid development stage. During the period, the whole level of fish disease research in China was obviously improved. Most parasites and their pathogenic mechanisms were discovered, and a great lot of valuable reference materials about the taxonomy and ultrastructure of parasites, *etc*. were collected. In addition, the studies of pathogenesis on fish diseases were also improved from organic level to cellular and molecule biology level, the studies of pathology on fish diseases were also enhanced to the level of ultrastructural histopathology, histochemistry and pathology physiology, and brought out a few academic books such as *Prevention manual of shrimp disease*[14], *Aquatic animal diseases*[15] , *etc*.. (6) The sixth period was from 2000 to nowadays, which was also the golden time. During the period, the level of fish disease research was further improved. Molecular biological techniques such as RT-PCR, multiplex PCR, suppression subtractive hybridization, cDNA microarray, *etc.* were widely used in the diagnosis of fish diseases, and the rapid diagnosis kits were also developed to diagnose the fish diseases, such as PCR diagnosis kit for fish iridovirus[16]. Furthermore, partial or complete sequences of functional genes of pathogenic viruses such as white spot syndrome virus (WSSV) were successfully tested, and partial genome cDNA library of pathogenic viruses were constructed, such as antiviral subtractive cDNA library of cultured fish cells[17]. Additionally, study on the biology and immune prevention have become hot issues in fish disease control research in China, and some biological agents have exhibited good application prospects, such as oral vaccines against *Aeromonas hydrophila*, DNA vaccine against lymphocystis disease virus, bivalent vaccine for turbot against *Vibrio anguillarum* and *Vibrio alginolyticus,* A3α-peptidoglycan, photosynthetic bacteria, *Bdelloviro bacteriovorus*, *etc.*[18～22].

2. Current Status of Research on Fish Disease in China

Fish disease research in China was started late, and was different from foreign developed aquaculture countries in the study objects, development trends and levels, *etc*.. However, there are more advantages in the fish disease research in China, and with the improvement of research conditions of fish disease, the disparity in fish disease research between China and foreign developed aquaculture countries is becoming gradually shrinking, and in some areas, China has already been close to or reached the top level of the world.

2.1. Research on Epidemiology of Fish Diseases

The epidemiological study on fish disease has great importance on illuminating the infectious sources and ways, reason of prevalence, *etc.* and fish disease investigators in China attached great importance to the epidemiology study of fish disease at the beginning. In recent years, fish disease investigators in China have spent most time making extensive surveys on a few serious fish diseases, of which the surveys on parasite and virus diseases were more extensive and comprehensive. The following are main features: (1) The prevalence of treamatodes, ciliates, helminthes, *etc.* was investigated. Of which the study on the epidemiology of treamatodes was the hot issue, such as the studies on the population dynamics of Saccocoelioides huidongensis in *Mugil cephalus*[23]and *Lenolopa huizhouensis* in *Scatophagus argus* [24], the population distribution and seasonal alternation of two species of monogeneans on the gills of *Hypophthalmichthys molitrix* and *Ctenopharyngodon idellus*[25]and the epidemiological investigation of *Azygia anguillae* in the host *Monopterus albus* from Dongting lake region[26], *etc.*; (2) The prevalence of virus diseases such as white spot syndrome, viral reddish body syndrome, infectious haematopoietic necrosis, and Taura syndrome, *etc.* was surveyed. Deng Deng (2005) investigated the status carrying WSSV in several spawning ground populations of *Fenneropenaeus chinensis*[27], Chengyin Shi (2005) reported the epidemic situation of viral reddish body syndrome among farmed *Scophthalmus maximus* in China [28]; (3) The prevalence of bacterial diseases in large-scale farming breeds was studied. Shan Jin (2005) investigated the epidemiology of vibriosis in large yellow croaker *Pseudosciaena crocea* (Richardson) in marine cage culture[29], Lei Ding (2000) made an epidemiology study on bacterial pustulae disease of *Ophiocephalus argus*[30]. The epidemiology study materials of the serious diseases mentioned above provided important evidence for the effective control on the outbreak of fish diseases.

2.2. Study on the Pathogen of Fish Disease

With vast aquaculture area and hundreds of farming breeds, China had formed an aquaculture pattern, whose farming breeds were mainly fishes, shrimps, crabs, and turtles, *etc.* in freshwater and shellfishes, algaes, fishes, prawns, crabs, *etc.* in seawater. However, the occurrence of fish diseases had also become more and more fierce. there were totally 207 kinds of fish diseases in China during the year 2005, which infected 62 kinds of fish, 11 kinds of shrimp, 3 kinds of crabs, 5 kinds of amphibious animals, and 8 kinds of shellfishes, *etc.*[2], and also provided new challenges and opportunities to the research on pathogeny of fish diseases. In recent years, China carried out broad studies on the pathogens of fish diseases, mainly in the following aspects:

2.2.1. Study on Fish Viruses

In 1914, Weissenberg announced the first fish virus disease in the world[31], which carved the milestone of fish virus research. From then on, all kinds of viruses were discovered one after another. In 1978, Institute of hydrobiology affiliated with China Academy of Science first confirmed the pathogen of hemorrhage of grass carps reovirus in China. That was the beginning of fish virus disease study. Up to now, investigators had discovered at least 33 kinds of viruses, of which at least 27 kinds of viruses were isolated from fish and shrimps (Table 1). Meanwhile, China made great progress in fish virus diseases, the following are main features:①the whole genomes of Taura syndrome virus （TSV） ZHZC3, infectious spleen and kidney necrosis virus(ISKNV), tiger frog virus(TFV) were completely sequenced[71], rapid detection methods of white spot syndrome virus (WSSV), Taura syndrome virus （TSV）, orange-spotted grouper nervous necrosis virus (OGNNV), lymphocystis disease virus (LCDV), infectious hypodermal and hematopoietic necrosis virus （IHHNV）, *etc*. were established, such as real-time PCR, PCR-dot blot hybridization, two-temperature and multiple PCR, *etc*.[72～74];②gene expression and its structure protein functions of OGNNV, ISKNV, GCRV, WSSV, LCDV were also studied, such as the expression of VP28 gene, VP19 gene and putative zinc finger protein gene of WSSV[75～77], the functions of the structure protein VP28 and VP26 of WSSV[78].

Table 1. Aquatic viruses reported in China

Virus	Morphological property	Disease	Reporter
Fish Reovirus	hexagonal in shape icosahedral symmetry, 68～72nm in diameter, with no capsula, double strand RNA		Yanyan Chen[32]
	spherical in shape, icosahedral symmetry,60～70nm in diameter, double strand RNA	hemorrhage of grass carp	Wuhan institute of virology[33]
	icosahedral in shape,60nm in diameter, double strand RNA		Shujian Mao [34]
	hexagonal in shape, icosahedral symmetry, 80～100nm in diameter, with no capsula, double strand RNA	hemorrhage of grass carp and black carp	Ziyu Zhai[35]
grass carp hemorrhage virus	spherical in shape icosahedral symmetry, 71nm in diameter, with no capsula,double strand RNA	hemorrhage of grass carp	Lihua Ke[36]

Table 1. (Continued)

Virus	Morphological property	Disease	Reporter
Picornavirus	hexagonal in shape, 24～30nm in diameter,with no capsula, single strand RNA	hemorrhage of grass carp	Shujian Mao[37]
spring viremia of carp virus	linear in shape, single strandRNA， 70nm ×180nm in size	spring Viremia of Carp	Yulin Jiang[38]
infectious pancreatic necrosis virus	hexagonal in shape, icosahedral symmetry, with no capsula, 55～65nm in diameter, double segment and double strand RNA	infectious pancreatic necrosis of *Salmo gairdnert*	Yulin Jiang[39]
infectious haematopoietic necrosis virus	rhabdovirus, 58～91nm in diameter, single strand RNA, with capsula	infectious haematopoietic necrosis of *Salmo gairdnert*	Zhizhuang Zhao[40]
Togavirus of eel	spherical or oval in shape, 70～85nm in diameter， single strand RNA, with capsula	mouth-opening disease of *Anguilla japonica*	Yanyan Chen[41]
coronavirus like virus	enlargement of top capsule, with flower arranged	mad swim disease of *Anguilla japonica*	Zengsi Tao[42]
lymphocystis disease virus-Chinese isolate	spherical in shape, icosahedral symmetry, 210nm in diameter ,with capsula	lymphocystic disease of *Paralichthys olivaceus*	Hongtao Xu [43]
lymphocystis disease virus	hexagonal in shape,180nm in diameter	lymphocystic disease of *Paralichthys olivaceus*	Yunkun Liu[44]
Iridovirus	hexagonal in shape,150nm in diameter	infective outbreaking disease of *Sciaenop ocellatus*	Kaibing Li[45]
infectious spleen and kidney necrosis virus	icosahedral symmetry, 150nm in diameter， with capsula, double strand DNA	infective outbreaking disease of *Siniperca chuatsi*	Shuqin Wu,[46];Jianguo He[47]
sphereovirus	spherical in shape,80～100nm in diameter,with capsula	- *Pagrosomus major*	Ming Jiang[48]
Baculovirus of *Penaeus chinensis*	rod in shape, 350nm×150nm in size, non-occluded, with capsula	Baculovirus disease of *Penaeus chinensis*	Wenbin Zhan[49]

Table 1. (Continued)

Virus	Morphological property	Disease	Reporter
	bacilliform in shape， with capsula, 440～500×170～180nm in size	acute sethal disease of prawns; Baculovirus disease of *Penaeus chinensis*	Baozheng Peng[50]; Bin Wang[51]
lymphoid cell nuclear baculovirus	thick rod in shape, 96～112nm in diameter, non-occluded	infective outbreaking disease of *Penaeus chinensis*，*Penaeus penicillatus*，*Penaeus japonicus*	Xifa Chen[52]
hepatopancreatic parvovirus	icosahedral symmetry, 23.3～29.8nm in diameter, single strand DNA		Qinggang Xue[53]
Reovirus	icosahedral symmetry, 64～66nm in diameter, with an capsule, double strand RNA	virus disease of *Penaeus chinensis*	
non-occluded baculovirus	short rod in shape, 330～400nm×110～130nm in size, with capsula and inclusion body	virus disease of *Penaeus chinensis*	Tuo Shi[54]
Penaeus monodon spherical virus	spherical in shape,28～35nm in diameter, spherical inclusion body	virus disease of *Penaeus monodon*	Xifa Chen[55]
Baculovirus of *Metapenacus eusis*	short rod in shape, 228nm×125nm in size, with capsula, non-occluded	virus disease of *Metapenacus eusis*	Shouming Feng[56]
Nodavirus	spherical in shape,24nm in diameter,single strand RNA	whitish muscle diseases of *Macrobrachium rosenbergii*	Dong Qian[57]
Taura syndrome virus	icosahedral in shape, with no capsula, 35nm in diameter	Taura syndrome of *Penaeus vannamei*	Liuji Wu[58]
Reovirus like virus	spherical in shape, with no capsula, 55nm in diameter	shiver disease of *Eriocheir sinensis*	Chengliang Gong[59]
freshwater mussel plague virus	spherical in shape, 46～85nm in diameter	freshwater mussel plague of *Hyriopsis cumingii*	Zhirong shen[60]
freshwater mussel herpersviroid virus	icosahedral symmetry, with capsula, 80～120nm in diameter	freshwater mussel plague of *Hyriopsis cumingii*	Yaofu Liu[61]

Table 1. (Continued)

Virus	Morphological property	Disease	Reporter
herpes-like virus	hexagonal in shape, 84～90nm in diameter	facial disc disintegration disease of *Argopecten irradians*	Jingyin Jiang[62]
virus of pig-frog	spherical in shape,with capsula, 150nm in diameter	lethal syndrome of pig-frog	Qiya Zhang[63]
soft shelled turtle iridovirus	spherical in shape, 120～160nm in diameter, with capsula	red neck disease of soft shelled turtle	Zaixian Chen64]
unidentified species	with an envelop,the core capsule is about 250 nm in diameter, te capsid is 200 nm in diameter. with no envelop,50nm in diameter, cluster in groups	hemorrhage disease of *Lateolabrax japonicus*	Liangyi Xue[65]
	spherical in shape, 50～80nm in diameter, with no capsula and tivesicular inclusion body	gonad atrophy disease of *Argopecten irradians*	Nin Yang[66]
	spherical in shape, 90～140nm in diameter,with double capsula, non-occluded	crack shell disease of *Haliotis discus hannai Ino*	Bin Wang[67] ,Xia Li [68]
	spherical in shape,50～80nm in diameter, with no capsula	red muscles disease of *Meretrix meretrix Linnaeus*	Sulian Ren[69]
	spherical in shape,150～200nm in diameter, with double capsula, non-occluded	crack shell disease of *Mytilus eduli*	Bin Wang[70]

2.2.2. Study on Fish Pathogenic Bacteria

Emmerich and Weibel (1894) first isolated and described the pathogenic bacterium of furunculosis[71], initiated the new porch of fish pathogenic bacteria research. In1956, Deming Wang (1956) first isolated the pathogenic bacterium from red skin disease in black carps in China[6], and now the species of the fish pathogenic bacteria isolated and identified in China had already reached at least 40, of which 35 species were isolated from fishes and shrimps (Table 2). And in recent years, new theories and techniques had already been applied to the fish pathogenic bacteria research, and new progress was achieved in the diagnosis and pathogenesis of the pathogenic bacteria, specially in the following areas: ① The methods to rapidly detect the familiar fish pathogenic bacteria and their pathogenic agents were set up, such as the ELISA technique to detect *Vibrio alginolyticus*[154], the indirect fluorescent antibody technique to detect *Vibrio fluvialis*[155], the dot immunobinding assay (DIA) to detect the exotoxin of *Vibrio mimicus*, and the PCR and multiplex PCR techniques to detect *Aeromonas hydrophila*[157, 158]. Furthermore, Sufang Shen (2000) developed the diagnosis kit for pathogenic *Aeromonas* on the principle of dot-ELISA technique[159]; ② The part of pathogenesis of familiar pathogenic bacteria was illuminated. Experiment results indicated that pilus and nonpilus adhesion were involved in the adhesions of *Aeromonas hydrophila*[160], elastase was the direct pathogenic agent of *Aeromonas hydrophila*[161], all strong virulent *Aeromonas hydrophila* strains were *aerA*+, *hlyA*+ and *ahpA*+ genotype, and all *Aeromonas hydrophila* strains with *ahpA+* genotype were virulent, and there was a positive correlation between the virulence of *Aeromonas hydrophila* and the presence of aerA and hlyA[162], furthermore, the virulence of *Aeromonas hydrophila* was related to water and fish body temperature, and water pH[163].

Table 2. Aquatic pathogenic bacteria reported in China

Bacteria	Morphological property	Disease	Reporter
Aeromonas hydrophila	Gram negative, short rod in shape, with one polar flagellum, without capsule and spore	red fin disease of eels	Huibo Chen[79]; Guoliang Wang[80]
		bacterial septicemia of *Channa maculate L.*	Changfu Chen[81]
		bacterial enteritis of *Silurus merdionalis*	Xilin Zhang[82]
		hemorrhagic disease of *Monopterus albus*	Jingyu Shen83]
		bacterial septicemia of *Channa asiastica* (Linnalus)	Aiyin Lei84]
		skin fester disease of grass carp	Xiaoyan Jiang85]
		skin ulceration disease of *Varicorhinus macrolepis*	Zengfu Wang86]
		bacterial septicemia of Acipenser gueldenstacdti; *Huso huso♀×Acipenser baeri♂*	Weihua Chu[87]
		bacterial septicemia syndrome of *Huso huso♀×Acipenser baeri♂*; *Huso dauricus*	Haipeng Cao[88]
		red gill disease of *Macrobrachium nipponense*	Jingyu Shen[89]
		red leg disease of bullfrog	Chengyu Hu[90]
		swollen abdomen of bullfrog	Lu He[91]
		black spot disease of *Trionyx sinensis*	Changyi Zhou[92]
Aeromonas punctata subsp. *caviae*	Gram negative, short rod in shape, with one polar flagellum, without capsule	bacterial septicemia of *Channa asiastica* (Linnalus)	Aiyin Lei[84]
		skin ulceration disease of *Varicorhinus macrolepis*	Zengfu Wang[86]

Table 2. (Continued)

Bacteria	Morphological property	Disease	Reporter
Aeromonas punctata subsp. *caviae*	Gram negative, short rod in shape, with one polar flagellum, without capsule	skin ulcer disease of *Salmo gairdner* Richerdson	Lei Ding[93]
		bacterial septicemia syndrome of *Huso huso*♀×*Acipenser baeri*♂; *Huso dauricus*	Haipeng Cao[88]
Aeromonas sobria	Gram negative, short rod in shape, with one polar flagellum, without capsule and spore	skin fester disease of grass carp	Xiaoyan Jiang[85]
		fulminant septicaemia of *Ictalurus punctatus*	Yingbing Su[94]
		bacterial septicemia of tilapias	Wanqi Cai[95]
		caverned disease of *Trionyx sinensis*	Peifang Sun[96]
Aeromonas salmonicida subsp. *salmonicida*	Gram negative, short rod in shape, without flagellum,capsule and spore	skin ulcer disease of *Salmo gairdner* Richerdson	Lei Ding[93]
Aeromonas punctata subsp. *punctata*	Gram negative, short rod in shape, without capsule and spore, one polar flagellum	stigmatosis of *Hypophthalmichthys molitrix; Aristichthys nobilis*	Bohai Xu[97]
		rotten skin disease of *Carassius auratus*	Haishen Wen[98]
Pseudomonas putrefaciens	Gram negative, rod in shape, with single flagellum	skin-ulcer disease of *Pseudosciaena crocea*	Kebing Lin[99]
Pseudomonas fluorescens	Gram negative, rod in shape, with one to three polar flagella	fester disease of *Discus abalone*	Lin Ye[100]
Pseudomonas alcaligenes	Gram negative, rod in shape, without capsule and spore	oral congestive disease of mud eels	Youzhi Ma[101]
Xanthomonas maltophilia	Gram negative, rod in shape, with 1～8 flagella, without capsule and spore	cystis disease of *Cistoclemmys flaciomarginata*	Bin Huang [102]
Edwardsiella tarda	Gram negative, rod in shape, peritrichous flagella	red-head disease of *Anguilla anguilla*	Kai Zhou[103]
		infection by *Edwardsiella tarda* of *Paralichthys olivaceus*	Xiaojun Zhang[104]
		hemorrhagic septicemia of *Scophthalmus maximus*	Bin Wang[105]

Table 2. (Continued)

Bacteria	Morphological property	Disease	Reporter
Edwardsiella tarda	Gram negative, rod in shape, peritrichous flagella	swollen abdomen of *Scophthalmus maximus*、*Paralichthys olivaceus*	Jun Li[106]; Shuxia Xue[107]
		white abdominal shell disease and skin fester disease of *Trionyx sinensis*	Lei Lei[108]
Vibrio harveyi	Gram negative, short rod in shape, with one polar flagellum,with lateral flagellum,without capsule	Vibriosis of *Epinephelus coioides*, *Lateolabrax japonicus*	Baokun Wang[109]; Xiangao Chen[110]
		tail-rot disease, skin-ulcer disease of *Pseudosciaena crocea*	GuozhongTong[111]; Kebing Lin[99]
		red leg disease of *Penaeus chinensis*	Bin Wang[112]
Vibrio alginolyticus	Gram negative, rod in shape, with one polar flagellum	Vibriosis of *Epinephelus malabaricus*	Xiaoyan Chen[113]
		ulcerative disease of *Epinephelus coioides*	Xuefeng Hu[114]
		swollen abdomen of *Scophthalmus maximus*	WeiniZhang[115]; Shuxia Xue[107]
		red leg disease of *Penaeus chinensis* red body disease of *Penaeus vannamei*	BinWang[112]; Yongcan Zhou[116]
		"mass mortality" , acute death and striping off disease of *Haliotis diversicolor*	Jing Zhou[117]; Guangfeng Liu[118]
Vibrio coralliilyticus	Gram negative, short rod in shape, with one polar flagellum	acute death and striping off disease of *Haliotis diversicolor*	Guangfeng Liu[118]
Vibrio parahaemolyticus	Gram negative, short rod in shape, with one polar flagellum	skin ulcer disease of *Pseudosciaena crocea*	Rongxing Ji[119]
Vibrio cholerae (Non-01)	Gram negative, arc in shape, without spore with one polar flagellum	intestine edema disease of *Penaeus chinensis*	Liping Wang[120]

Table 2. (Continued)

Bacteria	Morphological property	Disease	Reporter
Vibrio natriegen	Gram negative, short rod in shape, with one polar flagellum	red leg disease of *Penaeus chinensis*	Bin Wang[112]
		Vibriosis of *Argopecten irrsdins lamarck*	Xiaohua zhang[121]
Vibrio anguillarum	Gram negative, short rod or arc in shape, with one polar flagellum	rotted gill and rotted caudal fins of *Lateolabrax japonicu*	Hui Xiao[122]
		ulcerative fin disease of *Scophthalmus maximus*	Zheng Zhang[123]
Vibrio mimicus	Gram negative, short rod in shape, with one polar flagellum	"swollen abdomen" and"shivering disease" of *Eriocheir sinensis*	Jingyu Shen[124]
		ascites disease of *Sparus macrocephelus*	Guoliang Wang[125]
		ascites disease of *Eriocher sinensis*	Jinnian Li[126]
		red body syndrome of *Macrobrachium nipponense*	Wueping Ye[127]
Vibrio carchariae	Gram negative, short rod in shape, with one polar flagellum	Vibriosis of *Rachycentron canadum*	Mingyuan Guo[128]
Vibrio fluvialis	Gram negative, arc in shape, with one polar flagellum	infective outbreaking disease of freshwater fish	Bohai Xu[129]
		Vibriosis of *Paralichthys olivaceus*	Jianbo Chang[130]
		skin ulcer disease of *Sciaenops ocellatus*	Rongxing Ji[131]
Vibrio vulnificus	Gram negative, short rod in shape, with one polar flagellum	Vibriosis of *Rachycentron canadum*	Jichang Jian[132]
Vibrio splendidus	Gram negative, short rod in shape, with one polar flagellum	stomach ulcer disease of *Apostichopus japonicus*	Yingeng Wang[133]
Proteus mirabilis	Gram negative, rod in shape, with flagella from points all around the cell	rot-mouth disease of Trionyx sinensis	Keyu Xiao[134]
Proteus vulgaris	Gram negative, rod in shape, without capsule and spore	stomach and intestines' ulcer and bleed disease of *Trionyx sinensis*	Xiaofeng Chen[135]
Acinetobacter claeoaceticus	Gram negative, short rod in shape, without flagellum, capsule and spore	red leg disease of bullfrog	CHengyu Hu[90]
myxococcus piscicola	Gram negative, suppleness	gill-rot disease of *Aristichthys nobilis*	Weihao Huang[136]

Table 2. (Continued)

Bacteria	Morphological property	Disease	Reporter
Escherichia coli	Gram negative, rod in shape, with flagella from points all around the cell	hemorrhagic disease of *Oncorhynchus mykiss*	Bohai Xu[137]
Providencia rettgeri	Gram negative, rod in shape, with flagella from points all around the cell	septicemia of *Penaeus chinensis*	Wenbin Zhan[138]
Pasteurella anatipestifer	Gram negative, short rod in shape	bacterial enteritis of *Kareius bicoloratus*	JianboChang[139]
Leucothrix mucor	Gram negative, Filamentous in shape with no branches	filamentous bacterial disease of *Eriocheir sinensis*	Xihe Wan[140]
Shewamella algae	Gram negative	ulcer disease of *Scinenops ocellata*	Chang Chen[141]
Bacillus thuringiensis	Gram positive, rod in shape,with spore	appendage shivering disease of *Eriocheir sinensis*	Guozhang Zu[142]
Alcaligenes denitrificans	Gram negative, rod in shape		
Burkhol cepacid	Gram negative, short rod in shape, with multiple polar flagella,without capsule and spore	white cloud disease of *Micropterus salmoides*	Shan Jin[143]
Citrobacter freundii	Gram negative, short rod in shape	bacterial disease of *Cherax quadricarinatus*	Jingyu Shen[144]
Pseudoalteromonas nigrifaciens	Gram negative, short rod in shape, with one polar flagellum	skin ulcer syndrome of *Apostichopus japonicus*	Yingeng Wang[145]
Listonella anguillarum	Gram negative, rod in shape, without spore	bacterial septicemia of *Scophthalmus maximus*	Xiaojun Zhang[146]
Streptococcus iniae	Gram positive, spherical, Streptobacillus arrangement	bacterial fester disease of *Ophiocephalus argus*	Lei Ding[147]
		erythrodermatitis of *Tilapia nilotica*	Jiaqian Chai[148]; Qiaoqiu Wang[149]
		hemorrhagic septicemia of *Scortum bacoo*	Junqiang Qiu[150]
		Streptococcsiis of *Sciaenops ocellatus*	Zhihua Shen[151]
Morganella morganii	Gram negative, rod in shape, without capsule and spore	ulcerative disease of *Cyprinus carpio* L.	Xiaodan Lu[152]
Lactococcus garvieae	Gram positive, short rod or hexagonal in shape, Streptobacillus arrangement without spore	bacterial septicemia of *Paralichthys olivaceus* temminck et schlegel	Hai Fang[153]

2.2.3. Study on Fish Pathogenic Fungus

Fewer pathogenic fungus were reported in China, and up to now, the pathogenic fungus barely included *Saprolegnia* sp., *Aphanomyces invadans*, *Mucor* sp., *Candida oleophila* and

Torulopsis mogii (Table 3) , and the study on rapid detection of the pathogenic fungus was still sporadic. Xiaolin Tu (1996) once made the latex agglutination test for the detection of pathogenic *Torulopsis mogii* from *Macrobrachium rosenbergii*[169].

2.2.4. Study on Fish Parasites

In 1948, China began the study of Gyrodactylus, and up to now, there had been hundreds of species of parasites discovered in China (Table 4). The only Myxosporea were reported to reach 575 species by Qiliu Chen and Chenglun Ma[198]. In recent years, with the widely application of techniques in current immunology and molecular biology, great progress was made in the taxonomy and rapid detection of fish parasites, the following are main features: 1. DNA sequence analysis was applied to the sestematically taxonomic study of fish parasites, which exhibited more stable and credible traits at molecular and genetic levels. Yishan Lu (2004) made phylogenetic analysis of the myxosporeans parasitic in freshwater fish based on 18S ribosomal DNA[199]; 2. Immunological and molecular biological detection methods of fish parasites were developed, such as the indirect enzyme-linked immunosorbent assay (iELISA) to detect *Perkinsus marinus* in mollus[200], the ELISA against mature intact *Myxobolus rotundus* spores[201] and the PCR technique to detect *Thelohanellus kitauei* in *Cyprinus carpio*[202].

Table 3. Aquatic pathogenic fungus reported in China

Fungus	Morphological property	Disease	Reporter
Saprolegnia	slim mycelium,with no segregation, sparse branches and moving sporangium	Saprolegniasis of *Lateolabrax japonicus*	Aihua Li[164]
Aphanomyces invadans	—	ulcer disease of *Salvelinus alpinus*	Xiujie Shi[165]
Mucor sp.	Gram positive, thick and branching hyphae, no septum	mucormycosis of *Fugu obscurus*	YuanjieYang[166]
Candida oleophila	oval in shape, 2. 3～7. 8μm×1. 3～6. 5μm in size, cell arrayed in single or pair	muscle emulsification disease of *Portunus trituberculatus*	Wenjun Xu[167]
Torulopsis mogii	spherical or oval in shape, 2.5～5.0μm×3.75～6.25μm in size	explosive epidemic disease of *Macrobrachium rosenbergii*	Yuhua Sun[168]

Table 4. Aquatic pathogenic parasites reported in China

Parasite	Host	Parasitic location	Reporter
Ancistrum japonicum	*Dosinia japonica, Cyclina sinensis*	gill surface, mantle	
Ancistrum crassum	*Ruditapes philippinarum*、*Protothaca jedoensis*、*Caecella chinensis*、*Ruditapes variegata*	gill surface, mantle	Kuidong Xu[170]
Boveria subcylindrica	*Pinna pectinata*	gill surface, mantle	
Myxophyllum magnum n. sp.	*Sinonovacula constricta*	gill surface, mantle	Kuidong Xu[171]
Thigmophrya bivalviorum	*Saxidamus purpuratus*	gill surface, mantle	
Ancistrocoma primigenius	*Panope generosa*	gill surface	Kuidong Xu[172]
Ancistrocoma pelseneeri	*Mya arenaria*	gill surface	
Trichodina rectuncinata	*Lateolabrax japonicus*	gill	Kuidong Xu[173]
Trichodina paranigra sp.nov	*Carassius auratus*	gill	Fahui Tang[174]
Trichodina munnanica	*Acanthopagrus macrocephalus*	gill	Kuidong Xu[175]
Paratrichodina globonuclea	*Sebastes schlegeli*	gill	Kuidong Xu[176]
Neosentic celatus	*Monopteus albus*	intestine	Xiaoying Dong[177]
Cryptocaryon irritans	*Trachinotus blochii*	gill	Yue Ma[178]
Mesanophrys carcini	*Scophthalmus maximus*	head、operculare、base of fins	Yingeng Wang[179]
Pseodograffilla sp.	*Sciaenops ocellatus*	gill	Yanyin Wang[180]
Paralembus digitiformis	*Paralichthys olivaceus*	body surface	Li Zhou[181]
Microsporidian	*Penseus monodon*	muscle	Weina Wang[182]
Glugea epinephelus sp.nov.	*Epinephelus akaara*	abdomen	Houbo Wu[183]
Podospora yiduensis sp.nov.	*Squaliobarbus curricums* (Richardson)	gall bladder, kidney	Xiaoning Gong[184]
Myxobolus chinensis sp.nov.	*Hypseleotris swinhonis* (Gunther)	muscle of pack, between head an chest	

Table 4. Aquatic pathogenic parasites reported in China (continued)

Eustrongylides sp.	Protosalanx hyalocranius(Albbotl),Hemibarbus maculatus Bleeker,Channa argus(Canlor),Pelteobagrus fulvidraco(Richardson),Parasilurus asotus Linnaeus,and Odontobutis obscur(Temminck et Schlegel)	Celiac, swimbladder wall, stomach, muscle, ovary	Shan Jin[185]
Gyrodactylus donghuensis sp.nov.	Rhodeus sinensis	gill	Weijian Yao[186]
Gyrodactylus wuchangensis sp.nov.	Rhodeus sinensis	gill	
Gyrodactylus rhodei Zitnan	*Acanthorhodeus macropterus* Bleeker	gill	
Gyrodactylus macrorhodei Ergens et Yukhimenko	Rhodeus sinensis	gill	
Gyrodactylus coreiusi sp.nov.	*Coreius heterodon*	gill	Weijian Yao[187]
Gyrodactylus parabotia sp.nov.	*Parabotia fasciata* Dabry	gill	Weijian Yao[188]
Gyrodactylus lichuanensis sp.nov.	*Phoxinus lagowskii variegatus*	body surface	
Trichodina donghuensis sp.nov. *Trichodina ctenogobii* sp.nov.	*Ctenogobius giurnus* Rutter	body surface	Lianxiang Li[189]
Gyrodactylus gussevi	*Silurus meridionalis*	body surface	Qizhong Zhang[190]
Haliotrema austrosinense sp.nov *Haliotrema spirale* Yamaguti	*Sparusn macrocephalus*	gill	Haiyan Li[191]
Haliotrema hainanensis PanandZhang ,sp .nov	*Chaetodon wiebeli* Kaup	gill	Jun Pan[192]
Ancyrocephalus mogurndae	*Siniperca chuatsi*	gill	Qian Gao[193]
Neobenedeniella congeri *Allobenedenia convoluta*	*Epinephelus awoara*	body surface, filament	Wenchuan Yang[194]
Digenetic trematodes	*Meretrix meretrix*	gonad、gill, kidney, digestive diverticulum	Sulian Ren[195]
Waretrema piscicola Srivastava	*Scatophagus argus*	intestine	Shengfa Liu[196]
Rocinela sp.	*Cirrhina mrigala*	gill, bases of pectoral fins	Fei Zhao[197]

2.3. Study on the Pathology of Fish Diseases

Study on the pathology of fish diseases started late, but with the widely use of ultrathin sections and electron microscopy in the pathological study, the pathology was observed under submicroscopy and ultrastructure and rapid progress in the pathology of fish diseases was achieved, and up to now, the pathology of at least 40 species of fishes and shrimps was studied. In recent years, the pathological studies on fish diseases mainly focused on histopathological and cell pathological observations and haematological changes and the study results helped to elucidate the process and mechanism of occurrence of fish disease. The present studies of histopathology, cell pathology and haematology of fish diseases were embodied in the following areas:

2.3.1. Study on the Histopathology

The study subjects of the histopathology were mostly the fish diseases of economical valued fishes, shellfishes, crustacean zooplanktons, such as taura syndrome in *Litopenaeus vannamei,* viral nervous necrosis in *Epinephelus akaara,* lymphocystis disease in *Rachycentron canadum,* muscle whitish disease in *Macrobrachium rosenbergii,* "appendage shivering disease" in *Eriocheir sinensis, etc.* in virus diseases[203～210], external ulcerative disease in *Paralichthys olivaceus,* septicemia in *Megalobrania amblycephala*, septicemia in *Esox lucius,* shrimp black gill disease in *Macrobrachium nipponense*, ulcerate disease in *Haliotis diversicolor, etc.* in bacterial diseases[211～215], *Heteronchleidus* disease in *Macropodus opercularis*, *Ichthyophthirius Multifiliis* disease in juvenile *Leiocassis longirostris*, paradeontacvlix sinensis disease of *Fugu rubripe*, *Lateolabrax japonicus* disease in *Dactylogyrus inverse, Ichthyophthirius multifiliis fouquet* disease in *Coreius guichenoti*, *Perkinus* sp. disease in *Ruditapes philippinarum*, bucephalidae larvae-caused disease in *Meretrix meretrix*, "mantle hypertrophy disease" in *Meretirx meretrix*, *etc.* in parasite disease[216～222].

2.3.2. Study on the Cell Pathology

The study on the cell pathology of fish disease mainly focused on the cell pathological changes of the virus and bacterial diseases.The coinfection of fish cell line of grass carp fins (GCF) by Siniperca chuatsi spherical virus (SCSV) and Siniperca chuatsi rhabdovirus (SCRV) was revealed[223], the pathological changes of diseased cells were observed, such as the blood cells of Lateolabrax japonicus diseased with spleenomegaly and Litopenaeus vanammei infected by white spot syndrome virus[224, 225], the brain and heart muscle cells of Anguilla anguilla diseased with "mad swimming sysmptoms " [226], the cell line from grass carp (CIK) induced by fish reovirus[227], fish cell lines (EPC, EG, CO) infected by Trionyx sinesis virus[228], the epithelium of digestive diverticula of Meretrix meretrix diseased with red muscle and Argopecten irradians Lamarck diseased with gonad atrophy[229,230], the hormonal cell of adenohypophysis of Carassius auratus[231], the peripheral leucocytes coinfected by Vibrio anguillarum and Aeromonas hydrophila[232].

2.3.3. Study on the Haemolymph Pathology

The haemolymph pathology of fish diseases was a hot study in China now and a haematological change mode was established, whose indicator system involved the content of total plasma protein, SOD, ACP,AKP ,LDH and MPO, the number of white blood cells, antibacterial activity, and bacteriolytic activity. Shouming Feng (2006), Chengsong Zhang (2004) observed the number changes of blood cells of *Litopenaeus vannamei* and *Fenneropenaeus chinensis* diploid and triploid artificially challenged with WSSV[233,234], and point out that hemocyte cell was the main target cell of WSSV[233];and additionally, Shan Jin (2005) tested the change of serum enzymes of ALT, AST, LDH and the number of total protein, urea nitrogen, creatinine, K+, Na+, Cl- of *Micropterus salmoides* challenged with *Burkhol cepacid*[143], Jin Liu (2005), Xiaoying Dong (2006) studied the change of urea nitrogen, creatinine, K+, Na+, Cl-, LDT, GOT in the blood of *Monopterus albus* infected by *Eustronylides* sp. and *Pallisentis celatus*[235,236].

2.4. Study on the Immunology of Fish Diseases

Fish immunology research is a necessary and important part in fish disease defence study. In the 1950s, Deming Wang (1956) made active and passive immunization study with bacterin of red skin disease[6], and afterwards, many investigators in China carried out the studies of fish immunology one after another. The current research status in fish immunology focused on the following aspects:

2.4.1. Study on the Morphology, Structure, Distribution and Function of Immune System

Fish immune system included immune organs, immune cells and humoral immunity factor. Now the immune system research in China focused on the following aspects:

1. Morphological and structural observation of immune tissues and cells. Xiaochen Luo (2005) several kinds of immune cells existing in mucosal-associated tissues such as skin, eye's corn, gill, foregut, hindgut of *Epinephelus coioides*, including goblet cells, lymphocytes, macrophages, monocytes, eosinophils and nertrophils[237]; Honghu Wang (2006) testified experimentally that head kidney and spleen were the major hemopoietic organs in fish, and they were also important immune organs in *Lutjanus sanguineus*[238];

2. Study on the distribution and localization of endocrine cells in the digestive tract of fish, such as the endocrine cells of 5-hydroxytryptamine (5-HT)cells, somatostatin (SST) cells, gastrin (Gas) cells and glucagons (Glu) cells. The fish tested now mainly included *Anguilla japonicus, Pseudosciaena crocea, Sparus latus, Sparus macrocephalus, Bostrichthys sinensis, Paralichtys olivaceus, Paralichthys olivaceus*. The experiment results showed that 5—HT and SST cells were abundant in stomachus cardiacus and stomachus fundus, Glu cells were abundant in foregut and midgut, Gas cells were abundant in oesophagus, foregut and midgut[239～245], but the distribution of Gas cells was different in fishes[243， 245];

3. Study on the function of immune cell. Zhendong Zhang (2006) reported the observation of phagocytosis, deformability and immune adhesion of red cell for the first time, and pointed out that red cells played an important role not only in respiratory and circulation function[246];

4. Construction of cDNA library of immune cells. Qiang Lu (2004) successfully constructed the cDNA library of peripheral blood leucocytes in normal *Cyprinus carpio* L.[247], Guozhu Liu (2007) successfully constructed SMART cDNA library of kidney of large yellow croaker (*Pseudosciaena crocea*) stimulated with the inactivated trivalent bacteria vaccine[248].

2.4.2. Study on the Purification, Localization and Clone of Immunoglobulin (Ig)

It is of great importance of effect of the external factors on the development and function of immune system to the theory and application prospects of fish immunology research. Therefore, preparation for a polyclonal immunoglobulin or monoclonal antibodies, as a research tool, has demonstrated its increasingly important role. In recent years, China has conducted a lot of research, including the following: 1. Purification of Ig. Chinese investigators had already successfully purified Ig from the whole serum of *Pseudosciaena crocea, Anguilla anguilla,Lutjanus argentimaculatus, Epinephelus awoara, Lates calcarifer, Paralichthys olivaceus, Monopterus albus, Carassius auratus gibelio*, egg yolk of *Acipenser schrencki, Acipenser ruthenus, Acipenser baeri, Hucho taimen, Oncorhynchus mykiss*[249～255]; 2. Localization of Ig. Yueling Zhang (2005) found that do exist Ig-like proteins in the shrimp, which localized not only in hemolymph but also in haemocytes, lymphatic-like organ and hepatopancreas[256], Dajiang Li (2001) pointed out that maternal Ig of *Carassius auratus* began to appear in the early IV phase egg, during the development of egg, more Ig localized inside the internal membrane than at other sites, and more Ig existed in the internal membrane, embryo body and yolk duiring embryonic development[257]; 3. Clone and expression of Ig. Fan Wang (2006), Jinyu Shen (2006) successfully prepared monoclonal antibodies against *Siniperca chuatsi* Ig and *Carassius auratus gibelio* IgM[254,258],respectively; Hongbai Liu (2006) obtained cDNA of the Ig heavy chain variable regions from the spleen of *Acipenser schrenskii* by RT-PCR[259], Yan Yao (2006)inserted cDNA encoding the molt-inhibiting hormone 1 (Ers-MIHI) of the mitten crab *Eriocheir japonica sinensis* into pET28a (+) bacterial expression vector by recombinant DNA techniques and highly expressed the fusion protein in inclusion bodies in *E.coli* BL21[260].

2.4.3. Preparation and Detection of Antibody

Study on the preparation and detection of antibody was necessary to detect fish pathogens and fishery drug residues. Chinese investigators have successfully developed single-chain antibody fragment against white spot syndrome virus of Shrimp and monoclonal antibodies against *Macrobrachium rosenbergii* nodavirus, acute virus necrobiotic disease virus in *scallop Chlamys farreri, Aeromonas sobria, Aeromonas hydrophila* and its aerolysin and malachite green[262～267]; Meanwhile, Chinese investigators had also made great progress in the detection of antibodies. Qiang Li (2007) used ELISA to detect the antibody in the serum of Japanese flounder against lymphocystis disease virus (LCDV) by the monoclonal antibody 21-14[268].

2.5. Study on Pharmacokinetics of Fishery Drugs

Study on pharmacokinetics of fishery drugs played an important role in the accurate use of drugs, promotion of new fishery drugs and effective control of fish diseases. The progress of pharmacokinetics of fishery drugs obtained in recent years was reflected in the following aspects: 1. Drug administration and detection assay were diversified and simplified. The current drug residue determination assay had been improved from the familiar HPLC, RP-HPLC to more accurate and credible binding assays such as HPLC-GC-ECD, HPLC-MC, *etc.*, and more economical and convenient methods such as indirect competitive ELISA and microbiological assay; 2. The tested animals in the study was diversified, which were expanded from traditonal freshwater animals such as grass carps, crucian carps, *etc.* to the special economic and seawater species such as *acipenser schrenckii*, *Oreochromis niloticus, Sciaenops ocellatus*, *etc*., and the tested fishery drugs were also becoming abundant, which included nearly all the familiar fishery drugs such as enrofloxacin, norfloxacin, sarafloxacin, lomefloxacin, danofloxacin, oxolinic acid flumequine, sulfamethazine and sulfamethoxazole[269～275]; 3. New discoveries were made in the studies on the metabolism and elimination of fishery drugs. Xianle Yang (2005) found almost the same metabolism and elimination regularity of olaquindox in *Cyprinus carp* and *Carassius auraatus gibelio* bodies[274];Jun Tang (2006) discovered that plasma protein combination rate of enrofloxacin influence on the metabolism of enrofloxacin in *Eriocheir sinensis*[275].

2.6. Study on the Control of Fish Diseases

Drug prevention, immune control and biological control are three ways to treat fish deseases in China, which are also the orientations of fish disease control research in China. In the studies of fish diseases in China, virus disease control was the most difficult issue. Although Chinese investigators developed antiviral drugs such as fishery vaccines and immune polysaccharide, and some had bright application prospects. Guofu Chen (2005) found peptidoglycan could enhance the resistance of shrimp to WSSV[276], Zirong Xu (2005) proved that a specific immune response and protection could be induced in shrimp by oral vaccination of recombinant protein rVp28[277], Deshang Li (2004) developed fence-enclosing method to control white spot disease[278]; As for the control of bacterial diseases, immune and biological control measures were the present hot issues of fish bacterial diseases due to the serious drug resistance of pathogenic bacteria. Now Chinese investigators had already successfully developed some efficacious substance such as *Aeromonas hydrophila* biofilm vaccine, bivalent vaccine against *Vibrio anguillarum* and *Vibrio alginolyticus*, microencapsulated vaccine against *Vibrio anguillarum*, immune polysaccharide（yeast cell wall） and β-glucan[279～283]. Furthermore, progress was also achieved in the ecological control of fish bacterial disease. Xiaoming Gong (2006) proved that ozoned seawater could damnify and degrade Vibrio DNA greatly[284], Caihong Hu (2005) discovered that Cu beating montmorillonite（Cu-MMT） showed a strong ability to inhibit adhesion of pathogenic strains to fish epithelial cell[285]; As for the control of fungus diseases, the treatment of saprolegniasis was the hot issue in China. Xuehong Song (2007) discovered that the combination treatment with EM and deuterium sulfate prolonged the inhibition time of *Saprologenia* parasitia, and the stronger inhibition of *Saprologenia* parasitia was observed in the treatment of 8 μL/L EM and 50 μL/L deuterium sulfate[286], Caigen Yang (2000) proved that formaldehyde and neomycin had high efficacy to saprolegniasis[287]; additionally, as for the control of parasite diseases, drug control was the main treatment measure, and trichlorfon and copper sulfate were still the widely used insecticides. However, the superfluous use of the familiar insecticides tended to cause accumulation in soils and waters[288]. Chinese herb medicine as a novel insecticide gradually exhibited good effects on killing parasites in recent years. Gaoxue Wang (2006) proved that ginkgolic acids and *Cnidium monnieri* (L.) *Cuss.* extracts were efficient to kill *Dactylogyrus* with the killing rate of nearly 100%[289, 290].

3. Problems Existing in Fish Disease Research

Compared with fish disease research in foreign developed countries, there are more problems and difficulties in fish disease research in China, such as the weak research foundations, lagged research assays, ignorance of safety of fishery drug use. These problems are not harmonious with the current status of aquaculture industry in China. Therefore, some involved problems have to be handled urgently.

3.1. Study on Rapid, Sensitive and Accurate Detection Assays of Pathogens are Difficult to Meet the Demands of Industrialization

Rapid, sensitive and accurate detection assays are good tools to diagnose the pathogen in fish diseases and aquatic product. In foreign developed countries, immunological and molecular biological techniques were used to detect the fish pathogens long time ago, and most detection assays were successfully developed the commercial diagnosis kits. In 1996, William (1996) made use of a diagnosis kits for infectious pancreatic necrosis virus (IPNV) and obtained good results[291. While in China, even though a few diagnosis kits for pathogens were also developed, such as diagnosis kit for pathogenic *Aeromonas* sp. and fish iridovirus[292,293], most were seldom applied in aquaculture by reason of their high costs and technical difficulties; and the studies on the detection assays for parasites and *Saprolegnia* sp. were also few.

3.2. Pathogenic Mechanism of Pathogens Needs to be Clarified

Study on the pathogenic mechanism of pathogens is always the hot issue in fish disease research at home and abroad and now big progress was made in the pathogenic mechanism of some important pathogens such as WSSV, *Aeromonas hydrophila., etc*. However, more misunderstandings and errors still exist in the research. The types, components and adhesive mechanism of pilus in the pathogenic mechanism of *Vibrio* sp. such as *Vibrio vulnificus* and *Vibrio mimicus* were still puzzling[294], the physiological and biochemical properties of pathogenic-invited constituents of Rickettsia and Rickettsia-interference course with the normal metabolism of host cells were yet to be expained[295]. In such circumstances as a large variety of fish diseases and their pathogens in China, what a long way to go for the pathogenic mechanisms of pathogens?

3.3. Fish Immune-Related Genes Lack Solid Foundation

Study on the fish immune-related genes has become immune direction on an avant-garde at home and abroad. Boudinot(1998) and Lorenzen(2000) obtained glycoprotein genes of IHNV and VHSV respectively, which could introduce specific immune response of rainbow trout[296,297], Zou (1999) and Liu (2002) successfully cloned IL-1β cytokine and CC chemokine respectively[298,299], Yibing Zhang (2003) identified 23 antivirus-related genes from inactivated virus of grass carp hemorrhage[300].All these studies promoted the development of novel fishery vaccines. However, because the basic biological properties of pathogens such as pathogenicity, stability, variation, reproduction, storage still needed to be deeply understood and its basic theories and techniques were not yet fully grasped, the follow-up studies were like "castles in the air" and difficult to achieve an ideal result.

3.4. Progress in the Fishery Vaccine Development was Slow

Since Duff (1942) developed the world's first vaccine[301], nearly 30 foreign vaccine was put into the market and application[302], and the development of novel vaccines such as gene engineering vaccines, DNA vaccines also made significant headway, such as DNA immunization against the viral hemorrhagic septicemia virus, DNA vaccination against infectious hematopoietic necrosis virus, DNA vaccine against infectious hematopoietic necrosis virus[303～305]. However, in China, although the development of fishery vaccines become a hot research field and fishery vaccine development has also made significant headway, the Chinese fishery vaccine development was still greatly constrained by reason of variability of pathogens, diversity of antigen, weak research base of pathogen functional genes and lack of effective and convenient means to administer vaccines and enhance immune effects. So far, only cell cultured inactivated vaccine against grass carp hemorrhage and inactivated vaccine against *Aeromonas hydrophila* had access to new veterinary certificate of China, but the two fishery vaccines have not yet real industrial applications.

3.5. Exclusive Fishery Drugs were Scarce for Aquaculture

In the late 1980s, Bayer company in German developed enrofloxacin and put it into the market[306], Schering-Plough company in America developed florfenicol[307], both brought bright prospects to fish disease control. However, exclusive fishery drugs are still faced with the plight of scarcity, especially for antivirus drugs, there are still inefficient products available. On the one hand, independent research and development capacity of fishery drugs is weak, most fishery drugs directly or indirectly derived from human medicine, veterinary medicine or pesticides and had not yet formed independent product line[308]; On the other hand, research on the banned fishery drug substitutes also failed to keep up with aquaculture development, leading to the occurrence of the phenomenon that banned fishery drugs are still in widespread use. The "drug residue incident of eels" and "drug residue incident of *Scophthalmus maximus*" are very good reflections of the adverse phenomenon. Furthermore, with the widespread emergence of drug resistance of pathogens, new antibiotic development failed to resolve the increasingly complex issue of drug resistance.

3.6. Pharmacokinetics Study of Fishery Drugs can not Meet the Needs of Aquaculture

The study on fishery drugs pharmacokinetics began in the late 1990s, but had made substantial progress. Now Chinese investigators had studied pharmacokinetics of enrofloxacin, norfloxacin, sarafloxacin, lomefloxacin, danofloxacin, oxolinic acid flumequine, chloramphenicol, florfenicol in *Sciaenops ocellatus, Penaeus vannamei, Carassius auratus*, *Cyprinus carpio*, *Muraenesox cinereus,acipenser schrenckii, Oreochromis niloticus* , European eel, respectively, explored factors such as ways of administration, dosage, species differences, temperature, salinity, gender and age , etc. on the pharmacokinetics, and developed the corresponding residues of the highest standards and withdrawal period of fishery drugs[269～275，309～312].However, the type of drug and its metabolites, and animals tested in the pharmacokinetic studies still far failed to meet the needs of aquaculture , and pharmacokinetics of Chinese herbal medicines in aquatic animals was also seldom reported, and the "first pass effect" and "multi-drug crest" phenomenon of drug metabolism mechanism were also not well explained, remaining in the stage of description, derivation and analysis.

4. Prospects in Fish Disease Research

Taking an overview of the fish disease research history, fish disease research in China undoubtedly made many remarkable achievements. It is not difficult to find, however, previous studies generally focused on the basic application, prevention and treatment of fish diseases and the integration of existing application technology, studies in the rapid diagnosis of fish diseases, pathogenic mechanism of pathogens and biological control with innovative technology, *etc.* are very weak.In order to improve fish disease research in China on the overall level, the following recommendations to strengthen the research work are reflected in several aspects:

4.1. The Commercialization Process of Rapid Diagnosis Means Should be Accelerated

Study on the rapid, simple and accurate diagnosis means of pathogens is the hot issue in the fish disease control research. In recent years, studies on the elimination of drug residues, drug resistance of pathogens and deterioration of ecological environment were emphasized, which greatly promoted the molecular biology and immunology technology applying in fish pathogen detection. Therefore, how to integrate these new technologies into the development of rapid diagnosis kit for fish diseases, the commercialization process of diagnosis of fish pathogen was undoubtedly greatly accelerated. the main problems to be solved in future are purification of fish pathogens, removal of false-positive pathogens, improvement of sensitivity of pathogen detection method and strategy of the joint use of nucleic acid detection methods and immunoassay methods.

4.2. Study on the Pathogenic Mechanism Centered on Functional Genes (or Proteins) should be Deepened

Study on the functional genes (or protein) of pathogens has become the avant-garde direction in the pathogenic mechanism research. *Vibrio* sp., *Aeromonas* sp.and white spot syndrome virus are common pathogens of fish and extremely harmful. Therefore, how to strengthen pathogenic mechanism of these important pathogens is of great significance to the timely fish disease prevention. The future research should be carried out on the following: 1. From the start with the constraints and limitations producing pathogenic agents, pathogenic characteristics and its producing conditions of pathogenic agents are to be studied; 2. More efforts on the study of virulence factors of pathogens should be made, and new virulence factors and their corresponding gene regulation should also be determined; 3. the interactions between host resistance factors and pathogenic agents of pathogens should be studied.

4.3. Fish Pathology and its Study Means should be Further Strengthened

After long term research, most pathological changes of fish diseases have been clearly revealed from clinical and organic levels. However, pathological studies on the ultrastructural and molecular levels still remained mysteries, which promoted the conduction of fish pathology studies from the ultrastructural, cellular and molecular perspective and imported molecular biological and immunological technologies to fish pathological studies[313]. Therefore, some new technologies such as immuno-histochemical and molecular hybridization techniques and PCR, *etc.* will surely be widely used in the fish pathological study in China, which in turn promotes the rise and development of fish ultrastructural and cell pathology studies in China.

4.4. A Simple, Correct and Effective Evaluation Method of Pharmacodynamics on Fishery Drugs should be Established

Correct and comprehensive evaluation system of pharmacodynamics on fishery drugs is not only related to the research and development of fishery drugs, but also to the safe quality of aquatic products. China's current evaluation indicators of pharmacodynamics on fishery drugs are not perfect enough, effects of factors on the efficacy of antibiotics were less studied, the evaluation standards and detection indicators of some fishery immunostimulants were difficult to be determined, studies on the efficacy of Chinese herbal medicine were greatly limited for its multiple components, and still focused on the minimum inhibitory concentration (MIC), the minimum bactericidal concentration (MBC), MIC Index (FIC) [314]. In addition, most fishery drugs in China are in short of strict and comprehensive toxicological data, such as organic phosphorus, organochlorines, pyrethroid, heavy metal salt compounds and Chinese herb medicines. Therefore,

it is certainly the hot focus of research to perfect pharmacodynamic evaluation of fishery drugs and establish simple, correct and effective evaluation method of fishery drug pharmacodynamics.

4.5. New Ways to Effectively Prevent and Control Fish Diseases should be Explored

In recent years, drug resistance of pathogens and scarcity in efficient drugs have aroused extensive concern and anxiety. Therefore, how to explore new ways to effectively prevent and control fish diseases has become hot concern about fish disease control research. The directionS of future research are that: 1. By studies on pathogenic agent of pathogens, every effort will be made to find out the immune protective antigens and possible immune control methods and develop new vaccines; 2. The usage, dosage, *etc.*of immunostimulants should be deeply studied so as to make it an effective tool to defend pathogenic bacteria, viruses and parasites; 3. Probiotics with the function of inhibiting pathogens should be isolated and developed, and studies on the their administration routes and new formulations should be strengthened; 4. Chinese herbal medicine with low toxicity, low drug residue and small side effects should be vigorously developed so as to improve the fish immunity.

REFERENCES

[1] Xiangsheng Wu. Comparing the regional competitiveness of fishery areas in China. *Guangdong Feed*, 2006, 15(3): 3～5.

[2] Aiping Chen. Aquaculture disease monitoring report in China duiring 2005. *Scientific Fish Farming*, 2006, 8, 48～49.

[3] Wu Wang. Fish culture and populaotion increasing. Beijing: China Agriculture Press; 2000.

[4] Jianguo Wang. Twenty years of research on fish disease in China. *Fish Disease Research of China*, 2000, 22(1): 1～37.

[5] Dashu Ni. Fish prevention work report in 1953. *Acta Hydrobiologica Sinica*, 1955, 1: 7～24.

[6] Deming Wang. Primary study on the pathogen of red skin disease in black carp. *Acta Hydrobiologica Sinica*, 1956, 1: 1～17.

[7] Shanghai fisheries College. Fish disease (first edition). Beijing: China Agriculture Press; 1961.

[8] Institute of Hydrobiology. The manual for surveys of fish disease (first edition). Shanghai: Shanghai Science and Technology Press; 1961.

[9] Institute of Hydrobiology. The pathogeny graphy of fishes in Hubei province. Beijing: Science Press; 1973.

[10] Yangtze River fisheries research institute. Prevention manual of familiar fish disease (first edition). Beijing: China Agriculture Press; 1972.

[11] Wengong Zuo, Jingfu Chen. Prevention manual of familiar fish disease (third edition). Beijing: China Agriculture Press; 1987.

[12] Institute of Hydrobiology. The manual for surveys of fish disease (second edition). Shanghai: Shanghai Science and Technology Press; 1981.

[13] Jinpei Pan. Diagnosis and prevention manual of fish disease. Shanghai: Shanghai Science and Technology Press; 1988.

[14] Qingxian Meng. Prevention manual of shrimp disease. Qingdao: The press of Ocean University of Qingdao; 1991.

[15] Qiyan Huang. Aquatic animal diseases. Shanghai: Shanghai Science and Technology Press; 1993.

[16] Duan Lin, Yanhong Dong, Yin Yuan, *et al*. Application of PCR technique in molecular biology study and detection of Iridovirus. *Marine Environmental Science*, 2007, 26(2): 190～194.

[17] Yibing Zhang, Yaohua Shi, Jianfang Gui. Construction of antiviral subtractive cDNA librar y of cultured fish cells. *Acta Hydrobiologica Sinica*, 2003, 27(2): 113～119.

[18] Shihua Ding,Yiding Wang, Yuanyi Peng, *et al*. Immunogenecity and protective effects of an oral vaccine microp article against *Aeromonas hydrophila* in grass carp (*Ctenopharyngodon idella*). *Journal of Southwest Agricultural University*, 2005, 27(6): 888～893.

[19] Fengrong Zhen, Hongzhan Liu, Jinxing Zhang, *et al.* The role of liposome on immune efficacy of DNA vaccine against lymphocystis disease virus in *Paralichthys olivaceus*. *High Technology Letters*, 2006, 16(6): 643～648.

[20] Hongmei Cao, Jian Li, Wenbin Zhan. Immune efficacy of a bivalent vaccine for turbot against *Vibrio anguillarum* and *V. alginolyticus*. *Journal of Fisheries of Chin*a, 2006, 13(3): 397～403.

[21] Yujuan Guo, Xuenian Chen. The effect of A3α-peptidoglycan on immune response of Carassius auratus varpengze inoculated with *Aeromonas hydrophila* bacterin. *Chinese Journal of Preventive Veterinary Medicine*, 2006, 28(5): 562～566.

[22] Xianle Yang, Haipeng Cao, Yunyun Qian. *Bdellovibrio bacteriovorus*: A new prevention method against aquatic animal disease. *Freshwater Fisheries*, 2006, 36(2): 55～61.

[23] Jinying Wu, Junyi Lv, Hua Zeng, *et al*. Population dynamics of *Saccocoelioides huidongensis* in *Mugil cephalus*. *Journal of Fisheries of China*, 2000, 7(1): 46～51.

[24] Jinying Wu, Junyi Lv, Hua Zeng. Studies on the population dynamics of *Lenolopa huizhouensis* in *Scatophagus argus*. *Acta Hydrobiologica Sinica*, 2002, 26(1): 82～89.

[25] Weijian Yao, Pin Nie. Population distribution and seasonal alternation of two species of monogeneans on the gills of *Hypophthalmichthys molitrix* and *Ctenopharyngodon idellus*. *Acta Hydrobiologica Sinica*, 2004, 28(6): 664～668.

[26] Wenbing Wang, Buoping Zeng, Yushuang Luo, *et al*. Epidemiological investigation of *Azygia anguillae* in the host *Monopterus albus* from Dongting lake region. *Freshwater Fisheries*, 2006, 36(6): 36～40.

[27] Deng Deng, Qingwen Zhang, Weiji Wang, *et al*. Investigation on status carrying WSSV in several spawning ground populations of *Fenneropenaeus chinensis*. *Journal of Fisheries of China*, 2005, 29(1): 74～79.

[28] Chengyin Shi, Yingeng Wang, Lei Qin, *et al.* Investigation of pathogen and epidemiology of a novel disease, "viral reddish body syndrome", among farmed *Scophthalmus maximus* in China. *Marine Fisheries Research*, 2005, 26(1): 1～6.

[29] Shan Jin, Guoliang Wang, Qingsong Zhao, *et al.* Epidemiology of vibriosis in large yellow croaker *Pseudosciaena crocea* (Richardson) in marine cage culture. *Fisheries Science*, 2005, 24(1): 17～19.

[30] Lei Ding, Yongsheng Yue. Epidemiology study on bacterial pustulae disease of *Ophiocephalus argus*. *Journal of Fisheries of China*, 2000, 7(1): 121～123.

[31] Richard Weissenberg. Fifty years of research on the lymphocystis virus disease of fishes (1914~1964). *Annals of the New York Academy of Sciences,*1965, 126:1, 362～374.

[32] Yanyan Chen. Studies on the causative agent of hemorrhage of the grass carp (*Ctenopharyn-godon idellus*). *Acta Hydrobiologica Sinica*,1978, 6(3): 321～331.

[33] Wuhan institute of virology, Yangtze River fisheries research institute. Study on the nucleic acid characterization of the pathogen of hemorrhage of grass carp, fish reovirus. *Freshwater Fisheries*, 1984, 4: 6～8.

[34] Shujian Mao, Jianzhong Shao, Qi Hang, *et al.* On pathogen of hemorrhage of grass carp (*Ctenopharyngodon idellus*). *Journal of Fisheries of China*, 1989, 13(1): 1～5.

[35] Ziyu Zhai, Hongwen Ke, Haoxiang Yu, *et al.* Viral hemorrhagic disease of black carp (*Mylopharyngodon piceus*)[J]. Zoological Research,1987,8(4):379～388.

[36] Lihua Ke, Qin Fang, Yiquan Cai. Characteristics of a novel isolate of grass carp hemorrhagic virus. *Acta Hydrobiologica Sinica*, 1990, 14(2): 153～160.

[37] Shujian Mao, Qi Hang, Hanmin Chen, *et al.* Study on cytopathology of two pathogenic viruses of hemorrhage in grass carp, *Ctenopharynogodon idellus*. *Oceanologia Et Limnologia Sinica*, 1988, 19(5): 435～440.

[38] Yulin Jiang, Yan Li, Zhengqiu Li. Electron microscopic observation of pathogen of carp-pox disease. *Acta Hydrobiologica Sinica*, 1991, 15(2): 193～195.

[39] Yulin Jiang, Bohai Xu, Wei Li, *et al.* Isolation and identification of infectous pancreatic necrosis virus (IPNV) from imported rainbow trout (*Salmon gairdneri*) in P.R.China. *Acta Hydrobiologica Sinica*, 1989, 13(4): 353～359.

[40] Zhizhuang Zhao, Luqi Niu. Isolation and preliminary study of arhabdovirus from rainbow trout (*Oncorhynchus mykiss*) in Benxi, China. *Acta Hydrobiologica Sinica*, 1994, 18(4): 348～353.

[41] Yanyan Chen, Jun Xie, Wei Li. Charactization of atogavirus-like agent, isolated frome eels *Anguilla japonica* in China[J]. Acta Hydrobiologica Sinica, 1999,23(2):155～160.

[42] Zengsi Tao , Chengyao Li, Shenghua Yang, *et al.* Aetiological investigation of eel "mad swim disease. *Chinese Journal of Veterinary Science*, 1997, 17(1): 30～32.

[43] Hongtao Xu, Chunai Pu, Zhongliang Jiang, *et al.* Study on the causative agent of lymphocystic disease in cultured flounder, *Paralichthys Olivaceus*, in mainland China. *Chinese Journal of Virology*, 2000, 16(3): 223～226.

[44] Yunkun Liu, Xiuqin Sun, Jie Huang. Diagnosis of lymphocystis disease of cultured Japanese flounder *Paralichthys olivaceus*. *High Technology Letters*, 2002, 6: 92～95.

[45] Kaibing Li, Cunbing Li, Xinhui Li, *et al.* Pathogen of infective outbreaking disease in *Sciaenop ocellatus*. *Journal of Fisheries of China*, 2002, 9(1): 56～59.

[46] Shuqin Wu, Xinjun Li, Houjun Pan, *et al.* Research on the pathogen of the outbreak-infective disease of *Siniperca chuastsi*. *Journal of Fisheries of China*, 1997, 21: 56～60.

[47] Jianguo He, Shaoping Weng, Zhijian Huang, *et al.* Identification of outbreak and infectious diseases pathogen of *Siniperca chuatsi*. *Acta Scientiarum Naturalium Universitatis Sunyatseni*, 1998, 37(5): 74～77.

[48] Ming Jiang ,Ruiqing Fan, Shaoguo Ru, *et al.* Electron microscope obseration on the morphology of a sphereovirus and cell pathology of *Pagrosomus major*. *Journal of Fisheries of China*, 2000, 24(1): 52～57.

[49] Wenbin Zhan, Kaikang Yu, Qinxian Meng. Study on baculovirus disease of *Peanaeus chinensis*. *Journal of Fisheries of China*, 1995, 2(3): 22～28.

[50] Baozheng Peng, Jiaming Ren, Juying Shen, *et al.* The studies on baculovirus-caused disease of prawns (*Penacus chinesis osbeck*) in Shanghai suburb. *Chinese Journal of Virology*, 1995, 11(2): 151～156.

[51] Bing Wang, Jingyin Jiang, Hua Li, *et al.* Electron microscopic detection on pathogenic virus in chinese Shrimp (*Penaeus orientalis*). *Journal of Dalian Fisheries University*, 1996, 11(1): 22～27.

[52] Xifa Chen, Ping Chen, Dinghu Wu, *et al.* Study on a new bacilliform virus isolated from cultured prawns. *Science in China (series C)*, 1997, 27(5): 415～420.

[53] Qinggang Xue, Yunhao Guan, Wenxing Wang. Primary identification of a reovirus infected *Penaeus chinensis* together with hepatopancreatic parvovirus. *Oceanologia Et Limnologia Sinica*, 1996, 27(3): 314～318.

[54] Tuo Shi, Jie Kong, Zhengmin Bao. Studies on purification and ultrastructure of a baculoviral pathogen in *Penaeus chinesis*. *Acta Oceanologica Sinica*, 1998, 20(2): 60～64.

[55] Xifa Chen, Dinghu Wu, Huai Huang, *et al.* Studies of a new spherical viruses from *Penaeus monodon*. *Journal of Oceanography In Taiwan Strait*, 1996, 15(2): 159～163.

[56] Shouming Feng, Qin Zhang, Wei Deng, *et al.* Report of the baculoviruses disease found in *Metapenacus eusis*. *Animal Science andVeterinary Medicine*, 1999, 16(2): 24～26.

[57] Dong Qian, Guoliang Yang, Wen Liu, *et al.* Preliminary studies on the whitish muscle diseases of *Macrobrachium rosenbergii* post-larvae. *Acta Hydrobiologica Sinica*, 2002, 26(5): 472～476.

[58] Liuji Wu, Xinzhong Wu, Haisheng Xu, *et al.* Study on the pathogen of taura syndrome in *Panaeus chinensis* in Zhangjiang, Guangdong. *Chinese Journal of Veterinary Medicine*, 2006, 42(12): 52～53.

[59] Chengliang Gong, Renyu Xue, Guangli Cao, *et al.* Study on reovirus-like virus of *Eriocheir sinensis*. *Virologica Sinica*, 2000, 15(4): 395～399.

[60] Zhirong Shen, Xingyang Li, Luanfeng Pan, *et al.* The separation of an infectious virus to *Hyriopsis cumingi* and a preliminary research on its biological feature. *Journal of Zhejiang Ocean University*, 1986, 5(1): 95～97.

[61] Yaofu Liu, Zhuhui Huang, Shunhua He, *et al*. Preliminary study of the causative agent of the freshwater mussel *Hyriopsis cumingh plague*. *Freshwater Fisheries*, 1993, 23(1): 3～7.

[62] Jingyin Jiang, Dianlou Xing, Bin Wang, *et al*. Electron microscopic observation of a kinds of spherical viruses in *Eriocheir sinensis* for aquaculture in Ponds. *Journal of Dalian Fisheries University*, 1996, 11(1): 51～53.

[63] Qiya Zhang, Zhengqiu Li, Yulin Jiang, *et al*. Preliminary studies on virus isolation and cell infection from diseased frog *Rana grylio*. *Acta Hydrobiologica Sinica*, 1996, 20(4): 390～393.

[64] Zaixian Chen, Jianchuan Zhen, Yulin Jiang. An iridovirus isolated from sick soft shelled turtle with "red neck disease". *Chinese Journal of Veterinary Science*, 1998,18(2): 135～139.

[65] Liangyi Xue, Guoliang Wang, Xinglin Xu, *et al*. Electric microscopic observation on the hemorrhage disease of *Lateolabrax japonicus*. *Journal of Ningbo University (Natural Science and Engineering Edition)*, 1997, 10(2): 85～87.

[66] Nin Yang, Sulian Ren, Weibo Song. A spherical virus-like particle found in *Argopecten irradians Lamarck* and its cytopathology. *Marine Sciences*, 2005, 9: 56～59.

[67] Bin Wang, Xia Li, Chuanzhou Gao. Infection of spherical viruses from *Haliotis discus hannai Ino*. *Virologica Sinica*, 1997, 12(4): 360～363.

[68] Bin Wang, Xia Li, Shufan Liu, *et al*. Studies on pathogeny and histopathology of "crack shell disease" of *Haliotis discus hanni*. *Journal of Fisheries of China*, 1998, 22(1): 61～66.

[69] Sulian Ren, Dexiu Wang, Xiuzheng Sheng, *et al*. Cytopathology and morphogenesis of a spherical virus found in the *Meretrix meretrix* suffering "red muscle disease" (RMD). *Journal of Fisheries of China*, 2002, 26(3): 265～269.

[70] Bin Wang, Xia Li, Xidong Yang. Examination of spherical viruses from *Mytilus edulis*. *Virologica Sinica*, 2003, 18(3): 288～291.

[71] Emmerich R, Weibel E. Über eine durch Bakterien erregte seuche unter der forellen. *Arch Hyg Bakteriol* , 1894, 21: 1～21

[72] Zhixun Xie, Yaoshan Pan, Xianwen Deng, *et al*. Study on the detection of white spot syndrome virus and taura syndrome virus by multiplex PCR. *Acta Hydrobiologica Sinica*, 2005, 29(5): 595～597.

[73] Dongchun Yan, Shuanglin Dong, Jie Huang, *et al*. Trace white spot syndrome virus (WSSV) detection with polymerase chain reaction (PCR) –dot blot hybridization. *High technology letters*, 2004, 10: 33～36.

[74] Dongchun Yan, Shuanglin Dong, Jie Huang. PCR false negative result in white spot syndrome virus (WSSV) detection. *Oceanologia Et Limnologia Sinica*, 2007, 38(2): 146～150.

[75] Xiaolan Wang, Chengliang Gong, Yuxing Xue, *et al*. Gene clone and analysis of envelope protein VP28 of white spot syndrome virus of *Macrobrachium rosenbergii*. *Freshwater Fisheries*, 2007, 37(3): 12～16.

[76] Qijun Jia, Xiaolin Meng, Jinping Xu, *et al*. Expression of envelope protein VP19 of *Penaeus monodon* WSSV in *E.coli* and the effect against WSSV. *Virologica Sinica*, 2006, 21(6): 585～588.

[77] Yanbing Zhu, Feng Yang. Expression of three putative zinc finger protein genes wsv063,wsv069 and wsv477 in white spot syndrome virus (WSSV) genome and their products purification. *Journal of Xiamen University (Natural Science)*, 2005, 44(5): 697～700.

[78] Fei Liu, Shujian Cun, Kai Yang, *et al*. Functional analysis on structure proteins VP28 and VP26 of white spot syndrome virus. *Acta Scientiarum Naturalium Universitatis Sunyatseni*, 2006, 45(4): 86～89.

[79] Huibo Chen, Yangdong Lin, Yanling Weng, *et al*. Isolation, identification and drug resistance of the pathogens of red fin disease in eels. *Acta Hydrobiologica Sinica*, 1992, 16(1): 40～46.

[80] Guoliang Wang, Xinglin Xu, Zheng Lu. Study on the pathogens of red fin disease in eels. *Journal of Zhejiang Ocean University*, 1993, 12(3): 186～191.

[81] Changfu Chen, Yi Liu. Isolation and identification of the pathogen of bacterial septicemia in Taiwan snakehead (*Channa Maculata* L.). *Chinese Journal of Veterinary Science*, 1998, 18(2): 144～146.

[82] Xilin Zhang. Isolation of the pathogen of bacterial enteritis its pathology in *Silurus merdionalis*. *Freshwater Fisheries*, 1998, 28(3): 6～7.

[83] Jingyu Shen, Wen Liu, Dong Qian, *et al*. Studies on the pathogens of haemorrhagic disease of *Monopterus albus*. *Journal of Zhejiang Ocean University (Natural Science Edition)*, 2001, 20(2): 120～122.

[84] Aiying Lei, Weijian Huang, Aizheng Qin. Study on the pathogen (*Aeromonas* sp.) of bacterial septicemia in *Channa asiastica* (Linnalus). *Freshwater Fisheries*, 1999, 29(4): 16～19.

[85] Xiaoyan Jiang, Yueqiang bai, Shuying Xu. Primary study on the pathogen of skin fester disease in grass carp in cage culture. *Freshwater Fisheries*, 2003, 33(6): 8～11.

[86] Zengfu Wang, Hongmei Xie, Jing Zhang. Study on the pathogen of skin ulceration disease in *Varicorhinus macrolepis*. *Freshwater Fisheries*, 2001, 31(6): 40～42.

[87] Weihua Chu, Yong Yu. Isolation and identification of *Aeromonas hydrophila* in sturgens. *Freshwater Fisheries*, 2003, 33(2): 16～17.

[88] Haipeng Cao, Xianle Yang, Peng Gao, *et al*. Preliminary study of the pathogens isolated from bacterial septicaemia syndrome of sturgeons. *Freshwater Fisheries*, 2007, 37(2): 53～56.

[89] Jingyu Shen, Wen Liu, Dong Qian, *et al*. Studies on the pathogens of bacterial diseases of *Macrobrachium nipponense*. *Journal of Zhejiang Ocean University*, 2000, 19(3): 222～224.

[90] Chengyu Hu, Yijiang Hong. Study on pathogen of 'red-leg' disease in bullfrogs. *Journal of Fishery Sciences of China*, 2000, 7(2): 126～128.

[91] Lu He, Xiaohui Ai. Study on the pathogen isolated from ascitic fluid disease of bullfrog[J]. Freshwater Fisheries, 1995,25(3):16～18.

[92] Changyi Zhou, Xiaofeng Chen, Bin Luo, *et al*. Preliminary study on the bacterial pathogen of "black spot" disease of *Trionyx sinensis*. *Journal of Jimei University Natural Science*, 2001, 6(4): 308～311.

[93] Lei Ding, Yongsheng Lei, Jingyu Song. Study on pathogen of "skin ulcer"disease of rainbow trout. *Freshwater Fisheries*, 2002, 32(3): 28～30.

[94] Yingbing Su, Guiwei Zou, KepingYuan, *et al*. Isolation and identification of the pathogen of fulminant septicaemia from *Ictalurus punctatus*. *Freshwater Fisheries*, 2006, 36(5): 37～41.

[95] Wanqi Cai, Peifang Sun. A pathogen study on disease caused by *Aeromonas sobria* in tilapias and drug sensitive test. *Journal of Fishery Sciences of China*, 2002, 9(3): 243～246.

[96] Peifang Sun,Wanqi Cai. A pathogen study on disease caused by *Aeromonas sobria* in soft-shelled turtles. *Freshwater Fisheries*, 1998, 28(4): 3～5.

[97] Bohai Xu, Ruifang Xu, Mulin Xiong. Studies on the "stigmatosis" of silver carp (*Hypophthalmichthys molitrix*) and bighead (*Aristichthys nobilis*). *Oceanologia Et Limnologia Sinica*, 1980, 1: 3～8.

[98] Haishen Wen, Zhengqi Liu. A preliminary study on the pathogenic bacteria of"rotten skin"disease of crucian carp *Carassius auratus*. *Journal of Dalian Fisheries University*, 1995, 10(1): 49～55.

[99] Kebing Liu, Chen Zhou, Jiafu Liu. Studies on the pathogenic bacteria of *Pseudosciaena crocea* in marine cage culture. *Marine Sciences*, 1999, 4: 58～62.

[100] Lin Ye, Liyang Liu. The study on the pathogenic bacteria of the fester disease of cultured juvenile *Discus abalone*. *Journal of Fisheries of China*, 1997, 4(4): 43～48.

[101] Youzhi Ma, Miaoan Shu. Isolation and identification of *Pseudomonas alcaligenes* from mud eels. *Freshwater Fisheries*, 2000, 30(5): 33～34.

[102] Bin Huang, Shifeng Chen, Yong Chen. Study on the cys disease of *Cistoclemmys flaciomarginata*. *Freshwater Fisheries*, 2002, 34(5): 44～46.

[103] Kai Zhou, Qihuan Zhou. The pathogen of red-head disease in European eel (*Anguilla anguilla*)[J]. Acta Hydrobiologica Sinica, 1999,23(4):304～310.

[104] Xiaojun Zhang, Wenbin Zhang, Cuizheng Chen, *et al*. Studies on the infection and pathogen of *Edwardsiella tarda* in flounder *Paralichthys olivaceus*. *Acta Hydrobiologica Sinica*, 2005, 29(1): 31～37.

[105] Bin Wang, Qian Sun, Wei Fan, *et al*. Pathogenicity and identification of pathogenic bacterium causing hemorrhagic septicemia in cultured turbot (*Scophthalmus maximus*). *Journal of Dalian Fisheries University*, 2006, 21(2): 100～104.

[106] Jun Li, Xianhui Yan, Jixiang Chen, *et al*. Studies on the characteristics of pathogenic *Edwardsiella tarda* isolated from diseased *Scophthalmus maximus*. *Periodical of Ocean University of China*, 2006, 36(4): 649～654.

[107] Shuxia Xue, Shouming Feng, Jinsheng Sun. Isolation and identification of pathogenic bacteria in swollen abdomen of cultured turbot (*Scophthalmus maximus*) and flounder (*Paralichthys olivaceus*). *Oceanologia Et Limnologia Sinica*, 2006, 37(6): 548～554.

[108] Lei Ding, Yongsheng Yue, Jingyu Song. Study on the pathogen of white abdominal shell disease and skin fester disease of soft-shelled turtle (*Trionyx sinensis*) and its drug prevention and treatment. *Freshwater Fisheries*, 2001, 31(1): 46～48.

[109] Baokun Wang, Junhong Yu, Jun Li, *et al.* Isolation and identification of pathogen (*Vibrio harveyi*) from sea perch, *Lateolabrax japonicus*. *Journal of Fisheries of China*, 2002, 9(1): 52～55.

[110] Xiangao Chen, Shuqin Wu, Cunbin Shi, *et al.* Isolation and identification of pathogenic *Vibrio harveyi* from estuary cod *Epinephelus coioides*. *Journal of Fishery Sciences of China*, 2004, 11(4): 313～317.

[111] Guozhong Tong, Yasu Shi, Chaobo Xue. Analysis of the pathogenic bacteria and drug sensitivity test of the tail-rot *Pseudosciaena crocea* cultured in marine cage in Zhoushan, China. *Journal of Zhejiang Ocean University (Natural Science Edition)*, 2005, 24(2): 118～121.

[112] Bin Wang, Hua Li, *et al.* Study on two new pathogens of "red leg disease" found in *Penaeus chinensis*. *Journal of Dalian Fisheries University*, 1993, 8(2): 43～48.

[113] Xiaoyan Chen, Chaoqun Hu, Chang Chen, *et al.* Characteristics of pathogenic bacteria of vibriosis in caged malabar grouper malabar grouper, *Epinephelus malabaricus*. *Marine Sciences*, 2003, 27(6): 68～72.

[114] Xuefeng Hu, Cunbin Shi, Houjun Pan, *et al.* Preliminary studies on the pathogen (*Vibrio alginolyticus*) of the ulceration disease of maricultured estuary cod, *Epinephelus coioides*. *Periodical of Ocean University of China*, 2005, 35(2): 232～236.

[115] Weini Zhang, Li Zhou, Jing Xing, *et al.* Isolation and identification of pathogen SR1 associated with swollen abdomen of cultured turbot (*Scophthalmus maximus*). *Journal of Fisheries of China*, 2006, 13(4): 603～609.

[116] Yongcan Zhou, Ben Zhang, Xuefeng Chen, *et al.* Preliminary studies on the red body disease in *Penaeus vannamei*. *Marine Sciences*, 2003, 27(5): 61～65.

[117] Jing Zhou, Junpeng Cai, Hongzhi Yang. Isolation and characterization of *Vibrio alginolyticus* as a pathogen to the massive mortality of abalone (*Haliotis diversicolor*) postlarvae. *Transaction of Oceanology and Limnology*, 2006, 3: 54～59.

[118] Guangfeng Liu, Shinin Zhou, Liwen Xu, *et al.* Studies on the pathogenic bacteria of acute death and striping off disease in cultured postlarvae small abalone *Haliotis diversicolor* reeve. *Journal of Fisheries of China*, 2006, 13(4): 655～661.

[119] Rongxing Ji, Wenzheng Zou, Yinlong Li. Study on pathogen of "skin ulcer" disease of larve yellow croaker *Pseudosciaena crocea*. *Marine Sciences*, 2004, 28(11): 57～60.

[120] Liping Wang, Xiaoling Song. *Vibrio cholerae* (Non-01), a pathogen of *Penaeus chinensis* larvae (mysis stage) disease. *Journal of Fisheries of China*, 1997, 4(1): 45～51.

[121] Xiaohua Zhang, Shaoan Liao, Jun Li, *et al.* Studies on pathogenic bacteria (*Vibrio natriegen*) of *Argopecten irrsdins lamarck*. *Journal of Ocean University of Qingdao*, 1998, 28(3): 426～432.

[122] Hui Xiao, Jun Li. Studies on pathogens of rotted gill and rotted caudal fins of seaperch (*Lateolabrax japonicus*) fry. *Journal of Ocean University of Qingdao*, 1999, 29(1): 87～93.

[123] Zheng Zhang, Yingeng Wang, Wenjun Han, *et al.* Identification and phylogenetic study of pathogenic bacteria causing ulcerative fin of cultured turbot *Scophthalmus maximus*. *Advances in Mrine Science*, 2004, 22(2): 193～197.

[124] Jingyu Shen, Wenlin Yin. Studies on the pathogens of bacterial diseases of *Eriocheir sinensis*. *Journal of Fisheries of China*, 2000, 7(3): 89～92.

[125] Guoliang Wang, Tianlun Zhen. Preliminary studies on pathogen of ascites disease of cultured *Sparus macrocephelus*. *Chinese Journal of Veterinary Science*, 2003, 23(1): 33～35.

[126] Jinnian Li, Yuying Li, Shoukui Hu, *et al*. Analysis on pathogen producing ascitic fluid disease of *Eriocher sinensis*. *Journal of Fisheries of China*, 2005, 12(3): 267～274.

[127] Xueping Ye, Yizhi Luo, Guangzhi Yang, *et al*. A study on the pathogen of *Macrobrachium nipponense* red body syndrome[J]. Journal of Zhejiang Ocean University (Natural Science Edition), 2002,21(2):106～108,122.

[128] Mingyuan Guo, Guangfeng Liu, Juan Feng. Identification and phylogenetic analysis of a pathogenic *Vibrio* sp. isolated from *Rachycentron canadum*. *Journal of Fishery Science of China*, 2006, 13(5): 823～828.

[129] Bohai Xu, Zhan Yin, Yushen Wu, *et al*. Studies on the taxonomy of pathogenic bacteria of the bacterial hemorrhagic septicemia in cultured fishes in freshwater. *Acta Hydrobiologica Sinica*, 1993, 13(3): 5～11.

[130] Jianbo Chang, Xianghong Gong, Fengxian Sun, *et al*. Preliminary studies on a pathogen of vibriosis of cultured *Paralichthys olivaceus*. *Marine Fisheries Research*, 2001, 22(1): 37～41.

[131] Rongxing Ji, Shiliu Hu, Wenzheng Zou, *et al*. Preliminary studies on a pathogen of dermatosis of *Sciaenops ocellatus*. *Marine Sciences*, 2003, 27(8): 1～4.

[132] Jichang Jian, Zhaohe Wu, Gang Chen, *et al*. Isolation and characteristics of pathogen of vibriosis in cobia (*Rachycentron canadum*) maricultured in cage[J]. *Chinese Journal of Veterinary Science*, 2003,23(4):329～340.

[133] Yingeng Wang, Sufeng Sun, Xiaojun Rong. Stomach ulcer disease in auricularia of sea cucumber (*Apostichopus japonicus*) and its etiological identification. *Journal of Fisheries of China*, 2006, 13(6): 908～917.

[134] Keyu Xiao, Weimin Jiang, Xieli Jin. Study on the pathogen isolated from rot-mouth disease in soft-shelled turtles, Trionyx sinensis and its selective test for the curative medicine. *Freshwater Fisheries*, 1998, 28(1): 11～14.

[135] Xiaofeng Chen, Yafang Lu, Changyi Zhou. Studies on the pathogen of stomach and intestines' ulcer and bleed disease of *Trionyx sinensis*. *The Editorial Board of Jimei University (Natural Science)*, 2000, 5(3): 30～34.

[136] Weihao Huang, Jjihai Dong, Yueying Chen. Study on the bacterial "gill-rot disease" of big-head carp (*Aristichthys nobilis*) in cage culture. *Freshwater Fisheries*, 1980, 5: 10～15.

[137] Bohai Xu, Tao Ma. *Escherichia coli* as a pathogen in rainbow trout (*Oncorhynchus mykiss*). *Acta Hydrobiologica Sinica*, 1993, 17(4): 309～317.

[138] Wenbin Zhan, Li Zhou, Zhangqun Chen, *et al*. *Providencia rettgeri*, a new pathogen of *Penaeus chinensis*. *Journal of Fisheries of China*, 1997, 4(1): 38～44.

[139] Jianbo Chang, Xianghong Gong, Fengxian Sun, *et al*. Preliminary studies on a pathogen (*Pasteurella anatipestifer*) of cultured *Kareius bicoloratus*. *Transaction of Oceanology and Limnology*, 2002, 1: 57～61.

[140] Xihe Wan, Jun Li, Meiru Zhang, *et al.* A pathogen HL－1 isolated from *Eriocheir sinensis* larvae with filamentous bacterial disease. *Journal of Fisheries of China*, 2003, 10(5): 392～397.

[141] Chang Chen, Chaoqun Hu, Xiaoyan Chen, *et al.* Identification and characterization of *Shewamella algae* as a novel pathogen of ulcer disease of fish *Scinenops ocellata. Oceanologia Et Limnologia Sinica*, 2003, 34(1): 1～8.

[142] Guozhang Zu, Jinnian Li, Weiyi Yu, *et al.* Diagnosis and control of bacterial "appendage shivering disease" in *Eriocheir sinensis. Freshwater Fisheries*, 2004, 34(2): 27～30.

[143] Shan Jin, Guoliang Wang, Qingsong Zhao, *et al.* Preliminary studies on pathogen and hemopathology of the white cloud disease in *Micropterus salmoides. Acta Hydrobiologica Sinica*, 2005, 29(2): 184～188.

[144] Jinyu Shen, Zhimin Gu, Xiaoyi Pan, *et al.* Isolation and identification of *Citrobacter freundii* from *Cherax quadricarinatus. Journal of Fisheries of China*, 2005, 12(2): 197～200.

[145] Yingeng Wang, Bo Fang, Chunyun Zhang, *et al.* Etiology of skin ulcer syndrome in cultured juveniles of *Apostichopus japonicus* and analysis of reservoir of the pathogens. *Journal of Fishery Sciences of China*, 2006, 13(4): 610～616.

[146] Xiaojun Zhang, Cuizheng Zhang, Hai Fang, *et al.* Identification of pathogen *Listonella anguillarum* isolated from turbot *Scophthalmus maximus. Oceanologica Et Limnologica Sinica*, 2006, 37(5): 417～423.

[147] Lei Ding, Guangwei Xie. Study on the pathogen islated from in bacterial fester disease in *Ophiocephalus argus. Reservoir Fisheries*, 1999, 19(6): 27～28.

[148] Jiaqian Chai, Qiaoling Ding, Zhengguang Wang, *et al.* Isolation and identification of tilapia *Streptococcus. Chinese Journal of Preventive Veterinary Medicine*, 2002, 24(1): 18～20.

[149] Qiongqiu Wang. Isolation and identification of *Streptococcus* sp. from erythrodermatitis in *Tilapia nilotica. Chinese Journal of Preventive Veterinary Medicine*, 2001, 23(2): 150～152.

[150] Junqiang Qiu, Xianle Yang, Yi Le, *et al.* Preliminary study on the pathogen and histopathology of haemorrhagic blood poisoning in *Scortum bacoo. The Editorial Board For Journal of Shanghai Fisheries University*, 2004, 13(2): 115～120.

[151] Zhihua Shen, Dong Qian, Jun Qian, *et al.* Isolation, identification and pathogenicity of *Streptococcus iniae* isolated from red drum, *Sciaenops ocellatus. Acta Hydrobiologica Sinica*, 2005, 29(6): 678～683.

[152] Xiaodan Lu, Weimin Zou, Aiming Tan, *et al.* Identification and pathogenity of *Morganella morganii* isolated from *Cyprinus carpio* L.. *Freshwater Fisheries*, 2005, 35(2): 3～5.

[153] Hai Fang, Cuizheng Chen, Xiaojun Zhang. *Lactococcus garvieae* as a pathogen in flounder (*Paralichthys olivaceus* temminck et schlegel). *Journal of Fishery Sciences of China*, 2006, 13(3): 403～409.

[154] Qingpi Yan, Jun Wang, Yongquan Su, *et al.* Studies on ELISA method for detecting *Vibrio alginolyticus*, the pathogenic bacteria of *Pseudosciana crocea. Marine Sciences*, 2001, 25(9): 47～50.

[155] Qingpi Yan, Wenzheng Zou, Rongxing Ji , *et al*. Detection of *Vibrio fluviails* in *Paralichthys olivaceus* using indirect fluorescent antibody technique. *Marine Sciences*, 2006, 30(4): 16～19.

[156] Houbo Wu, Jinpei Pan. A dot immunobinding assay (DIA) to detect the exotoxin produced by the pathogenic bacteria of the vibriosis in marine cage-cultured *Pagrus major*. *Journal of Fisheries of China*, 2003, 27(6): 606～609.

[157] Weihua Chu, Chengping Lu. Detection of the serine-protease gene in *Aeromonas hydrophila* by PCR. *Chinese Journal of Veterinary Science*, 2003, 23(5): 463～464.

[158] Weihua Chu, Chengping Lu.PCR detection of pathogenic *Aeromonas hydrophila* by specific 16S rDNA and aerolysin gene. *Journal of Fisheries of China*, 2005, 29(1): 79～82.

[159] Sufang Shen, Weihua Chu. Preparation and application of diagnosis kit for pathogenic *Aeromonas* in fishes. *Chinese Journal of Veterinary Science*, 2000, 20(5): 462～464.

[160] Xingguo Zhu, Hongjie Fan, Chengping Lu, *et al*. The adhesions of *Aeromonas hydrophila* J-1 and their receptors. *Journal of Nanjing Agricultural University*, 2002, 25(2): 82～84.

[161] Yongjie Liu, Chengping Lu. Purification and characterization of elastase from *Aeromonas hydrophila* strain J－1[J]. *Chinese Journal of Veterinary Science*, 2006, 26(1): 54～56.

[162] Daling Zhu, Aihua Li, Jianguo Wang, *et al*. The correlation between the distribution pattern of virulence genes and the virulence of *Aeromonas hydrophila* strains. *Acta Scientiarum Naturalium Universitatis Sunyatseni*, 2006, 45(1): 82～85.

[163] Jing Hu, Aihua Li, Chengyu Hu, *et al*. Effects of temperature and pH on expression of virulence factors from *Aeromonas Hydrophila*. *Journal of Nanjing University of Science and Technology (Nature Science)*, 2006, 30(3): 375～380.

[164] Aihua Li, Pin Nie. A primary study of saprolegniasis and its pathogen of *Lateolabrax japonicus*. *Acta Hydrobiologica Sinica*, 1999,23(4): 388～390.

[165] Xiujie Shi, Hong Liu, Longying Gao, *et al*. Isolation and identification of causative agents of sick arctic charr (*Salvelinus alpinus*). *Journal of Huazhong Agricultural University*, 2007, 26(2): 223～227.

[166] Yuanjie Yang, Hui Chen, Yanhui He. Clinic and pathological observation of glasshouse cultured obscure puffer *Fugu obscurus* infected with mucormycosis. *Journal of Fishery Sciences of China*, 2006,13(2): 269～276.

[167] Wenjun Xu, Hanxiang Xu, Hui Shi , *et al*. Preliminary study on *Candida oleophila* disease in *Portunus trituberculatus*. *Journal of Fisheries of China*, 2005, 29(6): 831～836.

[168] Yuhua Sun, Qihuan Sun. Studies on pathogenic organism and prevention and cure for explosive epidemic disease of parent prawn of *Macrobrachium rosenberii*. *Journal of Fisheries of China*, 1998, 22(1): 56～60.

[169] Xiaolin Tu. Latex agglutination test for the detection of pathogenic *Torulopsis mogii* from *Macrobrachium rosenbergii*. *The Editorial Board For Journal of Shanghai Fisheries University*, 1996, 5(3): 201～204.

[170] Kuidong Xu, Weibo Song. Studies on pathogenetic ciliates from marine mollusks III. Thigmotrichine ciliates (Protozoa, Ciliophora, Scuticocillatida). *Journal of Ocean University of Qingdao*, 2000, 30(2): 230～236.

[171] Kuidong Xu, Weibo Song. Studies on pathogenetic ciliates from marine mollusks II. Philasterine ciliates (Protozoa, Ciliophora, Scuticocillatida). *Journal of Ocean University of Qingdao*, 2000, 30(2): 224～229.

[172] Kuidong Xu, Weibo Song. Studies on pathogenetic ciliates from marine mollusks I .Rhynchodida ciliates (Protozoa, Ciliophora). *Journal of Ocean University of Qingdao*, 2000, 30(2): 217～223.

[173] Kuidong Xu, Fanlin Meng, Weibo Song. Scanning electron microscopic observations on the histopathology of trichodiniasis of the mariculture fish, *Lateolabrax japonicus*. *Journal of Ocean University of Qingdao*, 2000, 30(3): 418～422.

[174] Fahui Tang, Yuanjun Zhao, Hui Chen. Trichodinid ectoparasites from golden carp, with a description of *Trichodina paranigra* sp. nov.. *Acta Hydrobiologica Sinica*, 2005, 29(1): 75～80.

[175] Kuidong Xu, Weibo Song. Parasitic trichodinid ciliates (Cilophora, Peritrichida) from mariculture fishes I . Genus*Trichodina*. *Journal of Ocean University of Qingdao*, 2000, 30(3): 406～412.

[176] Kuidong Xu, Weibo Song. Parasitic trichodinid ciliates (Cilophora, Peritrichida) from mariculture fishes II. Genera. *Paratrichodina*, *Trichodinella* and *Dipartiela*. *Journal of Ocean University of Qingdao*, 2000, 30(3): 413～417.

[177] Xiaoying Dong, Haiyun Li, Shengqiu Tang. Scanning electron microscopic observation on *Neosentis celatus* in *Monopterus albus*. *Reservoir Fisheries*, 2005, 25(3): 75～76.

[178] Yue Ma, Anxing Li, Mingquan Xie, *et al.* Cortical ultrastructures of *Cryptocaryon irritans* GD1 strain infecting *Trachinotus blochi* experimentally. *Acta Zoologica Sinica*, 2006, 50(2): 396～405.

[179] Yingeng Wang, Jiejun Chen, Lei Qin. *Mesanphrys carcini* causing severe scuticociliatosis in farmed turbot *Scophthalmus maximus* in China. *Journal of Fisheries of China*, 2005, 12(5): 594～601.

[180] Yanyin Wang, Guowei Guo, Rucheng Gao, *et al.* Turbellarian parasitised in *Sciaenops ocellatus* and its control. *Journal of Fisheries of China*, 2002, 26(4): 379～381.

[181] Li Zhou, Kuidong Xu, Wenbin Zhan, *et al* . Morphological studies of a ciliate from the external ulcerative Japanese flounder, *Paralichthys olivaceus*. *Journal of Ocean University of Qingdao*, 2001, 31(2): 190～194.

[182] Weina Wang, Anli Wang, Li Chen, *et al.* Ultrastructure of the microsporidian from Penaeus monodon. *Acta Zoologica Sinica*, 2001, 47: 78～81.

[183] Houbo Wu, Yingsong Wu, Zhaohe Wu. Occurrence of a new microsporidium in the abdominal cavity of *Epinephelus akaara*. *Acta Hydrobiologica Sinica*, 2005, 29(2): 150～154.

[184] Xiaonin Gong, Xingren Xie. Two new species of *Myxosporidia* of the fish, *Squaliobarbus curriculus* and *Hypseleotris swinhonis*. *Acta Hydrobiologica Sinic*a, 2000, 24(4): 380～383.

[185] Shan Jin, Guoliang Wang, Qingsong Zhao, *et al.* Scanning electron microscopic observation on the pathogen of nematode disease in freshwater fish. *Chinese Journal of Veterinary Science*, 2005, 25(2): 159～161.

[186] Weijian Yao. Four species of *Gyrodactylus* from fish of the family *Acheilognthinae*. *Acta Hydrobiologica Sinica*, 2002, 26(5): 513～516.

[187] Weijian Yao. A new species of *Gyrodactylus nordmann* from *Coreius heterodon bleeker*. *Acta Hydrobiologica Sinica*, 2007, 31(1): 104～106.

[188] Weijian Yao. Three new species of *Gyrodactylus* of fishes from Wuling mountains region, China. *Acta Hydrobiologica Sinica*, 2002, 26(5): 517～520.

[189] Lianxiang Li. Descriptions of two new species of Trichodinids freshwater fishes, *Ctenogobius giurnus retter*. *Acta Hydrobiologica Sinica*, 2001, 25(5): 5.

[190] Qizhong Zhang. Studies on *Gyrodactyliasis* of the southern catfish (*Silurus meridionalis*). *Journal of Southwest China Normal University (Natural Science)*, 1999, 24(4): 468～473.

[191] Haiyan Li, Yinjie Chen. Description of two new species of the genus *Haliotrema* from marine fishes in Guangdong, China. *Marine Sciences*, 2005, 29(5): 50～53.

[192] Jun Pan. Monogenoidea of marine fishes from Hainan Ⅳ. Two new species of *Haliotrema* from south China sea. *Tropic oceanology*, 2000, 19(2): 33～38.

[193] Qian Gao, Pin Nie, Weijian Yao. Observation of *Ancyrocephalus mogurndae* (Monogenea: Ancyrocephalidae) from the the gills of the mandarin fish, *Siniperca chuatsi* by scanning electron microscopy. *Acta Hydrobiologica Sinica*, 2001, 25(6): 597～604.

[194] Wenchuan Li, Liwei Li, Lei Shi, *et al.* Description on two species of Benedenian from Epinephelus awoara in Xiamen , Fujian, China (Monogenea: Capsalidae). *Journal of Xiamen University (Natural Science)*, 2001, 40(5): 1133～1137.

[195] Sulian Ren, Xinchun Yang, Weibo Song. A treamatode larva parasitilized in cultured marine bivalve, *Meretrix meretrix* linnaeus and its histopathology. *Periodical of Ocean University of China*, 2005, 35(3): 387～391.

[196] Shengfa Liu, Yanbo Yang. A redescription and a new geographical record of *Watertrema piscicola* srivastava, 1939,(Trematoda: watertrematidae) in south China sea. *Journal of Tropical Oceanography*, 2003, 22(5): 82～86.

[197] Fei Zhao, Xiaodan Lu, Weimin Zou, *et al.* Parasitic behaviors of *Rocinela* sp.—— a newly found freshwater isopoda. *Journal of Huazhong Agricultural University*, 2006, 25(3): 270～273.

[198] Qiliu Chen, Chenglun Ma. Myxosporean record. Beijing: Science Press; 1998.

[199] Yishan Lu, Pin Nie. Phylogenetic analysis of the myxosporeans parasitic in freshwater fish based on 18S ribosomal DNA. *Acta Hydrobiologica Sinica*, 2004, 28(6): 587～591.

[200] Jingfeng Fan, Yubo Liang, Lichao Song, *et al* . Establishment of indirect enzyme-linked immunosorbent assay (iELISA) of detecting *Perkinsus marinus* in mollusc. *Marine Envronmental Science*, 2006, 25(2): 62～64.

[201] Jingyong Zhang, Jianguo Zhang, Yingsong Wu, *et al.* Establishment of an ELISA against mature intact *Myxobolus rotundus* spores. *Journal of Fisheries of China*, 2005, 12(2): 133～137.

[202] Xianwei Feng, Zhengguo Yang, Guangmou Yan, *et al*. Detection of *Thelohanellus kitauei* in *Cyprinus carpio* by PCR. *Progress in Veterinary Medicine*, 2002, 23(5): 62～63.

[203] Yanqing Gong, Xinzhong Chen, Yinyao Huang. Study on the pathogen and histopathology of taura syndrome in cultured prawn, *Litopenaeus vannamei*. Marine Fisheries Research, 2004, 25(2): 18～23.

[204] Li Lin, Jiannan Huang, Shaoping Weng. Histopathological examination and electron microscopy observation of viral nervous necrosis in red-spotted grouper, *Epinephelus akaara*. *Journal of Fisheries of China*, 2005, 29(4): 519～523.

[205] Ouqin Chang, Cunbin Chang, Hong Ma, *et al*. Histopathological study on lymphocystis disease of *Rachycentron canadum*. *Journal of Fisheries of China*, 2006, 13(6): 973～979.

[206] Wenjun Xu, Xiuzheng Sheng, Hui Shi, *et al*. Artificial infection for *Portunus trituberculatus* by WSSV and histopathological observation. *The Editorial Board For Journal of Shanghai Fisheries University*, 2007, 16(1): 33～39.

[207] Hongda Lu, Haiqing Chen. Pathogen and histopathology of muscle whitish disease in *Macrobrachium rosenbergii*. *Journal of Fisheries of China*, 2003, 10(2): 126～132.

[208] Min Fang, Weizhong Jin, Linsheng Song, *et al*. Histopathological study on "appendage shivering disease" of *Eriocheir sinensis*. *Oceanologia Et Limnologia Sinica*, 2003, 34(3): 322～328.

[209] Chongluo Fu, Weibo Song, Yun Li, *et al*. Histopathological research and immunofluorescence of AVND virus infection in cultured Scallop *Chlamys farreri*. *Acta Microbiologica Sinica*, 2004, 44(6): 741～744.

[210] Li Zhou, Wenbin Zhan, Weibo Song. Observations on the histopathology of the external ulcerative disease of Japanese flounder, *Paralichthys olivaceus*. *Journal of Ocean University of Qingdao*, 2000, 30(4): 593～597.

[211] Lijun He, Likun Liao, Jinfeng Yuan, *et al*. Pathological observation of bacterial septicemia in *Megalobrania amblycephala*. *Journal of Southwest Agricultural University*, 2006, 28(3): 483～486.

[212] Gaoxue Wang, Zengrong Huang. Studies on the pathogen and histopathology of septicemia in *Esox lucius*. *Journal of Fisheries of China*, 2006, 30(3): 383～389.

[213] Huixian Wu, JUnzeng Xue, Liuya Tang. Studies on the histopathology of *Macrobrachium nipponense* with shrimp black gill disease. *Bulletin of Science and Technology*, 2002, 18(1): 73～75.

[214] Zhisheng Chen, Junyi Lv, Jinying Wu. Pathological studies on ulcerate disease in *Haliotis diversicolor*. *Chinese Journal of Zology*, 2004, 39(2): 6～9.

[215] Haiyan Li, Yongyu Xiang. Gill histopathology of *Heteronchleidus* in *Macropodus opercularis* (Monogenea: Ancyrocephalidae) infection. *Journal of Fisheries of China*, 2006, 13(5): 829～833.

[216] Dali Chen, Qizhong Zhang, Zhijian Wang, *et al*. Histological observation of juvenile *Leiocassis longirostris* infected by *Ichthyophthirius Multifiliis*. *Journal of Southwest China Normal University (Natural Science)*, 2004, 29(2): 278～281.

[217] Zhengyong Liu, Guomin Zhang, Youquan Xie, *et al.* Tissue pathological observation on the paradeontacvlix sinensis disease of *Fugu rubripe*. *The Editorial Board of Jimei University (Natural Science)*, 2006, 11(1): 24～27.

[218] Liwei Li, Wenchuan Yang. Microhabitats and histopathology of *Dactylogyrus inversus* (Monogenea: Dactylogyridae) infection in perch, *Lateolabrax japonicus*. *Journal of Xiamen University (Natural Science)*, 2002, 41(2): 234～238.

[219] Qizhong Zhang, Dali Chen. Histopathological observation of *Coreius guichenoti* infected by *Ichthyophthirius multifiliis fouquet*, 1876. *Journal of Southwest China Normal University (Natural Science)*, 2005, 30(6): 1112～1115.

[220] Xia Li, Jingbo Wang, Bin Wang, *et al.* Histopathological study on *Perkinus* sp. disease of *Ruditapes philippinarum*. *Journal of Dalian Fisheries University*, 2002, 17(3): 215～219.

[221] Sulian Ren, Weibo Song. Histopathology of bucephalidae larvae-caused disease in *Meretrix meretrix*. *Journal of Fisheries of China*, 2002, 26(5): 459～454.

[222] Sulian Ren, Xinchun Yang, Weibo Song. Study on histopathology of the cultured clam *Meretirx meretrix* suffered from the "mantle hypertrophy disease". *Periodical of Ocean University of China*, 2005, 35(6): 949～954.

[223] Qiya Zhang, jianjun Tao, Xiaohong Huang, *et al.* Coinfection of two siniperca chuatsi viruses and morphologica changes of the infected cells. *Acta Hydrobiologica Sinica*, 2005, 29(5): 473～478.

[224] Guoliang Wang, Shan Jin, Huiqun Chen. Pathogen and cytopathology of spleenomegaly disease in *Lateolabrax japonicus*. *Journal of Fisheries of China*, 2003, 27(2): 158～162.

[225] Zhidong Zhang, Wenbin Zhan, Yanhong Xue, *et al.* Variations of granular hamaecytes when shrimp *Litopenaeus vanammei* infected by white spot syndrome virus (WSSV). *Oceanologica et Limnologica Sinica*, 2005, 36(1): 67～71.

[226] Jianghua Yan, Qing Yan, Yinyao Huang, *et al.* Cellular distribution and pathological ultramicro-structural change of "mad swimming sysmptoms " of *Anguilla anguilla*. *Journal of Oceanography in Taiwan Strait*, 2001, 20(1): 54～57.

[227] Sanpu Qu, Xiaorong Zhang, Congyi Zhen, *et al.* Apoptosis of the cell line from grass carp (CIK) induced by fish reovirus. *Acta Hydrobiologica Sinica*, 2000, 24(6): 616～621.

[228] Qiya Zhang, Zhengqiu Li. Susceptibility of fish cell lines of *Trionyx sinesis* virus. *Chinese Journal of Veterinary Science*, 2000, 20(1): 15～18.

[229] Sulian Ren, Dexiu Wang, Xiuzheng Sheng, *et al.* Cytopathology and morphogenesis of a spherical virus found in the *Meretrix meretrix* suffering from "red muscle disease" (RMD). *Journal of Fisheries of China*, 2002, 26(3): 265～269.

[230] Nin Yang, Sulian Ren, Weibo Song. A spherical virus-like particle found in *Argopecten irradians Lamarck* and its cytopathology. *Marine Sciences*, 2005, 29(9): 55～58.

[231] Ming Jiang, Shaoguo Ru, Xiuqin Sun, *et al.* Study on the ultrastructure of sphereovirus and cell pathology in hormonal cell of adenohypophysis of *Carassius auratus*. *High Technology Letters*, 2003,12: 85～88.

[232] Jiahua Xie, Jjianjun Yuan, Liping Wang, *et al.* Changes in number and structure eel infected with two species of peripheral leucocytes of Pike of fish bacterial pathogens. *Chinese Journal of Zoology*, 2006, 41(5): 88～91.

[233] Shouming Feng, Xianle Yang, Jun Li, *et al.* Haemolymph pathological research on white spot syndrome (WSS) of *Litopenaeus vannamei*. *Journal of Fisheries of China*, 2006, 30(1): 108～112.

[234] Chengsong Zhang, Fuhua Li, Kuijie Yu, *et al.* Haematological changes of sibling diploid and triploid *Fenneropenaeus chinensis* after being challenged with pathogens. *Jouranl of Fisheries of China*, 2004, 28(5): 535～540.

[235] Jin Liu, Boping Zeng, Hengmei Yan, *et al.* The effect of infection *Monopterus albus* by two species of chief parasites on some biochemical values in the blood. *Journal of Natural Science of Hunan Normal University*, 2005, 28(1): 67～69.

[236] Xiaoying Dong, Haiyun Li, Shengqiu Tang, *et al.* Study on hemopathology and histopathology of *Monopterus albus* infected by *Neosentis celatus*. *Reservoir Fisheries*, 2006, 26(1): 86～89.

[237] Xiaochun Luo, Anxing Li, Mingquan Xie, *et al.* Studies on the morphology of grouper (*Epinephelus coioides*) mucosal immune system. *Acta Hydrobiologica Sinica*, 2005, 29(2): 193～198.

[238] Honghu Wang, Changling Li, Fujun Cao, *et al.* Microstructure of head-kidney and spleen in *Lutjanus sanguineus*. *Journal of Oceanography in Taiwan Strait*, 2006, 25(3): 360～367.

[239] Shugen Lin, Defeng Wu, Shoukun Wang, *et al.* Location and identification of endocrine cells in the gastro-entero-pancreas of *Anguilla japonicus*. *Journal of Fisheries of China*, 2003, 27(6): 513～518.

[240] Shugen Lin, Quanxi Wang, Shoukun Wang. Immunohistochemical localization of endocrine cells in the digestive tract of *Pseudosciaena crocea*. *Fujian Journal of Agricultural Sciences*, 2005, 20(4): 238～242.

[241] Qing Yang, Haihui Ye, Huiyang Huang, *et al.* Immunohistochemical studies on endocrine cells in the digestive tract of *Sparus latus*. *Marine Science Bulletin*, 2005, 24(4): 87～90.

[242] Shugen Lin, Zhifeng Yang, Shoukun Wang. Localization of immunohistochemistry of endocrine cells in the digestive tract of *Sparus macrocephalus*. *Journal of Fujian Agricultural and Forestry University*, 2006, 35(1): 77～81.

[243] Haihui Ye, Aijia Zhu, Huiyang Huang, *et al.* Identification and localization of endocrine cells in the digestive tract of *Bostrichthys sinensis*. *Journal of Xiamen University (Natural Science)*, 2006, 45(4): 545～548.

[244] Jingfeng Wang, Yang Liu, Qin Zhao, *et al.* Immunocytochemistry study on somatostatin secreting cells in *Paralichtys olivaceus*. *Journal of Fishery Sciences of China*, 2007, 14(3): 397～402.

[245] Shoukun Wang, Junqiu Li, Jianbin Lin, *et al.* Localization on immunohistochemistry of endocrine cells in the digestive tract of *Paralichthys olivaceus*. *Journal of Fujian Agricultural and Forestry University*, 2007, 36(2): 171～175.

[246] Zhengdong Zhang, Peijun Zhang, Zhaolan Mo. Preliminary study on red cell immune function of left-eyed flounders, *Paralichthys olivaceus*. *High Technology Letters*, 2006, 16(12): 1311～1314.

[247] Qiang Lu, Peijin Feng, Lianrui Li , *et al* . Construction of cDNA library of peripheral blood leucocytes in normal *Cyprinus carpio* L.. *Journal of Fisheries of China*, 2004, 28(5): 585～588.

[248] Guozhu Liu, Wenbiao Zhen, Xinhua Chen. Construction of SMART cDNA library from kidney of large yellow croaker (*Pseudosciaena crocea*) stimulated with the inactivated trivalent bacteria vaccine. *Acta Oceanologica sinica*, 2007, 29(2): 161～166.

[249] Juan Feng, Chaoqun Hu. Purification and characteristics of serum immunoglobulins of four major cultured marine fishes in china. *Journal of Tropical Oceanography*, 2002, 21(4): 8～13.

[250] Yun Liu, Feng Sun, Guoliang Jiang. Purification and partial characterization of serum immunoglobulins from Japanese flounder (*Paralichthys olivaceus*). *Journal of Fishery Sciences of China*, 2007, 14(4): 547～553.

[251] Juanjuan Lin, Jinxian Yang, Qiang Chen, *et al*. Purification and structure analysis of serum immunoglobulin in *Anguilla anguilla*. *Journal of Fisheries of China*, 2006, 30(6): 806～811.

[252] Qingpi Yan, Yifan Han, Tianxiang Gao, *et al*. Purification of serum IgM from large yellow croaker (*Pseudosciaena crocea*) and preparation of rabbit sera anti-IgM. *Journal of Fishery Sciences of China*, 2006, 13(3): 475～480.

[253] Jing Zhang, Liguo An, Fumiao Zhang, *et al*. Isolation purification and partial characterization of serum and intestine mucus IgM from rice field eel(*Monopterus albus*). *Fisheries Science*, 2007, 26(3): 130～133.

[254] Jingyu Shen,Thompson K D,Adams A , *et al*. Production and characterization of monoclonal antibodies against *Carassius auratus gibelio* IgM. *Journal of Fisheries of China*, 2006, 30(3): 421～423.

[255] Ying Zhang, Qiuzhi Qu, Tangbin Huo, *et al*. Comparison of immunoglobulins in egg yolk of three species of sturgeons and two species of salmonids. *Acta Zoological Sinica*, 2007, 53(2): 264～269.

[256] Yueling Zhang, Wenjie Liu, Sanying Wang, *et al*. On immunolocalization of Ig-like proteins in shrimp *Litopenaeus vannamei*. *Marine Sciences*, 2005, 29(9): 31～35.

[257] Dajiang Li, Mingchao Zhong, Haoran Lin. Location of maternal immunoglobulin in the egg and embryo of crucion carp(*Carassius auratus*). *Acta Scientiarum Naturalium Universitatis Sunyatseni*, 2001, 40(2): 82～85.

[258] Fan Wang, Tianlong Lin, Houjun Pan, *et al*. Production and characterization of monoclonal antibodies against *Siniperca chuatsi* Ig. *Journal of Fisheries of China*, 2006, 30(2): 285～288.

[259] Hongbo Liu, Di Wang. Ig heavy chain of amur sturgeon *Acipenser schrenskii*: cDNA sequence and diversity. *Acta Zoologica Sinica*, 2006, 52(3): 557～563.

[260] Yan Yao, Kaiya Zhou, Daxiang Song. Expression and polyclonal antibody preparation of molt-inhibiting hormone 1 (MIH1) from the mitten crab *Eriocheir japonica sinensis*. *Acta Zoologica Sinica*, 2006, 52(1): 209～214.

[261] Juan Qian, Yipeng Qi. Generation and characterization of a single-chain antibody fragment against white spot syndrome virus of Shrimp from phage display antibody library. *Chinese Journal of Virology*, 2005, 21(6): 461～467.

[262] Wen Liu, Dong Qian, Jianxiang Wu, *et al.* Production and application of monoclonal antibodies to *Macrobrachium rosenbergii* Nodavirus. *Journal of Fisheries of China*, 2005, 29(4): 529～533.

[263] Chongluo Fu, Weibo Song, Yun Li, *et al.* Monoclonal antibodies prepared and used for detection of acute virus necrobiotic disease virus in scallop Chlamys farreri by indirect ELISA. *Acta Microbiologica Sinica*, 2005, 45(1): 116～120.

[264] Jun Huang, Minglong Liang, Yun Zhu. Preparation and characterization of monoclonal antibodies against *Aeromonas sobria*. *Journal of Henan Agricultural Sciences*, 2006, 2: 108～111.

[265] Hongyan Chen, Tianlong Lin, Risheng Chen, *et al.* Preparation and characterization of monoclonal antibodies to *Aeromonas hydrophila*. *Journal of Fishery Sciences of China*, 2003, 10(3): 121～125.

[266] Chuang Guo, Ping Fang, Lixin Guo, *et al.* *Aeromonas hydrophila*: Developing its monoclonal antibody against aerolysin and preliminary application. *Fisheris Science*, 2007, 26(3): 167～170.

[267] Pengfei Gong, Quan Wang, Yongjun Chen. Preparation and identification of monoclonal antibody against malachite green. *Veterinary Science in China*, 2007, 37(4): 359～362.

[268] Qiang Li, Wenbin Zhan, Xiuzheng Sheng, *et al.* Detection of specific antibody against lymphocystis disease virus in Japanese flounder (*Paralichthys olivaceus*) by using monoclonal antibody technology. *Periodical of Ocean University of China*, 2007, 37(2): 243～246.

[269] Tongyan Lu, Yuhui Yang, Lianwei Xu, *et al.* The Pharmacokinetic of danofloxacin in healthy and diseased *acipenser schrenckii* infected by *Aeromonas hydrophila*. *Acta Hydrobiologica Sinica*, 2006, 30(3): 349～355.

[270] Xianle Yang, Jia Zhan, Kun Hu. The comparative pharmacokinetics of chloramphenicol in *Oreochromis niloticus*[J]. *Journal of Fisheries of China*, 2005,29(1):60～65.

[271] Peijian Yu, Zutong Weng, Haiping Fan, *et al.* Pharmacokinetics of florfenicol in European eel. *Journal of Fujian Fisheries*, 2005, 4: 72～57

[272] Jichang Jian, Zhaohe Wu, Gang Chen.Pharmacokinetics of enrofloxacin in *Sciaenops ocellatus*. Chinese *Journal of Veterinary Science*, 2005, 25(2): 195～197.

[273] Liwen Xu, ChangRong Liao, Ruixuan Wang, *et al.* A preliminary study on pharmacokinetics of florfenicol in *Haliotis diversicolor supertexta*. *Journal of Oceanography in Taiwan*, 2006, 25(2): 216～221.

[274] Xianle Yang, Kun Hu, Junqiang Qiu, *et al.* Studies on accumulation and toxicity of olaquindox in fishes. *Acta Hydrobiologica Sinica*, 2005, 29(1): 13～19.

[275] Jun Tang, Xianle Yang, Zonglin Zhen, *et al.* Pharmacokinetics and the active metabolite of enrofloxacin in Chinese mitten-handed crab (*Eriocheir sinensis*). *Aquaculture*, 2006, 260: 69～76.

[276] Guofu Chen, Xiaoling Song, Jie Huang, *et al.* Influence of A3α-peptidoglycan on growth, immune capability and anti-infectivity of *Litopenaeus vannamei*. *High Technology Letters*, 2005, 15(8): 100～106.

[277] Zirong Xu, Keqiang Wei. Oral vaccination trail with *Cambarus proclarkii* to induce protection against white spot syndrome virus. *Chinese Journal of Veterinary Science*, 2005, 25(6): 588～591.

[278] Deshang Li, Jifang Li, Guo Qiang. A seawater quick treating device for preventing viral disease in shrimp culture. *Marine Fisheries Research*, 2004, 25(4): 78～64.

[279] Jihong Zhang, Weihua Chu, Chengping Lu. Evaluation of biofilm of *Aeromonas hydrophila* for oral vaccination. *Journal of Fishery Sciences of China*, 2004, 11(5): 404～407.

[280] Hongmei Cao, Jian Li, Wenbin Zhan. Immune efficacy of a bivalent vaccine for turbot against *Vibrio anguillarum* and *V. alginolyticus*. *Journal of Fishery Sciences of China*, 2006, 13(3): 397～402.

[281] Junhong Yu, Jihong Shen, Xianghong Wang Oral vaccination of sea perch, Lateolabrax japonicus, against *vibrio anguillarum* using microencapsulated vaccine. *Journal of Fishery Sciences of China*, 2001, 8(2): 76～79.

[282] Zheng Xu, Juan Yao, Changfu Chen , *et al.* Enhancement of disease resistance in *Eriocheri sinensis* by oral administration of immune polysaccharide (yeast cell wall). *Journal of Huazhong Agricultural University*, 2005, 24(4): 383～386.

[283] Shuyan Chi, Qicun Zhou, Jianbin Zhou, *et al.* Effects of dietary β-glucan on growth performance and disease resistance of juvenile hybrid tilapia (*Oreochromis aureus*♂×*Oreochromis niloticus*♀). *Journal of Fishery Sciences of China*, 2006, 13(5): 767～774.

[284] Xiaoming Gong, Jun Sun, Ping Liu, *et al.* Damage of ozone-produced oxidant to *Vibrio* DNA. *Marine Sciences*, 2006, 30(10): 1～3.

[285] Caihong Hu, Meisheng Xia, Li Xiong, *et al.* Effects of Cu bearing montmorillonite on *Aeromonas hydrophila* adhesion to epithelial cells of *Nile tilapia*. *Journal of Fisheries of China*, 2005, 29(5): 619～623.

[286] Xuehong Song, Kui Chen, Yongling Wang, *et al.* Effects of EM and deuterium sulfate treatment on the inhibition rate of *Saprologenia parasitia* during the hatching of *Pseudobagrus fulvidraco* fertilized eggs. *Freshwater Fisheries*, 2007, 37(1): 9～13.

[287] Caigen Yang, Chunfang Cai, Zhanfang Guo. Effect of antibacterial drugs on the control of saprolegniasis of salangidae eggs. *Reservoir Fisheries*, 2000, 20(5): 38～39.

[288] Weina Xu, Xin Zhang, Wenbin Liu. Toxicity of trichlorfon against *Carassais auratus gibebio* and its impacting factors. *Journal of Agro-Environment Science*, 2007, 26(B03): 68～71.

[289] Gaoxue Wang, Yunkui Zhao, Jiayun Yao, *et al.* Study on killing activity of ginkgolic acids to pathogen and dactylogyrus of fish. *Journal of Northwest Sci-Tech Univ of Agr And*, 2006, 34(10): 11～15.

[290] Gaoxue Wang, Chao Chen, Anliang Chen, *et al.* Dactylogyrus killing efficacies of 22 plants extracts and its 6 compounds. *Acta Botanica Boreali-Occidentalia Sinica*, 2006, 26(12): 2567～2573.

[291] William H. Carr, James N. Sweeney, Linda Nunan, *et al.* The use of an infectious hypodermal and hematopoietic necrosis virus gene probe serodiagnostic field kit for the screening of candidate specific pathogen-free *Penaeus vannamei* broodstock. *Aquaculture*, 1996, 147(1—2): 1～8.

[292] Sufang Shen, Weihua Chu, Chengping Lu. Preparation and application of diagnosis kit for pathogenic *Aeromonas* in fishes. *Chinese Journal of Veterinary Science*, 2000, 20(5): 462～464.

[293] Duan Lin, Yanhong Dong, Yin Yuan, *et al.* Application of PCR technique in molecular biology study and detection of Iridovirus. *Marine Environmental Science*, 2007, 26(2): 190～193.

[294] Uchimura M, Yamamoto T. Production of hemagglutinins and pili by *Vibrio minicus* and its adherence to human and small intestines *in vitro*. *FEMS Microbiol Lett*, 1992, 91: 73～78.

[295] Xinzhong Wu, Jinpei Pan. Studies on the rickettsia-like organism disease of tropical marine pearl oyster V. Ultrastructural pathology and pathogenesis of rickettsia-like organism disease. *Acta Oceanologica Sinica*, 1999, 21(3): 113～121.

[296] Boudinot P,Blanco M,Kinkelin P. Combined DNA immunization with the glycoprotein gene of the viral hermorraghic septicemia virus and infectious hematopoietic necrosis virus induces double-specific protective immunity and non-specific response in rainbow trout. *Virol*, 1998, 249: 297～306.

[297] Lorenzen N,Lorenzen E,Einer-Jensen K,*et al.* DNA vaccines as a tool for analyzing the protective immune response against rhabdoviruses in rainbow trout. *Fish and Shellfish Immunol*, 2000, 12: 439～453.

[298] Zou J,Grabowski P S,Cunnigham C, *et al.* Molecular cloning of interleukin I beta from rainbow trout oncorhynchus mykiss reveals no evidence of an ice cut sit. *Cytokine*, 1999, 11: 552～560.

[299] Liu L,Fujiki K,Dixon B,*et al.* Molecular cloning of a novel rainbow trout (*Oncorhynchus mykiss*) CC chemokine with a fractalkine-like stalk and a TNT decoy receptor using cDNA fragments containing AU-rich elements. *Cytokine*,2002,2:71～81.

[300] Yibing Zhang, Qiya Zhang, Dequan Xu, *et al.* Identification of fish antiviral related genes from culture cells induced with inactive virus. *Chinese Science Bulletin*, 2003, 27(2): 581～588.

[301] Duff D C B. The oral immunization of trout against *Bacterium Salmonicida*. *J Immun*, 1942, 44: 87～93.

[302] Xianle Yang, Haipeng Cao. The development of aquaculture vaccines in China. *Journal of Fisheries of China*, 2006, 30(2): 264～271.

[303] Katrin Utke, Holger Kock, Heike Schuetze, *et al.* Cell-mediated immune responses in rainbow trout after DNA immunization against the viral hemorrhagic septicemia virus. *Developmental and Comparative Immunology, In Press, Corrected Proof*, 2007, 1～17.

[304] Gael Kurath, Kyle A. Garver, Serge Corbeil,*et al.* Protective immunity and lack of histopathological damage two years after DNA vaccination against infectious hematopoietic necrosis virus in trout. *Vaccine*, 2006, 24(3): 345～354.

[305] Serge Corbeil, Gael Kurath and Scott E. Lapatra. Fish DNA vaccine against infectious hematopoietic necrosis virus: efficacy of various routes of immunisation. *Fish and Shellfish Immunology*, 2000, 10(8): 711～723.

[306] Kaiyong Liu, Kaiyu Wang. Study and practice on enrofloxacin on fishery. *Chinese Journal of Veterinary Drug*, 2004, 38(10): 32～34.

[307] Liwen Xu, Changrong Liao, Guangfeng Liu. Security about florfenicol used in aquaculture. *Journal of Fishery Sciences of China*, 2005, 12(4): 512～518.

[308] Xianle Yang. Fisheries medicine manual (second edition). Beijing: China Agriculture Press; 2005.

[309] Xiangling Wang, Zhiping Fang, Jiyue Cao, *et al*. Studies on residue and depletion of sarafloxacin hydrochloride in goldfish, *Carassius auratus*. *Acta Hydrobiologica Sinica*, 2006, 30(2): 198～204.

[310] Wenhong Fang, Guoxing Zhen. Pharmacokinetics of norfloxacin in hemolymph from whiteleg shrimip, *Penaeus vannamei* following intramuscle injection and oral administration. *Acta Hydrobiologica Sinica*, 2006, 30(5): 541～546.

[311] Zhiping Fang, Yufang Peng, Jiyue Cao, *et al*. Pharmacokinetics and residue of lomefloxacin in *Cyprinus carpio*. *Freshwater Fisheries*, 2005, 35(1): 16～18.

[312] Zenghui Liang, Liming Lin, JIngjing Liu, *et al*. Pharmacokinetic study of oxolinic acid and flumequine in *Muraenesox cinereus*. *Marine Fisheries Research*, 2006, 27(3): 86～92.

[313] Adams M B,Nowak B F. Amoebic gill disease: sequential pathology in cultured Atlantic salmon, *Salmo salar L .Journal of Fish Diseases*, 2003, 26: 601～614.

[314] Jinming Ye,Xianxiang Yang, Zenghua Jiang, *et al*. Current situation, problems and reasonal advice of the fishery drugs in China. *China Fisheries*, 2007, 5: 65～69.

In: Aquaculture Research Trends
Editor: Stephen H. Schwartz, pp.59-112
ISBN 978-1-60456-217-0

Chapter 2

ESTIMATING PARENTAGE RELATIONSHIPS USING MOLECULAR MARKERS IN AQUACULTURE

Paulino Martínez[1] and Jesús Fernández[2]
[1] Departamento de Genética, Facultad de Veterinaria,
Universidad de Santiago de Compostela, 27002 Lugo, Spain
[2] Departamento de Mejora Genética Animal,
Instituto Nacional de Investigación y Tecnología Agraria y Alimentaria,
28040 Madrid, Spain

ABSTRACT

The inference of parentage relationships between individuals from their molecular resemblance represents a useful methodology for genetic improvement in Aquaculture. This makes possible the reorganization of broodstock in groups with low relatedness, as well as the identification of families in selection programs to avoid the harmful effects of inbreeding and to maintain the highest diversity as possible across generations. The choice of the appropriate genetic marker and the statistical methodology is essential to get the best solution for the questions considered. Depending on the availability of previous information on the genealogy, paternity or kinship analysis will be applied. The simplest approach to paternity inference, involves parent identification through the exclusion of the remaining candidates. However, this can lead to more than one solution, solvable after application of maximum likelihood procedures. The ability to identify the parents of each offspring depends on the sampling scenario (number of candidate parents and the fraction sampled), on the potential of the markers used (polymorphism, genotyping errors), and on the conformance to theoretical assumptions of the statistical models applied. A large number of paternity programs are freely available. They display usually complementary performances, and there is not the best software for all situations.

When we deal with a single group of individuals belonging to the same generation or that can not be separated into known generations, the aim is just to estimate the degree of genetic relationship between them, usually expressed as the coancestry coefficient. The basic idea is to determine how much of the molecular similarity (Identity By State) is due to the Identity By Descent (the really important parameter). There are two groups of relationship estimators from the molecular information. One of them includes methods directed at estimating coancestry for only a pair of individuals at a time, usually relaying on the knowledge of the allele frequencies of the reference population. Estimators within this group can be further divided into those called Method of Moments Estimators (MME) and those based on Maximum Likelihood (MLE). The other group of methods uses jointly the information of all individuals to determine the more probable population/familiar structure. They perform an explicit reconstruction of the genealogy leading to the observed population (at least for one generation above). Depending on the type of estimator and the assumptions used in their development, each of them presents some advantages and/or limitations which should be taken into account when choosing the one to use. In the present study, the most relevant estimators are presented and, also, a list of free software available for their application.

1. Introduction

The study of relatedness using molecular markers begins in humans by means of blood groups (Weir *et al.* 2006) and later in Drosophila applying chromosome polymorphisms (Milkman and Zeitler 1974) and, especially, allozymes (Meagher 1986). With the arrival of recombinant DNA technology, a wide range of molecular markers are developed, which combine the specificity of PCR with the analytical capacity of automatic sequencers for fragment analysis. Then, microsatellite loci, a type of markers ideal for studying parentage relationships, are discovered. The development, in parallel, of specific statistical methods for this purpose (Luikart and England 1999), gives rise to a large amount of theoretical and empirical works in this field, currently constituting a very active area of research (Anderson and Garza 2006; Hadfield *et al.* 2006; Wang 2006; Weir *et al.* 2006).

A great deal of statistical developments associated to parentage relationship studies emerge in close connection with forensic genetics and molecular ecology (Weir *et al.* 2006). Forensic genetics has mostly focussed in paternity analysis and individual identification. Statistical developments for paternity inference proceeded initially from this field (Chakraborty *et al.* 1988; Pena and Chakraborty, 1994; Jin and Chakraborty, 1995). On the other hand, molecular ecology is more interested in estimating population parameters (gene flow, effective population size, breeding system and care of progeny, etc.) than relatedness between individuals itself (Wilson and Ferguson 2002; Jones and Ardren 2003). In livestock production, the availability of pedigree information has determined that statistical tools for covering its requirements have been mainly taken from forensics. Aquaculture shows particular characteristics, such as the capture of genetic diversity from natural populations for genetic breeding programs, the high fecundity of aquatic species, the complexity of broodstock management and finally, the production of thousands of larvae and fry from different families in individual tanks. This shapes a particular scenario, which requires from specific solutions. The revision of publications on parentage relationships in Aquaculture from databases reveals that most of these studies have been focussed on fish, and only very recently there appear studies on crustaceans and molluscs (Taris *et al.* 2005; Jerry *et al.* 2006a, 2006b; Lucas *et al.* 2006; Johnson *et al.* 2007). Consequently, the problematic presented in this revision, as well as the references cited, will be mainly focussed on fish farming.

The relevance of parentage relationship analysis in aquaculture is mainly connected with genetic improvement and to a minor extent, with food and environmental safety (Hastein *et al.* 2001; Liu and Cordes 2004). The high fecundity of aquatic organisms has frequently determined the application of simple mass selection programs resulting in genetic diversity losses and inbreeding depression problems (Hulata 2001; Gjedrem 2005). The identification of familiar relationships is therefore a key point to apply familiar breeding programs, which combine inter and intrafamily genetic diversity components avoiding these harmful effects (Falconer and McKay 1996). Also, quite often, parentage relationships among breeders are unknown, because there is no detailed information about their origin or because they proceed from other companies. Usually, commercial batches show a high relatedness among them, because they have been obtained from a few breeders.

Genetic breeding programmes are being implemented in no more than 40 aquatic species (Gjedrem 2005). For their development, it is necessary to check the families founded in species with planned crosses like turbot, salmonids or catfish. The complex management of larvae and fry at farm facilities, which involves the movement of fishes between tanks and their classification by size to accommodate feeding and growth, increases the probability of error, and consequently that actual families do not correspond to those planned. Also, it is common from a certain size, the mixture of families in the same tank to avoid environmental interactions for heritability estimation or for the selection programme itself (Castro *et al.* 2004). In this context, the appropriate weight of families regarding the selection intensity applied, requires the knowledge of the relationships of the individuals selected. On the other hand, in species with mass spawning, such as black sea bream, Senegal sole or European sea bass, it is essential the identification of families which are reared together in the same tank, both to ascertain the breeding structure as well as to apply a familiar selection programme (De León *et al.* 1998; Boudry *et al.* 2002; Fessehaye *et al.* 2006; Castro *et al.* 2006). Furthermore, in these species heritability estimation to find out the best selection strategy also depends upon family identification (Vandeputte *et al.* 2005; Lucas *et al.* 2006; Wang *et al.* 2006). Finally, an important issue related with genetic breeding programs is the maintenance of genetic diversity across generations. In addition to maintaining a broodstock of appropriate size, several authors have proposed the application of methods based in relatedness coefficient to minimize the loss of genetic diversity across generations (Ballou and Lacy 1995; Caballero and Toro 2000; Sekino *et al.* 2004).

Other collateral issues related with Aquaculture are fish traceability and the control of releases from farms, which can contaminate natural populations (Wilson and Ferguson 2002). The identification of parentage relationships through molecular markers can permit to trace back individuals with market problems (e.g. toxins) or released in natural populations, to breeders in the farm which have produced them (Hastein *et al.* 2001; Hayes *et al.* 2005). A similar situation can also occur in fattening farms, which receive fry from other companies, and where the presence of some anomaly or pathology can be trace back to the company of origin through paternity analysis (Martínez *et* al. unpublished data).

2. General Considerations on Genetic Markers and their Potential for Relatedness Analysis

2.1. Properties of Genetic Markers

The study of parentage relationships through molecular markers is based on the genetic similarity existing between individuals, in turn determined by Mendelian genetic transmission laws (Figure 1). Thus, a parent shares an allele *identical by descent* (IBD) with its offspring, and the allocation of parents to a particular offspring is known as paternity inference. Therefore, for this analysis, it will be necessary individuals of two consecutive generations, the parental and the filial ones. On the other hand, the study of kinship between whatever two individuals (dyad) pertaining or not to a particular generation, will be based on the proportion of IBD alleles shared between them. So, full-sibs share on average one IBD allele, though at some loci they will not share any allele (25%) and in another 25%, both alleles will be IBD. Statistical procedures for allocating parents (paternity analysis) will differ from those devoted to estimate the relationship at dyads (kinship/coancestry analysis), and will be treated in two different sections in this revision.

On the other hand, genetic resemblance between individuals will be estimated from the allelic similarity in the loci analyzed, and this will be obtained through genotyping. Consequently, two alleles will be *identical by state* (IBS), if they show the same genotypic value, measured for example in base pairs (bp) at microsatellite loci, or as relative electrophoretic migration at allozymes. Two individuals can share IBS alleles in a particular

	M-1	M-2	M-3	M-4	M-5	M-8	M-14	M-125
MOTHER	165183	095097	165167	090134	147151	192198	116138	132136
FATHER	173181	095095	165165	116120	149151	192194	132132	120136
O1	173183	095095	165167	120134	147151	192198	116132	136136
O2	165173	095097	165165	120134	147151	194198	116132	136136
O3	173183	095097	165167	090120	151151	194198	132138	136136
O4	173183	095097	165165	120134	147151	192192	116132	136136
O5	173183	095095	165165	120134	147151	192198	116132	136136
O6	173183	095097	165165	090116	149151	192198	132138	120136
O7	165173	095095	165165	120134	147151	194198	116132	120132
O8	165173	095097	165165	090120	151151	192194	132138	120132
O9	165173	095097	165165	120134	147151	192192	132138	136136
O10	173183	095095	165167	090116	149151	192194	116132	120132
O11	167169	107125	177179	090122	147151	198202	132132	120128

Notice that all offspring share an allele with their parents at all loci, and that full-sibs can share 0, 1 or 2 alleles per locus.

Genotypes are given in base pairs (165183: heterozygote for 165 and 183 bp alleles)

Microsatellite (M) loci are presented in the upper row from M-1 until M-125

Mendelian incompatibilities with one or both parents of the false offspring O11 appeared marked in yellow and red, respectively.

Figure 1. Microsatellite genotypes in a family from a breeding selection programme in turbot (*Scophthalmus maximus*).

population, though they are not related, depending on the frequency of this allele in the population under study. The key for paternity, and especially for kinship analysis, will be on the 'goodness' of the statistical methods for estimating IBD from IBS and the allelic frequencies of the population studied (Blouin 2003; Jones and Ardren, 2003; Weir *et al.* 2006).

In this context, the choice of the best genetic marker for achieving the appropriate power is essential. This will be dependent on their polymorphism, accurate genotyping, technical drawbacks, expression model and cost. As a rule of thumb, the higher the polymorphism of a locus, measured as the number of alleles or as the expected heterozygosity, the greater its power for the analysis. Genotyping accuracy is other essential element, since mistakes can lead to false exclusions in paternity analysis or to decrease genetic similarity at specific dyads when estimating kinship (Sancristobal and Chevalet 1997; Hoffman and Amos 2005; Castro *et al.* 2006; Johnson *et al.* 2007). The model of expression is other relevant point, since with dominant markers it is not possible to distinguish between homozygous or heterozygous individuals of dominant phenotype, and, consequently, the capacity of estimating relationships notably decreases. Codominant markers can be up to one order more resolutive than dominant ones (Gerber *et al.* 2000). Finally, cost and technical drawbacks are usually two issues closely connected, the later being related with the amount of genotyping errors.

Since their discovering, microsatellites have been the most popular markers used for parentage analysis (Hulata 2001; Wilson and Ferguson 2002; Jones and Ardren 2003; Chistiakov *et al.* 2006). Their high polymorphism, codominant expression, technical simplicity and reliability, as well as their genomic abundance and ubiquity, have determined their appropriateness for this purpose. However, microsatellites are not free from important drawbacks, including their development, sometimes associated to tandem repeat enriched library construction (Zane *et al.* 2002), whose efficiency is usually below 5% (Carreras-Carbonell *et al.* 2004; Pardo *et al.* 2007). Alternatively, it is possible to cross-amplify microsatellites from close related species, though success is highly variable, and sometimes the effort and time consumed is finally greater than constructing enriched libraries. In fish, a high interspecific microsatellite conservation has been observed in salmonids (Angers and Bernatchez 1997), important in sparids (Brown *et al.* 2005; Pinera *et al.* 2006; Castro *et al.* in press), but very low in pleuronectids (Bouza *et al.* 2002; Castro *et al.* 2006) and in gadids (Castro *et al.*, unpublished data).

Other relevant problem for parentage analysis associated to microsatellites, is the presence of genotyping errors (Marshall *et al.* 1998; Castro *et al.* 2004, 2006; Hoffman and Amos 2005; Pompanon *et al.* 2005). Genotyping errors are due to the presence of adjacent alleles differing by 1 bp; to "dropping out", related to misamplification of some alleles, particularly the longest ones in heterozygotes of divergent allelic sizes; to confusion with internal standard peaks in automatic sequencers with only one fluorophore; and to stuttering, related with strand slippage during PCR reaction, more common in dinucleotide than in longer microsatellite motifs (Figure 2). All these sources of error can be increased by the tendency of Taq polymerase to add an extranucleotide after each PCR cycle appearing an echo pike/band at each peak produced in the reaction (Pompanon *et al.* 2005).

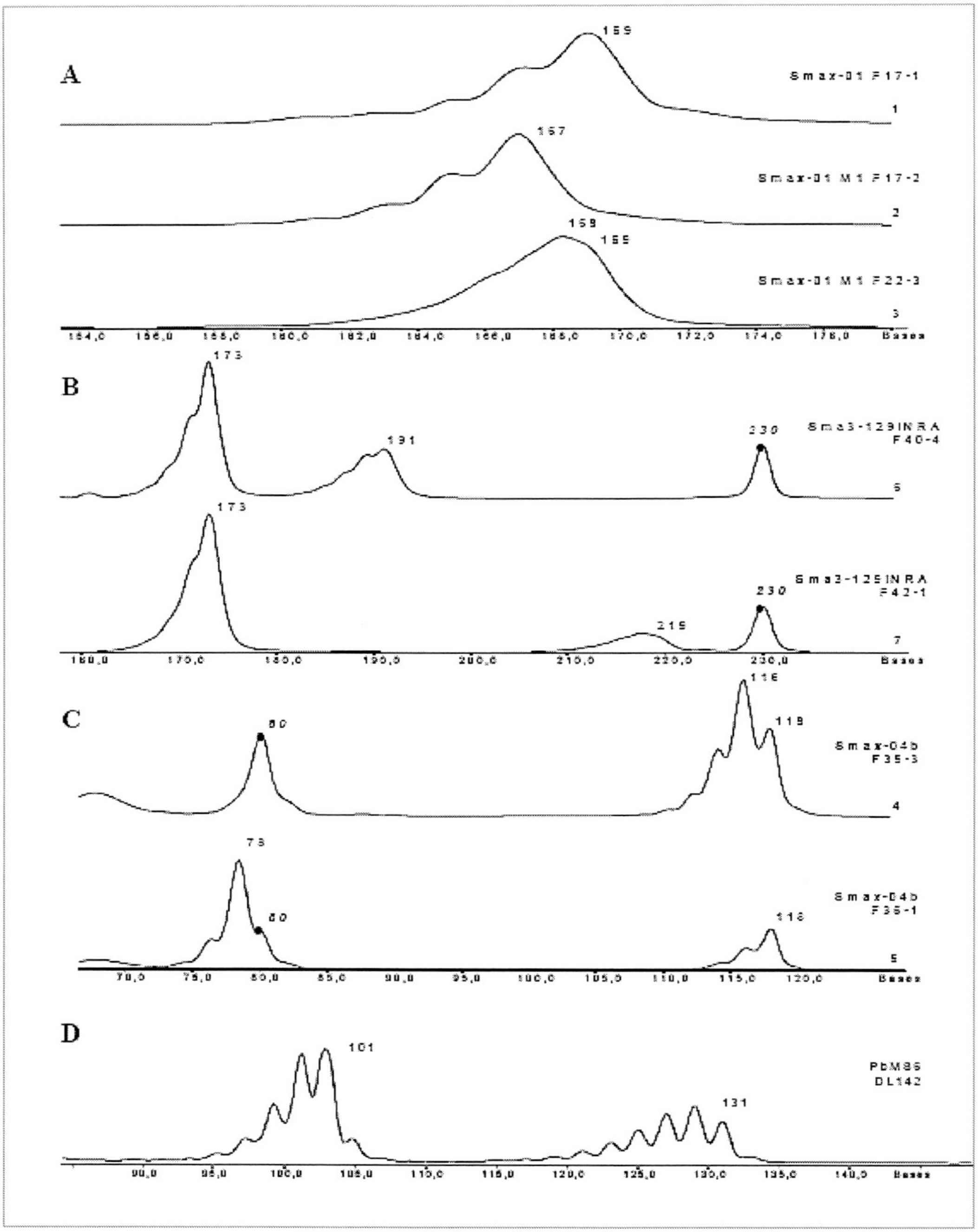

Modified from Castro *et al.* (2005)

A) Intermediate alleles; B) Drop out; C) Confusion with an internal marker; D) "Stuttering"

The three first cases proceed from turbot (*Scophthalmus maximus*) and the last one from black sea bream (*Sparus aurata*). This one has been obtained by cross-amplification from red sea bream (*Pagellus bogaraveo*).

Figure 2. Sources of genotyping error at microsatellite loci.

Null alleles are due to the presence of mutations in the regions where primers anneal and can be present at high frequencies in some loci (Figure 3; Dakin and Avise 2004; Jerry *et al.* 2004; Castro *et al.* in press). This annealing failing leads to non-amplification of the allele,

FAMILY 1	M-1	M-2	M-3	M-4	M-5	M-6	M-7	M-8	M-9
FATHER	169169	095095	165165	090090	147147	092096	132136	171189	097097
MOTHER	165167	125137	165165	116120	147147	088092	136144	171177	099103
O-1	**165169**	**095125**	**165165**	**090116**	**147147**	**092092**	**132136**	**171189**	**097103**
O-2	**167169**	**095125**	**165165**	**090116**	**147147**	**092096**	**136144**	**171189**	**097103**
O-3	**167169**	**095137**	165179	**090116**	**147147**	**092096**	**136136**	**171177**	**097103**
O-4	**165169**	**095137**	**165165**	**090120**	**147147**	**088096**	**132136**	**171177**	**097103**
O-5	**165169**	**095137**	**165165**	**090120**	**147147**	**088092**	**136136**	**171177**	**097099**
O-6	**167169**	**095137**	**165165**	**090120**	**147147**	**092092**	**132136**	**177189**	**097103**

FAMILY 2	**M-1**	**M-2**	**M-3**	**M-4**	**M-5**	**M-12**	**M-125**	**M-129**	**M-152**
Not analyzed	***167/***	***095119***	***179/***	***088090***	***151/***	***092/***	***128132***	***167/187***	***097405***
MOTHER	**152169**	**095109**	**165179**	**116122**	**147147**	**090092**	**128148**	**173173**	**095099**
O1	**152167**	**095119**	**165179**	**088116**	**147151**	**090106**	**128132**	**187187**	**095097**
O2	**167169**	**095095**	**165179**	**090122**	**147151**	**092098**	**128148**	**167173**	**095105**
O3	**152167**	**109119**	**165179**	**090122**	**147151**	**092098**	**128128**	**167173**	**097099**
O4	**152167**	**095119**	**165179**	**088116**	**147151**	**090106**	**128132**	**187187**	**095097**
O5	**152167**	**095095**	**179179**	**088122**	**147151**	**092098**	**128148**	**187187**	**095105**
O6	**167169**	**095109**	**179179**	**088122**	**147151**	**090098**	**132148**	**173187**	**099105**

Microsatellites (M) analyzed in the upper row
Parent-offspring incompatible genotypes highlighted in red
Family 2 illustrates the possibility of parental reconstruction starting from offspring data (italics genotypes).

Figure 3. Mutation (family 1) and null alleles (family 2), as a source of paternity incompatibility in turbot .

being then genotyped as homozygote, when actually is a heterozygote genotype. Finally, though there are unstable microsatellites with mutation rates above 10^{-2} (Primmer *et al.* 1998; Bacon *et al.* 2001), these usually range between 10^{-3} and 10^{-5} (Schlotterer 2000), which represents a negligible source of mismatches in the context of parentage analysis (Jones and Ardren 2003).

Other genetic markers used for parentage relationship inference are the AFLPs ('Amplified Fragment Length Polymorphism'; Gerber *et al.* 2000; Ezaz *et al.* 2004), and more recently the SNPs ('Single Nucleotide Polymorphism'). SNPs are due to nucleotide substitutions (Anderson and Garza 2006; Weir *et al.* 2006), while AFLPs detect a more complex source of molecular variation. SNPs are codominant markers between 10-20 times more frequent across the genome, with higher genotyping accuracy and 5 fold lower cost than microsatellites (Hayes *et al.* 2005). However, they are usually biallelic, and consequently, exhibit a lower polymorphism than microsatellites. SNPs represent a sound alternative to microsatellites in those species where large genetic resources are available (Anderson and

Garza 2006; Weir *et al.* 2006). Though genomic resources are being largely increased in the last years in aquaculture species, only in salmonids and in some other model species SNPs have been deposited in public databases. Finally, AFLPs have been occasionally used for parentage analysis in aquatic species. Their main advantage is the low cost per locus and their easy development, especially valuable in species with low genomic resources. However, they are essentially dominant and show lower genotyping confidence than SNPs and microsatellites. Specific statistical methods have been developed for kinship analysis with AFLPs (Gerber *et al.* 2000), and their performances have been evaluated in comparison with microsatellites in aquatic species (Ezaz *et al.* 2004). These authors have emphasized that a selection of AFLP loci, where the dominant allele is at frequencies between 0.1 and 0.4, would result in a significant improvement for their allocation capacity.

Most studies on parentage relationships in Aquaculture have been addressed with microsatellite loci. In spite of the potential of SNPs, it will take several years to see their wide application in Aquaculture. Therefore, most examples, references and statistical tools across this revision will be referred to microsatellites.

2.2. Statistical power of genetic markers for estimating parentage relationships

Though the potential of markers for parentage analysis is closely associated with their polymorphism (number of alleles, heterozygosity), there are more accurate statistical approaches for its assessment. This is a central question to obtain the most powerful set of markers involving the minimum effort for the analysis (Wang 2006). The most popular procedures for this aim are associated with paternity inference derived from forensic genetics (Chakarborty *et al.* 1988). In this sense, it has been defined the potential of exclusion of a particular locus, as its capacity for excluding a false parent taken at random from the reference population (usually the population of potential parents), when the other parent is unknown (Excl1) or when it is known (Excl2). The Excl2 probability will be always higher than Excl1, since Mendelian restrictions are higher (Table 1). The analysis is based on the calculus of the proportion of genotypes impossible for the true parent. The probability of exclusion of a locus will be equal to the sum of the frequencies of all incompatible genotypes averaged over all possible parents (Chakraborty *et al.* 1988; Sancristobal and Chevalet 1997). The sampling of genotypes for its calculus is performed using allele frequencies of the reference population assuming Hardy-Weinberg equilibrium. The combined exclusion potential over several loci (*Pe*(C)) is obtained as the product of the probabilities of assignment (1 - *Pe*) for the *L* loci analyzed assuming independent segregation, according to the following equation:

$$P_E(C) = 1 - \prod_{l=1}^{L} (1 - Pe_l)$$

where *Pe* is the exclusion probability for the *l*th locus (Marshall *et al.* 1998).

Table 1. Genetic diversity estimators and exclusionary power of 10 microsatellite loci from European sea bass (*Dicentrarchus labrax*)

Locus	Na	He	PIC	Excl1	Excl2
Labrax3	28	0.881	0.870	0.625	0.769
Labrax6	10	0.683	0.635	0.269	0.442
Labrax8	25	0.851	0.838	0.563	0.723
Labrax13	31	0.921	0.914	0.727	0.841
Labrax17	18	0.855	0.837	0.548	0.710
Labrax29	31	0.876	0.866	0.622	0.767
Dla11	16	0.743	0.724	0.381	0.569
Dla12	22	0.920	0.912	0.716	0.835
Dla20	53	0.940	0.935	0.783	0.877
Mean/Total	26	0.852	0.837	0.99978	0.999997

Bouza *et al.* (unpublished data)
Na: number of alleles
He: expected heterozygosity
PIC: polymorphic information content
Excl: exclusionary potential 1 (when no parent is known)
Exc2: exclusionary potential 2 (when one parent is known)

However, the probability of exclusion exhibits certain limitations, because it is averaged at sample level and simulation is based in a predefined family representation (Taggart 2007). Additionally, the simulation is performed with a limited number of offspring and does not take into account the existence of genotyping errors and mutations at specific loci (Wang 2006). Finally, the method is based on paternity exclusion and does not consider neither other possible relationships, nor the genotypic information of compatible parents (Wang 2006). Goodnight and Queller (1999) proposed a more powerful method fitted to each relationship context analyzed, though computationally more complex. The method evaluates the potential of a set of loci to discriminate between two alternative hypotheses (primary and null), based in the distribution of the probability ratio between the primary and the null hypotheses obtained through simulation starting from allele frequencies of the reference population. Wang (2006) went forward and introduced the possibility of genotyping errors and mutations, and used a more rapid and accurate method to determine the statistical confidence and potential without a simulation support. Finally, Taggart (2007) has developed an exhaustive and efficient algorithm to evaluate the discriminatory power of a set of markers at family, group or sample level.

3. General Considerations on Statistical Methodology

3.1. Maximum Likelihood Estimation

Parametric statistics assumes that results from a particular process (IBS for markers in this context) are samples from a random variable with a known distribution (the model), which depends on one or several parameters θ (IBD and the allelic frequencies of the reference population). When those parameters are known a density function f(x | θ) can be constructed and thus, we may *a priori* calculate the probability of obtaining a particular value. However, when the estimation is pursued (like determining the degree of relatedness from molecular data), parameters are unknown and we actually take samples of observations to infer their value. Now, we have a fixed data set (X_0: IBSs) and we can define the likelihood of a particular value of the parameter (e. g. full-sibs) as the probability of obtaining the results if the true value of θ were the one considered, $f(X_0 \mid \theta)$. Likelihood function l(θ | x) o l(θ) is the opposite concept to density function, as it yield the probability of different values of the parameters for a fixed set of observations.

Likelihood allows for the comparison between alternate hypothesis about the true value of the parameters (for example, H_1: one of the individuals being parent of the other, against H_2: they are unrelated). Consequently, the parameter's value with the highest likelihood will correspond with the more probable alternative. Commonly, comparisons are made through the likelihood *ratio*, so the interpretation is simpler, since $l(\theta_1) / l(\theta_2) = 3$ will mean that θ_1 value will be three times more likely than its alternate. Usually, the logarithm of the likelihood is computed [L(θ) = ln l(θ)].

To find the maximum likelihood estimator of the parameter we can calculate the likelihood of a broad range of values (ideally the whole range) and use the one with the highest likelihood. But if the likelihood function can be constructed and derived, we know that the maximum value can be obtained by equating the first derivative to zero and solving for the parameter (if the maximum is not at the extreme of the distribution).

To obtain a measure of the accuracy of our estimate confidence intervals can be constructed around the maximum likelihood estimator. More thoroughly explanation on this methodology and its comparison with other types of estimator can be found in any general book of statistics.

3.2. Bayesian Inference

If we consider parameters as random variables which depends on fixed observations, then its distribution $f(\theta \mid X_0)$ can be obtained. The conditional probability of event *A* given *B* is the probability that both events occur, provided that event *B* has actually occurred. Bayes theorem allows for the calculation of such probability following the expression

$$p(A \mid B) = \frac{p(B \mid A) \cdot p(A)}{p(B)}$$

If we translate this formula to the estimation of parameters from the sampled data we have:

$$f(\theta \mid X_0) = \frac{f(X_0 \mid \theta) \cdot f(\theta)}{f(X_0)} = \frac{f(X_0 \mid \theta) \cdot f(\theta)}{\int f(X_0 \mid \theta) \cdot f(\theta) d\theta}$$

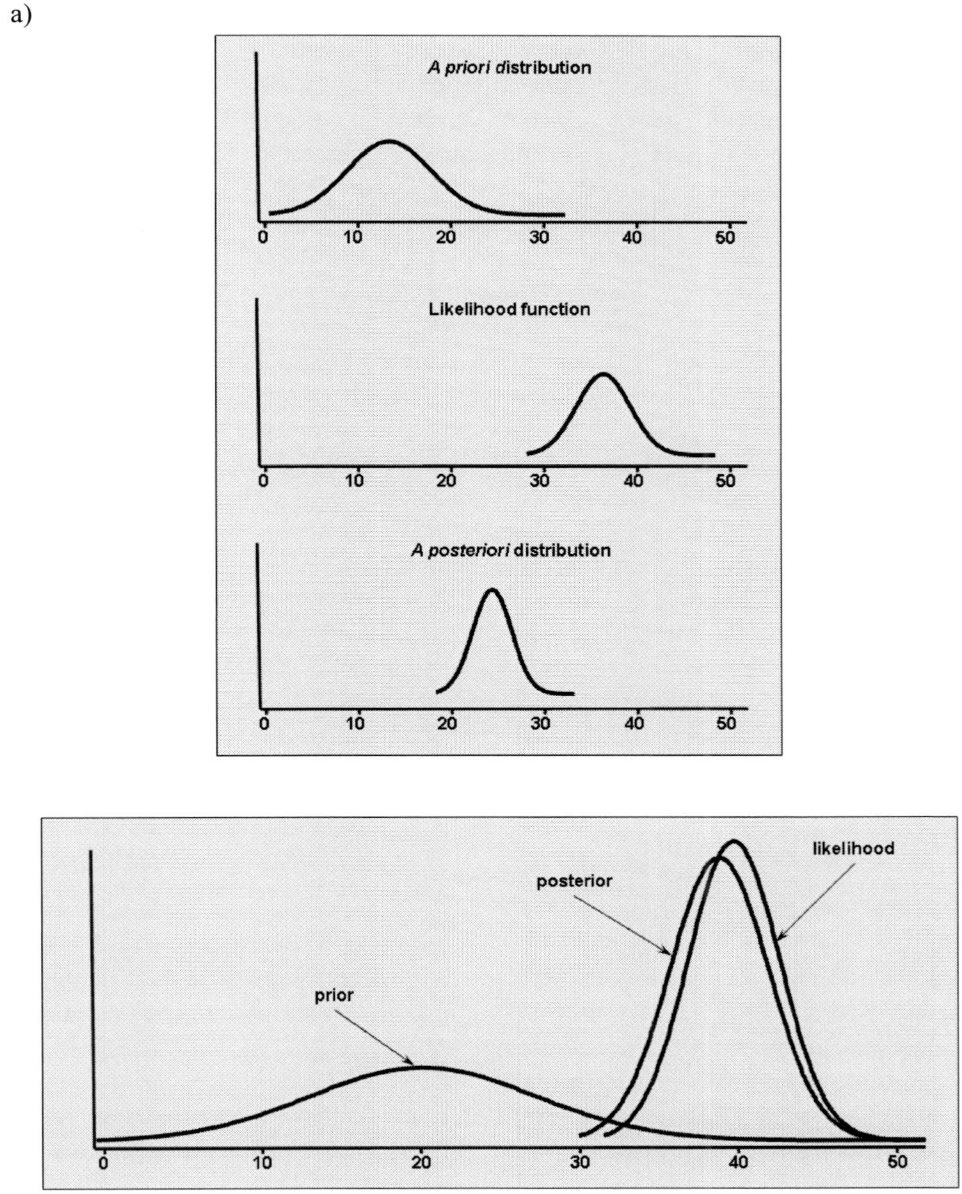

Figure 4. a) Posterior distribution as a combination of the previous information (*a priori*) and the information of data (likelihood). b) With scarce previous information (little informative prior) posterior is very similar to likelihood.

A careful reading of this expression tell us that the final or *a posteriori* distribution depends on the probability of the data given the value of the parameter (the likelihood if we consider it as a function of θ) and the *a priori* distribution of the parameter itself. This in turn is related with the previous information about the process we have. The denominator does not depend on θ, and does not affect the parameter estimation. Thus, the posterior distribution is in some way the product of the prior by the likelihood and it will range between them, as shown in Figure 4a.

The main objection against the Bayesian inference is that the result depends on previous assumptions (which may include the researcher prejudices) through the *a priori* probability. To avoid this, a non-informative *a priori* distribution may be used (e. g. a uniform distribution along the definition range of the parameter) and the results will only depend on the likelihood. Notwithstanding, this feature is appealing because it permits the Bayesian methodology to easily process information in a sequential pattern. If after a first estimation we obtain a second set of independent observations, we can repeat the analysis using as *a priori* the posterior distribution obtained with the first sample and the likelihood of the second one.

Another advantage is that estimation consists of a whole distribution, instead of single point estimation and a confidence interval as it happens with maximum likelihood estimation. When a single value is required, commonly the mode of the posterior distribution is the chosen value (it is the most probable value). If we use a criterion minimising the risk of the estimator of giving erroneous values (usually a quadratic function) then the chosen value as a punctual estimator will be the mean.

The smaller the amount of initial information, the closer the posterior distribution will be to the likelihood function, and its mode will be more similar to the maximum likelihood estimate (see Figure 4b). When sample sizes are large, Bayesian estimators converge to maximum likelihood estimators, because the importance of the data is greater than the previous information we may have. Moreover, the mean and the mode of the posterior distribution will be quite close due to the asymptotic normality of this distribution.

Deeper insights on the Bayesian methodology can be found in the book by Bernardo and Smith (1994).

3.3. MCMC Methods

Many times the process we want to characterize depends on more than one parameter and, consequently, the posterior distributions also do (joint posterior distribution). But obtaining an estimate for each parameter requires the calculation of the posterior distribution depending on a single parameter, integrating for the rest (marginal posterior distribution). This integration is difficult to perform (if not impossible), so we have to implement algorithms which allows for the construction of the marginal distributions without an explicit integration (Gelfand and Smith 1990). All MCMC methods (Markov Chain Monte Carlo) arise from the Metropolis-Hastings algorithm, the oldest one, being the most popular the *Gibbs sampling*. Markov chains are stochastic successions of events or states in which the probability of a particular state is only function of the previous state. *Gibbs sampling* is based on this kind of chains and basically consists in correlatively sampling from the complete distribution of each parameter, in cycles, converging this chain to the marginal distribution for each of them.

MCMC methods are also useful in searching across a huge space of feasible solutions (for example, all possible combinations of individuals in the population into full-sibs groups) to find the one with highest probability.

3.4. Simulation

As we pointed out in the previous section, some operations in the field of the estimation of genetic relationships are hardly manageable in an arithmetic way and require the use of other approaches. In these situations one of the available tools is the computer simulation. Some problems where simulation can be implemented, as explained later, include the determination of significance values for likelihood ratios, the calculation of confidence intervals for some estimators or the establishment of thresholds between adjacent relationship classes (e. g. separation between half-sibs and full-sibs).

The basic idea consists in generating a series of virtual individuals carrying alleles in each locus with a probability equal to the frequency, known or estimated, in the same population. From them, dyads (or groups) of offspring are generated following the Mendelian transmission rules and, therefore, relationship between any pair of individuals are known. From the alleles at the marker loci we can estimate relationships and, with many dyads of the same type, construct a distribution of the estimator which allows for the calculation of confidence intervals and other parameters.

If available, it would be better to simulate virtual parents from the genotypic or haplotypic frequencies, because in this way the possible disequilibrium between or within loci can be taken into account in the process.

4. Paternity Analysis

4.1. Methods for Parentage Allocation

The fact that each parent necessarily shares one allele with its offspring makes possible an efficient approach to paternity analysis based on exclusion. This is not possible when the relationship between two individuals not directly related through consecutive generations is evaluated. This circumstance also permits to go back from offspring to dead or non-sampled parents for genotyping reconstruction, and makes also possible to estimate the minimum number of parents responsible of a particular progeny. But, as in kinship analysis, efficient Bayesian and maximum likelihood methods have been developed for paternity inference.

4.1.1. Allocation by Exclusion

At first sight, this is the simplest and most appealing solution for parentage allocation (Figure 5). In summary, the true parent is identified by excluding the remaining candidates based on the detection of Mendelian incompatibilities at the loci used. Obviously, the potential of exclusion will be dependent on the number of candidates, the number of loci and their power of exclusion (Wilson and Ferguson 2002; Jones and Ardren 2003). The higher the number of candidate parents, the larger the number of loci necessary for excluding all false parents. However, this relationship is not linear, and the inclusion of an additional locus compensates for a proportionally greater increase of new candidates (Bernatchez and Duchesne 2000). Finally, if one parent is known the allocation of the other parent will be much more straightforward (Marshall *et al.* 1998).

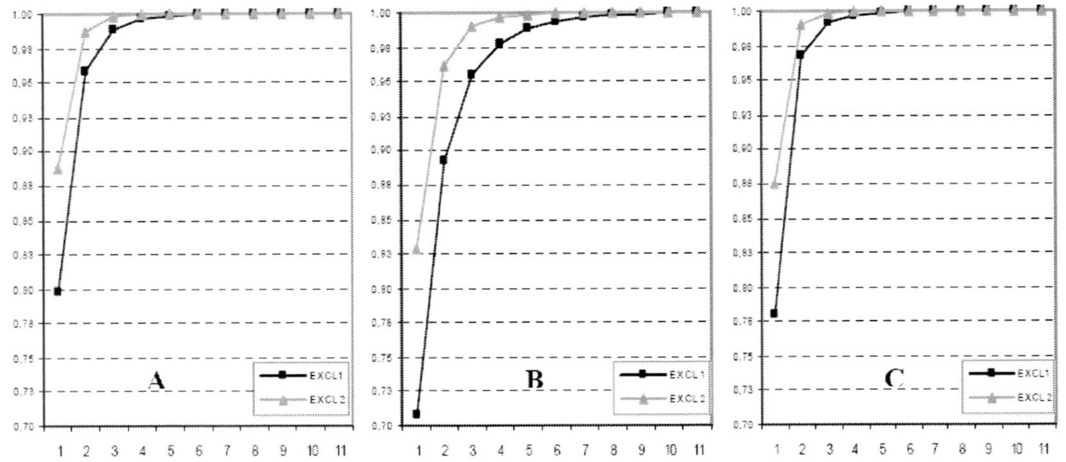

Modified from Castro *et al.* (2005)

A: black sea bream (*Sparus aurata*); B: turbot (*Scophthalmus maximus*); C: Senegal sole (*Solea senegalensis*)

Abscises: number of loci analyzed; Ordinates: exclusionary probabilities when no parent is known (Exlc1) and when one parent is known (Excl2).

Figure 5. Exclusionary potential variation of microsatellites in three Aquaculture species.

However, this approach presents some drawbacks and can perform better at specific scenarios. Firstly, and depending on the number of candidates and loci managed, it is possible that not all candidates but the true parents are excluded (Wilson and Ferguson 2002). The application of the minimum possible number of loci to alleviate the effort and cost in Aquaculture, increases undefined situations, appearing trios or higher numerous groups of putative parents (Estoup *et al.* 1998; Castro *et al.* 2004, 2006, 2007; Johnson *et al.* 2007). Secondly, the strict exclusion principle is valid when no genotyping errors or null alleles exist at the markers used, which is not a realistic assumption. The percentage of genotyping errors for microsatellite loci, the most popular markers for paternity inference, ranges between 0.5-2% (Estoup *et al.* 1998; Hoffman and Amos 2001; Castro *et al.* 2004, 2006, 2007; Pompanon *et al.* 2005). The proportion of null alleles is variable, but can affect to 10% of the loci, and though their frequencies are usually below 10%, they can reach 20% or even higher values (Dakin and Avise 2004; McCartney *et al.* 2004; Castro *et al.* 2007). Both circumstances would lead to false exclusions, but also, though less frequently, to erroneous assignments. The existence of genotyping errors and null alleles leads to an epistemological paradox of the exclusionary approach. The increase in the number of markers, necessary to gain allocation potential, unavoidably leads to a higher number of false exclusions (Marshall *et al.* 1998; Duchesne *et al.* 2005). The solution is to introduce certain flexibility in the system permitting a predetermined number of incompatibilities (Vandeputte *et al.* 2006; Taggart 2007). Obviously, this solution decreases the potential of exclusion. Additionally, and since the tolerance is homogeneously applied at all managed loci, the existence of an important error rate at specific loci, will lead to a higher flexibility in the whole system, increasing incorrect allocations. In this sense, the previous checking of loci for null alleles and genotyping errors using family or population data will facilitate the adjustment of error rate at the sample managed (Marshall *et al.* 1998; Taggart 2007; Castro *et al.* 2007).

4.1.2. Maximum Likelihood Allocation

Maximum likelihood methods do not intend to exclude false parents, but allocating the most probable one taking into account the genetic frequencies of a reference population in addition to the genotypes of candidate parents and offspring (Meagher and Thompson, 1986; Marshall *et al.* 1998; Jones and Ardren, 2003). In this way, the uncertainty associated to several compatible parents is overcome (Anderson and Garza 2006). On the other hand, the existence of genotyping errors will not preclude the identification of the true parent, which can still be the most probable solution. For this analysis, as in all methods of paternity inference, it should be considered if the other parent is known, and if one is searching for the most probable single parent or couple. The likelihood of each candidate parent (or couple) of the offspring tested will be obtained as the log of the likelihood ratio ('LOD score') of that candidate being the true parent respect to one individual taken at random from the reference population. These random individual is obtained from allelic frequencies of the reference population assuming HW equilibrium. This likelihood will be estimated as the proportion of alleles shared per locus. As pointed out by Jones and Ardren (2003) the method will introduce a bias, since homozygotes will have a higher probability of sharing the compatible allele with its offspring. If loci segregate at random, a usual and reasonable assumption in spite of no mapping information available, it is possible to obtain a combined estimation over all loci analyzed by summing the LOD scores at individual loci.

Knowing the most likely parent is not enough to soundly establish paternity. It is necessary to know the statistical confidence, since there can be putative parents with close likelihoods. A relevant contribution to solve this question has been the simulation approach developed by Marshall *et al.* (1998), applied with some modifications by other authors (Goodnight and Queller 1999). Essentially, the method estimates the difference in LOD scores between the two most probable candidates (Δ) and establishes confidence thresholds through simulation from allele frequencies of the reference population (Table 2).

Table 2. Paternity assignment output from CERVUS following a maximum likelihood approach

Offspring	P non-exclusion	Candidates	N loci	N incomp	LOD score	Δ	Confid
O2	1,59E-04	P1	11	3	2,84E+00	2,69E-01	+
O2	1,59E-04	P2	11	4	2,57E+00	0,00E+00	
O2	1,59E-04	P3	11	2	2,44E+00	0,00E+00	
O3	5,54E-03	P4	11	0	5,22E+00	2,42E+00	*
O3	5,54E-03	P5	11	3	2,80E+00	0,00E+00	
O3	5,54E-03	P6	11	3	2,40E+00	0,00E+00	
O9	5,20E-04	P7	11	1	3,10E+00	2,92E-01	+
O9	5,20E-04	P8	11	3	2,81E+00	0,00E+00	
O9	5,20E-04	P9	11	2	2,09E+00	0,00E+00	

The analysis of three potential offspring (O2, O3 and O9) and 9 candidate parents (P1a P9) based in 11 microsatellite loci is shown

P: probability

N incomp: number of loci incompatible

Confid.: statistical confidence (+ P>80%; * P>95%)

An alternative Bayesian approach which takes into account the information of all compatible pairs has been proposed by different authors (Cercueil *et al.* 2002; Signorovitch and Nielsen 2002). However, the interpretation of the posterior probabilities is not clear, because no methods to obtain critical values for significance have been suggested.

Maximum likelihood allocation can be categorical or fractional (Jones and Ardren 2003). In the former, the goal is to identify the true parents of the offspring tested. Fractional allocation distributes the probability of each offspring among all parents with LOD score > 0. This approach is meaningful when one is interested in allocating the parentage success of groups of individuals and can perform better than the categorical one for this task (Nielsen *et al.* 2001; Duchesne *et al.* 2002; Araki and Blouin 2005). This approach can have some interest in species with mass spawning in Aquaculture when breeders of different ages or origins are managed. However, the categorical allocation is much more widely applied, since the applications in this field are mostly focused at individual level.

4.1.3. Full Probability Methods

The maximum likelihood models presented do not take into account previous information which could be relevant for parentage assignment. In molecular ecology studies, but also in the aquaculture context, not all individuals have the same *a priori* probability to sire a particular offspring. This information, if appropriately introduced in the system, can improve paternity inference. At farm facilities, there is no random mating and broodstock is structured following spatial-temporal variables. Neef *et al.* (2001) developed a Bayesian method to introduce *a priori* information, and so, to appropriately weight the search of potential parents to obtain the posterior probabilities. More recently, Hadfield *et al.* (2006) proposed the so called full probability method in which parent allocation is analyzed simultaneously with the searching of population parameters following an interactive model.

4.1.4. Genotypic Reconstruction of Parents

This approach, rather than identifying putative candidates, intends to reconstruct the genotypes of parents in a highly specific scenario (Jones 2001, 2005). The fact that each parent shares one allele with each offspring makes possible the reconstruction of parental genotypes in full/half-sibs mixtures, when a single female or male is responsible of the progeny. Unfortunately, the method is computationally complex, and no more than six parents of the other sex can be searched. The method works well even when the genotype of the single female (or male) responsible for the progeny is not known (Jones 2005). This procedure could have application in highly specific scenarios in Aquaculture, such as the reconstruction of parents in commercial batches where a single female and several males or *vice versa* are crossed. The estimation of the number of parents responsible of a particular progeny is a relevant point for application of paternity programs, since many of them require both their number and sampled fraction. However, there exist more appropriate methods to solve this question. Thus, the minimum number of parents can be grossly estimated as half the number of alleles of the highest polymorphic locus analyzed. The multilocus minimum method uses the information of all loci and divides all possible haplotypes across loci by 2^L, where L is the number of loci (Fiumera *et al.* 2001). A computational more sophisticated method has been suggested by Duchesne *et al.* (2005), which uses the cumulative allocation probability of loci to obtain the proportion of collected and uncollected parents responsible of a particular progeny.

4.2. Statistical Methodology Problems

4.2.1. Sampling Scenarios in Paternity Analysis

The strategy for paternity inference is greatly dependent on the sampling scenario, i. e., the number of candidates, the proportion of sampled parents and previous allocation of candidates. The situation in Aquaculture is much simpler than that described in natural populations (Jones and Ardren 2003). The most favourable scenario is when crosses can be planned, such in salmonids, turbot or catfish. In these species, if families are reared in independent tanks, a primary goal is to confirm that the families actually correspond to that planned. The management of a complex broodstock, the movement and classification of fry to encompass growth, the jumping of fish between adjacent tanks, and so on, makes recommendable the checking of short samples per family to confirm their family origin (Castro *et al.* 2004). If this checking is not carried out, an increasing proportion of errors will be accumulated across the selection programme diminishing the expected progress according to heritability estimates. In these cases, it is not necessary the application of statistical tools, the Mendelian congruence being checked manually, always depending on the amount of data managed. Also, the availability of several individuals per family makes the analysis much more consistent, since genotyping errors or null alleles at specific loci are more easily detected, and the assignation is much more powerful (Jones and Ardren, 2003; Castro *et al.* 2004, 2006). No more than 2 - 4 microsatellites are usually applied for this task (Castro *et al.* 2004).

In mass spawning species, the situation is more complex and will be dependent on the number of breeders per tank. Contrarily to what happens in studies in natural populations, here all breeders are usually available for analysis, which makes easier the allocation of parents. Additionally, breeders are subdivided in several tanks and photoperiods, which makes possible to restrict allocation to small subgroups. In these species both parents are unknown, which can suggest allocation by couples, as several statistical programs perform (Cercueil *et al.* 2002; Kalinowski *et al.* 2007). Though more powerful, this option should be taken with caution, since genotyping errors at one parent would automatically discard the identification of both parents. A low number of microsatellite markers (3-5) can be enough to achieve a potential assignment above 95% in these cases (Herbinger *et al.* 1995; Pérez-Enríquez *et al.* 1999; Hara and Sekino 2003; Dong *et al.* 2006; Castro *et al.* 2006, 2007).

As outlined before, acquisition of fish from other commercial sources is a common practice in Aquaculture for renewing genetic diversity and taking advantage of selection gain. Usually, these batches are founded with 1 - 2 females and several males. The identification of kinship among these potential breeders is essential to be incorporated in the selection programme. Here, the disadvantage is the lack of parents, but the low number of founders and the mixture of half/full-sib families make this task affordable through parental reconstruction methods (Jones 2001, 2005).

4.2.2. Treatment of Technical and Genetic Drawbacks of Molecular Markers

At the beginning of this work, it was indicated the types and frequency of genotyping errors associated to each genetic marker. In this section, it will be described and discussed the appropriate way for treating these errors for limiting their impact on paternity inference. Null alleles are one of the main sources of error associated to microsatellites (Wilson and Ferguson

2002) (Figure 6). More than 10% of the microsatellites reported exhibited null allele frequencies around or above 5%, a figure which can seriously compromise paternity inference (Marshall *et al.* 1998). In some cases, they can be eliminated by redesigning the primers starting from the library clone where they were identified (Dakin and Avise 2004). A first statistical approach to this problem is to use a tool which alerts us about their presence from population data. Null alleles give rise to heterozygote deficit in panmictic populations. Unless this deficit is caused by population structure which affect all loci in a similar way (inbreeding, Wahlund effect), microsatellites with null alleles will exhibit heterozygote deficit (Brookfield 1996). Some programmes designed for paternity inference include algorithms to identify those loci suspicious of harbouring null alleles, to be discarded from the analysis or to be taken with caution for their application (Marshall *et al.* 1998; Amos 1999). On the other hand, detection of null alleles will be much more accurate if family data are available, or if, as outlined before, the data managed include full-sib families, a situation common in the aquaculture context. In these cases, null alleles will be identified by the presence of homozygote-homozygote mismatches between parent and offspring at specific loci (Figure 3). If enough family data are available, an accurate estimate of null allele frequency can be obtained at each locus (Primmer *et al.* 1998; Dakin and Avise 2004; Castro *et al.* 2004, 2006, 2007). An alternative to avoid the exclusion of true parents due to null allele mismatches is the transformation of all homozygote genotypes to heterozygotes for a null allele, at least at those suspicious loci, and especially if an approximation of exclusion for allocation is followed (Danzmann 1997).

It has already been mentioned, that, in spite of exhibiting mutation rates highly above than single nucleotide substitutions false exclusions due to mutations will be exceptional when using microsatellite loci. Thousands of genotypes will not include more than a few incompatibilities due to mutation (Ellegren 2000). Only highly unstable microsatellites could

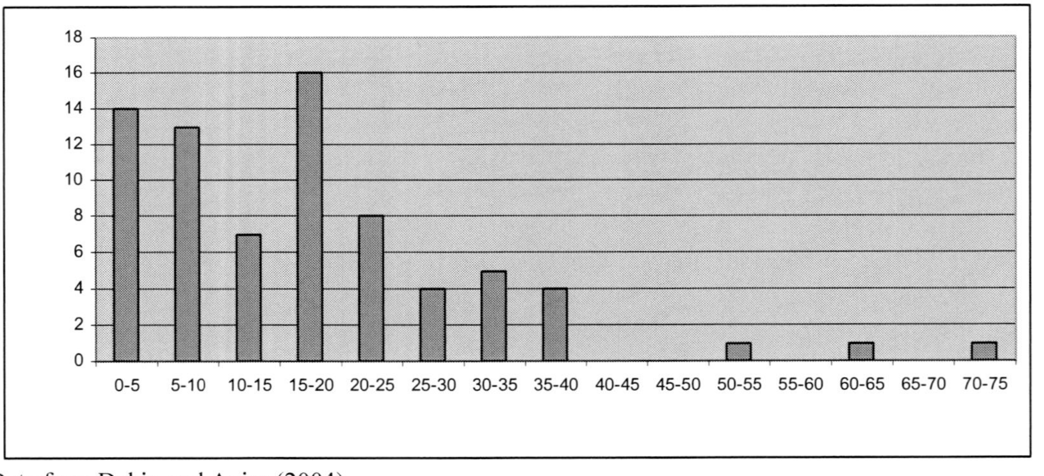

Data from Dakin and Avise (2004)

Number of loci with null alleles (ordinates) within a particular frequency range (abscises; in percentage).

Figure 6. Null allele frequency at microsatellite loci.

introduce a proportion of mismatches to be taken into consideration (Primmer *et al.* 1998). However, genotyping errors, as null alleles, can produce much higher impact. Particularly, 'drop out' and 'stuttering' are the cause for systematic errors in microsatellite genotyping, which can notably affect paternity inference (Castro *et al.* 2004, 2006; Hoffman and Amos 2004; Johnson *et al.* 2007). Their particular profiles, and especially their iteration, makes easier their detection, especially when family data are available. This obligates to reanalyze electropherograms and facilitates their elimination for further analysis. Most genotyping errors of microsatellites leave a particular signature, which facilitates their detection starting from population data. In this sense, the analysis of expected and observed heterozygote distributions can be used to detect drop out (low frequency of heterozygotes for highly divergent alleles) or stuttering (excess of heterozygotes for shortly distant alleles). The program MICRO-CHECKER (Van Oosterhout *et al.* 2004) uses this approach to identify and quantify the presence of null alleles, drop out and stuttering.

4.2.3. Departure from Theoretical Assumptions

Most approaches for paternity inference, excluding strict exclusion methods, rely on several theoretical assumptions, usually the conformance to HW equilibrium and independent segregation of loci (Wilson and Ferguson 2002). HW is usually assumed when simulating genotypes of parents and offspring for calculating exclusion probabilities or when looking for confidence in paternity inference. The lack of HW equilibrium in the reference population will determine an overestimation of exclusion probabilities, especially when heterozygote deficit is the cause (Estoup *et al.* 1998; Castro *et al.* 2006). Also, independent segregation of loci is assumed to calculate the combined potential of exclusion or the combined assignment for paternity. This information is usually unavailable, but when few loci are managed the probability of strong linkage between them which could originate a severe distortion in the analysis is really low. However, linkage has been observed in different studies (Ardren *et al.* 1999), mostly due to non-independent isolation of clones from enriched libraries. The use of different tandem repeated sequences from the same clone to increase the efficiency of libraries for microsatellite isolation has been observed in several aquaculture species (Iyengar *et al.* 2000; Bouza *et al.* 2002; Castro *et al.* 2006, 2007). The diminishing power for assignation of these linked loci will be associated to the lack of gametic equilibrium, which, in general, will be more intense the closer the linkage between loci. However, if gametic equilibrium exists in spite of linkage, there will not be a departure of parentage assignment from expectations, at least for the whole sample. If linkage phase is known both linked loci would even constitute a superlocus, providing a higher potential for assignment (Jones *et al.* 1998). However, this is not frequent and this possibility represents more a theoretical exercise rather than a practical tool for paternity analysis. Additionally, though expectations for the whole sample do not change if gametic equilibrium exists, this situation would be variable depending on the particular genotypes of the families analyzed (Anderson and Garza 2006; Castro *et al.* 2006). Therefore, as a rule it should be recommended the application of unlinked loci for parentage inference.

A final element which notably influences paternity allocation is the presence of a strong family structure among candidates, or especially, if among candidates are included half or full-sibs of the offspring to be allocated (Estoup *et al.* 1998; Marshall *et al.* 1998; Castro *et al.* 2007). This circumstance would be unlike in aquaculture since ages of breeders are usually known, at least when the genetic breeding programme is working. However, it could occur when founding the broodstock from native individuals or when individuals come from unknown commercial source. In these cases, the actual parent could show LOD scores very close to their relatives or even lower when it competes with a full-sib of the offspring analyzed. The best solution would be to increase the number of loci. As a general rule maximum likelihood approaches would be more affected than exclusion ones by this problem (Marshall *et al.* 1998; Jones and Ardren 2003). These situations would be avoided if breeders were stocked according to minimum relatedness (Castro *et al.* 2007).

4.3. Computer Programs for Paternity Inference and their Performances

A relevant question in paternity inference is the choice of the appropriate program among the many existing ones (Table 3). However, there is not the best software for all situations, and many of them present complementary performances as emphasized across the previous sections in this revision and also by different authors (Kasumovic *et al.* 2003). The most common scenario in aquaculture breeding programmes is the knowledge of most, if not all, parents of the progeny to be sired. This limits the utility of programmes such as PAPA and PASOS, which offer performances related with the identification of the fraction of parents sampled, more useful for natural populations. Also the requirement of these variables by some programmes, such as CERVUS and FAMOZ (Marshall *et al.* 1998; Gerber *et al.* 2000), would not represent a handicap. In fact, as seen later, CERVUS is the most popular program among aquaculture researchers. Also, the problem related with the lack of some parents when an exclusive searching by couples approach is performed (PROBMAX, Danzmann 1997; FAP, Taggart 2007; PAPA, Duchesne *et al.* 2002) would not represent a drawback in the aquaculture context. In Aquaculture, parental reconstruction provided by GERUD (Jones 2001, 2005) could aid to identify the small fraction of parents which have not been sampled or died (Castro *et al.* 2004).

Regarding the allocation method, again the availability of all or most parents in aquaculture is a relevant point to take into consideration. The exclusion approaches with variable tolerance implemented in several programs (NEWPAT, Worthington Wilmer *et al.* 1999; PARENTE, Cercueil *et al.* 2002; FAP, Taggart 2007) will show here their best performance. Most offspring could be allocated depending on the genotyping accuracy and the number of loci applied. An error rate below 3 - 4% combined with an appropriate tolerance could be enough to solve correctly most paternities (Vandeputte *et al.* 2006). However, the use of a low number or low polymorphic markers, or the existence of relatedness among candidates could lead to multiple allocation (Castro *et al.* 2004, 2006). Maximum likelihood methods could perform better when error rate is high and additionally, they could solve multiple allocation paternities. Castro *et al.* (2007) demonstrated a better performance of the exclusion over maximum likelihood methods when all parents were available in *Sparus aurata*, but after an exhaustive correction of errors and in a scenario with relevant relatedness among parents. This is the worst situation for maximum likelihood methods. More investigation using actual data will be necessary in the future to clarify this point.

Table 3. Paternity inference programs of interest in Aquaculture

Program	Method	Markers	SP	PP	PR	SC	EP	Required information	Management of technical/ genetic sources of incongruence	Comments
PROBMAX[1]	Exclusion	Codominant/ dominant		X				Parent and offspring genotypes. Candidate subgroups if wished	Permits retesting of data assuming SMM errors and recodifing homozygotes for NA	Evaluates exclusion capacity of markers
NEWPAT[2]	Exclusion	Codominant/sex linked	X			X		Offspring and sexed parents genotypes	NA frequency estimates. Exclusion tolerance to be decided by the user	Estimates the minimum number of loci for exclusion
PARENTE[3]	Exclusion	Codominant	X	X		X		Parent and offspring genotypes.	It admits error tolerance to be introduced by the user	Age of candidates can be introduced
FAP[4]	Exclusion	Codominant		X				Parent and offspring genotypes.	Introduces tolerance according to user. Identifies systematic genotyping errors or NAs	It permits to identify the best loci set of loci for assignment
CERVUS[5]	Categorical allocation/ Max likelih	Codominant	X	X		X	X	Offspring and sexed parents genotypes	Proportion of errors defined by the user. It estimates genotyping errors when true parents available	It permits exclusion approach with variable tolerance
PAPA[6]	Categorical allocation/ Max likelih	Codominant		X		X		Offspring and parents genotypes	Error rate introduced by the user using a SMM model.	Identifies the best set of loci for allocation
FAMOZ[7]	Categorical allocation/ Max likelih.	Codominant, dominant and uniparental	X	X		X	X	Parent and offspring genotypes	It can be introduced proportion of errors and deviations from panmixia by the user	Estimates probability of genotypic identity
PASOS[8]	Categorical allocation/ Max likelih.	Codominant	X	X				Parent and offspring genotypes	Transmission errors according to SMM. It admits error tolerance within an exclusion approach	Estimates number of candidates from offspring genotypes

Table 3. Paternity inference programs of interest in Aquaculture (Continued

Program	Method	Markers	SP	PP	PR	SC	EP	Required information	Management of technical/ genetic sources of incongruence	Comments
PATRI[9]	Fractional allocation/ Bayesian	Codominant	X			X		Offspring-parent pairs and sexed parents genotypes	Evaluation of genotyping errors using offspring-parent pairs data	Candidates subgroups can be defined for fractional allocation
GERUD[10,11]	Parental reconstruction	Codominant			X	X	X	Offspring and known parent genotypes		Estimates minimum parent number

SP: Single parental allocation

PP: Couple allocation

PP: Parental reconstruction of non-genotyped candidates

SC: Statistical confidence for assignment

EP: It provides exclusion probabilities as an estimation of assignment potential per locus

SMM: Stepwise mutation model

NA: Null allele

[1]Danzmann (1997); in: http://www.uoguelph.ca/~rdanzman/software/probmax/

[2]Worthington Wilmer *et al.* (1999); in:http://www.zoo.cam.ac.uk/zoostaff/amos/newpat.html

[3]Cercueil *et al.* (2002); in: http://www2.ujf-grenoble.fr/leca/membres/manel.html

[4]Taggart (2006); in: http://www.aqua.stir.ac.uk/rep-gen/Downloads.html

[5]Marshall *et al.* (1998); in: http://helios.bto.ed.ac.uk/evolgen/cervus/cervus.html

[6]Duchesne *et al.* (2002); in: http://www.bio.ulaval.ca/contenu-fra/professeurs/Prof-l-bernatchez.html

[7]Gerber *et al.* (2000); in: http://www.pierroton.inra.fr/genetics/labo/Software/Famoz/

[8] Duchesne *et al.* (2005) ; in : http://www2.bio.ulaval.ca/louisbernatchez/downloads.htm

[9]Signorovitch and Nielsen (2002); in: http://www.biom.cornell.edu/Homepages/Rasmus_Nielsen/files.html

[10]Jones (2001); in: http://www.biology.gatech.edu/professors/labsites/jones/parentage.html

[11]Jones (2005); in: http://www.bio.tamu.edu/USERS/ajones/parentage.htm

Finally, as outlined before, fractional allocation programmes, such as PATRI (Signorovitch and Nielsen 2002) could have some interest to evaluate the relative contribution of different parental groups or strains in species with mass-spawning, since they perform better than categorical allocation approaches for this purpose (Jones and Ardren 2003).

A relevant point to be considered in paternity inference is the treatment of genotyping errors or null alleles, which could induce to false assignations (Jones and Ardren 2003). Regarding this point, the availability of population data would be of great value, since, as outlined before, it would permit the estimation of null allele frequency and the presence of genotyping errors at the loci used. CERVUS or NEWPAT include algorithms to estimate null allele frequencies based on heterozygote deficit, and therefore, users can be alerted against those loci with moderate-high null allele frequencies with reasonable accuracy (Castro *et al.* 2007). PROBMAX uses a highly conservative option to avoid false exclusions due to null alleles, which implies the recoding of all homozygotes as heterozygotes for a null allele.

The availability of family data is of great value to identify systematic genotyping errors or accurately estimate null allele frequencies (Castro *et al.* 2004, 2006, 2007). The existence of large progenies in Aquaculture with a limited number of parents permits to identify systematic errors using the progenies analyzed itself, such as implemented in the program FAP (Taggart 2007). This is probably the most appropriate option in an Aquaculture scenario. The remaining programmes use automatic options which treat all mismatches in the same way, irrespective if they are null alleles or genotyping errors. Additionally, they treat all loci in a homogeneous way, leading to an error rate overestimation (Jones and Ardren 2003; Kalinowski *et al.* 2007). Thus, CERVUS and PARENTE interpret mismatches as replacements by genotypes or alleles taken at random from allele frequencies of the reference population. In these cases, assignations will be more probable if the error involves a frequent genotype or allele in the reference population. Other programmes, such as PAPA and PASOS calculate a probability depending on a particular mutational model selected by the user, commonly the stepwise model applied to microsatellites. Most programmes based in allocation by exclusion use variable tolerance to be decided by the user (FAP, NEWPAT) and/or permit the recoding of mismatches according to the stepwise mutation model (PROBMAX).

4.4. Paternity Inference in Aquaculture

A representative sample of some of the most relevant studies on parentage inference in Aquaculture is presented in Table 4. Firstly, it should be noticed their progressive increase during last years, so in 2006-2007, there appeared as many publications as in previous years jointly. Secondly, though most works have been applied on fish, a remarkable increment on molluscs and crustaceans is evident in the last years. Microsatellites are the markers usually applied for paternity inference in Aquaculture, but in some species with large genomic resources, such as salmonids, SNPs are beginning to be used. The increase in genomic resources (markers, genetic maps) in some aquaculture species is determining the development of more sophisticated molecular tools for throughput parentage allocation by checking exhaustively null alleles, genotyping errors and linkage among markers (Johnson *et al.* 2007; Li *et al.* 2007; Castro *et al.* 2007). CERVUS is the most popular program among aquaculture users, though manual assignation appears still quite common. The new version of

CERVUS (Kalinowski *et al.* 2007) provides several new options over the previous one, such as exclusion approach with tolerance or searching by couples, which could enhance its application. In the future, it will be necessary to evaluate the use of exclusionary-based programmes, such as FAP, specifically designed for aquaculture context. Finally, many publications detected in this survey are related with microsatellite implementation, which ensures a future application of paternity inference and their applications in these species. Heritability estimation, reproductive structure and support in familiar breeding programmes are the most sophisticated applications used in Aquaculture.

Table 4. Representative references for paternity inference in Aquaculture

Species	Taxonomic group	Genetic marker	Program	Application	Reference
Sparus aurata	Fish	Microsatellites	CERVUS	Parentage assignment	Castro *et al.* in press
Morone spp. hybrids	Fish	Microsatellites	Own program	Heritability estimation	Wang *et al.* 2007
Oncorhynchus mykiss	Fish	Microsatellites	-	Parentage assignment	Johnson *et al.* 2007
Penaeus monodon	Crustaceans	Microsatellites	-	Genetic diversity and fingerprinting	Li *et al.* 2007
Acipenser ruthenus	Fish	Microsatellites	Manual	Gynogenetic evaluation	Fopp-Bayat *et al.* 2007
Gadus morhua	Fish	Microsatellites	Manual	Parentage assignment	Wesmajervi *et al.* 2006
Pagellus bogaraveo	Fish	Microsatellites	CERVUS	Parentage assignment	Lemos *et al.* 2006
Oreochromis niloticus	Fish	Microsatellites	PAPA	Reproductive structure	Fessehaye *et al.* 2006
Morone sazatilis y Morone chrysops	Fish	Microsatellites	Manual	Heritability estimation	Wang *et al.* 2006
Haliotis asinina	Molluscs	Microsatellites	Manual	Heritability estimation	Lucas *et al.* 2006
Fenneropenaeus chinensis	Crustaceans	Microsatellites	CERVUS	Parentage assignment	Dong *et al.* 2006
Penaeus monodon	Crustaceans	Microsatellites	CERVUS	Parentage assignment	Jerry *et al.* 2006a
Penaeus japonicus	Crustaceans	Microsatellites	CERVUS	Interfamiliar growth evaluation	Jerry *et al.* 2006b
Solea senegalensis	Fish	Microsatellites	CERVUS	Parentage assignment	Castro *et al.* 2006
Salmo salar	Fish	Microsatellites/ SNPs	Manual	Parentage assignment and genetic diversity	Rengmark *et al.* 2006

Species	Taxonomic group	Genetic marker	Program	Application	Reference
Crassostrea gigas	Molluscs	Microsatellites	PAPA	Breeding programme	Taris *et al.* 2005
Salmo salar	Fish	Microsatellites/ SNPs	Own program	Parentage assignment and traceability	Hayes *et al.* 2005
Cyprinus carpio	Fish	Microsatellites	Own program	Heritability estimation	Vandeputte *et al.* 2004
Scophthalmus maximus	Fish	Microsatellites	CERVUS	Parentage assignment	Castro *et al.* 2004
Oreochromis niloticus	Fish	Microsatellites/ AFLPs	-	Gynogenetic evaluation	Ezaz *et al.* 2004
Scophthalmus maximus	Fish	Microsatellites	CERVUS PROBMAX	Parentage assignment	Borrell *et al.* 2004
Scophthalmus maximus	Fish	Microsatellites	CERVUS	Gynogenetic evaluation	Castro *et al.* 2003
Paralichthys olivaceus	Fish	Microsatellites	Manual	Parentage assignment and genetic diversity	Sekino *et al.* 2003
Crassostrea gigas	Molluscs	Microsatellites	Manual	Reproductive structure	Boudry *et al.* 2002
Salmo salar	Fish	Microsatellites	KINSHIP	Parentage assignment	Norris *et al.* 2000
Clarias gariepinus	Fish	Microsatellites	FAMILY	Breeding programme	Volckaert and Hellemans 1999
Dicentrarchus labrax	Fish	Microsatellites	Manual	Breeding programme	De León *et al.* 1998
Oncorhynchus mykiss	Fish	Microsatellites	Manual	Breeding programme	Herbinger *et al.* 1995

Note: The references are a representative sample taken from some relevant publications in Aquaculture.

5. Kinship Estimation

In the context of genetics, when we talk about kinship or coancestry we refer, in general, to the common origin of alleles carried by two individuals in a particular locus. What we really try to detect or predict is the probability of those alleles being *Identical By Descent* (IBD). This means that both alleles are copies of a single allele in a common ancestor, in opposition to the concept of *Identical By State* or *Alike In State* (IBS or AIS) when alleles are equal although they derive from different alleles in the reference population. A graphical representation of both concepts can be found in Figure 7.

The more general and exhaustive way of describing the relationship between two individuals is in terms of the nine models of identity by descent presented by Jacquard (1974) (see Figure 8). Following this methodology several coefficients $\Delta = (\Delta_1, \Delta_2, \ldots, \Delta_9)$ can be calculated corresponding to the probability of the four alleles of the dyad conforming to that pattern of IBD. Another way of mathematically representing the degree of genealogical relationship is through the probability of the genotypes of two individuals sharing 2 (k_2), 1 (k_1) or 0 (k_0) alleles (Thompson 1991).

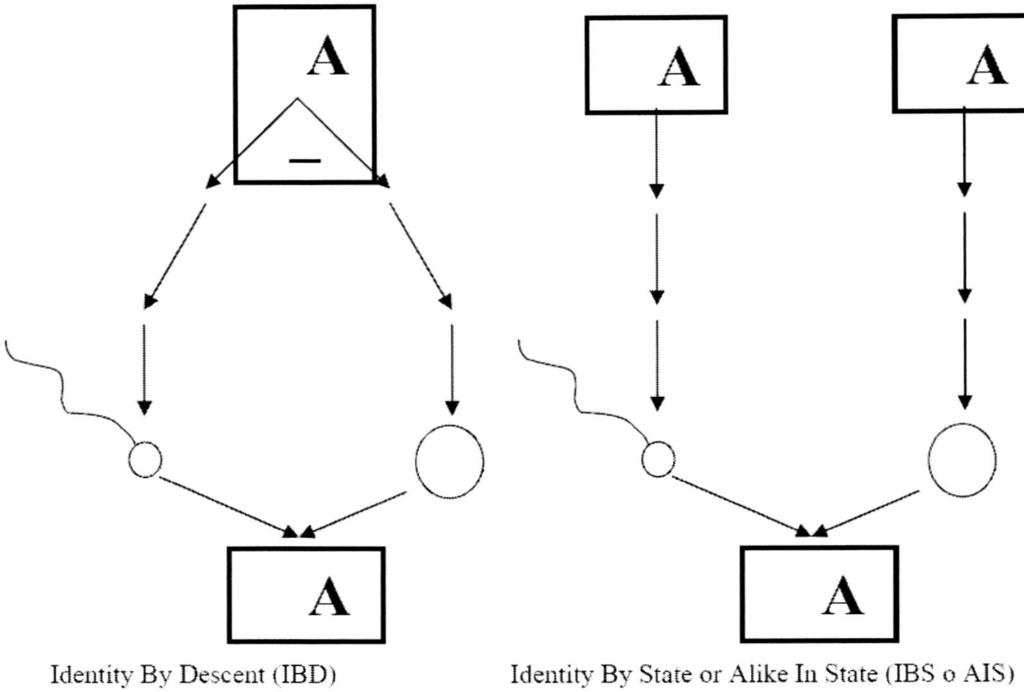

Figure 7. Graphical representation of the difference between alleles identical by descent or by state.

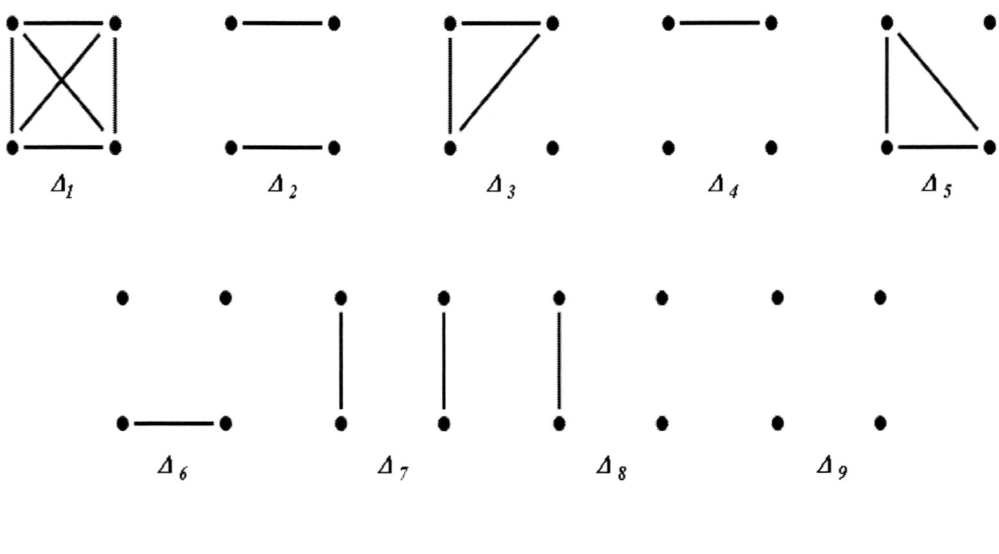

Figure 8. Identity by descent modes between two individuals.

Commonly, the degree of genetic relationship between two individuals is summarised using the coancestry coefficient (f or θ, depending on the authors). This was defined by Malecot (1948) as the probability of sampling two alleles identical by descent taken at random one from each individual in the same locus. The coancestry coefficient for non inbred individuals is half the additive genetic correlation or relatedness coefficient (r). Some methodologies have been developed for the estimation of r, but since in most situations the transformation between parameters is straightforward ($r = 2f$), we shall refer to f estimators thereafter. When the complete genealogy is known, the coancestry coefficient can be easily calculated following classical rules (Falconer and Mackay 1996) or implementing tabular methods (Emik and Terrill 1949). We should also remember that when comparing two alleles from the same individual the probability of being IBD is the inbreeding coefficient (F), equal to the genealogical coancestry of the parents of that individual.

5.1. Molecular Information

When the genealogy of the population under study is not available, information from molecular markers developed should be used to estimate kinship. The basic idea is to define a measure of the molecular resemblance between individuals, assuming that those sharing the highest number of alleles for the markers will be those with the highest probability of carrying IBD alleles (Blouin 2003; Weir *et al.* 2006). Different estimation strategies and procedures exist which can be classified in the groups described below.

5.2. Pairwise Estimators

The main characteristic of these methods is that relationships are determined for just one couple of individuals (dyad) at a time based on their molecular information. Data corresponding to the remaining individuals in the population are not used but for the estimation of allelic frequencies.

5.2.1. Molecular Coancestry

This coefficient arise from the application of Malecot (1948) definition to marker loci. Accordingly, molecular coancestry (f_M) between two individuals is the probability of sampling the same allele (whatever IBD or not) when taking one allele from the same locus of each individual. Molecular coancestry ranges from 0 (when none of the four possible comparisons is between IBS alleles) to 1 (when all four alleles are IBS), with only two intermediate values corresponding to 0.25 and 0.5. The relative contribution of each allele from a particular locus to the molecular coancestry can be calculated separately and then summed up, as shown in Table 5 for a locus with three alleles. If information from several loci is available, molecular coancestry is calculated by averaging their particular value over loci.

The main problem with this measure is that only reflects IBS and, thus, the parameter of interest (IBD) will be inflated.

Table 5. Contribution of each allele to the molecular coancestry between two individuals depending on their genotype

Genotypes	f_{Mi}	f_{Mj}	f_{Mk}	f_M
$A_i A_i , A_i A_i$	1.00	0.00	0.00	1.00
$A_i A_i , A_i A_j$	0.50	0.00	0.00	0.50
$A_i A_i , A_j A_k$	0.00	0.00	0.00	0.00
$A_i A_j , A_i A_j$	0.25	0.25	0.00	0.50
$A_i A_j , A_i A_k$	0.25	0.00	0.00	0.25

5.2.2. Similarity Indices

Originally developed for DNA fingerprinting, these methods are applicable for any electrophoretic pattern where different bands can be observed. If comparing an individual X where n_X bands are detected with another individual Y with n_Y bands, n_{XY} being common, a simple similarity measure can be constructed as

$$S = \frac{2n_{xy}}{n_x + n_y}$$

which corresponds to the Dice's coefficient (Sneath and Sokal 1973). Its value will range from 0 (no bands shared) to 1 (all common bands).

When similarity is calculated based on codominant markers, different modifications have been proposed, for example the one by Li *et al.* (1993):

$$S = \frac{n_{xy}}{2}\left(\frac{1}{n_x} + \frac{1}{n_y}\right)$$

These authors pointed out that the expectation of such similarity index is

$$E(S) = 2f + (1-2f)S_0, \text{ being } S_0 = \sum_{i=1}^{n} p_i^2 (2 - p_i)$$

for a locus with n alleles, where p_i is the frequency of allele i in the reference population. This equation allows for the construction of a first estimator of the genealogical coancestry, expressed as

$$\hat{f} = \frac{S - S_0}{2(1 - S_0)}$$

A different approach was proposed by Bowcok *et al.* (1994), who suggested an index for multilocus analysis consisting in the total number of shared alleles (a maximum of two per locus) divided by twice the number of genotyped loci.

5.2.3. Methods of Moments Estimators (MME)

In an attempt to estimate IBD from the molecular coancestry, we can use the expectation of that coancestry for a particular allele, that is

$$E(f_{Mi}) = f \cdot p_i + (1 - p_i)p_i^2$$

where p_i is the frequency of the allele *i* in the reference population and *f* is the genealogical coancestry (i.e. the IBD probability). The first term in the right hand side of the equation is the probability of finding two alleles *i* identical by descent, while the second one is the probability of sampling two equal alleles in a population with an allele frequency p_i. Solving for *f*, a simple estimator of the IBD due to a single allele can be obtained as:

$$\hat{f} = \frac{f_{Mi} - p_i^2}{p_i(1 - p_i)} \qquad \text{(Ritland 1996)}$$

Estimates for each allele from the same locus can be averaged in different ways depending on the weighting factor (w_i) assumed (Li and Horvitz 1953; Ritland 1996).

Using a different approach, Lynch and Ritland (1999) developed a coancestry estimator based on the probability of finding a particular genotype in a 'proband' individual (*Y* with alleles *k* and *m*), conditional on the genotype of another 'reference' individual (*X* with alleles *i* and *j*). This value is proportional to the probability of sharing one or two alleles and, therefore, an estimation of the coancestry coefficient can be calculated defining a variable S_{ij}. This will take a value 1 when *i* and *j* are equal (irrespective if they are in the same or different individuals) and 0 if they are different. Due to the underlying idea of predicting the genotype of *Y* through the genotype of *X*, these methods are commonly called *regression estimators*. In concordance with the methodology used in their construction, these estimators are 'asymmetric' being f_{XY} (*X* is the reference and *Y* the proband) not necessarily equal to f_{YX}. The arithmetic mean of both values is used as the estimate of the coancestry.

Related with the previous method, although originally developed for the estimation of average coancestry of groups of individuals, the Queller and Goodnight (1989) estimator is based on the same parameters as the one by Lynch and Ritland (1999). A drawback of this method is that the estimator is not defined for biallelic loci when individuals under study are heterozygous, limiting its use to multiallelic loci.

Wang (2002) developed a regression estimator recalculating the probabilities of finding dyads of individuals with different similarity indices (S = 1, ¾, ½ or 0 following the definition by Li *et al.* 1993 in the previous section). In his particular formulation, P_1 is equivalent to S = 1 and P_2 to S = ¾. Although in certain situations this estimator coincides with the Queller and Goodnight (1989) and Lynch and Ritland (1999) estimators, it exhibits better performance for some situations, as tested by simulation using actual turbot data (Pino-Querido *et al.* submitted).

The mathematical expressions of the estimators described above when dealing with single locus information can be found in Table 6a.

Table 6a. Different estimators for single locus with their corresponding weights

Assumption	Estimator	Reference
$w_i = p_i(1-p_i)$	$\hat{f} = \dfrac{f_{Mi} - \sum_{i=1}^{n} p_i^2}{1 - \sum_{i=1}^{n} p_i}$	Li and Horvitz (1953)
$w_i = \dfrac{1-p_i}{n-1}$	$\hat{f} = \dfrac{1}{n-1}\sum_{i=1}^{n}\dfrac{f_{Mi} - p_i^2}{p_i}$	Ritland (1996)
	$\hat{f}_{XY} = \dfrac{p_i(S_{jk}+S_{jm}) + p_j(S_{ik}+S_{im}) - 4p_ip_j}{2(1+S_{ij}) + S_{km}(p_i+p_j) - 8p_ip_j}$	Lynch and Ritland (1999)
	$\hat{f}_{XY} = \dfrac{(S_{ik}+S_{im}+S_{jk}+S_{jm}) - p_i - p_j - p_k - p_m}{2 + S_{ij} + S_{km} - p_i - p_j - p_k - p_m}$	Queller and Goodnight (1989)
	$\hat{f}_{XY} = \dfrac{4P_1 + 3P_2 - 2\left(1 + \sum_{i=1}^{n} p_i^2\right)}{4\left(1 - \sum_{i=1}^{n} p_i^2\right)}$	Wang (2002)

Different weights to average the estimates obtained from different loci have been proposed. These will essentially depend on the criteria about the relationship chosen among loci and the weight given to each estimate. Oliehoek *et al.* (2006), besides including a different weighting factor for each locus, suggested a further correction on the estimation of the probability of IBS. They argued that, although different allele frequencies at the loci used would imply different probabilities of IBS, the same coancestry estimation should be obtained from all of them for a particular dyad of individuals. Their WEDS estimator (Weighted Equal Drift Similarity) try to equalise the increase in coancestry from the base population for all loci by estimating the probability of IBS at each locus in the following way:

$$\widehat{S}_l = \frac{\sum_{i=1}^{n_l} p_i^2 - S_{\min}}{1 - S_{\min}}$$

where $S_{\min} = \min(\sum_{i=1}^{n_l} p_i^2;\quad l = 1, ..., L)$. In this way we set $\widehat{S}_l = 0$ for the locus having the lowest expected similarity (*S*).

Table 6b summarises some of the multilocus MME estimators that have been proposed.

Table 6b. Different multilocus estimators with their implicit assumptions and corresponding weights

Single locus estimator	Assumption	Multilocus estimator	Reference
Li and Horvitz (1953)		$\hat{f} = \frac{\sum_{l=1}^{N} \hat{f}_l}{N}$	Toro *et al.* (2002)
Li and Horvitz (1953)	Correlation between loci = 1	$\hat{f} = \frac{\sum_{l=1}^{N} f_{Ml} - \sum_{l=1}^{N}\sum_{i=1}^{n} p_{li}^2}{N - \sum_{l=1}^{N}\sum_{i=1}^{n} p_{li}^2}$	Ritland (1996)
Ritland (1996)	Correlation between loci = 0 ($w_l = n_l - 1$)	$\hat{f} = \frac{\sum_{l=1}^{N}\sum_{i=1}^{n} \frac{f_{Mli} p_{li}^2}{p_{li}^2}}{\sum_{l=1}^{N} (n_l - 1)}$	Ritland (1996)
Lynch and Ritland (1999)	$w_l = \frac{(1+S_{ij})(p_i + p_j) - 4p_i p_j}{8 p_i p_j}$	$\hat{f} = \frac{1}{\sum_{l=1}^{N} w_l} \sum_{l=1}^{N} w_l \hat{f}_l$	Lynch and Ritland (1999)
Queller and Goodnight (1989)		Sum of the numerators of single locus estimators divided by the sum of denominators	Toro *et al.* (2002)
	$w_l^{-1} = \frac{\sum_{i=1}^{n_l} p_i^2 \left(1 - \sum_{i=1}^{n_l} p_i^2\right)}{(1 - \hat{s}_l)^2}$	$\hat{f} = \frac{1}{\sum_{l=1}^{L} w_l} \sum_{l=1}^{L} w_l \frac{S_l - \hat{s}_l}{1 - \hat{s}_l}$	Oliehoek *et al.* (2006)

Comparisons of the precision for different MME estimators can be found, for example, in Van de Casteele *et al.* (2001), Toro *et al.* (2002), Wang (2002), Csilléry *et al.* (2006) and Pino-Querido *et al.* (submitted).

5.2.4. Maximum Likelihood Estimators (MLE)

Another way to deal with the problem of the estimation of coancestry is to calculate the probability of observing the genotypes of the dyad under study ($\mathcal{S}_i$ in the Milligan 2003 notation; see also Herbinger *et al.* 1997; Kalinowski *et al.* 2006) depending on a particular type of relationship (Δ). This is actually the likelihood of Δ, $L(\Delta) = \Pr(\mathcal{S}_i|\Delta) = \Sigma \Pr(\mathcal{S}_i \mid S_j)\Delta_j$, being S_j each of the nine identity modes explained in a previous section. Recalling that these modes can be summarised into the coefficients k_0, k_1 and k_2, and knowing that the coancestry coefficient in non inbred populations is $f = \frac{1}{4} k_1 + \frac{1}{2} k_2$, we can calculate the likelihood of a particular degree of coancestry from the expressions on Table 7, which depend on the frequencies of each allele in the population (p_i, p_j, p_k, p_m).

Table 7. Probability of the nine possible combinations of genotypes in a dyad given identity by descent modes k_0, k_1, k_2.

X	Y	k_0	k_1	k_2
$A_i A_i$	$A_i A_i$	p_i^4	p_i^3	p_i^2
$A_i A_i$	$A_i A_j$	$2p_i^3 p_j$	$p_i^2 p_j$	
$A_i A_i$	$A_j A_j$	$p_i^2 p_j^2$		
$A_i A_j$	$A_i A_i$	$2p_i^3 p_j$	$p_i^2 p_j$	
$A_i A_j$	$A_i A_j$	$4 p_i^2 p_j^2$	$p_i p_j (p_i + p_j)$	$2 p_i p_j$
$A_i A_i$	$A_j A_k$	$2 p_i^2 p_j p_k$		
$A_j A_k$	$A_i A_i$	$2 p_i^2 p_j p_k$		
$A_i A_j$	$A_i A_k$	$4 p_i^2 p_j p_k$	$p_i p_j p_k$	
$A_i A_j$	$A_k A_m$	$4 p_i^2 p_j p_k p_m$		

When information on more than one locus is available, the maximum likelihood estimator will be obtained by multiplying the likelihoods of the different loci. A revision of the factors affecting the precision of the MLE estimators as well as a comparison with the performance of the MME estimators can be found in Milligan (2003).

5.2.5. Problems of the Pairwise Estimators

Several are the factors that influence the efficiency of pairwise estimators. One of them is that these methods assume Hardy-Weinberg (HW) and gametic equilibrium at all loci analyzed. Although not formally proven, all authors agree that the departure from these assumptions may yield biased estimates.

Another major problem both of the MME and MLE methods is that they require the knowledge of the allelic frequencies in the base or reference population (Wang 2002; Toro *et al.* 2002; Thomas 2005). When the number of generations elapsed from the base is small and the population size large, no relevant changes in the frequencies are expected and the estimates obtained from the individuals we actually want to determine their coancestries will be acceptable. But if there had been enough margin for the genetic drift to act (many generations and/or reduced population sizes) frequencies could have changed and even some alleles lost in the process. Under this scenario, as shown by Toro *et al.* (2002), underestimations are systematically obtained, because of the magnified probability of IBS of the surviving alleles. MME methods may even provide negative values (out of the definition range, as it is a probability) making difficult the interpretation of the results. The correlation between estimates and true values of coancestry is low when wrong frequencies are used in the estimation, even with a large number of markers genotyped. When the true frequencies are known the performances of the estimators greatly improves (Toro *et al.* 2002; Pino-Querido *et al.* submitted).

In practical scenarios, the estimation of the coancestry between individuals reduces to choose between a few types of fixed relationships. For example, we may want to decide if two individuals are full-sibs or unrelated. The MLE methods, due to its nature, can deal with this situation just by looking for the degree of relationship with the highest likelihood and classifying the dyad in that group. If the appropriate choice is between two particular relationships (usually comparing a degree of relationship against being unrelated) a likelihood ratio can be calculated. The level of significance can be obtained by simulating a great number of dyads of the less likely relationship (using the population allelic frequencies) and then calculating the likelihood ratio for each of them under both hypotheses. The percentage of replicates whose value is above the observed ratio will be the probability of misclassification of the dyad (*p*-value).

Contrarily, the MME methods provide estimates on a continuous range. Therefore, it is necessary to determine the thresholds where assignation will change between fixed relationships (Figure 9). If no *a priori* information exists, values for those thresholds are usually the midpoints between actual genealogical values of each type of relation (Blouin *et al.* 1996), for example: 0.0625 between unrelated and half-sibs and 0.1875 between half-sibs and full-sibs (see Figure 9 and 10). The reason for these figures is to equalise the probabilities of under- and overestimating coancestry (type I and type II errors). The expected genealogical values may be substituted by the mean values of a large number of estimates from simulated dyads according to the observed population frequencies.

When the true relationships to be obtained appear in different frequencies, a common situation in Aquaculture where unrelated is the most common relationship, thresholds should be shifted to the less represented category to ensure the symmetry in the classification errors (Pino-Querido *et al.* submitted). But the relative proportion of each type of relationship usually is unknown *a priori*. These authors have proposed two methods to estimate the proportion of dyads belonging to each category using the properties of the normal curve. With their approach, the thresholds could be shifted to balance type I and II errors, and the resulting relatedness classification would improve.

When pairwise estimates are transformed to fixed genealogical relationships these methods can be affected by another problem. The consideration of each dyad independently of the rest of individuals may lead to incongruous assignations and aberrant coancestry matrices (Thomas and Hill 2000; Fernández and Toro 2006). For example, molecular information could be consistent with individuals A and B being full-sibs and similarly for individuals B and C. But this does not guarantee that individuals A and C will be correctly classified as full-sibs. Rodríguez-Ramilo *et al.* (in press) testing different pairwise estimators in a turbot population comprising 560 individuals with known relationships and genotyped

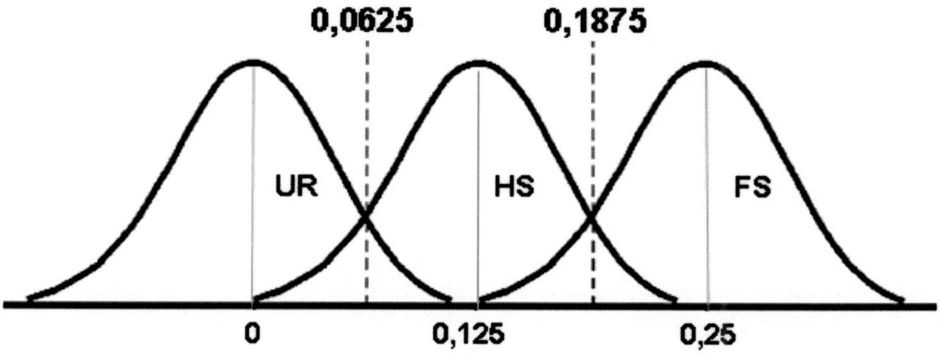

Figure 9. Distributions of the estimates for dyads of full-sibs (FS), half-sibs (HS) and unrelated (UR), respectively, with the thresholds commonly used for classification.

	R3	R5	R6	R8	R10	R13	R14	R19	R21	R23	R1	R4	R7
R3													
R5	-0,0260												
R6	0,1004	0,0907											
R8	0,0153	0,1548	0,1092										
R10	0,1146	0,0461	0,1256	0,1248									
R13	0,3352	0,1020	0,2131	-0,0015	0,1166								
R14	0,1028	0,2278	0,0287	0,2577	0,1031	0,0805							
R19	0,0898	0,1036	0,0603	0,3503	0,1186	0,0678	0,2052						
R21	-0,0305	0,0065	-0,0375	-0,0050	-0,0420	0,0444	0,1209	-0,0135					
R23	0,2921	0,0142	0,1009	-0,0915	0,0832	0,3354	-0,0555	0,0084	0,0033				
R1	0,2725	0,0419	0,0462	-0,1205	-0,0205	0,2891	-0,0090	-0,0250	0,0266	0,3012			
R4	0,0705	0,0335	-0,0475	0,0577	0,0067	-0,0040	0,1415	0,0774	0,0548	-0,0150	-0,1035		
R7	-0,0265	-0,0700	-0,1070	0,0529	-0,0705	-0,0995	0,0056	0,0173	0,0487	-0,0880	-0,1210	0,0830	
R9	-0,0015	0,1077	0,0837	0,2460	0,0583	0,0098	0,1394	0,1845	0,1133	-0,0945	-0,0680	-0,0120	0,0093

Note.- Unrelated dyads are shown in green, half-sib dyads in yellow and full-sib dyads in red. The limits considered for classification were 0.0625 between unrelated and halfsibs, and 0.1875 between halfsibs and fullsibs.

Figure 10. Kinship classification of pairs of individuals (dyads) in a turbot broodstock

for 11 microsatellites, have observed that the proportion of non reciprocal triplets of full-sibs is more than 70% of the triplets constructed from the molecular information.

Thomas and Hill (2002) pointed out that inferring coancestries in a pairwise mode may lead to reconstruct full-sib groups incompatible with Mendelian segregation rules. For example, three individuals with genotypes AA, BB and CC, respectively, cannot be full-sibs at the same time, but in pairwise comparisons this possibility will not be discarded. In the study by Rodríguez-Ramilo *et al.* (2007) the proportion of such incorrect groups was 60% of the total number of full-sibs families reconstructed with more than three individuals.

5.3. Pedigree Reconstruction

Methods grouped in this category were developed as an alternative to pairwise methods, to take advantage of the information from the rest of individuals on the population. A first approach was presented by Browning and Thompson (1999) who proposed to work with triplets of individuals instead of pairs. In this way, some genotype incompatibilities could be detected and coancestries assigned with more precision. A refinement of the method was developed by Sieberts *et al.* (2002) for joint estimation of triplets with mixed full-sibs, half-sibs, twin or unrelated, using Markov Chains to determine the posterior probability of each type of relationship.

Generalising, the ideal procedure would use the information from all the individuals in the population and would estimate every relationship simultaneously. Under this approach what we obtain is the global structure of relationships and, thus, the explicit reconstruction of the genealogy (at least for one generation). The negative side of these methods is that they require more complex calculations being computationally more demanding. Consequently, most of them have been limited to simple population structures like full-sib families.

Almudevar and Field (1999) presented an enumeration algorithm for feasible full-sib families, being the final objective to find the families with the highest number of members. When two alternate configurations comprising the same number of individuals exist, the one with the highest partial likelihood is chosen. Another method to obtain solutions involving the largest families is maximising the *Simpson's index* (Butler *et al.* 2004) defined for a population with N individuals structured in g groups comprising N_i elements as:

$$\frac{\sum_{i=1}^{g} N_i(N_i - 1)}{N(N-1)}$$

In this method, as in Almudevar and Field (1999), the compatibility of the proposed families with Mendelian segregation rules (e. g. all members could arise from a single couple) has to be tested to make sure they yield congruent solutions. The search for the optimum configuration is performed in both methods by Monte Carlo techniques.

In the literature, other estimators can be found relying on MCMC methods to search across the space of feasible solutions (compatible full-sib groups) but using different criteria to determine the more likely configuration. The driving parameter may be a combination of the likelihood ratios of all possible pairs of individuals (Smith *et al.* 2001) or the joint likelihood of each of the possible configurations (Painter 1997; Thomas and Hill 2000; Emery *et al.* 2001; Smith *et al.* 2001). The likelihood calculation of a particular family will depend, as in the pairwise MLE methods, on the allelic frequencies and the number of individuals of each genotype grouped together, as shown in Table 8.

Thus, a family with five of its members with genotype AA and another five with genotype BB, although possible, is much less likely than a family comprising heterozygous as well.

Thomas and Hill (2002) and Wang (2004b) extended these methods to populations with a hierarchical structure involving full-sib families and nested half-sib families.

Although the pedigree reconstruction methods have clear advantages over the pairwise methods, they still present some drawbacks. First, its use is limited to very simple structures (full- or half-sibs) and requires more or less detailed knowledge of the population structure. Second, the likelihood calculations need the estimation of the true frequencies on the population (except for the estimator based on the Simpson's Index and the one by Almudevar and Field 1999, based on the partial likelihoods). Moreover, all formulas are also developed under the assumption of the existence of HW equilibrium for the alleles at each locus and linkage equilibrium between loci.

Table 8. Likelihood of full-sibs families depending on the observed genotypes for one locus

Genotypes	Likelihood
AA	$p_A^4 + 4\,p_A^3\,p_X\,(1/2)^n + 4\,p_A^2\,p_X^2\,(1/4)^n$
AB	$2[p_A^2\,p_B^2 + (1/2)^{nAB} 2(p_A^3\,p_B + p_A^2\,p_B\,p_X + p_A\,p_B^3 + p_A\,p_B^2\,p_X) + (1/4)^{nAB}$
	$4(p_A^2\,p_B\,p_X + p_A\,p_B^2\,p_X + p_A\,p_B\,p_X^2)] + (1/2)^{nAB} 4\,p_A^2\,p_B^2$
AA BB	$4\,p_A^2\,p_B^2\,(1/4)^n$
AA AB	$4\,p_A^3\,p_B\,(1/2)^n + 4\,p_A^2\,p_B^2\,(1/4)^{nAA}(1/2)^{nAB} + 8\,p_A^2\,p_B\,p_X(1/2)^n$
AA AB BB	$4\,p_A^2\,p_B^2\,(1/4)^{nAA+nBB}(1/2)^{nAB}$
AA BC	$8\,p_A^2\,p_B\,p_C(1/4)^n$
AB AC	$4\,p_A^2\,p_B\,p_C(1/2)^n + 8\,p_A^2\,p_B\,p_C(1/4)^n + 8\,p_A\,p_B\,p_C\,p_D(1/4)^n$
AA AB AC	$8\,p_A^2\,p_B\,p_C(1/4)^n$
AA AB BC	$8\,p_A^2\,p_B\,p_C(1/4)^n$
AB AC BC	$8\,p_A^2\,p_B\,p_C(1 - p_X)(1/4)^n$
AA AB AC BC	$8\,p_A^2\,p_B\,p_C(1/4)$
AC BD	$8\,p_A\,p_B\,p_C\,p_D(1/4)^n$
AD AC BC	$8\,p_A\,p_B\,p_C\,p_D(1/4)^n$
AC AD BC BD	$8\,p_A\,p_B\,p_C\,p_D(1/4)^n$

p_X is the frequency of all alleles not present in the family considered

nAB is the number of individuals in a family carrying genotype *AB* (from a total of *n* full-sibs)

Recently, Fernández and Toro (2006) have developed a new method based on the observation by Toro *et al.* (2002) that, although pairwise molecular coancestry is an overestimation of the true value, the molecular coancestry matrix may exhibit a high correlation with the genealogical coancestry matrix. Consequently, these authors proposed a method which implements a *simulated annealing* algorithm to search within the space of all possible configurations of the population using the correlation between the molecular coancestry matrix and the genealogical matrix corresponding to the proposed structure, as the decision criterion. Obviously, only the compatible full-sibs families are considered. The advantage of this method is that it does not require the knowledge of the allelic frequencies in the reference population, and it makes no assumptions about the equilibrium within and between loci. Moreover, as calculations are simple, it allows for the estimation of more complex structures (e.g. non nested families) or even higher degree relationships like cousins or brothers from inbred parents, by reconstructing genealogies of more than one generation. The efficiency of some of the pedigree reconstruction estimators has been tested under different scenarios by Butler *et al.* (2004) and Fernández and Toro (2006).

A novel way of dealing with the problem of the joint estimation of all coancestry relationships was proposed by Beyer and May (2003). Their method is based on graphs theory representing the population as nodes (each individual) which can be linked by lines if they are full-sibs. Figure 11 shows different groups of connected nodes representing full-sibs families.

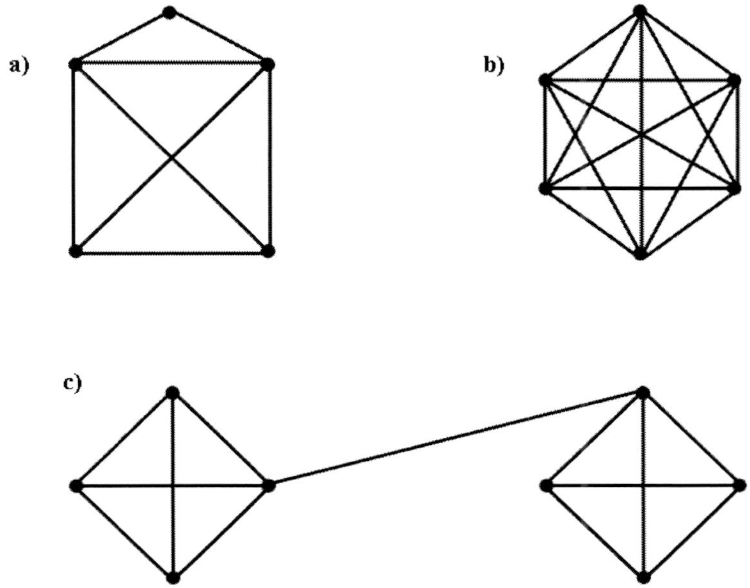

Modified from Beyer and May (2003)
a) Improper linked component (some individuals are not connected)
b) Proper/complete linked component (all elements connected with each other)
c) A wrong connection between two proper linked components (families).

Figure 11. Graphical representation of groups of individual and their sib-ship relation (linked points).

From an initial configuration based on the pairwise likelihood ratios (Goodnight and Queller 1999), we have to move to another structure where all individuals of the same family are connected to each other but disconnected from the nodes of other families, through graphical algorithms like the *minimum-cut*.

A final strategy to define groups of relatives consists in using the matrix of genetic distances (or relatedness coefficients) between individuals (e.g. allele sharing coefficient) to construct the subsequent phylogenetic trees using different algorithms (Bentzen *et al.* 2001). Those individuals clustering together will be assumed to belong to the same family. These methods, as suggested by Blouin (2003), appear to work reasonably well in situations where simple pedigree data are analyzed.

5.4. Dominant Markers

All previously explained estimators (except for the first similarity indices) have been developed for codominant markers. However, there are situations where it is much easy and cheaper to apply dominant markers, especially for species with low genomic resources. Therefore, it would be a good alternative to estimate coancestry from such type of markers.

Hardy (2003) proposed an estimator based on the fact that coancestry between two individuals is proportional to the correlation of the phenotypic frequencies for each allele in both individuals. For dominant loci those frequencies can only be 0 (lack of band) or 1 (presence of band, dominant allele present).

Table 9. Probability of the three possible combinations of phenotypes in a dyad given *identity by descent modes* k_0, k_1, k_2.

X	Y	k_0	k_1	k_2
Recessive	Recessive	q^4	q^3	q^2
Dominant	Recessive	$2(1-q^2)q^2$	$2q^2 - 2q^3$	
Dominant	Dominant	$2(1-q^2)^2$	$1 - 2q^2 + q^3$	$1 - q^2$

q is the frequency of the recessive allele

The advantage of this estimator, besides the nature of the markers it uses, is that it can be applied to populations not in HW equilibrium. However, to make use of this feature, the degree of deviation must be known (we have to calculate, through an independent estimate, the mean inbreeding coefficient of the population).

Wang (2004a) and Ritland (2005) developed MME estimators from the expected values of the number of IBD alleles shared by two individuals (k_2, k_1, k_0) conditional on the phenotype of the dyad for the marker (Table 9). This table is equivalent to Table 7 with the difference that now the only three possibilities are both individuals presenting band, only one individual presenting band, or both individuals without band.

From these probabilities the reasoning is similar to that used for the MME methods on codominant markers (dependence on the knowledge of the frequencies, equilibrium within- and between-loci required, ...).

5.5. Genotyping Errors

A major problem that cannot be denied is that inferences in any of the studied estimators are performed on molecular data that we assume correct. But molecular information may contain errors. The types of common errors were explained in a previous section. Some estimates of the rate of appearance of different kind of errors can be found in Talbot *et al.* (1995) and Ewen *et al.* (2000).

In the case of pairwise estimators, the incorrect assignation of a genotype to an individual affects their molecular similarity with the rest, but its influence will be relative depending on the amount of information from the other loci correctly genotyped. Epstein *et al.* (2000) included in their maximum likelihood estimator the possibility of errors in the genotype, assuming that when identical alleles are observed in two individuals the probability of that similarity corresponding to an error is ε and, consequently, the probability of both alleles being really equal would be $(1 - \varepsilon)$. Then, probabilities of being IBD conditional on the observed genotypes can be recalculated. Obviously, an estimate of the error rate has to be available.

However, methods involving pedigree reconstruction are more sensitive to point errors, as they may cause the individual bearing the erroneous genotype to be incompatible with its actual family. This effect is independent on the amount of error-free markers we have genotyped. In fact, the larger the number of loci genotyped, the higher the probability of including errors. A possible solution to the problem was proposed by Fernández and Toro (2006). The idea is relaxing the compatibility restrictions in the creation of full-sibs groups allowing for a number of loci where incompatibilities can exist. The expectation is that individuals with errors can be included in their correct family, at the risk of accepting erroneous assignations of individuals correctly genotyped.

Wang (2004b) extended a model similar to the one by Epstein *et al.* (2000) to calculate the likelihoods of different configurations of the population incorporating the probabilities of different type of errors (actually two types are distinguished). From the configuration with the highest likelihood, this author proposed to calculate the likelihood ratios for each family and locus assuming that one of the error rates is zero or both are null. In this way, it is possible to detect those loci more prone to bear an error and the more frequent kinds of error.

The use of any *a priori* information (e.g. genealogy) may be crucial to purge data before implementing any estimator. This practice will greatly improve the accuracy of the estimators.

5.6. Measures of Accuracy

The selection of the estimator to be used, besides the considerations regarding its nature (pairwise or not, dependent on frequencies, ...), requires any measure of the accuracy or reliability which allows for comparison between different methods. These parameters can be used when genealogical data are also available or when simulation studies are conducted.

If we consider the coancestry as a value associated to each dyad, it seems appropriate to use the bias (difference between the estimated value and the true value) to determine the 'goodness' of the estimator. Although the expected value provided by an estimator may be unbiased, a great dispersion of estimates around the expectation may turn it into a low efficient estimator. One way to measure dispersion is the mean squared error (or its squared root; Milligan 2003).

Another proposed measure of fit is the correlation between the estimated coancestry matrix and the genealogical relationships for all possible dyads in the population (Toro *et al.* 2002; Fernández and Toro 2006). High correlation values mean that, although estimates may be biased, the ranking of pairs remains constant and also the relative values for different dyads.

When relationships between individuals are classified in a few types of coancestry (commonly full-sibs, half-sibs or non-related), accuracy of the method can be measured through the proportion of dyads correctly assigned to their actual group and the different errors of classification (Blouin *et al.* 1996; Thomas and Hill 2002; Fernández and Toro 2006). Figure 12 shows the nine possible combinations of estimated and real relationships when only three types are considered.

Using this kind of tables the tendency of a method to underestimate or overestimate coancestry can be determined, as well as its power for the detection of particular degrees of relationship. Obviously, before using this measure of fitness on MME pairwise estimators, we should transform the continuous values to qualitative data with the associated drawbacks already pointed out.

A more general comparison can be done through a Chi-square test, where the number of estimated and actual full-sibs, half-sibs and unrelated dyads are combined to calculate the statistic. The greater the value obtained, the worse the fit between estimates and true coancestries.

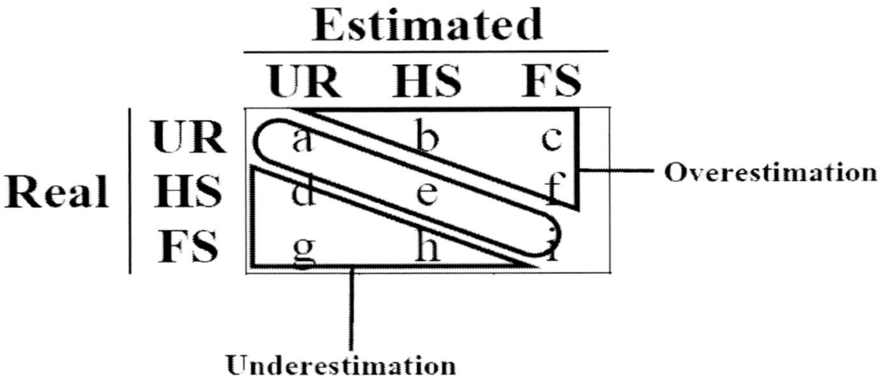

Figure 12. All types of possible assignations considering three kind of relationships: unrelated (UR), half-sibs (HS) or full-sibs (FS).

Smith *et al.* (2001), in the case of full-sibs family reconstruction, proposed a measure of accuracy which consisted in determining the lowest number of movements (i.e. exchanges of individuals between families) required to convert the estimated population structure into the true one. A refinement of this method distinguishes between movement of single individuals and the movement of blocks of individuals. For example: it is not equivalent to erroneously classify *N* individuals into *N* different families than splitting a correct family into two, one of the groups containing the *N* individuals.

Notwithstanding, as stated by different authors (Fernández and Toro 2006; Oliehoek *et al.* 2006), the appropriate comparison should be made in practical grounds, i. e.: coancestry overestimates may be less important for the arrangement of the mating scheme under the minimum coancestry strategy, while underestimates are preferred if the coancestries have to be included in the estimation of breeding values using BLUP methodology. Oliehoek *et al.* (2006) found out that the ranking of estimators, according to the levels of genetic diversity preserved in a population managed under minimum coancestry contributions, changed depending on the population structure but not with the number of markers used or their polymorphism. In the aquaculture field, Porta and Fernández (unpublished data) have found that differences in accuracy between some coancestry estimators nearly vanished when values are used to organise individuals of a stock into different tanks as to yield the minimum average within-tank coancestry (mating between relatives is avoided and the genetic representation of each tank increases).

5.7. Software

A large number of computer applications devoted to the estimation of coancestry from molecular information are available. Most of them implement pairwise methodologies, although several tools have been already developed for the use of group estimation. Table 10 summarises the most relevant programs and their characteristics.

Table 10. Available software for the estimation of coancestry

Program	Drawbacks / Comments	Reference	Web site
Pairwise (ML)			
RELPAIR	Allows for linkage / Accounts for genotyping errors / Allows for chromosome X markers/ Optional input of probable genealogies (checking pedigree errors)	Epstein *et al.* (2000)	http://www.sph.umich.edu/statgen/boehnke/relpair.html
PREST	Allows for linkage / Accounts for genotyping errors (some cases) / Optional input of probable genealogies (checking pedigree errors)	McPeek and Sun (2000)	http://galton.uchicago.edu/~mcpeek/software/prest/
ECLIPSE (in PANGAEA)	Analyses triplets of individuals / Allows for linkage / Accounts for genotyping errors (some cases) / Optional input of probable genealogies (checking pedigree errors)	Sieberts *et al.* (2002)	http://www.stat.washington.edu/thompson/Genepi/pangaea.shtml
ML-RELATE		Kalinowski *et al.* (2006)	http://www.montana.edu/kalinowski/kalinowski_software.htm
Pairwise (MM)			
RELATEDNESS	Only for Macintosh / Assumes unlinked loci	Queller and Goodnight (1989)	http://www.gsoftnet.us/GSoft.html
KINSHIP	Only for Macintosh / Assumes unlinked loci	Goodnight and Queller (1999)	http://www.gsoftnet.us/GSoft.html
DELRIOUS	Requires software Mathematica / Assumes unlinked loci	Stone and Björklund (2001)	http://www.zoo.utoronto.ca/stone/delrious/delrious.htm
MER	Dominant or codominant markers	Wang (2002); Wang (2004a)	http://www.zoo.cam.ac.uk/ioz/software.htm
SPAGEDI	Dominant or codominant markers / Compares genetic differentiation with geographical distance	Hardy and Vekemans (2002); Hardy (2003)	http://www.ulb.ac.be/sciences/ecoevol/spagedi.html
Genealogy reconstruction			
BOREL (in PANGAEA)	FS families (likelihood evaluation)	Painter (1997)	http://www.stat.washington.edu/thompson/Genepi/pangaea.shtml
PRT	FS families (enumeration of compatible groups)	Almudevar and Field (1999)	http://www.urmc.rochester.edu/smd/biostat/people/faculty/almudevar.html
PARENTAGE	FS and non-nested HS families / Allows for previous information / Accounts for genotyping errors / Slow and little accurate with large populations	Emery *et al.* (2001)	http://www.mas.ncl.ac.uk/~nijw/

Table 10. (Continued)

Program	Drawbacks / Comments	Reference	Web site
Pairwise (ML)			
COLONY	FS families nested in HS families / Accounts for genotyping errors / Accounts for known parents genotypes	Wang (2004b)	http://www.zoo.cam.ac.uk/ioz/software.htm
MOLCOAN	Not dependent on frequencies knowledge / FS and non-nested HS families (1 generation) or other relationships (< 1 generation) / Accounts for known genealogy / Accounts for genotyping errors	Fernández and Toro (2006)	http://www.uvigo.es/webs/c03/webc03/XENETICA/XB2/Jesus/Fernandez.htm

5.8. Kinship Estimation in Aquaculture

Molecular estimation of kinship in aquaculture has been mostly applied to evaluate family structure within a broodstock or at specific generations in a breeding programme to avoid inbreeding or loss of genetic diversity across generations of selection (Table 11). The main drawback for this approach is the large error associated to the estimators with the number of microsatellites commonly applied (around 10), which determines only enough power to discriminate with reasonable accuracy between quite divergent relationships (unrelated *vs.* full-sibs). This is obviously an important objective when beginning a genetic breeding programme with a broodstock of unknown origin, and probably, the application of a broader battery of microsatellites could be profitable to increase the accuracy of the estimations. The uncertainty associated to this analysis is on the basis of an important amount of works devoted to evaluate the accuracy of microsatellites comparing molecular estimations with actual pedigree data. As in paternity analysis kinship has been mainly estimated in fish and only using microsatellites.

Regarding the methodology, at all cases pairwise estimators were applied and usually categorical classification was made by applying thresholds between predetermined categories. The more recent development of pedigree reconstruction methods, and specially, their higher complexity for computation has probably limited their application. The Queller and Goodnight (1989) estimator of relatedness is the most popular and the programs by these authors, Relatedness and Kinship, have been widely applied. Pairwise distance matrices has also been used to reveal kinship through graphical methods using different algorithms for dendrogram reconstruction (UPGMA, Neigbor-joining, ..).

Table 11. Representative references for kinship evaluation in Aquaculture

Species	Taxonomic group	Pairwise estimator	Program	Analysis	Reference
Sparus aurata	Fish	r_{QG}	RELATEDNESS	Family structure in broodstock	Castro *et al.* in press
Pangasianodon gigas	Fish	r_R	Own program	Family structure in broodstock	Sriphairoj *et al.* 2007
Salvelinus alpinus	Fish	r_{QG}	KINSHIP	Family structure in broodstock	Ditlecadet *et al.* 2007
Solea senegalensis	Fish	r_{QG}	KINSHIP; GENETIX[4]	Family structure in broodstock and F1	Porta *et al.* 2006
Solea senegalensis	Fish	r_{QG}	RELATEDNESS	Family structure in broodstock	Castro *et al.* 2006
Haliotis discus hannai	Molluscs	r_{LR}	IDENTIX[2]	Assessing programmes for population enhancement	Sekino *et al.* 2005
Paralichthys olivaceus	Fish	r_{QG}	RELATEDNESS; PHYLIP[2]	Heritability estimation through relatedness estimation	Shikano 2005
Scophthalmus maximus	Fish	r_{QG}	RELATEDNESS	Accuracy of relatedness estimates *vs* real data	Borrell *et al.* 2004
Haliotis spp.	Molluscs	r_{QG}	RELATEDNESS	Family structure in broodstock *vs* natural populations	Evans *et al.* 2004
Acipenser transmontanus	Fish	*r* for dominant markers[3]	Own program; PHYLIP[1]	Evaluation of microsatellites as dominant markers for estimating relatedness	Rodzen *et al.* 2004

Table 11. (Continued)

Species	Taxonomic group	Pairwise estimator	Program	Analysis	Reference
Paralichthys olivaceus	Fish	r_{QG}	Own program	Using relatedness for enhancing genetic diversity in breeding programmes	Sekino *et al.* 2004
Oncorhynchus mykiss	Fish	r_{QG}	RELATEDNESS	Accuracy of relatedness estimates *vs* real data	McDonald *et al.* 2004
Haliotis discus hannai	Molluscs	*D*	PHYLIP[1]	Family structure in broodstock and F1	Park and Kijima 2003
Salmo salar	Fish	r_{QG}	KINSHIP	Accuracy of relatedness estimates *vs* real data	Norris *et al.* 2000

The references are a representative sample taken relevant publications in Aquaculture
All studies were performed using microsatellite loci
[1]Clustering methods from relatedness matrices to identify full-sib families (PHYLIP program: Felsenstein, 1993).
[2]Belkhir *et al.* (2002)
[3]Lynch and Milligan (1994)
[4]Belkhir *et al.* (2003)
Pairwise estimators: QG (Queller and Goodnight, 1989); R (Ritland, 1996); LR (Lynch and Ritland , 1999); D (genetic distance)

Acknowledgements

Many examples and data presented in this article have been obtained by the Group of Genetics for Aquaculture and Conservation of the University of Santiago de Compostela (ACUIGEN). Our acknowledgement to Jaime Castro, Ania Pino-Querido, Miguel Hermida, Belén G. Pardo and Carmen Bouza, and to the technicians Lucía Insua, Sonia Gómez, María Portela, Susana Sánchez y María López.

References

Almudevar A. and Field C. (1999) Estimation of single-generation sibling relationships based on DNA markers. *Journal of Agricultural Biological and Environmental Statistics* 4, 136–165.

Amos W. (1999). A comparative approach to the study of microsatellite. In: *Microsatellites. Evolution and Applications* (ed. by D.B. Goldstein and C. Schlötterer), pp. 66-79. Oxford University Press, Oxford.

Anderson E.C. and Garza J.C. (2006) The power of single-nucleotide polymorphisms for large-scale parentage inference. *Genetics* 172, 2567-2582.

Angers B. and Bernatchez L. (1997) Complex evolution of a salmonid microsatellite locus and its consequences in inferring allelic divergence from size information. *Molecular Biology and Evolution* 14, 230-238.

Araki H. and Blouin M.S. (2005) Unbiased estimation of relative reproductive success of different groups: evaluation and correction of bias caused by parentage assignment errors. *Molecular Ecology* 14, 4097-4109.

Ardren W.R., Borer S., Thrower F., Joyce J.E. and Kapuscinski A.R. (1999) Inheritance of 12 microsatellite loci in *Oncorhynchus mykiss*. *Journal of Heredity* 90, 529-536.

Bacon A.L., Dunlop M.G. and Farrington S.M. (2001) Hypermutability at a poly(A/T) tract in the human germline. *Nucleic Acids Research* 29, 4405-4413.

Ballou J.D. and Lacy R.C. (1995) Identifying genetically important individuals for management of genetic variation in pedigreed populations. In: *Population Management for Survival and Recovery*, (ed. by J.D. Ballou, M. Gilpin and T.J. Foose), pp. 76-111. Columbia University Press, New York.

Belkhir K., Castric V. and Bonhomme F. (2002) IDENTIX, a software to test for relatedness in a population using permutation methods. *Molecular Ecology Notes* 2, 611-614.

Belkhir K., Borsa P., Chikhi L., Raufaste N. and Bonhomme F. (2003) *GENETIX 4.04, logiciel sous Windows TM pour la génétique des populations.* Laboratoire Génome, Populations, Interactions, CNRS UMR 5000, Université de Montpellier II, Montpellier, France (in French).

Bernardo J.M. and Smith A.F.M. (1994) *Bayesian Theory*. John Wiley and Sons.

Bernatchez L. and Duchesne P. (2000) Individual-based genotype analysis in studies of parentage and population assignment: how many loci, how many alleles? *Canadian Journal of Fisheries and Aquatic Sciences* 57, 1-12.

Beyer J. and May B. (2003) A graph-theoretic approach to the partition of individuals into full-sib families. *Molecular Ecology* 12, 2243-2250.

Blouin M.S. (2003) DNA-based methods for pedigree reconstruction and kinship analysis in natural populations. *Trends in Ecology and Evolution* 18, 503-511.

Blouin M.S., Parsons M., Lacaille V. and Lotz S. (1996) Use of microsatellite loci to classify individuals by relatedness. *Molecular Ecology* 5, 393–401

Borrell Y.J., Alvarez J., Vázquez E., Pato C.F., Tapia C.M., Sánchez J.A. and Blanco G. (2004) Applying microsatellites to the management of farmed turbot stocks (*Scophthalmus maximus* L.) in hatcheries. *Aquaculture* 241, 133-150.

Boudry P., Collet B., Cornette F., Hervouet V. and Bonhomme F. (2002) High variance in reproductive success of the Pacific oyster (*Crassostrea gigas*, Thunberg) revealed by microsatellite-based parentage analysis of multifactorial crosses. *Aquaculture* 204, 283-296.

Bouza C., Castro J., Presa P., Sánchez L. and Martínez P. (2002) Allozyme and microsatellite diversity in natural and domestic populations of turbot (*Scophthalmus maximus*) in comparison with other Pleuronectiformes. *Canadian Journal of Fisheries and Aquatic Sciences* 59, 1460-1473.

Bowcock A.M., Ruiz-Linares A., Tomfohrde J., Minch E., Kidd J.R. and Cavalli-Sforza L.L. (1994) High resolution of human evolutionary trees with polymorphic microsatellites. *Nature* 368, 455-457.

Brookfield J.F.Y. (1996) A simple new method for estimating null allele frequency from heterozygote deficiency. *Molecular Ecology* 5, 453-455.

Brown R.C., Tsalavouta M., Terzoglou V., Magoulas A. and Mcandrew B.J. (2005) Additional microsatellites for *Sparus aurata* and cross-species amplification within the Sparidae family. *Molecular Ecology Notes* 5, 605-607.

Browning S. and Thompson E.A. (1999) Interference in the analysis of genetic marker data. *American Journal of Human Genetics* Suppl 65, A244

Butler K., Field C., Herbinger C.M. and Smith B.R. (2004) Accuracy, efficiency and robustness of four algorithms allowing full sibship reconstruction from DNA marker data. *Molecular Ecology* 13, 1589–1600.

Caballero A. and Toro M.A. (2000) Interrelations between effective population size and other pedigree tools for the management of conserved populations. *Genetical Research* 75, 331-343.

Carreras-Carbonell J., Macpherson E. and Pascual M. (2004) Isolation and characterization of microsatellite loci in *Tripterygion delaisi*. *Molecular Ecology Notes* 4, 438-439.

Castro J., Bouza C., Sánchez L., Cal R.M., Piferrer F. and Martínez P. (2003) Gynogenesis assessment by using microsatellite genetic markers in turbot (*Scophthalmus maximus*). *Marine Biotechnology* 5, 584-592.

Castro J., Bouza C., Presa P., Pino-Querido A., Riaza A., Ferreiro I., Sánchez L. and Martínez P. (2004) Potential sources of error in parentage assessment of turbot (*Scophthalmus maximus*) using microsatellite loci. *Aquaculture* 242, 119-135.

Castro J., Pino-Querido A., Chavarrías D., Merino P., Ferreiro I., Riaza A., Hermida M., Bouza C., Sánchez L. and Martínez P. (2005) Asignación de parentescos mediante marcadores microsatélite en rodaballo (*Scophthalmus maximus*), lenguado (*Solea* senegalensis) y dorada (*Sparus aurata*). pp. 456-457 *X Congreso Nacional de Acuicultura*. Gandía, Valencia.

Castro J., Pino A., Hermida M., Bouza C., Riaza A., Ferreiro I., Sánchez L. and Martínez P. (2006) A microsatellite marker tool for parentage analysis in Senegal sole (*Solea senegalensis*): Genotyping errors, null alleles and conformance to theoretical assumptions. *Aquaculture* 261, 1194-1203.

Castro J., Pino A., Hermida M., Bouza C., Chavarrías D., Merino P., Sánchez L. and Martínez P. (2007) A microsatellite marker tool for parentage assessment in gilthead sea bream (*Sparus aurata*). *Aquaculture* 272S1, S210-S216.

Cercueil A., Bellemain E. and Manel S. (2002) PARENTE: computer program for parentage analysis. *Journal of Heredity* 93, 458-459.

Chakraborty R., Meagher T.R. and Smouse P.E. (1988) Parentage analysis with genetic markers in natural populations. I. The expected proportion of offspring with unambiguous paternity. *Genetics* 118, 527-536.

Chistiakov D.A., Hellemans B. and Volckaert F.A.M. (2006) Microsatellites and their genomic distribution, evolution, function and applications: A review with special reference to fish genetics. *Aquaculture* 255, 1-29.

Csilléry K., Johnson T., Beraldi D., Clutton-Brock T., Coltman D., Hansson B., Spong G. and Pemberton J.M. (2006) Performance of marker-based relatedness estimators in natural populations of outbred vertebrates. *Genetics* 173, 2091–2101

Dakin E.E. and Avise J.C. (2004) Microsatellite null alleles in parentage analysis. *Heredity* 93, 504-509.

Danzmann R.G. (1997) PROBMAX: a computer program for assigning unknown parentage in pedigree analysis from known genotypic pools of parents and progeny. *Journal of Heredity* 88, 333.

De León F.J.G., Canonne M., Quillet E., Bonhomme F. and Chatain B. (1998) The application of microsatellite markers to breeding programmes in the sea bass, *Dicentrarchus labrax*. *Aquaculture* 159, 303-316.

Ditlecadet D., Dufresne F., Le Francois N.R. and Blier P.U. (2006) Applying microsatellites in two commercial strains of Arctic charr (*Salvelinus alpinus*): Potential for a selective breeding program. *Aquaculture* 257, 37-43.

Dong S.R., Kong J., Zhang T.S., Meng X.H. and Wang R.C. (2006) Parentage determination of Chinese shrimp (*Fenneropenaeus chinensis*) based on microsatellite DNA markers. *Aquaculture* 258, 283-288.

Duchesne P., Godbout M.H. and Bernatchez L. (2002) PAPA (package for the analysis of parental allocation): a computer program for simulated and real parental allocation. *Molecular Ecology Notes* 2, 191-193.

Duchesne P., Castric T. and Bernatchez L. (2005) PASOS (parental allocation of singles in open systems): a computer program for individual parental allocation with missing parents. *Molecular Ecology Notes* 5, 701-704.

Ellegren H. (2000) Microsatellite mutations in the germline: implications for evolutionary inference. *Trends in Genetics* 16, 551-558.

Emery A.M., Wilson I.J., Craig S., Boyle P.R. and Noble L.R. (2001) Assignment of paternity groups without access to parental genotypes: multiple mating and developmental plasticity in squid. *Molecular Ecology* 10, 1265–1278

Emik L.O. and Terrill C.E. (1949) Systematic procedures for calculating inbreeding coefficients. *Journal of Heredity* 40, 51–55.

Epstein M.P., Duren W.L. and Boehnke M. (2000) Improved inference of relationship for pairs of individuals. *American Journal of Human Genetics* 67, 1219–1231

Estoup A., Gharbi K., San Cristobal M., Chevalet C., Haffray P. and Guyomard R. (1998) Parentage assignment using microsatellites in turbot (*Scophthalmus maximus*) and rainbow trout (*Oncorhynchus mykiss*) hatchery populations. *Canadian Journal of Fisheries and Aquatic Sciences* 55, 715-725.

Evans B., Bartlett J., Sweijd N., Cook P. and Elliott N.G. (2004) Loss of genetic variation at microsatellite loci in hatchery produced abalone in Australia (*Haliotis rubra*) and South Africa (*Haliotis midae*). *Aquaculture* 233, 109-127.

Ewen K.R., Bahlo M., Treloar S.A., Levinson D.F., Mowry B., Barlow J.W. and Foote S.J. (2000) Identification and analysis of error types in highthroughput genotyping. *American Journal of Human Genetics* 67, 727–736

Ezaz M.T., Sayeed S., McAndrew B.J. and Penman D.J. (2004) Use of microsatellite loci and AFLP markers to verify gynogenesis and clonal lines in Nile tilapia *Oreochromis niloticus* L. *Aquaculture Research* 35, 1472-1481.

Falconer D.S. and Mackay T.F. (1996) *Introduction to Quantitative Genetics*. 4th ed. Longman, Harlow, Essex.

Felsenstein J. (1993) *PHYLIP (Phylogeny Inference Package) version 3.5c*. Distributed by the author. Department of Genetics, University of Washington, Seattle.

Fernández J. and Toro M.A. (2006) A new method to estimate relatedness from molecular markers. *Molecular Ecology* 15, 1657–1667.

Fessehaye Y., El-Bialy Z., Rezk M.A., Crooijmans R., Bovenhuis H. and Komen H. (2006) Mating systems and male reproductive success in Nile tilapia (*Oreochromis niloticus*) in breeding hapas: a microsatellite analysis. *Aquaculture* 256, 148-158.

Fiumera A.C., DeWoody Y.D., DeWoody J.A., Asmussen M.A. and Avise J.C. (2001) Accuracy and precision of methods to estimate the number of parents contributing to a half-sib progeny array. *Journal of Heredity* 92, 120–126.

Fopp-Bayat D., Kolman R. and Woznicki P. (2007) Induction of meiotic gynogenesis in sterlet (*Acipenser ruthenus*) using UV-irradiated bester sperm. *Aquaculture* 264, 54-58.

Gelfand A.E. and Smith A.F.M. (1990) Sampling based approaches to calculating marginal densities. *Journal of the American Statistical Association* 85, 398-409.

Gerber S., Mariette S., Streiff R., Bodenes C. and Kremer A. (2000) Comparison of microsatellites and amplified fragment length polymorphism markers for parentage analysis. *Molecular Ecology* 9, 1037-1048.

Gjedrem T. (2005) *Selection and Breeding Programs in Aquaculture*. Springer, Dordrecht.

Goodnight K. F. and Queller D.C. (1999) Computer software for performing likelihood tests of pedigree relationship using genetic markers. *Molecular Ecology* 8, 1231–1234.

Hadfield J.D., Richardson D.S. and Burke T. (2006) Towards unbiased parentage assignment: combining genetic, behavioural and spatial data in a Bayesian framework. *Molecular Ecology* 15, 3715-3730.

Hara M. and Sekino M. (2003) Efficient detection of parentage in a cultured Japanese flounder *Paralichthys olivaceus* using microsatellite DNA marker. *Aquaculture* 217, 107-14.

Hardy O. (2003) Estimation of pairwise relatedness between individuals and characterization of isolation-by-distance processes using dominant genetic markers. *Molecular Ecology* 12, 1577–1588.

Hardy O. and Vekemans X. (2002) SPAGeDi: a versatile computer program to analyse spatial genetic structure at the individual or population levels. *Molecular Ecology Notes* 2, 618–620.

Hastein T., Hill B.J., Berthe F. and Lightner D.V. (2001) Traceability of aquatic animals. *Rev. Sci. Tech. – Off. Int. Épizoot.* 20, 564-583.

Hayes B., Sonesson A.K. and Gjerde B. (2005) Evaluation of three strategies using DNA markers for traceability in aquaculture species. *Aquaculture* 250, 70-81.

Herbinger C.M., Doyle R.W., Pitman E.R., Paquet D., Mesa K.A., Morris D.B., Wright J.M. and Cook D. (1995) DNA fingerprint based analysis of paternal and maternal effects on offspring growth and survival in communally reared rainbow trout. *Aquaculture* 137, 245-256.

Herbinger C.M., Doyle R.W., Taggart C.T., Lochman S.E., Brooker A.L., Wright J.M. and Cook D. (1997) Family relationships and effective population size in a natural cohort of Atlantic cod (*Gadus morhua*) larvae. *Canadian Journal of Fisheries and Aquatic Sciences* 54, 11–18.

Hoffman J.I. and Amos W. (2005) Does kin selection influence fostering behaviour in Antarctic fur seals (*Arctocephalus gazella*)? *Proceedings of the Royal Society B-Biological Sciences* 272: 2017-2022.

Hulata G. (2001) Genetic manipulations in aquaculture: a review of stock improvement by classical and modern technologies. *Genetica* 111, 155-173.

Iyengar A., Piyapattanakorn S., Heipel D.A., Stone D.M., Howell B.R., Child A.R. and MaClean N. (2000) A suite of highly polymorphic microsatellite markers in turbot (*Scophthalmus maximus* L.) with potential for use across several flatfish species. *Molecular Ecology* 9, 368-371.

Jerry D.R., Preston N.P., Crocos P.J., Keys S., Meadows J.R.S. and Li Y.T. (2004) Parentage determination of Kuruma shrimp *Penaeus (Marsupenaeus) japonicus* using microsatellite markers (Bate). *Aquaculture* 235, 237-247.

Jerry D.R., Evans B.S., Kenway M. and Wilson K. (2006a) Development of a microsatellite DNA parentage marker suite for black tiger shrimp *Penaeus monodon. Aquaculture* 255, 542-547.

Jerry D.R., Preston N.P., Crocos P.J., Keys S., Meadows J.R.S. and Li Y.T. (2006b) Application of DNA parentage analysis for determining relative growth rates of *Penaeus japonicus* families reared in commercial ponds. *Aquaculture* 254, 171-181.

Jin L. and Chakraborty R. (1995) Population structure, stepwise mutations, heterozygote deficiency and their implications in DNA forensics. *Heredity* 74, 274–285.

Johnson N.A., Rexroad C.E., Hallerman E.M., Vallejo R.L. and Palti Y. (2007) Development and evaluation of a new microsatellite multiplex system for parental allocation and management of rainbow trout (*Oncorhynchus mykiss*) broodstocks. *Aquaculture* 266, 53-62.

Jones A.G. (2001) GERUD 1.0: a computer program for the reconstruction of parental genotypes from progeny arrays using multilocus DNA data. *Molecular Ecology Notes* 1, 215-218.

Jones A.G. (2005) GERUD 2.0: a computer program for the reconstruction of parental genotypes from half-sib progeny arrays with known or unknown parents. *Molecular Ecology Notes* 5, 708-711.

Jones A.G. and Ardren W.R. (2003) Methods of parentage analysis in natural populations. *Molecular Ecology* 12, 2511-2523.

Jones A.G., Kvarnemo C., Moore G.I., Simmons L.W. and Avise J.C. (1998) Microsatellite evidence for monogamy and sex-biased recombination in the Western Australian seahorse, *Hippocampus angustus*. *Molecular Ecology* 7, 1497-1505.

Kalinowski S.T., Wagner A.P. and Taper M.L. (2006) ML-RELATE: a computer program for maximum likelihood estimation of relatedness and relationship. *Molecular Ecology Notes* 6, 576–579.

Kalinowski S.T., Taper M.L. and Marshall T.C. (2007) Revising how the computer program CERVUS accommodates genotyping error increases success in paternity assignment. *Molecular Ecology* 16, 1099-1106.

Kasumovic M.M., Ratcliffe L.M. and Boag P.T. (2003) A method to improve confidence in paternity assignment in an open mating system. *Canadian Journal of Zoology* 81, 2073-2076.

Lemos A., Freitas A.I., Fernandes A.T., Goncalves R., Jesus J., Andrade C. and Brehm A. (2006) Microsatellite variability in natural populations of the blackspot seabream *Pagellus bogaraveo* (Brunnick, 1768): a database to access parentage assignment in aquaculture. *Aquaculture Research* 37, 1028-1033.

Li Q., Park C. and Kijima A. (2003) Allelic transmission of microsatellites and application to kinship analysis in newly hatched Pacific abalone larvae. *Fisheries Science* 69, 883-889.

Li Y.T., Wongprasert K., Shekhar M., Ryan J., Dierens L., Meadows j., Preston N., Coman G. and Lyons R.E. (2007) Development of two microsatellite multiplex systems for black tiger shrimp *Penaeus monodon* and its application in genetic diversity study for two populations. *Aquaculture* 266, 279-288.

Li C.C. and D.G. Horvitz (1953) Some methods of estimating the inbreeding coefficient. *American Journal of Human Genetics* 5, 107-117.

Li C.C., Weeks D.E. and Chakrabarti A. (1993) Similarity of DNA fingerprints due to chance and relatedness. *Human Heredity* 43, 45-52.

Liu Z.J. and Cordes J.F. (2004) DNA marker technologies and their applications in aquaculture genetics. *Aquaculture* 238, 1-37.

Lucas T., Macbeth M., Degnan S.M., Knibb W. and Degnan B.M. (2006) Heritability estimates for growth in the tropical abalone *Haliotis asinina* using microsatellites to assign parentage. *Aquaculture* 259, 146-152.

Luikart G. and England P.R. (1999) Statistical analysis of microsatellite DNA data. *Trends in Ecology and Evolution* 14, 253– 256.

Lynch M. and Ritland K. (1999) Estimation of pairwise relatedness with molecular markers. *Genetics* 152, 1753-1766.

Malécot G. (1948) *Les Mathématiques de l'Hérédité*. Masson et Cie, Paris.

Marshall T.C., Slate J., Kruuk L.E.B. and Pemberton J.M. (1998) Statistical confidence for likelihood-based paternity inference in natural populations. *Molecular Ecology* 7, 639-655.

McCartney M.A., Brayer K. and Levitan D.R. (2004) Polymorphic microsatellite loci from the red urchin, *Strongylocentrotus franciscanus*, with comments on heterozygote deficit. *Molecular Ecology Notes* 4, 226-228.

McDonald G.J., Danzmann R.G. and Ferguson M.M. (2004) Relatedness determination in the absence of pedigree information in three cultured strains of rainbow trout (*Oncorhynchus mykiss*). *Aquaculture* 233, 65–78.

McPeek M.S. and Sun L. (2000) Statistical tests for detection of misspecified relationships by use of genome screen data. *American Journal of Human Genetics* 66, 1076–1094.

Meagher T.R. (1986) Analysis of paternity within a natural population of *Chamaelirium luteum*. I. Identification of most-likely male parents. *American Naturalist* 128, 199-215.

Meagher T.R. and Thompson E.A. (1986) The relationship between single and parent pair genetic likelihoods in genealogy reconstruction. *Theoretical Population Biology* 29, 87–106.

Milkman R. and Zeitler R.R. (1974) Concurrent multiple paternity in natural and laboratory populations of *Drosophila melanogaster*. *Genetics* 78, 1191-1193.

Milligan B.G. (2003) Maximum-likelihood estimation of relatedness. *Genetics* 163, 1153–1167.

Neef B.D., Repka J. and Gross M.R. (2001) A Bayesian framework for parentage analysis: The value of genetic and other biological data. *Theoretical Population Biology* 59, 315-331.

Nielsen R., Mattila D.K., Clapham P.J. and Palsboll P.J. (2001) Statistical approaches to paternity analysis in natural populations and applications to the North Atlantic humpback whale. *Genetics* 157, 1673-1682.

Norris A.T., Bradley D.G. and Cunningham E.P. (2000) Parentage and relatedness determination in farmed Atlantic salmon (*Salmon salar*) using microsatellite markers. *Aquaculture* 182, 73-83.

Oliehoek P.A., Windig J.J., van Arendonk J.A.M. and Bijma P. (2006) Estimating relatedness between individuals in general populations with a focus on their use in conservation programs. *Genetics* 173, 483–496.

Painter I. (1997) Sibship reconstruction without parental information. *Journal of Agricultural Biological and Environmental Statistics* 2, 212-229.

Pardo B.G., Fernández C., Hermida M., Vázquez A., Pérez M., Presa P., Calaza M., Alvarez-Dios J.A., Comesaña A.S., Raposo-Guillán J., Bouza C. and Martínez P. (2007) Development and characterization of 248 novel microsatellite markers in turbot (*Scophthalmus maximus*). *Genome* 50, 329-332.

Pena S.D.J. and Chakraborty R. (1994) Paternity testing in the DNA era. *Trends in Genetics* 10, 204-209.

Pérez-Enríquez R., Takagi M. and Taniguchi N. (1999) Genetic variability and pedigree tracing of a hatchery-reared stock of red sea bream (*Pagrus major*) used for stock enhancement, based on microsatellite DNA markers. *Aquaculture* 173 411-421.

Pinera J.A., Bernardo D., Blanco G., Vázquez E. and Sánchez J.A. (2006) Isolation and characterization of polymorphic microsatellite markers in *Pagellus bogaraveo*, and cross-species amplification in *Sparus aurata* and *Dicentrarchus labrax*. *Molecular Ecology Notes* 6, 33-35.

Pompanon F., Bonin A., Bellemain E. and Taberlet P. (2005) Genotyping errors: causes, consequences and solutions. *Nature Reviews Genetics* 6, 847-859.

Porta J., Porta J.M., Martínez-Rodríguez G. and Alvarez M.C. (2006) Genetic structure and genetic relatedness of a hatchery stock of Senegal sole (*Solea senegalensis*) inferred by microsatellites. *Aquaculture* 251, 46-55.

Primmer C.R., Saino N., Moller A.P. and Ellegren H. (1998) Unravelling the processes of microsatellite evolution through analysis of germ line mutations in barn swallows *Hirundo rustica*. *Molecular Biology Evolution* 15, 1047-1054.

Queller D.C. and Goodnight K.F. (1989) Estimating relatedness using molecular markers. *Evolution* 43, 258-274.

Rengmark A.H., Slettan A., Skaala O., Lie O. and Lingaas F. (2006) Genetic variability in wild and farmed Atlantic salmon (*Salmon salar*) strains estimated by SNP and microsatellites. *Aquaculture* 253, 229-237.

Ritland K. (1996) Estimators for pairwise relatedness and inbreeding coefficients. *Genetical Research* 67, 175-186.

Ritland K. (2005) Multilocus estimation of pairwise relatedness with dominant markers. *Molecular Ecology* 14, 3157–3165.

Rodríguez-Ramilo S., Toro M.A., Martínez P., Castro J., Bouza C. and Fernández J. (2007) Accuracy of the pairwise methods in the reconstruction of family relationships using molecular information of turbot (*Scophthalmus maximus*). *Aquaculture*.273, 434-442.

Rodzen J.A., Famula T.R. and May B. (2004) Estimation of parentage and relatedness in the polyploid white sturgeon (*Acipenser transmontanus*) using a dominant marker approach for duplicated microsatellite loci. *Aquaculture* 232, 165-182.

Sancristobal M. and Chevalet C. (1997) Error tolerant parent identification from a finite set of individuals. *Genetical Research* 70, 53-62.

Schlötterer C. (2000) Evolutionary dynamics of microsatellite DNA. *Chromosoma* 109, 365-371.

Sekino M., Saido T., Fujita T., Kobayashi R. and Takami H. (2005) Microsatellite DNA markers of Ezo abalone (*Haliotis discus hannai*): a preliminary assessment of natural populations sampled from heavily stocked areas. *Aquaculture* 243, 33-47.

Sekino M., Saitoh K., Yamada T., Kumagai A., Hara M. and Yamashita Y. (2003) Microsatellite-based pedigree tracing in a Japanese flounder *Paralichthys olivaceus* hatchery strain: implications for hatchery management related to stock enhancement program. *Aquaculture* 221, 255-263.

Sekino M., Sugaya T., Hara M. and Taniguchi N. (2004) Relatedness inferred from microsatellite genotypes as a tool for broodstock management of Japanese flounder *Paralichthys olivaceus*. *Aquaculture* 233, 163-172.

Shikano T. (2005) Marker-based estimation of heritability for body color variation in Japanese flounder *Paralichthys olivaceus*. *Aquaculture* 249, 95-105.

Sieberts S.K., Wijsman E.M. and Thompson E.A. (2002) Relationship inference from trios of individuals, in the presence of typing error. *American Journal of Human Genetics* 70, 170-180

Signorovitch J. and Nielsen R. (2002) PATRI- paternity inference using genetic data. *Bioinformatics* 18, 341-342.

Smith B.R., Herbinger C.M. and Merry H.R. (2001) Accurate partition of individuals into full-sib families from genetic data without parental information. *Genetics* 158, 1329-1338

Sneath P.H.A. and Sokal R.R. (1973) *The Principles and Practice of Numerical Classifications*. W.H. Freeman, San Francisco.

Sriphairoj K., Kamonrat W. and Na-Nakorn U. (2007) Genetic aspect in broodstock management of the critically endangered Mekong giant catfish, *Pangasianodon gigas* in Thailand. *Aquaculture* 264, 36-46.

Stone J. and Björklund M. (2001) DELRIOUS: a computer program designed to analyse molecular marker data and calculate delta and relatedness estimates with confidence. *Molecular Ecology Notes* 1, 209–212.

Taggart J.B. (2007) FAP: an exclusion-based parental assignment program with enhanced predictive functions. *Molecular Ecology Notes* 7, 412-415.

Talbot C.C. Jr, Avramopouls D., Gerken S., Chakravarti A., Armour J.A., Matsunami N., White R. and Antonarakis S.E. (1995) The tetranucleotide repeat polymorphism D21S1245 demonstrates hypermutability in germline and somatic cells. *Human Molecular Genetics* 4, 1193-1199.

Taris N., Baron S., Sharbel T.F., Sauvage C. and Boudry P. (2005) A combined microsatellite multiplexing and boiling DNA extraction method for high-throughput parentage analyses in the Pacific oyster (*Crassostrea gigas*). *Aquaculture Research* 36, 516-518.

Thomas S.C. (2005) The estimation of genetic relationships using molecular markers and their efficiency in estimating heritability in natural populations. *Philosophical Transactions of the Royal Society of London. Series B, Biological Sciences* 360, 1457-1467.

Thomas S.C. and Hill W.G. (2000) Estimating quantitative genetic parameters using sibships reconstructed from marker data. *Genetics* 155, 1961–1972.

Thomas S.C. and Hill W.G. (2002) Sibship reconstruction in hierarchical population structures using Markov chain Monte Carlo techniques. *Genetical Research* 79, 227–234.

Thompson E.A. (1991) Estimation of relationships from genetic data. In: *Handbook of Statistics* (ed. by C.R. Rao and R. Chakraborty), vol. 8, pp. 255-269. Elsevier Science Publishers, Amsterdam.

Toro M.A., Barragán C., Óvilo C., Rodrigañez J., Rodríguez C. and Silió L. (2002) Estimation of coancestry in Iberian pigs using molecular markers. *Conservation Genetics* 3, 309-320.

van de Caastele T., Galbusera P. and Matthysen E. (2001) A comparison of microsatellite-based pairwise relatedness estimators. *Molecular Ecology* 10, 1539-1549.

Van Oosterhout C., Hutchinson W.F., Wills D.P.M. and Shipley P. (2004) micro-checker: software for identifying and correcting genotyping errors in microsatellite data. *Molecular Ecology Notes* 4, 535-538.

Vandeputte M., Kocour M., Mauger S., Dupont-Nivet M., De Guerry D., Gela D., Vallod D., Linhart O. and Chevassus B. (2005) Heritability estimates for growth-related traits using microsatellite parentage assignment in juvenile common carp (*Cyprinus carpio* L.). *Aquaculture* 247, 31-32.

Vandeputte M., Mauger S. and Dupont-Nivet M. (2006) An evaluation of allowing for mismatches as a way to manage genotyping errors in parentage assignment by exclusion. *Molecular Ecology Notes* 6, 265-267.

Volckaert F.A.M. and Hellemans B. (1999) Survival, growth and selection in a communally reared multifactorial cross of African catfish (*Clarias gariepinus*). *Aquaculture* 171, 49-64.

Wang X.X., Ross K.E., Saillant E., Gatlin D.M. and Gold J.R. (2007) Genetic effects on carcass-quality traits in hybrid striped bass (*Morone chrysops* female x *Morone saxatilis* male). *Aquaculture Research* 38, 973-980.

Wang J. (2002) An estimator for pairwise relatedness using molecular markers. *Genetics* 160, 1203–1215.

Wang J. (2004a) Estimating pairwise relatedness from dominant genetic markers. *Molecular Ecology* 13: 3169–3178.

Wang J. (2004b) Sibship reconstruction from genetic data with typing errors. *Genetics* 166, 1963-1979.

Wang J. (2006) Informativeness of genetic markers for pairwise relationship and relatedness inference. *Theoretical Population Biology* 70, 300-321.

Wang X.X., Ross K.E., Saillant E., Gatlin D.M. and Gold J.R. (2006) Quantitative genetics and heritability of growth-related traits in hybrid striped bass (*Morone chrysops* female x *Morone saxatilis* male). *Aquaculture* 261, 535-545.

Weir B.S., Anderson A.D. and Heppler G.M. 2006 Genetic relatedness analysis: modern data and new challenges. *Nature Reviews Genetics* 7, 771-780.

Wesmajervi M.S., Westgaard J.I. and Delghandi M. (2006) Evaluation of a novel pentaplex microsatellite marker system for paternity studies in Atlantic cod (*Gadus morhua* L.). *Aquaculture Research* 37, 1195-1201.

Wilson A.J. and Ferguson M.M. (2002) Molecular pedigree analysis in natural populations of fishes: approaches, applications, and practical considerations. *Canadian Journal of Fisheries and Aquatic Sciences* 59, 1696-1707.

Worthington Wilmer J., Allen P.J., Pomeroy P.P., Twiss S.D. and Amos W. (1999) Where have all the fathers gone? An extensive microsatellite analysis of paternity in the grey seal (*Halichoerus grypus*). *Molecular Ecology* 8, 1417-1429.

Zane L., Bargelloni L. and Patarnello T. (2002) Strategies for microsatellite isolation: a review. *Molecular Ecology* 11, 1-16.

In: Aquaculture Research Trends
Editor: Stephen H. Schwartz, pp.113-148

ISBN 978-1-60456-217-0

Chapter 3

LARVAL FISH NUTRITION AND REARING TECHNOLOGIES: STATE OF THE ART AND FUTURE

Jian G. Qin[1]
School of Biological Sciences, Flinders University,
GPO Box 2100, Adelaide, SA, Australia

ABSTRACT

Marine finfish larviculture techniques have been quickly developed in the past two decades. Although hatchery production technology still follows similar protocols worldwide, feeding regimes and culture systems have been greatly modified to improve hatchery performance and efficiency. Marine fish larvae are reared more often in green water intensive systems than in clear water systems, while extensive culture is less commonly used. The use of live food organisms, especially at first feeding, is still obligatory in most marine fish larvae mass-culture. Rotifers *Brachionus* spp. and brine shrimp *Artemia* nauplii are the main live food items for early life stages of the major finfish species, with inert feed provided at a later stage. Super-intensive rotifer culture using algal paste has improved cost-effectiveness of hatchery production, whereas *Artemia* nauplii have been used as biocapsules to deliver vitamins and a variety of nutrition (e.g., EPA, DAH and ARA) from different sources. Recently, the use of copepods and cladocerans as live food has been explored. However, low production of copepods is the major hurdle of mass production, but there is a growing interest in using saline water cladocerans as live food in hatchery. Until recently, it is considered impossible to feed most newly hatched marine fish species with a compound diet. Substituting a compound diet for live prey is performed several weeks after hatching, depending on the species. Ontogenetic studies suggest that most digestive enzymes are present in young larvae, but the quantity is not sufficient to digest formulated feed. Therefore, it is essential to provide live feed for the first feeding of marine finfish species. Weaning of fish larvae to inert feed at a later stage of development is easily achieved by co-feeding inert feed with live food because live food organisms consumed by the larvae assist the digestion process by donating their digestive enzymes to the gut of fish larvae. In larviculture rearing, high mortalities are likely to occur during the diet

[1] Email: Jian.Qin@Flinders.edu.au.

transition from endogenous to exogenous nutrition and the period of weaning, and cannibalism could contribute greatly to mortality when great size heterogeneity exists. Reduction of larval deformity is a challenge in some fish species and further understanding of the role of nutrition in fish development, especially at the molecular level will allow improving the quality of hatchery reared larvae.

1. Introduction

Aquaculture continues to grow more rapidly than all other animal food-producing sectors, with an average annual growth rate for the world of 8.8 percent per year since 1970, compared with only 1.2 percent for capture fisheries and 2.8 percent for terrestrial farmed meat production systems (FAO, 2006). Comparing the difference between the demand of aquaculture products and the supply from the capture, a great potential exists for aquaculture to meet the needs of seafood market, as well as to reduce the threat to the nature resources and the endangered species. For marine finfish aquaculture, the bottleneck is mass-production of high quality fry for farming. At present, the larviculture of suitable marine finfish species is considered the main limiting factor for commercial fish farming industrial development (Liao et al., 2001; Shields, 2001).

Nutritional requirements centre the research and hatchery management of larval fish rearing and production (Cahu et al., 2003; Lee, 2003). Currently, the seed supply of fish juveniles in commercial hatcheries relies on the successful supply of live zooplankton species such as rotifers, copepods and *Artemia* nauplii, during the larval stage (Hagiwara et al., 2001; Lee et al., 2005; Sorgeloos et al., 2001; Støttrup and Norsker, 1997). To ensure a sufficient live food supply, the fish hatcheries need to establish a food chain supply from algae to zooplankton. The high cost of infrastructure and maintenance of live food culture have made researchers search for a replacement of live food (Southgate and Partridge, 1998). In the past decade, intensive research has focus on the replacements for live food organisms with compound diets (Kolkovski, 2001). Although significant improvement has been made in co-feeding live and compound diets, especially at the *Artemia* feeding stage, compound diets alone have not matched the growth and survival of fish fed with live feeds (Lee, 2003). The use of live food organisms, especially at first feeding, is still obligatory in marine fish larvae mass-culture. The fundamental reason for the reliance on live food is because the digestive tract in most fish species at first feeding contains limited enzymes for digestion, absorption and assimilation of large molecules of proteins, lipids and glycogen (Cara et al., 2003; Chen et al., 2006b). The live food organisms consumed by the larvae assist the digestion process by donating their digestive enzymes to the gut of fish larvae (Dabrowski and Glogowski, 1977; Kolkovski et al., 1993). The nutritional requirements of fish larvae change during the course of ontogenetic development during early life history (Oozeki and Bailey, 1995). The variations of nutritional need depend on the morphology, physiology and feeding behaviour of different fish larvae.

This chapter will first review the digestive biology of fish larvae covering ontogenetic development of digestive systems, nutritional requirements and food digestions. Subsequently, the current methods for live food culture and inert food formation will be examined. Then, the larviculture systems and operations will be compared and feeding practices and major issues associated with larval rearing will be presented. Last, future research perspectives will be outlined. This review covers the state of art of larval fish rearing concept, technology and practices. It is my hope that this review will shed a light toward the improvement of hatchery technology and management strategies for larval fish rearing.

2. LARVAL FISH BIOLOGY

2.1. Ontogenetic Development Digestive Systems

The changes in morphology and physiology of marine fish larvae in the early life development lead to a number of important variations in feeding and nutritional requirements through the larvae to juveniles. These variations occur in the morphology of the digestive organs, the digestive processes, the nutrient requirements and the feeding behavior (De Silva and Anderson, 1995).

The morphology of the digestive tract is generally very simple in swim-up larvae. Dabrowski (1984) summarized the important changes in the digestive tract during ontogenesis of fish larvae (Figure 1). At hatching, the digestive tract is a straight tube closed at the mouth and histologically undifferentiated along its length. It remains quite unchanged from mouth opening until the completion of yolk absorption, and then becomes segmented into a buccopharynx, foregut, midgut and hindgut. The larval period ends with the development of a stomach with gastric glands and pyloric caeca. The liver and the pancreas are formed at hatching and are functional at first feeding.

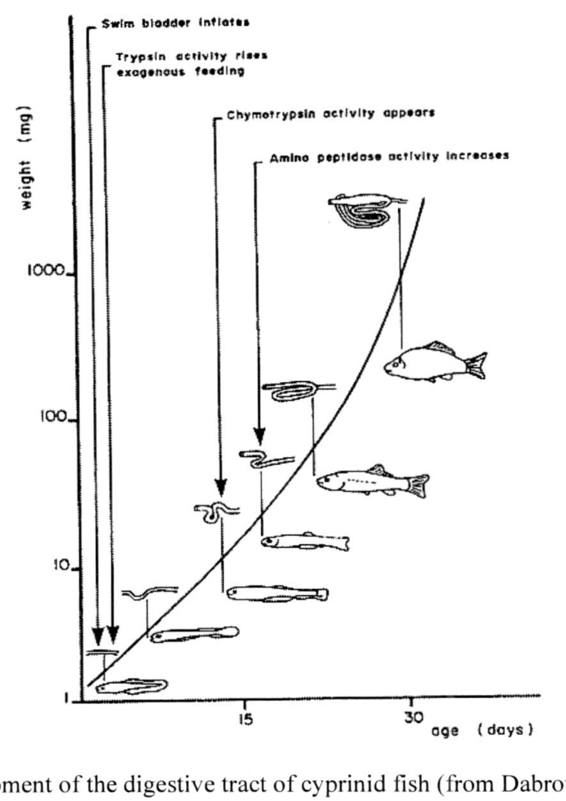

Figure 1. Ontogenic development of the digestive tract of cyprinid fish (from Dabrowski, 1984).

Ontogenetic development of the digestive system in most teleost fish is generally divided into three major phases (Bisbal and Bengtson, 1995a; Boulhic and Gabaudan, 1992; Chen et al., 2006b). The first phase starts from hatching and ends at the completion of endogenous feeding. During this period, larvae depend on energy reserves in the yolk sac and oil globules. Toward the end of the first phase, fish experience a transition of endogenous to exogenous feeding before exclusively feeding on external food. The second phase starts from the onset of exogenous feeding and ends before the formation of gastric glands in the stomach, characterized by the lack of sufficient digestive capabilities; (Boulhic and Gabaudan, 1992; Buddington, 1985). During this phase, larval fish mainly depend on pinocytosis and intracellular digestion and absorption (Watanabe, 1982). As a result, fish larvae usually feed on live food such as rotifers that can be easily ingested and digested. The third phase starts from the presence of gastric glands and pyloric caeca to metamorphosis onward, indicating the functional maturation of the digestive system (Bisbal and Bengtson, 1995a; Tanaka, 1971).

Changes in mouth size are extremely important in considering larval nutrition as they affect the capacity of the fish to ingest food organisms (Qin and Fast, 1997). Larval fish provided with food particles that are too large will be unable to consume them and starvation is inevitable. In contrast, if the particles are too small the energy required to gather sufficient quantities of food for their needs will be too great and will result in poor growth. Therefore, one of the most important features of larval nutrition is the suitability of food particle size to mouth size (Cunha and Planas, 1999).

After first exogenous feeding, a number of changes occur in yellowtail kingfish (Chen et al., 2006a). Mucosal folds gradually develop and the intestine develops regional differentiation. Absorption of nutrients in these animals is characterized by pinocytosis of macromolecules (e.g., protein and lipid) in which the surface membrane of the epithelial cells engulfs the molecule. Materials absorbed in this way are subsequently broken down intracellularly. In later development, the stomach and pyloric caeca form and the intestine lengthens relative to body length and begins to coil. With the development of the mucosa, greater amounts of digestive enzymes are produced, which facilitate digestion within the gas-intestinal tract (Stroband and Kroon, 1981).

Considerable variation in larval mouth size occurs both between species and throughout the larval stage. Larvae of first-feeding snakehead larvae of 6-7 mm total length (TL) with a mouth opening of 0.55 mm selected for *Artemia* nauplii, and against formulated feed (Qin and Fast, 1997). However, larvae of a flounder *Rhombosolea tapirina* have extremely small mouths and cannot accept *Artemia* nauplii or even nauplii of rotifers which are smaller (95-350 μm in cross-section). For this species it may be more appropriate to provide ciliate protozoans or mollusc velligers, both of which may be as small as 35 μm in cross-section.

2.2. Nutritional Requirement

Nutritional requirement is one of the major factors determining the survival and growth of marine fish larvae (Kolkovski et al., 1997b). Larval nutrition relies on feeding with live foods such as rotifers *Brachionus plicafilis* and brine shrimp *Artemia* during the first days (until 20-25 days after hatching), followed by a combination of live and formulated feed until they are completely weaned (Barahona-Fernandes and Girin, 1976; Kolkovski et al., 1997b).

Most research effort has been focused on meeting the nutritional requirements of marine fish larvae using cultured live prey and inert replacement diets for larvae. Much progress has been made in defining larval digestive capabilities (Chen et al., 2006b; Kolkovski, 2001) and nutrient requirements (Izquierdo et al., 1992). New techniques for administering essential nutrients via enriched live prey have been investigated (McEvoy et al., 1997; Sargent et al., 1997) and progress has been made in formulating and presenting inert feeds to larvae (Cahu and Zambonion Infante, 2001; Kelly et al., 2000). The global shortage in mid 1990s of *Artemia* cysts has provided impetus for diet replacement and a growing range of commercial microparticulate feeds is now offered to partially replace *Artemia* and to enable earlier weaning of fry.

Progresses have been made on the total dietary protein and lipid requirements of marine fish larvae (Øie et al., 1997). It has been documented that the total quantity and types of free amino acids (FAA) present in rotifers and *Artemia* differ from those in marine fish eggs/larvae and from their natural planktonic prey (Harel et al., 1999; Ronnestad et al., 1998). Delivery of FAA to larvae via formulated microdiets has proven difficult due to rapid leaching of amino acids to the environment).

The essential fatty acid (EFA) deficiencies of rotifers and *Artemia* as prey for marine fish larvae are well documented (Copeman et al., 2002; Sargent et al., 1997). Recent research on the dietary-essential fatty acid requirements of marine fish larvae has considered the interactions of all three EFAs, docosahexaenoic acid (DHA, 22:6n−3), eicosapentaenoic acid (EPA, 20:5n−3) and arachidonic acid (ARA, 20:4n−6)(Sargent et al., 1999a). The importance of high dietary DHA levels for the correct neural development of larvae has been shown to be unequivocal; however, the appropriate dietary ratio of EPA:ARA, in terms of eicosanoid interactions, appears to vary according to species (Sargent et al., 1999a).

The dietary requirements of larval fish are distinctly different from those of juvenile and adult fish. Nutritional requirements of fish larvae cannot be determined using traditional nutritional approaches, and thus the formulation of inert feeds remains problematic (Cahu and Zambonino Infante, 2001). In general, larval fish need 50–70% protein (for growth), with 28% lipids (for energy), and 3% highly unsaturated fatty acids (HUFAs) (for growth, survival, and neural development, among other parameters of fish health) (Cahu and Zambonino Infante, 2001).

There have been more available data on lipid requirements of larvae than other nutrients, because the profiles of fatty acids in live food can be manipulated, while altering other nutrient contents has not yet been successful. Lipid content of the feed provides the main energy source for the larvae, although the required lipid levels vary among different fish species. Sea bream (Salhi et al., 1994) and red drum (Brinkmeyer and Holt, 1998) have been reported to require 18% lipid in formulated diets, while sea bass (Zambonino Infante and Cahu, 1999) requires 30% lipid. Sea bream has been reported to require 29–37% lipid in *Artemia* (Koven et al., 1992) and Japanese flounder requires 25% (Furuita et al., 1998).

Phospholipids, the major component of cell membranes in fish larvae, have dual roles in the diet. They have been shown to improve cohesiveness of particles and to reduce leaching of the soluble nutrients in the microdiet (Cahu and Zambonino Infante, 2001). Since marine fish eggs and yolk sac larvae contain 10% phospholipids, it is hypothesized that larval diets must include these nutrients because larvae cannot synthesize them fast enough (Sargent et al., 1999b). Phospholipids can be provided from soybean lecithin or from fish meal. The optimal level of HUFAs is proposed at 3%, based on studies using live organisms and microdiets (Watanabe and Kiron, 1994; Zambonino Infante and Cahu, 1999).

In microdiets for larvae, the protein content is usually adjusted to about 50%, based on the results of various studies on juveniles (Brinkmeyer and Holt, 1998; Peres et al., 1997; Salhi et al., 1994). The main sources of protein have been fish meal and squid meal (Cahu and Zambonino Infante, 2001). Protein hydrolysates have attracted interest as additives in early larval nutritional studies. Fish protein hydrolysate added to the diet, for instance, has effectively increased growth in salmon fry (Berge and Storebakken, 1996) and carp larvae (Carvalho et al., 1997).

The effectiveness of protein hydrolysates, however, is limited to larvae, because peptidase activity decreases in juveniles, which are less dependent upon this mode of digestion (Cahu and Zambonino Infante, 2001). Kolkovski (2001) suggested that free amino acids or hydrolysates can only partially (30%) replace protein sources in the diets, because the short peptides and free amino acids flow too fast through the gut wall. Nevertheless, protein hydrolysates have high nutritive value for larvae, and may allow earlier weaning of fish larvae to inert feeds (Cahu and Zambonino Infante, 2001).

Although fish larvae have sufficient digestive enzymes for live food organisms at first feeding (Cahu and Zambonino Infante, 2001), the levels may not be sufficient to fully digest microparticle diets (Chen et al., 2006b; Kolkovski, 2001). Dietary enzyme supplements have been added to microdiets, with variable results, indicating that their effectiveness depends on the age and food habits of the particular species, the development of the brush border enzymes in the intestine, and the type of dietary enzymes used (Kolkovski, 2001).

2.3. Food Digestion

The development of digestive enzymes in marine fish larvae has been considered to be genetically programmed (Cahu and Zambonino Infante, 2001) and the development of the digestive processes of fish larvae goes concomitantly with increasing morphological complexity (De Silva and Anderson, 1995). Majority of the epithelial cells of the digestive tract are absorptive in swim-up larvae indicates that the secretion of digestive enzymes is likely to be limited early in development (Chen et al. 2006).

Baragi and Lovell (1986) described the activities of a number of digestive enzymes in the gut of striped bass *Morone saxatilis.* Four enzymes, trypsin, chymotrypsin, carboxypeptidase and pepsin, are for protein-degrading, and their activities can be correlated with developmental changes in the gastrointestinal tract associated with the change from intracellular to intragastrointestinal digestion of protein. Trypsin increases through larval development to age 12 days, but decreases on day 16. However, by day 25 it has increased again. Trypsin is most active at neutral pH. Pepsin, active at low pH and present only in the stomach, increases from day 16. This increase is probably due to the development of the stomach and the glands that secrete the enzyme and acid to produce the low-pH environment. Chymotrypsin shows no great variation in activity over development and carboxypeptidase shows a decline at day 16, but like trypsin carboxypeptidase activities increase at day 25. Amylase activity also remains largely unchanged in striped bass through larval development.

Most digestive enzymes are present before the onset of the exogenous feeding (Cara et al., 2003; Ribeiro et al., 1999; Zambonino Infante and Cahu, 1994) and the development of these enzymes determines the timing to digest exogenous food (Gawlicka et al., 2000). In a recent study, Chen et al. (2006a) found that the specific activities of trypsin, amylase, lipase and alkaline phosphatase in 2-day-old yellowtail king fish are significantly stronger than those in the newly hatched larvae. In white sea bream, the activity of digestive enzymes such as acid and alkaline proteases, amylase, lipase, and acid and alkaline phosphatases is also detected at mouth opening (Cara et al., 2003). Similarly, in Eurasian perch larvae, the specific activities of trypsin and amylase were detected soon after hatching and increased substantially in the first few days after exogenous feeding (Cuvier-Peres and Kestemont, 2001). In Pacific threadfin the amylase activity was not measurable at hatching, but its activity was detected prior to first feeding (Kim et al., 2001). These studies suggest that digestive enzymes are present at a low level upon hatching and their activities increase to a high level prior to the onset of exogenous feeding. Chen et al. (2006a) suggested that the onset of digestive enzymes is triggered by internal mechanisms, rather than by dietary intake. The developmental pattern of amylase, lipase, and alkaline phosphatase is closely related to the structural development of the digestive system identified yellowtail king fish.

Chen et al (2006a) reported that the increase in specific activities of these digestive enzymes took place before exogenous feeding. The fluctuations in enzyme activities covered the period of morphological differentiation in the digestive tract and associated glands. After the formation of gastric glands, the digestive system became functional and the specific activities of these digestive enzymes remained constant, while the total enzyme activities increased gradually with age. In walleye pollock larvae, high specific activities of the trypsin and amylase were coincident with the transition period from endogenous to exogenous feeding (Oozeki and Bailey, 1995). Similarly, Zambonino Infante and Cahu (1994) suggested that the increase of the digestive enzyme activities in the early larval stage was genetically programmed but was not induced by the external food. All these studies clearly show the concurrence between the digestive enzyme activities and the morphological development of the digestive system in larval fish. The understanding of the digestive mechanisms is fundamental for diet formulation and provision to fish larvae in he early developmental stage.

Most fish larvae have sufficient levels of digestive enzymes to cope with live food organisms at first feeding, but there may not be enough for digestion of formulated feed. Formulated diets contain proteins and other ingredients that are difficult for larvae to digest, since they are much harder to break down than live organisms (Lindner et al., 1995). A practical method was first used by Dabrowski and Glogowski (1977) to overcome this problem. In their study, they supplemented a diet for common carp juveniles with bovine trypsin and reported an increase in proteolytic activity correlated with the bovine trypsin proportion. Similarly, Kolkovski et al. (1993) added porcine pancreatic extract to the diet for gilthead seabream and found that supplementation of 0.05% (dry weight) pancreatin enhanced the food assimilation by 30% and larval growth by 200%, compared with the larvae fed on the control feed without enzyme additives. However, other studies assessed the effect of dietary enzyme supplementation in formulated feed to different fish species are not conclusive (Kolkovski, 2001). It can be concluded that the effect of dietary digestive enzymes depends on fish age, species and the type of the dietary enzymes as well as food habits.

3. Larval Fish Diets

Currently, the culture of most marine fish larvae still depends on live food during the larval stages. To ensure a reliable, a series of culture including algae and various zooplankton species has to be established. Live feed supply can be continued until metamorphosis when the fish are then weaned onto dry formulated diets. During larval rearing, the reliable supply of some live food species could be a challenge. For example, in mid 1990s, the production of *Artemia* cysts in the Great Salt Lake, UT, USA, was disrupted by 'El-Niño' climatic condition changes. Total production of *Artemia* cysts was only sufficient for no more than 25% of the world demand. This is one of the most dramatic incentives to develop alternative live food and starter diets in aquaculture industry.

3.1. Live Food

3.1.1. Rotifers

Rotifers, which have been widely used in marine finfish hatcheries throughout the world, are commonly referred to as one of three types: L-type, S-type, or SS-type, based on their size initially *B. plicatilis* (130–340 μm in lorica length) is referred as an L-type rotifer in cold water (18–25°C) and *B. rotundiformis* (100–210 μm) as an S-type in warmwater (28–35°C, Figure 2). The super small rotifers referred to as SS-type rotifers (90–150 μm) are found in subtropical and tropical waters (Hagiwara et al., 1995).

Both species are common in brackish waters, but show strong tolerance to high salinity. In Japan, these rotifers are previously known as harmful organisms because they consume a large amount of oxygen and cause a rapid change of water color in eel culture ponds. Before 1980, the major developments of rotifer mass culture include: (1) introduction of *Nannochloropsis oculata* and baker's yeast as rotifer diets; and (2) establishment of an enrichment protocol before feeding fish larvae (Watanabe et al., 1978). Since 1990, the major advances in rotifer culture are (1) development of condensed freshwater *Chlorella* for rotifer food (Maruyama and Hirayama, 1993), (2) the development of high density rotifer mass culture technology using condensed phytoplankton products and (3) techniques for rotifer preservation (Yoshimura et al., 1997).

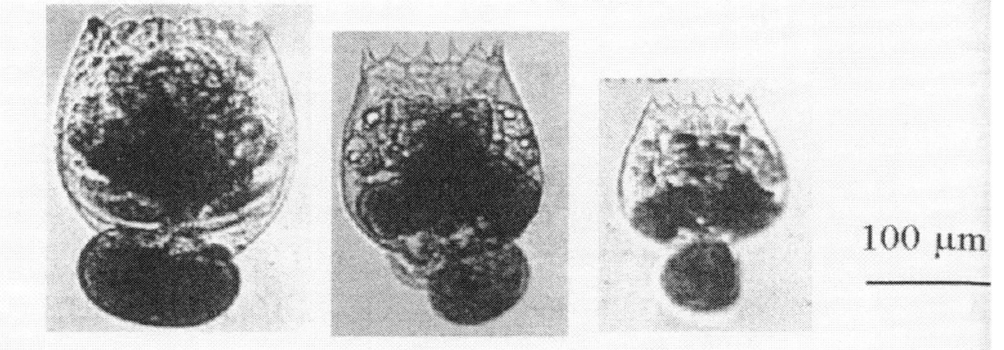

Figure 2. Euryhaline *Brachionus* species used in aquaculture as live food: *B. plicatilis* (left), *B. rotundiformis* S-type (middle) and *B. rotundiformis* SS-type (right, from Hagiwara et al. 2001).

Chlorella-Based Rotifer Culture

A method of condensed phytoplankton culture has been developed for feeding mass-cultured rotifers to reduce the space need for growing microalgae in the hatchery in Japan. Among several freshwater algal species, freshwater *Chlorella* is the most popular for the condense culture. It is available in an 18-liter volume with the density of 20 billion cells/ml at a cost of US$140–150. During the manufacturing process, *Chlorella* cells receive enrichment with vitamin B, which is essential for rotifers (Hirayama and Funamoto, 1983; Scott, 1981). The feature of *Chlorella* culture is that it can be grown in an organic carbon source, such as glucose or acetic acid, which significantly reduces the costs for lighting (Hirayama et al., 1989). Another feature of this culture is the reduction of *Chlorella* doubling time. In comparison, the doubling time of *N. oculata* is 16-20 h with light, but the doubling time for *Chlorella* is reduced to 3 h with an organic carbon source without light. More importantly, *n*-3 HUFA has been successfully incorporated in *Chlorella* cells (Hayashi et al., 1993). The *Chlorella* product can be used for mass culture of rotifers, and the harvested rotifers can be directly fed to fish larvae, which eliminates the procedure for enrichment culture. In addition, fortified DHA and vitamin C also enhanced the growth rate of rotifer populations (Hayashi et al., 2001). By using such an algal product, rotifer culture technique has also been significantly improved.

High-Density Culture of Rotifers

The high-density rotifer culture system can significantly reduce the spatial need at the hatchery facility for live food production. Conventionally, a 55-m^3 mass-culture tank is needed to produce 10 billion *B. rotundiformis* (180 ind/ml) and the culture of *N. oculata* requires a 200 m^3 tank volume to feed these rotifers. In comparison, by using the condensed *Chlorella* product, the density of *B. rotundiformis* can be increased to 5,000-8,000 ind/ml, and only one 4 m^3 tank can produce 10 billion rotifers. Under aeration and pH control, it is possible to increase the rotifer density up to 20,000 ind/ml, and to decrease the culture volume to 1 m^3 (Yoshimura et al., 1996). At such as density, the removal of foam and solid substances is essential (Yoshimura et al., 1997). The Fukuoka Mariculture operates two 1-m^3 tank rotifer tanks with a 2-day batch culture, producing about 1×10^{10} rotifers daily.

Rotifers can also be produced at high densities in a continuous culture system (Fu et al., 1997). With this system, the density of *B. rotundiformis* ranged from 3,000 to 6,000 ind/ml, and *B. plicatilis* from 1,100 to 2,200 ind/ml. The continuous culture system is more advantageous for maintaining the environment of rotifer culture (James et al., 1989), and it seems to have a great potential to adopt this style for rotifer mass-culture.

Rotifer Storage

For a short-term storage, rotifers can be preserved for periods varying from a few days to 3-4 weeks at temperatures ranging between -2 and +8°C, while resting eggs can be used for long-term preservation (Lubzens et al., 1990). For preserving *B. plicatilis* and *B. rotundiformis* at 4°C, attentions should be paid to culture medium exchange, and temperature acclimation at 15°C and strain selection (Lubzens et al., 1990). For *B. plicatilis*, the percent survival after 30 days of preservation at 4°C was up to 60%. The S- and SS-type *B. rotundiformis* are weak under 4°C, and can not survive for more than 15 days. In comparison, on day 5 at 4°C, the temperate strains showed 75% survival, but tropical SS type rotifers exhibited a very low level.

For long-term storage of rotifers, the significant factors for mass production of resting eggs include strain selection and environment regulation (Hagiwara et al., 1994; Hagiwara et al., 1997; Lubzens et al., 1989). In 10 days, 10 billion resting eggs were produced in 50-m^3 tanks (Hagiwara et al., 1993) and the resting eggs can be stored in cans (Balompapueng et al., 1997).

Resting eggs are the product of rotifer sexual reproduction, which is initiated by the appearance of mictic females (mixis). The mixis is induced by accumulated substances in the rotifer culture water (Hino and Hirano, 1976). By conducting axenic rotifer cultures, Hagiwara et al. (1994) showed that mixis is induced in the presence of certain bacteria and with the addition of a water-soluble extract of rotifers. It appears that an excess amount of rotifer extract inhibits mixis induction. A semi-continuous culture system was used to maintain the concentration of mixis-inducing substance, resulting in efficient resting egg production (Balompapueng et al., 1997).

Recent progress in resting egg production has been achieved by temperature manipulation. For *B. plicatilis*, the temperature was first adjusted to a low level to enhance sexual reproduction, then adjusted to high to promote fertilized rotifers to produce resting eggs (Kogane et al., 1997). For *B. rotundiformis*, the temperature was first set at a high level, but decreased thereafter to avoid the decline in water quality. For *B. plicatilis*, resting egg formation was slow at 15°C, and very few resting eggs were produced after 7 days. However, by increasing the temperature to 25°C on day 5, 15-20 million resting eggs were harvested on day 7. For *B. rotundiformis*, the number of resting eggs produced was also improved by decreasing the temperature as the culture progressed.

3.1.2. Copepods

A workshop was conducted at the Oceanic Institute, Hawaii with the focus of copepods culture as a live food for aquaculture (Lee et al., 2005). Like rotifers and brine shrimp naupplii, copepods could be additional desirable food for fish larvae. Especially, nauplii of some copepod species can be used to raise fish larvae that require starter food smaller than rotifers. At the workshop, a number of promising candidate genera was identified for culture. Currently, mass production technology is not well established. Among the calanoid copepods, the genera *Acartia, Pseudodiaptomus, Sinocalanus, Eurytemora, Gladioferens, Parvocalanus, Bestiolina, Temora, Centropages,* and *Labidocera* were proposed for first feeding. Nauplii of some *Acartia* spp. are as small as 100 μm in length and 50-60 μm in width, making them suitable for first feeding as well. Harpacticoid copepod species tend to have high production rates, are generally not cannibalistic, and are found all over the world. In addition, it is likely that many genera among this group can be raised on formulated feeds. Two species that can be mass cultured at high densities are *Tisbe holothuriae* and *Nitokra lacustris.* Among the cyclopoid copepods, the genera *Oithona* and *Dioithona* were listed as candidates for first-feeding larvae.

The use of copepods as live food for larval fish rearing is largely restricted to extensive systems, where wild zooplankton are collected, using different filters to obtain specific size-ranges, and fed directly or allowed proliferate prior to being fed as food for fish larvae (Støttrup and Norsker, 1997). Attempts have been made to culture several species such as

Acartia spp., *Tigriopus juponicus, Oithona* spp., *Paracalanus* spp., and *Euryternora* spp., for feeding fish larvae at stages in intensive systems (Støttrup et al., 1986).

Cultivation of copepods has rarely extended beyond experimental scale (Lee et al., 2005). For the aquaculture industry, mass rearing of copepods in intensive systems had little appeal in the light of the ready availability of *Artemia eggs.* As long as these are easily harvested from vast natural resources, a direct economic comparison, based on costs per biomass unit, will not favor copepod cultivation. However, the decline of *Artemia* production in late 1990s makes the study of copepods as larval fish food more attractive. There are three main groups of copepods, i.e., calanoids, cyclopoids and harpacticoids. Like rotifers, calanoid copepods are planktonic and relatively easy to operate and to scale up. Calanoids, however, can only be cultivated at very low densities. Harpacticoids, on the other hand, may be produced in volumetrically much denser cultures, but being benthic, their proliferation depends on the area of a solid substratum. Such an environment is not homogenous and hence is considerably more difficult to manage and scale up. Cylopoids are pelagic, but most species are carnivorous and are not easy to reach high biomass.

Among the calanoids, *A. tonsa* has been used in mass cultivation for feeding purposes and has been cultured successfully for many generations in the laboratory (Støttrup et al., 1986), but the low culture densities make it an inappropriate species for commercial intensive mass cultivation. Mass cultures of *A. tsuensis* in outdoor tanks have been used in extensive rearing systems for marine fish (Ohno et al., 1990). Other harpacticoids like *Tigriopus* sp. have been part of commercial rearing practices (Fukusho, 1980). As a model species, Støttrup and Norsker (1997) used harpacticoid *T. holothuriae* to illustrate how batch culture and continuous cultivation could be achieved. In assessing the feeding status of the copepods, they examined the ammonia excretion to control the diet level.

Batch Culture System

Flat trays, 40 × 60 cm, were filled with about 3 1 of filtered (0.22 μm) seawater and 40,000 adult *Tisbe* copepods consisting predominantly of oviger females. Each tray was filtered daily through two filters, one of mesh size of 220 μm retaining almost only oviger females and one of 40 μm to maintain nauplii hatched in the interim period. The adults collected on the filter were replaced in the rearing trays with fresh seawater and fresh algae were added. To each tray about 400 ml *Rhodomonas* was added at concentrations of one million cells/ml. Temperature was around 18°C.

Cultivation in daily harvested trays proved reliable and feasible. On average, 2×10^6 nauplii could be repeatedly produced from 4-5 trays. Handling time of the cultures was estimated to be l-l.5 h per day. Scaling up the production using the tray system probably would increase the labor proportionately. Larger shallow trays (1 × 1 m) do not offer immediate benefits as they are more difficult to manipulate and less productive, possibly due to an uneven food distribution.

Continuous Culture System

The continuous cultivation system consisted of a closed copepod bioreactor (Figure 3) connected to a continuous microalgal cultivation system. The production of *Rhodomonas* was around 20 1 daily corresponding to a biomass of 2 g dry matter, and was collected in a closed, agitated collector (C), which was emptied twice daily into the continuous copepod cultivation unit.

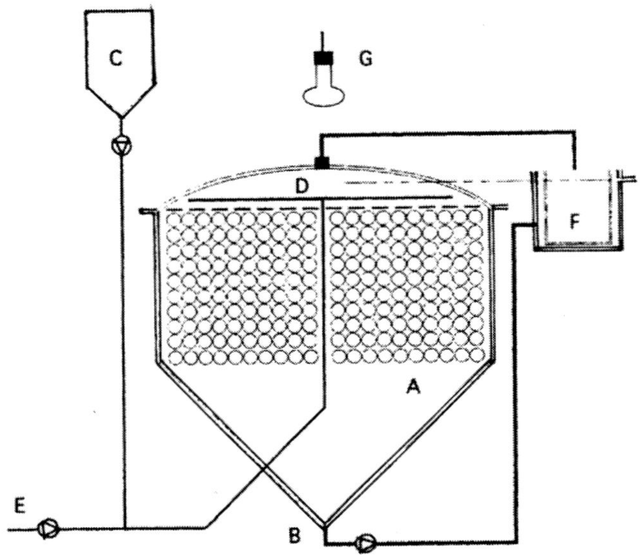

Figure 3. A schematic diagram of the continuous copepod cultivation unit (A), bottom outlet (B), agitated algal collector (C), water distributor (D), water inlet (E), harvest-filter (F) and light bulb (G, from Støttrup and Norsker 1997).

The copepod cultivation unit (A) consisted of a 150 l PVC tank with a conical bottom and semi-spherical translucent top. The vertical section of the unit was filled with 4.5 cm polypropylene balls suppressed by a grid below the water-distributor (D), which had four arms with holes at regular intervals injecting water in horizontal-tangential rays. The distributor was also used for feeding and dimensioned to administer the suspended feed in a thin layer that subsequently would be drawn into the ball bed of the copepod bioreactor. Solenoid valves controlled the addition of algae and harvest of copepods. A remote computer system controlled the operation of the valves on a predetermined time schedule and was also used to control the microalgal cultivation process. The algal production unit was set to a dilution rate of 0.33/day.

Filtered sea water pumped through the water distributor left the unit through the bottom outlet (B) and flowed through a second 40 μm filter, placed after the harvest-filter (F). At harvest, the bottom outlet (B) was closed and at the same time a light bulb (G), placed directly over the unit, was switched on. In contrast to adults and copepodites which are generally negatively phototactic, *Tisbe* nauplii remain in the free water volume in the semi-hemisphere and collect in the upper water layers. With the bottom outlet closed, the water was forced through the top outlet and through the harvest-filter (F). Total numbers of copepods (nauplii and some copepodites) were retained on the filter. After harvest, normal flow was regained and the copepods were fed. The system including the algal and copepod unit was kept at ambient laboratory temperature conditions at approximately 18°C. The continuous system produced on average 2.3×10^5 nauplii and 2.7×10^5 copepodites daily when fed on the algal supply from the algal bioreactor.

3.1.3. Cladocerans

Cladocera, as a general rule, appear to be restricted to freshwater environments of salinity <1 ppt (Hart et al., 1991). Among cladocerans, a great majority is exclusive freshwater species (Pennak, 1989), and only few species such as *Moina hutchinsoni, M. salina* and *Daphniopsis pusilla* live in saline lakes (He et al., 1989; Hebert and Wilson, 2000; Martinez-Jeronimo et al., 2004). *Moina mongolica* (Daday) is one of few halophilic cladocerans in the genus *Moina* found in inland saline water (Goulden, 1968). In China, *M. mongolica* has been found in brackish lakes scattered in the northwestern arid region and can easily adapt to brackish and hypersaline waters (He et al., 1989). *Moina mongolica* is 0.6–0.7 mm long and 0.5 mm wide, and the adult is 1.0–1.4 mm long and 0.8 mm wide (Figures 4 and 5). Its size is about 5–10 times larger than *B. plicatilis* and 1.2–3.5 times larger than *Artemia* nauplii (He et al., 1989). The size dimension of *M. mongolica* fits the mouth opening of most fish postlarvae (He et al., 2001). If the rotifer is a suitable starter food for marine fish larvae, then juvenile *M. mongolica* may serve as live food for fish postlarvae after rotifer, and adult *M. mongolica* can be a transitional food for fish fingerlings between live food and formulated feed.

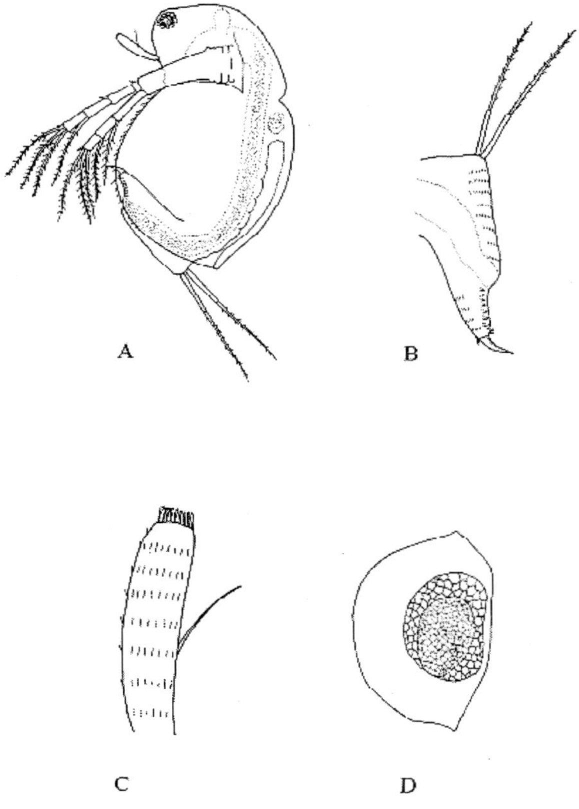

Figure 4. Illustration of a saline water cladoceran *Moina mongolica* (female): whole body (A), postabdomen (B), first antenna (C), and ephippium (D, from He et al. 2001).

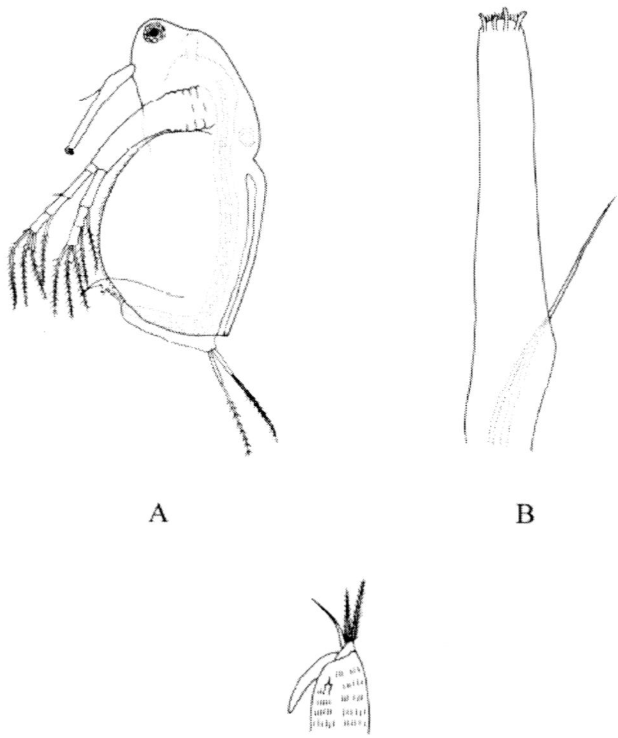

Figure 5. Illustration of a saline water cladoceran *Moina monglica* (male): whole body (A), first antenna (B) and last segment of the first leg (C, from He et al. 2001).

In most marine fish hatcheries, the rotifer *Brachionus plicatilis* has been widely accepted as a starter food for larval fish (Fulks and Main, 1991). As larval fish grow, the size of rotifers becomes too small for fish to ingest, and subsequently, rotifers have been replaced by brine shrimp *Artemia*. Although it has been used for many years, the resource crisis of brine shrimp in Great Salt Lake alarmed researchers to seek alternative food to substitute *Artemia* in larval fish rearing. Development of the culture technology in marine copepods for larval fish has been making progress but because the long generation time in most copepod species makes it difficult to produce mass copepods in intensive culture systems (Schipp et al., 1999; Støttrup and Norsker, 1997). Between cladoceran and copepods, most fish larvae prefer the former because the jerky movement of cladocerans makes them more visible to fish while copepods swim too fast for fish larvae to catch (Mayer and Wahl, 1997). In freshwater aquaculture, cladocerans such as *Daphnia* and *Moina* have been widely used as larval fish food in pond culture (Qin and Culver, 1995, 1996). In contrast, the use saline cladoceran species as food for fish larvae has rarely been explored partially because only few species of Cladocera are available in the world ocean (Nybakken and Bertness, 2005).

Moina mongolica could be potential species for marine finfish larviculture due to its fast population growth. Tong et al. (1988) reported that content of 20:5*n*3 (eicosapentaenoic acid) in *M. mongolica* was 6-7 times higher than it in *Artemia* and *B. plicatilis*. He *et al.* (1997) successfully substituted *Artemia* nauplii with *Moina mongolica* in larviculture of red seabream *Pagrosomus major* sea perch *Lateolabrax japonicus* and achieved satisfactory growth and survival in marine fish hatcheries. Recently, He et al. (2001) summarized four methods (static batch method, static continuous method, aerated static method and flow-through method) for the mass production of *M. mongolica*. In each culture method, salinity ranges from 30‰ to 32‰ and temperature varies from 19.2°C to 26.0°C in the culture.

Static Batch Method

In this method, *M. mongolica* are harvested at the end of the culture. *Nannochloropsis oculata* was used as live food and inoculated into 500 l fibreglass tanks up to 1/5–1/4 of the tank volume at $2–5\times10^6$ cells/ml. *Moina mongolica* were then introduced into each tank at 250 ind/l. During the culture, the density of *N. oculata* was maintained at $>2\times10^6$/ml. The density of *M. mongolica* in each tank reached 540–3,500 ind/1 during the culture period. The biomass yielded 9.8–47.4 g /m^3/d after 28 days. He et al. (1998) found that food supply could be a limiting factor in this type of culture due to the high filtering rate of *M. mongolica* (3.0 ml/d/ind at 25°C. It is difficult to maintain the *M. mongolica* density >1,000 ind/l by using algae as a sole food supply. Harvesting is recommended when the animal density reaches 500–1,000 ind/l.

Static Continuous Method

Animals are harvested periodically after the biomass reaches a target level in this culture. *Nannochloropsis oculata* was used as live food for *M. mongolica* and inoculated into 500 l tanks in a way similar to the static batch method, but beer yeast was added as supplemental feed together with *N. oculata*. Dry yeast powders were soaked in water for 1–2 h, and 25 mg yeast cells/l were added to the culture twice a day. *Moina mongolica* were harvested when the population density reached 500 /l. At each time of harvesting, 10–50% of the tank volume was drained and collected with a fine mesh filter. The tank was then topped up with *N. oculata* ($2–5\times10^6$ cells ml/1) after each harvest and the remaining *M. mongolica* population in the tank was served as seed for further production. When the density of *M. mongolica* was in the range of 1,225–5,400/l, the wet weight production could reach 14.3–45.8 g /m^3/d. Due to high food consumption by *M. mongolica*, He et al. (1998) suggested that the addition of supplemental food such as yeast is a practical method to increase the unit production, but dissolved oxygen can be a limiting factor for further increase of production. The oxygen consumption rate of *M. mongolica* is 0.008 mg ind/1/d and the lowest oxygen that the animal can tolerate is 0.14–0.93 mg /l. In static culture, the minimum level of dissolved oxygen in the culture vessels should be >0.5 mg/l. When the animal density reaches 1,000/l, the daily oxygen consumption can reach 8 mg/l. Therefore, when the density of *M. mongolica* is more than 1,000 ind/l, aeration should be supplied. The recommended level of dissolved oxygen for *M. mongolica* culture is 6–8 mg/l.

Aerated Static Method

This method is similar to the static continuous culture, but aeration is provided when dissolved oxygen drops to a threshold level of 6 mg/l. *Nannochloropsis oculata* and *M. mongolica* were inoculated into 500 l tanks in the way similar to the static continuous culture, but the culture was daily aerated 2–4 times when dissolved oxygen dropped <6 mg/l. Each aeration period lasted 1–3 h. The yeast cells were added 2 times a day and 40 mg/l was added each time. The density of *M. mongolica* reached 4,735–10,520/l in the tanks and daily biomass production reached 43.3–103.3 g/m^3. The animals in tanks were harvested between 5 and 24 d. Dissolved oxygen was maintained 5–8 mg/l in all culture tanks.

Flow-Through Method

This method is an innovation of the static continuous culture, but algae are continuously supplied to *M. mongolica*. At the beginning, *N. oculata* and *M. mongolica* were inoculated into 100 l tanks. The food was supplied through a 30 l algal tank in which *N. oculata* (2–5$\times 10^6$ cells/ml) was cultured. The algal tank was placed at an elevated position so that algae could be supplied to the *Moina* tank by gravity. In the *Moina* tank, the water exchange rate was 0.9–1.5 times/d. In this method, the density of *M. mongolica* reached 6,723–7,890/l in the tank and daily biomass production reached 33.5–88.6 g/m^3. The production level in the flow-thought system is higher than that in both static batch culture and static continuous culture, but is lower than that in the aerated static culture.

3.1.4. Artemia

Since 1970s, the development of commercial marine fish culture has gradually relied on *Artemia* cysts as live feeds for finfish in early stages (Sorgeloos et al., 2001). During the last three decades, the Great Salt Lake has been the premier supplier of *Artemia* cysts to the world aquaculture market and the subject of numerous speculations regarding its capacity to sustain a growing aquaculture industry (Lavens and Sorgeloos, 2000). Due to an intense El Niño in 1982–1984, the heavy snowfall and a huge amount of melting water from the bordering mountains resulted in a considerable drop in salinity. The 1999–2000 harvest was a record low. Total GSL harvest will not even satisfy 20% of the global demand and previous year's stocks have been sold. Since no artificial feed formulation is yet available to completely substitute for *Artemia*, feeding live prey to young fish larvae still remains essential in commercial hatchery operations. The nutritional quality of commercially available *Artemia* strains being relatively poor in eicosapentaenoic acid (EPA, 20:5*n*-3) and especially docosahexaenoic acid (DHA, 22:6*n*-3), it is essential and common practice to enrich these live prey with emulsions of marine oils.

Biology and Living Environments

Artemia cysts are harvested from various locations in the world. The criteria for cyst quality are good hatching synchrony (less than 7 h between hatching of first and last nauplii) and high hatching efficiency (more than 200,000 nauplii per gram product). Several factors are critical for the hatching conditions. These include cyst decapsulation prior to incubation, and hatching under the following optimal conditions: constant temperature of 25–28 °C, 15–35 ppt salinity, minimum pH of 8.0, near saturated oxygen levels, maximum cyst densities of 2 g/l, and strong illumination of 2,000 lx (Lavens and Sorgeloos, 1996).

After hatching, *Artemia* nauplii do not feed in their first stage of development (Instar I), but consume their own energy reserves (Lavens and Sorgeloos, 2000). At high incubation water, freshly hatched *Artemia* nauplii develop into the second larval stage (Instar II metanauplii) within 6–8 h. It is important to use first-instar nauplii for feeding, rather than starved second-instar metanauplii, which are transparent and less visible. Instar II metanauplii are about 50% larger in length and swim faster than first instars. As a result, they are less acceptable as prey. Furthermore, they contain lower amounts of free amino acids, so they are less digestible and their lower individual dry weights and energy content reduce the energy uptake by the predator per unit of hunting effort (Léger et al., 1986).

The unique feature of feeding biology makes it possible to manipulate the nutritional value of *Artemia*. About 8 h following hatching, brine shrimp nauplii will molt into the second instar stage and become non-selective particle feeders. Simple methods have been developed to pack different kinds of products into the *Artemia* prior to feeding to predator larvae. This method of nutrient packing is called *Artemia* enrichment or boosting, which is widely applied in marine fish and crustacean hatcheries for enhancing the nutritional value of *Artemia* with essential fatty acids.

Nutritional Quality

(1) Fatty acids

In the late 1960s and early 1970s, several authors reported low success in larviculture with marine fish and crustacean species when using *Artemia* as food (Persoone and Sorgeloos, 1980). Léger et al. (1986) suggested that the main factor affecting the nutritional value of *Artemia* was the content of the highly unsaturated fatty acid (HUFA) eicosapentaenoic acid, 20:5*n*-3(EPA). In the 1980s, most attention was dedicated to the presence of EPA in *Artemia* as a guarantee for successful production of marine fish larvae (Watanabe et al., 1983). In the late 1980s and early 1990s, more attention was paid to the level of DHA; good survival appeared to be correlated with EPA, but DHA improved larval quality and growth (Kraul, 1993). The importance of DHA, more particularly the requirement for high DHA/EPA ratios in promoting growth, stress resistance, and pigmentation, was revealed (Mourente et al., 1993; Reitan et al., 1994). While in the past, satisfactory results were obtained with DHA/EPA ratios of less than 1, the emphasis now is on attaining levels of 2 and higher (Sorgeloos et al., 2001).

The enrichment products include selected microalgae, micro-encapsulated products, yeast, emulsified preparations, self-emulsifying concentrates, and micro-particulate products, either singly or in various combinations (Léger et al., 1986). The highest enrichment levels are obtained from emulsified concentrates through the following procedure. Freshly hatched nauplii are transferred to the enrichment tank at a density of 100–300 nauplii/ml for enrichment periods of 24 or 48 h. The enrichment medium consists of hypochlorite-disinfected and neutralized seawater maintained at 25°C. The enrichment emulsion is added in consecutive doses of 0.3 g/l every 12 h. Strong aeration using air stones or pure oxygen is required to maintain dissolved oxygen levels above 4 ppm. Enriched nauplii are harvested after 24 or 48 h, thoroughly rinsed and stored at temperatures below 10°C to assure that HUFAs are not metabolized during storage. Enrichment levels of 50–60 mg/g DW *n*-3 HUFAs are obtained after 24-h enrichment with the emulsified concentrates. Nauplii should be transferred or exposed to the enrichment medium as soon as possible before first feeding, so they begin feeding immediately after the opening of the alimentary tract (instar II stage). As a result, the increase of nauplius size during enrichment can be minimized, i.e., after 24-h enrichment, *Artemia* nauplii will reach about 660 μm, and after 48-h enrichment, about 790 μm (Lavens and Sorgeloos, 1996).

Besides DHA and EPA, Koven et al. (2003) showed that arachidonic acid ARA (20:4*n*-6) may play a significant role. ARA may improve larval growth and pigmentation in several marine fish species since it provides precursors for eicosanoid production (Castell et al., 1994; Estevez et al., 1997). The requirement of ARA in fish, however, seems to depend on the fish species and larval development, and needs to be dosed with extreme care since it may act in a different way depending on the DHA concentration (Castell et al., 1994; Koven et al., 2003).

(2) Vitamins

The effectiveness of *Artemia* nauplii as a dietary carrier system could be tested for various other nutritional components administered via an emulsion, water-soluble compounds via liposomes (Hontoria et al., 1994), and microcapsule delivery (Sakamoto et al., 1982). This can be applied to study the role of minor nutrients such as vitamins in larval fish nutrition.

Vitamin C, or ascorbic acid (AA), is generally considered to be an essential dietary component for the various stages of aquaculture organisms (Merchie et al., 1997). Several biological (e.g., skeletal development, growth, survival) and physiological functions (e.g., resistance to toxicants and stress, immunoactivity) are enhanced in larvae from supplemental dietary ascorbate (Dabrowski, 1992). Tests have been conducted to incorporate extra AA into *Artemia* nauplii in a stable and bioavailable form. When vitamin c-enriched *Artemia* were fed to turbot larvae, no differences in growth or overall survival could be detected with the non-enriched live food (Merchie et al., 1995). The larvae of the high AA treatment, however, showed a better pigmentation rate compared to the control group. Evaluation of the physiological condition applying a salinity stress test revealed an improvement by feeding extra AA. Cumulative mortalities after challenge with *Vibrio anguillarum* reached 50% for the control versus 40% for the AA-supplemented fish, with a slower onset of mortality for the AA-fed fish (Merchie et al., 1996).

High levels of vitamin E, α-tocopherol, can be bioaccumulated and maintained in *Artemia* nauplii, making this live food delivery system useful for studying dietary requirements as well as the antioxydative effects of vitamin E (Huo et al., 1996). Vitamin A levels in *Artemia* nauplii could be raised from 1.3 to 1283 IU/g DW over an 18-h period through the addition of vitamin A palmitate to an egg-yolk-based emulsion (Dedi et al., 1995). However, for each nutrient, the usefulness of the *Artemia* bioencapsulation method remains to be verified.

3.2. Formulated Feed

The high cost of live food production in finfish hatcheries could be minimized by cheaper production of live food organisms and earlier weaning onto formulated feeds. However, Up to the present, it has been essential to provide live feed for the first feeding of marine finfish species. Limited success has been achieved when inert feed is provided as first food for species such as red drum (Holt, 1993), sea bass (Cahu et al., 1998), red sea bream (Takeuchi, 2001), and others (Southgate and Partridge, 1998). Complete or significant partial replacement of live food with a suitable artificial or inert feed is the ultimate goal of research in this field. The majority of artificial diets for finfish larvae can be categorised as either microencapsulated diets or microbound diets (Southgate and Partride, 1998).

Microencapsulated Diets

Microencapsulated diets (MED) are composed of dietary materials encapsulated within a membrane or capsule wall (Figure 2). MED may have limited use for presenting artificial diets to marine finfish larvae due to their poor digestibility. Yufera et al. (1995) showed that gilthead seabream larvae *Sparus aurata* can be transferred directly from live to microencapsulated food without appreciable variation in the incidence of feeding or in the rate of food consumption from first feeding onwards. Walford et al. (1991) attempted to rear newly hatched *L. calcarifer* with a protein-walled microcapsule; however, when fed alone these microcapsules resulted in 100% larval mortality by day 10. Because MED have a capsule wall or membrane separating the bulk of the diet from the water in which the MED is suspended, leaching of dietary components is limited. This helps maintain the integrity of the MED until eaten and assists in maintaining high water quality. However, some nutrient leaching from artificial food particles is considered to promote diet attractiveness (Lopez-Alvarado et al., 1994).

Microbound Diets

Microbound diets (MBD) consist of dietary components held within a gelatinised matrix or binder (Figure 2) (Lopez-Alvarado et al., 1994) and differ from MED in that they lack a capsule wall. It has been suggested that the lack of a capsule wall enhances digestion of the artificial food particles and, because of this, MBD may be more suitable than MED for presenting artificial diets to marine fish larvae (Southgate and Lee, 1993).

Materials for MBD vary considerably in their source, properties and nutritional value. Commonly used materials include polysaccharides from seaweeds such as agar, carrageenan and alginate and proteins such as zein and gelatin (Hashim and Saat, 1992).

Status of Artificial Diets for Marine Finfish Larvae

Southgate and Partridge (1998) summarised the use of experimental artificial diets for first-feeding marine fish larvae and found that total replacement of live prey is still not a viable option for most marine fish as it results in substantially lower growth rates and survival and often leads to a much higher incidence of deformities.

Given the difficulties in feeding artificial diets to first-feeding larvae, partial replacement of live foods by artificial feeds is possible and can result in considerable cost savings in live feed production (Jones et al., 1993). Results obtained so far with fish larvae fed MBD have been promising. Kanazawa et al. (1989) reported that artificial feeds could substitute live food for newly hatched red seabrearn *P. major*. In a similar study, growth of day 8 *S. aurata* larvae fed a ration in which 80% of rotifers were replaced with MBD was similar to the treatment receiving 100% live food (Tandler and Kolkovski, 1992).

Weaning at an early stage would be an effective way to reduce the cost of live food (Lavens and Sorgeloos, 2000). Chen et al (2007) reported weaning yellowtail kingfish can start as early as 15-18 day after hatching when the fundic stomach is formed. A technique for weaning *D. labrax* larvae 15 days earlier has enabled savings in *Artemia* production of up to 80% (Person-Le Ruyet, 1989). In fat snook *Centropomus parallelus*, larvae were successfully weaned by 35 DPH, but larvae weaned by 40 DPH displayed higher growth rates and were significantly larger by the end of the experiment (Alves et al., 2006).

4. Larviculture Systems

The marine finfish larviculture techniques have been developed quickly in the past three decades. The objective and function of larvciculture technique is to mass-produce healthy seed sufficiently and economically in a certain period of time for aquaculture industries. The general larviculture techniques for commercially farmed marine finfish can be classified into three schemes as (1) intensive rearing in clear water, (2) intensive rearing in greenwater, (3) extensive rearing (Shields, 2001). Sea bass *Dicentrarchus labrax* and sea bream larvae are used as examples to illustrate this culture system as described by Shields (2001).

4.1. Intensive Rearing in Clear Water

The rearing system is carried out closed water systems without microalgae. Water quality in the closed circuit is maintained by means of mechanical filtration followed by pressurized biological filtration and UV sterilization (Chatain, 1997). Water renewal rate is adjusted to maintain total ammonia nitrogen levels below 0.5 ppm and pH in the range 6.5–8.5. In order to equilibrate gas saturation levels, each larval rearing tank is equipped with a packed desaturation column. Fish eggs are incubated in upwelling tanks at a temperature of circa 13–15 °C. Newly hatched larvae are stocked at a density of 100/l into small rearing tanks (~4 m^3) containing clear water. Water temperature is controlled to approximately 18–20 °C over the following 10 days, during which time the larvae subsist in darkness on their endogenous reserves. *Artemia* nauplii (<5/ml) are introduced as a first prey item, at days 8–10, when overhead lighting (intensity <1500 lx) is switched on. According to Coves (1986), provision of *Artemia* nauplii continues until day 20, with enriched metanauplii being introduced from day 15 until weaning (day 45) when the larvae have reached a weight of at least 40 mg.

While this method of sea bass larviculture offers simplicity in terms of live feed production, it is not suitable for the smaller larvae of gilthead sea bream or other marine fish larvae, due to inappropriate size of the *Artemia* prey. In response, French researchers modified the closed circuit/clear water protocol, by incorporating rotifers during first feeding (Chatain, 1997). Warmer water temperatures (circa 17 °C) are applied during egg incubation and the sea bream larvae are fed enriched rotifers at a density of 5–10/ml from days 4–20 post-hatch (Coves, 1986). Small *Artemia* nauplii are introduced from circa Day 18, being replaced by enriched metanauplii (Utah strain) from day 26. Weaning on to formulated diets starts progressively around day 36 and should be complete by about day 60 post-hatch.

In practice, clear-water rearing techniques are little used for commercial production, due to the complexity of the system and high variability in performance. The absence of microalgae is considered especially critical in performance terms and most commercial sea bass and sea bream hatcheries undertake algae production, despite the additional workload involved.

4.2. Intensive Rearing in Green Water

In this system, fish larvae are cultured in is open-circuit green water. Some hatcheries operate part or all of their water circuits on a semi-closed basis, as a means of saving energy, excluding pathogens, or applying specific water quality conditions. For both open and semi-closed rearing systems, incoming hatchery water is mechanically filtered then sterilized using UV or ozone. In semi-closed systems, the process water is additionally filtered to remove suspended solids (pressurized sand filter, or gravity-fed disc filter) and passed across a biofilter, before being UV-sterilized and oxygenated on return to the rearing tank. Tank daily exchange volume typically exceeds replenishment volume by 10–100 times in semi-closed systems.

Exact system configurations and husbandry protocols for intensive larval production vary considerably among hatcheries. In particular, the types and densities of microalgae used and the timing of diet transitions differ according to local conditions and operator preference. Larval rearing tanks are typically circular, ranging in volume from 1 to 20 m^3, with a slightly sloping floor. Eggs may be incubated separately and stocked after hatching, or stocked directly in the larval rearing tanks at a density of approximately 50–100/l. During early rearing stages, water is commonly introduced via a surface inflow and exits via a surface-mounted "banjo" filter on the opposing side of the tank. Low water exchange rates (~1 exchange/day) are applied during the rotifer/green water phase and care is taken to avoid directional flow, in order to prevent spinal deformities. The surface water inlet/outlet arrangement may be replaced during later stages by a sub-surface, directional inflow and central drain.

Commonly cultured species of microalgae include *Nannochloris atomus*, *Nannochloropsis oculata*, *Chlorella minutissima*, *Isochrysis* sp., and *Tetraselmis* sp. Algae is added daily to the rearing tanks to maintain a density of 50–200 cells/µl. Rotifers are typically supplied 2–3 times per day, with stocking density increasing from 2-7 rotifers/ml as the larvae develop. Small mouth larvae (e.g., sea bream) larvae may be offered rotifers for only a few days, while large mouth larvae (e.g., sea bass) are commonly retained for more than 20 days on this prey. Diet transitions are performed gradually, both species receiving an intermediate diet of small strain *Artemia* nauplii, before transferring to enriched metanauplii. *Artemia* are presented up to eight times per day, to maintain a density of 1–2/ml in the rearing tank.

Sea bass larvae are conventionally weaned onto formulated feeds by days 60–65 post-hatch, although earlier weaning can be achieved using currently available microparticulate feeds. Depending on system configuration, the larvae may be partially or completely weaned in the hatchery before transfer to nursery tanks.

4.3. Extensive Larval Rearing

Larval fish rearing has been widely used in freshwater ponds (Qin and Culver, 1992; Qin et al., 1995), but it less common for marine fish larvae. Lee and Ostowski (2001) described the extensive method of larval rearing in outdoor ponds. Ponds are fertilized to increase primary productivity before the stocking of hatched larvae. This extensive larval rearing method has been practiced in species such as red drum, spotted seatrout and southern flounder (Lee and Ostrowski 2001). Unlike intensive larval rearing methods, extensive larval rearing methods usually solely depend on natural productivity to provide for the nutritional needs of fish larvae. Live feed production facilities are not normally part of the facilities in extensive larval rearing systems. Failure to sustain the wild plankton in the rearing pond will cause mass mortality of fish larvae from food deprivation. One strategy to overcome unexpected food shortages is to supplement the available food with cultured live feed or formulated feed.

Production by an extensive larval rearing system depends on management skills and therefore the number of juveniles produced can fluctuate considerably. Production costs, however, can be lower than in an intensive larval rearing system, due to lower operating costs, a larger volume of water, and lower costs for labor (Lee et al., 1997). Pond preparation for larval rearing before stocking is critical to the success of outdoor hatchery operations. The fertilization strategy of culture ponds should be modified according to the water sources, pond conditions, and production goals. For instance, red drum larvae were stocked in outdoor ponds 36-h post-hatch (Lyon and Fisher, 1998). In this study, the success of stocking was judged according to the number of larvae from ichthyoplankton tows: 50 or more larvae/tow was considered successful. Stocking success was improved by maintaining water temperature at 25–30 °C at a maximum pH below 9.4. Southern flounder, stocked in a 0.1-ha pond at a density of 74 larvae 4-day post-hatch/m^2, had an average survival of 3.8% after 74 days (Jenkins and Smith, 1999).

5. Feeding Practice and Problems

5.1. Starter Feed

The feeding regimes begin with live prey, usually rotifers *Brachionus* spp. and brine shrimp *Artemia* nauplii for the early life stages, and then larvae are weaned to inert feeds (Lee, 2003). The best feeding regime provides fish larvae with food organisms of a suitable particle size and, with the proper nutrients, at the right time.

Brine shrimp nauplii are widely used in hatcheries throughout the world. Nauplii of *Artemia* can be easily obtained from the hatching of resting eggs. Also, nauplii grow to a larger size and provide more biomass to larvae. However, the availability of the cysts in nature fluctuates (Lavens and Sorgeloos, 2000; Sorgeloos et al., 2001). The annual available brine shrimp cysts were 800 metric tonnes or 40% of the total demand from aquaculture (Sorgeloos et al., 2001). Fluctuations in the supply of cysts have resulted in unstable prices and higher production costs for the hatchery.

Nutritional deficiencies have been major concern when using brine shrimp. Most Artemia nauplii stocks have shown a deficiency in eicosapentaenoic acid (EPA; 20:5*n*-3) and docosahexaenoic acid (DHA; 22:6*n*-3)(Watanabe et al., 1978). *Artemia* nauplii can be enriched with fatty acids, phospholipids, and vitamins (Sorgeloos et al., 2001). Initially, most studies focused on the enrichment of EPA (Watanabe et al., 1983). Lately, DHA, the DHA/EPA ratio (Bell et al., 2003; Menoyo et al., 2007; Nevejan et al., 2003), and arachidonic acid (20:4*n*-6)(Koven et al., 2003; Van Anholt et al., 2004) have become the topics for enrichment studies. Control of bacterial input from *Artemia* cysts has become another concern of using *Artemia* cysts. Decapsulation procedures that have been described by (Sorgeloos et al., 1977) improve the percentage of hatching and provide the additional benefit of disinfecting the cysts. Commercially available products for disinfecting *Artemia* have reduced vibrio-load during hatching and the enrichment process (Tolomei et al., 2004).

5.2. Co-Feeding and Weaning

Advancements in microdiet development have been reviewed in several reports (Kolkovski et al., 1997a; Kolkovski et al., 1997b; Lee, 2003; Yufera et al., 2005). Up to the present, it has been essential to provide live feed for the first feeding of marine finfish species. Limited success has been achieved when inert feed is provided as first food for species such as red drum (Holt, 1993), sea bass (Cahu et al., 1998), red sea bream (Takeuchi, 2001), and others (Southgate and Partridge, 1998). Weaning of fish larvae to inert feed at a later stage of development, however, is easily achieved by co-feeding inert feed with live food.

Weaning to dry diets is done gradually, starting with co-feeding and reducing the amount of live feeds (usually *Artemia*) until replacement with dry diets is completed (Jones et al., 1993; Kolkovski, 2001). Feeding prey organisms, such as *Artemia* nauplii, a short time before dry diets are offered may increase digestive activity and increase the ability of larvae to digest the dry diets (Kolkovski et al., 1997b). Many works have described the advantages of co-feeding dry and live feed to fish larvae (Ehrlich et al., 1989; Fermin and Bolivar, 1991; Qin et al., 1997; Walford et al., 1991). For example, Tandler and Kolkovski (1992) achieved 80% survival with no growth problems using the co-feeding technique for rearing 10-day-old gilthead seabream larvae.

5.3. Larval Mortalities

During marine finfish larviculture rearing, high mortalities are likely to occur in two phases. One is during the transition from endogenous to exogenous nutritional stage (Green and McCormick, 2001; Sarasquete et al., 1995), and the other occurs during the weaning period from live feeds to formulated feeds (Hamlin and Kling, 2001). The characteristic of the two critical periods is that diet transition causes mortality (Shields, 2001).

In the first critical period, fish larvae only have a partly developed digestive tract and inadequate digestive enzymes (Dhert et al., 1998). Fish larvae need to rely on a mixture of endogenous and exogenous nutrition for growth. The knowledge of ontogenetic development of the digestive system in larval fish and the feeding performance such as gut content, gut evacuation in certain stage of larval development is the key to improve fish survival (Bisbal and Bengtson, 1995b; Canino and Bailey, 1995).

In the second critical period, the fish larvae totally rely on exogenous nutrition. The function of the digestive system changes from larval to juvenile stage in this period in characteristics of fully developed digestive enzymes and functional stomach. Commercially, the juvenile fish must be weaned to inert diet before sold. However, in this critical period, the survival rate is low comparing to the other stages. Further research on development of the digestive tract will help to solve the bottleneck of high mortalities in the weaning period (Bisbal and Bengtson, 1995a; Gordon and Hecht, 2002).

Cannibalism is a common and widespread phenomenon causing fish mortality especially during larval period (Hecht and Pienaar, 1993). In fish larvae, cannibalism is usually associated with heterogeneous size variation, limited food availability, high population densities, limited refuge areas, and light conditions (Qin and Fast, 1996). Among these variables, size variation and food availability are considered the primary causes of cannibalism (Hecht and Appelbaum, 1988; Katavic et al., 1989). Heterogeneous size distributions often lead to social dominance, which in turn results in aggressive behavior and cannibalistic responses (Hecht and Appelbaum, 1988). An inadequate food supply can initially induce cannibalism in striped bass larvae; and subsequently, cannibalistic fish may rely solely on cannibalism and never learn to accept artificial food (Paller and Lewis, 1987). Qin and Fast (1996) reported size grading and feeding manipulation could reduced the incident of cannibalism. In their study, cannibalism rate was 100% during a 5 day trial when the total length ratio of smaller snakehead *Channa striatus* to larger fish was 0.35; but the cannibalism rate decreased to 43% when the total length ratio of smaller fish to larger ones increased to 0.64. Increased availability of formulated feed also reduced cannibalism. With no feed offered, cannibalism was 83%; but decreased to 43% when daily feeding rate was 15% of the larger fish's body weight. Cannibalism is unavoidable in most carnivorous fish during the larval period, but can be greatly reduced among juvenile fish by size grading and *ad libitum* feeding.

5.4. Larval Deformity

Marine fish larvae undergo major functional and morphological changes during the developmental stages and several factors can interfere with the normal development of larvae and affect fry quality. Skeletal malformations, such as spinal malformation—scoliosis, lordosis, coiled vertebral column, missing or additional fin rays, bending opercle or jaw malformations, are frequently observed in hatchery-reared larvae (Cahu et al., 2003). Morphological deformities occur in a wide range of aquaculture fish species including the sea bass, Dicentrarchus labrax (Daoulas et al., 1991; Fraser et al., 2004), red sea bream *Pagrus major* (Kihara et al., 2002), the white sea bream *Diplodus sargus* (Sfakianakis et al., 2003), the common Dentex *Dentex dentex* (Koumoundouros et al., 2001), rainbow trout *Salmo gairdneri* (Dabrowski et al., 1990), milkfish *Chanos chanos* (Gapasin et al., 1998), senegal sole *Slea senegalensis* (Gavaia et al., 2002) and tilapia *Oreochromis niloticus* (Soliman et al., 1986). These deformities are often associated with intensive hatchery production and are a significant problem for aquaculture production (Battaglene and Cobcroft, 2007).

Several factors have been evoked to explain skeletal deformities in fish larvae reared in hatcheries. Lordosis in sea bass and sea bream, mainly affecting vertebrae 14–15, has been related to the hydrodynamics in tanks, particularly the water current intensity (Chatain, 1994). Inadequate light intensity as well as temperature and salinity in rearing water have been reported as sources of skeletal malformations (Johnson and Katavic, 1984). Pathology in very young larvae, and especially development of some pathogenic bacteria can induce high rates of malformations (Madsen and Dalsgaard, 1999). Moreover, it has been shown that prevalence of skeletal malformations, including jaw deformities, is significantly higher in triploid farmed fry of Atlantic salmon (Sadler et al., 2001). Recently more data become available about nutritional requirements of fish larvae, sustaining optimal growth and survival. Compound diets are now being formulated, usable from first feeding onwards in some marine fish larvae. Further understanding of the role of nutrition in fish development, especially at the molecular level will allow improving the quality of hatchery reared larvae.

6. Future Perspectives

Despite the recent technology advancement in fish hatchery production, the improvement for both quantity and quality of juvenile fish is still the major goal in hatcheries management. At present, live food is indispensable for most species of larval fish rearing in hatcheries. Research in live food production should focus on improving production efficiency and finding more species that are more suitable for the feeding requirement of a diverse fish species. Because live feed production is not cost-effective, technology to reduce the reliance on live feed in hatchery operations is highly desired. The replacement of live feed with inert diet should be based on the knowledge of nutritional needs of fish larvae, micro-diet processing for better ingestion and digestion, and protocols co-feeding applications.

In rotifer culture, there is a need to select for rotifer strains that are resistant to environmental stress during mass culture or during preservation process. The selection of desirable rotifer strains for larviculture could be done through inbreeding or appropriate biotechnological methods using strains with useful characteristics. Development of techniques that can hybridize between strains is desirable for larviculture practices. Technology of high density rotifer culture should be further developed to find small alternative preys or inert diet for rotifers to manipulate the nutritional value of rotifers. In mixed culture as some organisms such as *Artemia* outcompete rotifers for food, development of effective chemical treatments to eliminate harmful competitors is desirable.

Copepods are raised as live feed by a variety of methods and under different culture conditions. Like rotifer production, continuous or batch culture methods can be used to produce copepods. Part of the strategy for culturing copepods is growing the appropriate phytoplankton. Harpacticoids are comparatively easy to culture. Among the calanoid copepods, *Acartia* spp. are much easier to cultivate than *Calanus* spp. or *Temora* spp. Culture techniques may be considerably different between even closely related species.

An ideal copepod species for aquaculture was not identified at the workshop, but several potential species were recommended for further study. More research is needed to explore their potential. At present, the production level of copepods is unlikely to match that of rotifers. The potential contribution of copepods to hatchery production exists and can become increasingly important especially in rearing of new aquaculture species.

There are several cladoceran species in saline water. The potential of *M. mongolica* for finfish larviculture has been demonstrated in terms of its adaptability to environmental change, rapid reproduction, wide choice for food, applicability to mass culture, nutritional composition and suitable size for fish ingestion. Its body size is suitable as a food for finfish larviculture when the size of rotifer becomes too small. In the feeding trials, it is possible that *M. mongolica* can substitute *Artemia* cysts as a new food source in marine larviculture. The formation of resting eggs makes it easier to transport to a new location. Future research on *M. mongolica* should focus on its commercial use for other aquaculture species including shrimp, crab larvae in various geographic locations.

At present, no artificial feed formulation is yet available to completely substitute for *Artemia*, and feeding live prey to young fish larvae is still essential in commercial hatchery operations. The nutritional quality of commercially available *Artemia* strains being relatively poor in eicosapentaenoic acid (EPA, 20:5*n*-3) and especially docosahexaenoic acid (DHA, 22:6*n*-3), it is essential and common practice to enrich these live prey with emulsions of marine oils. Various enrichment emulsions have been formulated differing in the fatty acid composition of their triglycerides. The traditional formulations rich in EPA have been replaced by new products rich in DHA and arachidonic acid. To reduce the risks for oxidation of these fatty acids, higher concentrations of vitamin E are incorporated into the emulsions. Also, vitamin C has been incorporated in booster formulations that increase the level of ascorbic acid in *Artemia* to 2000 ppm. All these changes in the formulation of the enrichment diets offer more possibilities to cover the needs of different species and help to reduce problems related to diseases, stress resistance, malformation, and pigmentation in numerous fish species.

Many achievements have been made in the development of artificial diets for marine finfish larvae over the past two decades, but it is still unlikely to completely replace live food with artificial diet for larval rearing. Future research on artificial diets for marine fish larvae should be conducted concurrently with research to define specific nutritional requirements of finfish larvae. Artificial food particles are generally ingested at a lower rate than live food organisms. As a result, artificial diets are usually fed in excess to provide a sufficiently high food density. This strategy can lead to water-quality problems from uneaten food building up on the bottom of larval culture tanks and from nutrients leaching from the food particles. Improving diet attractability may solve this problem. Further progress on artificial diet for fish larvae will reduce the reliance of marine fish hatcheries on live foods and lead to more cost-effective hatchery practices.

REFERENCES

Alves, J.T.T., V.R. Cerqueira and J.A. Brown, 2006. Early weaning of fat snook (*Centropomus parallelus* Poey 1864) larvae. *Aquaculture* 253, 334-342.

Balompapueng, M.D., A. Hagiwara, Y. Nozaki and K. Hirayama, 1997. Preservation of resting eggs of the euryhaline rotifer *Brachionus plicatilis* O. F. Muller by canning. *Hydrobiologia* 358, 163-166.

Baragi, V. and R.T. Lovell, 1986. Digestive enzyme activities in striped bass from first feeding through larva development. *Transactions of the American Fisheries Society* 115, 478-484.

Barahona-Fernandes, M.-H. and M. Girin, 1976. Preliminary tests on the optimal pellet-adaptation age for sea bass larvae (Pisces, *Dicentrarchus labrax* L. 1758). *Aquaculture* 8, 283-290.

Battaglene, S.C. and J.M. Cobcroft, 2007. Advances in the culture of striped trumpeter larvae: A review. *Aquaculture* 268, 195-208.

Bell, J.G., L.A. McEvoy, A. Estevez, R.J. Shields and J.R. Sargent, 2003. Optimising lipid nutrition in first-feeding flatfish larvae. *Aquaculture* 227, 211-220.

Berge, G.M. and T. Storebakken, 1996. Fish protein hydrolyzate in starter diets for Atlantic salmon (*Salmo salar*) fry. *Aquaculture* 145, 205-212.

Bisbal, G.A. and D.A. Bengtson, 1995a. Development of the digestive tract in larval summer flounder. *Journal of Fish Biology* 47, 277-291

Bisbal, G.A. and D.A. Bengtson, 1995b. Effects of delayed feeding on survival and growth of summer flounder *Paralichthys dentatus* larvae. *Marine Ecology-Progress Series* 121, 301-306.

Boulhic, M. and J. Gabaudan, 1992. Histological study of the organogenesis of the digestive system and swim bladder of the Dover sole, *Solea solea* (Linnaeus 1758). *Aquaculture* 102, 373-396.

Brinkmeyer, R.L. and G.J. Holt, 1998. Highly unsaturated fatty acids in diets for red drum (*Sciaenops ocellatus*) larvae. *Aquaculture* 161, 253-268.

Buddington, R.K., 1985. Digestive secretions of lake sturgeon, *Acipenser fulvescens*, during early development. *Journal of Fish Biolology* 26, 715-723.

Cahu, C. and J.Z. Zambonion Infante, 2001. Substitution of live food by formulated diets in marine fish larvae. *Aquaculture* 200, 161-180.

Cahu, C. and J. Zambonino Infante, 2001. Substitution of live food by formulated diets in marine fish larvae. *Aquaculture* 200, 161-180.

Cahu, C., J. Zambonino Infante and T. Takeuchi, 2003. Nutritional components affecting skeletal development in fish larvae. *Aquaculture* 227, 245-258.

Cahu, C., J. Zambonino Infante, A.M. Escaffre, P. Bergot and S. Kaushik, 1998. Preliminary results on sea bass (*Dicentrarchus labrax*) larvae rearing with compound diet from first feeding. Comparison with carp (*Cyprinus carpio*) larvae. *Aquaculture* 169, 1-7.

Canino, M.F. and K.M. Bailey, 1995. Gut evacuation of walleye pollock larvae in response to feeding conditions. *Journal of Fish Biology* 46, 389-403.

Cara, J.B., F.J. Moyano, S. Cardenas, C. Fernandez-Diaz and M. Yufera, 2003. Assessment of digestive enzyme activities during larval development of white bream. *Journal of Fish Biology* 63, 48-58.

Carvalho, A.P., A.M. Escaffre, A.O. Teles and P. Bergot, 1997. First feeding of common carp larvae on diets with high levels of protein hydrolysates. *Aquaculture International* 5, 361-367.

Castell, J.D., J.G. Bell, D.R. Tocher and J.R. Sargent, 1994. Effects of purified diets containing different combinations of arachidonic and docosahexaenoic acid on survival, growth and fatty acid composition of juvenile turbot (*Scophthalmus maximus*). *Aquaculture* 128, 315-333.

Chatain, B., 1994. Abnormal swimbladder development and lordosis in sea bass (*Dicentrarchus labrax*) and sea bream (*Sparus auratus*). Aquaculture 119, 371-379.

Chatain, B., 1997. Development and achievements of marine fish-rearing technology in France over the last 15 years. *Hydrobiologia* 358, 7-11.

Chen, B.N., J.G. Qin, M.S. Kumar, W. Hutchinson and S. Clarke, 2006a. Ontogenetic development of the digestive system in yellowtail kingfish *Seriola lalandi* larvae. *Aquaculture* 256, 489-501.

Chen, B.N., J.G. Qin, M.S. Kumar, W.G. Hutchinson and S.M. Clarke, 2006b. Ontogenetic development of digestive enzymes in yellowtail kingfish *Seriola lalandi* larvae. *Aquaculture* 260, 264-271.

Copeman, L.A., C.C. Parrish, J.A. Brown and M. Harel, 2002. Effects of docosahexaenoic, eicosapentaenoic, and arachidonic acids on the early growth, survival, lipid composition and pigmentation of yellowtail flounder (*Limanda ferruginea*): a live food enrichment experiment. *Aquaculture* 210, 285-304.

Coves, D., 1986. New rearing techniques and larvae quality improvements in sea-bass (*Dicentrarchus labrax*) hatcheries. In: Technics used for intensive rearing and alimentation of fish and shellfish. FAO/UNDP Mediterranean Reginal Aquaculture Project. Tunis (Tunisia). pp. 342-356.

Cunha, I. and M. Planas, 1999. Optimal prey size for early turbot larvae (*Scophthalmus maximus* L.) based on mouth and ingested prey size. *Aquaculture* 175, 103-110.

Cuvier-Peres, A. and P. Kestemont, 2001. Development of some digestive enzymes in Eurasian perch larvae *Perca fluviatilis*. *Fish Physiology and Biochemistry* 24, 279-285.

Dabrowski, K., 1984. The feeding of fish larvae: present "state of the art" and perspectives. *Reprodution Nutrition and Development* 24, 807-833

Dabrowski, K., 1992. Ascorbate concentration in fish ontogeny. *Journal of Fish Biology* 40, 273-279.

Dabrowski, K. and J. Glogowski, 1977. Studies on the role of exogenous proteolytic enzymes in digestion processes in fish. *Hydrobiologia* 54, 129-134.

Daoulas, C., A.N. Economou and I. Bantavas, 1991. Osteological abnormalities in laboratory reared sea-bass (*Dicentrarchus labrax*) fingerlings. *Aquaculture* 97, 169-180.

De Silva, S.S. and T.A. Anderson, 1995. Fish Nutrition in Aquaculture. Chapman and Hall, London, xix, 319 p. pp.

Dedi, J., T. Takeuchi, T. Seikai and T. Watanabe, 1995. Hypervitaminosis and safe levels of vitamin A for larval flounder *Paralichthys olivaceus* fed *Artemia* nauplii. *Aquaculture* 133, 135-146.

Dhert, P., P. Divanach, M. Kentouri and P. Sorgeloos, 1998. Rearing techniques for difficult marine fish larvae. *World Aquaculture* 29, 48-55.

Ehrlich, K.F., M.C. Cantin, M.B. Rust and B. Grant, 1989. Growth and survival of larval and postlarval smallmouth bass fed a commercially prepared dry feed and/or *Artemia* nauplii. *Journal of the World Aquaculture Society* 20, 1-6.

Estevez, A., M. Ishikawa and A. Kanazawa, 1997. Effects of arachidonic acid on pigmentation and fatty acid composition of Japanese flounder, *Paralichthys olivaceus* (Temminck an Schlegel). *Aquaculture Research* 28, 279-289.

FAO, 2006. The State of World Fisheries and Aquaculture, Rome, 180 pp.

Fermin, A.C. and M.E.C. Bolivar, 1991. Larval rearing of the Philippine freshwater catfish, *Clarias macrocephalus* (Gunther), fed live zooplankton and artificial diet: A preliminary study. *Israeli Journal of Aquaculture /Bamidgeh* 43, 87-94.

Fraser, M.R., T.A. Anderson and R. de Nys, 2004. Ontogenic development of the spine and spinal deformities in larval barramundi (*Lates calcarifer*) culture. *Aquaculture* 242, 697-711.

Fu, Y., A. Hada, T. Yamashita, Y. Yoshida and A. Hino, 1997. Development of a continuous culture system for stable mass production of the marine rotifer *Brachionus*. *Hydrobiologia* 358, 145-151.

Fukusho, K., 1980. Mass production of a copepod, *Tigriopus japonicus* in combination culture with a rotifer *Brachionus plicatilis* fed w-yeast as a food source. *Bulletin of Japanese Society of Scientific Fisheries* 46, 625-629.

Fulks, W. and K.L. Main, 1991. Rotifer and microalgae culture systems: proceedings of a U.S. - Asia Workshop held in Honolulu, Hawaii, January 28-31, 1991. Oceanic Institute, Honolulu, Hawaii, xi, 364 p. pp.

Furuita, H., T. Takeuchi and K. Uematsu, 1998. Effects of eicosapentaenoic and docosahexaenoic acids on growth, survival and brain development of larval Japanese flounder (*Paralichthys olivaceus*). *Aquaculture* 161, 269-279.

Gavaia, P.J., M.T. Dinis and M.L. Cancela, 2002. Osteological development and abnormalities of the vertebral column and caudal skeleton in larval and juvenile stages of hatchery-reared Senegal sole (*Solea senegalensis*). *Aquaculture* 211, 305-323.

Gawlicka, A., B. Parent, M.H. Horn, N. Ross, I. Opstad and O.J. Torrissen, 2000. Activity of digestive enzymes in yolk-sac larvae of Atlantic halibut (*Hippoglossus hippoglossus*): indication of readiness for first feeding. *Aquaculture* 184, 303-314.

Gordon, A.K. and T. Hecht, 2002. Histological studies on the development of the digestive system of the clownfish *Amphiprion percual* and the time of weaning. *Journal of Applied Ichthyology* 18, 113-117.

Goulden, C.E., 1968. The systematics and evolution of the Moinidae. *Transaction of the American Philosophicial Society New Series* 58, 3-101.

Green, B.S. and M.I. McCormick, 2001. Ontogeny of the digestive and feeding systems in the anemonefish *Amphiprion melanopus*. *Environmental Biolology of Fishes* 61, 73-83.

Hagiwara, A., K. Hamada, S. Hori and K. Hirayama, 1994. Increased sexual reproduction in *Brachionus plicatilis* (Rotifera) with the addition of bacteria and rotifer extracts. *Journal of Experimental Marine Biology and Ecology* 181, 1-8.

Hagiwara, A., M.D. Balompapueng, N. Munuswamy and K. Hirayama, 1997. Mass production and preservation of the resting eggs of the euryhaline rotifer *Brachionus plicatilis* and *B. rotundiformis*. *Aquaculture* 155, 223-230.

Hagiwara, A., K. Hamada, A. Nishi, K. Imaizumi and K. Hirayama, 1993 Mass production of rotifer *Brachionus plicatilis* resting eggs in 50 m^3 tanks. *Nippon Suisan Gakkaishi* 59, 93-98.

Hagiwara, A., T. Kotani, T.W. Snell, M. AssavaAree and K. Hirayama, 1995. Morphology, reproduction, genetics, and mating behavior of small, tropical marine *Brachionus* strains (rotifera). *Journal of Experimental Marine Biology and Ecology* 194, 25-37.

Hagiwara, A., W.G. Gallardo, M. Assavaaree, T. Kotani and A.B. de Araujo, 2001. Live food production in Japan: recent progress and future aspects. *Aquaculture* 200, 111-127.

Hamlin, H.J. and L.J. Kling, 2001. The culture and early weaning of larval haddock (*Melanogrammus aeglefinus*) using a microparticulate diet. *Aquaculture* 201, 61-72.

Harel, M., S. Ozkizilcik, E. Lund, P. Behrens and A.R. Place, 1999. Enhanced absorption of docosahexaenoic acid (DHA, 22 : 6n-3) in Artemia nauplii using a dietary combination of DHA-rich phospholipids and DHA-sodium salts. *Comparative Biochemistry and Physiology B-Biochemistry and Molecular Biology* 124, 169-176.

Hart, B.T., P. Bailey, R. Edwards, K. Hortle, K. James, A. McMahon, C. Meredith and K. Swadling, 1991. A review of the salt sensitivity of the Australian freshwater biota. *Hydrobiologia* 210, 105-144.

Hashim, R. and M.A.M. Saat, 1992. The utilization of seaweed meals as binding agents in pelleted feeds for snakehead (*Channa striatus*) fry and their effects on growth. *Aquaculture* 108, 299-308.

Hayashi, M., K. Toda and S. Kitaoka, 1993. Enriching *Euglena* with unsaturated fatty-acids. *Bioscience Biotechnology and Biochemistry* 57, 352-353.

Hayashi, M., T. Yukino, I. Maruyama, S. Kido and S. Kitaoka, 2001. Uptake and accumulation of exogenous docosahexaenoic acid by *Chlorella*. *Bioscience Biotechnology and Biochemistry* 65, 202-204.

He, Z.H., J.G. Qin, H.Q. Wang, Z.Y. Wamg and X. Xia, 1989. Saline and supersaline zooplankton in west Shanxi Province and Yinchuan Region, China. *Acta Hydrobiology Sinica* 13, 24-37.

He, Z.H., H. Jiang, Z.Q. Jiang, and L.L. Xue, 1997. Use *Moina* as live food for marine fish larvae. *Journal of Dalian Fisheries University* 12: 1-7.

He, Z.H., Y. Wang, H. Cui, L.Z. Guo and H. Qian, 1998. Studies on the mass culture of *Moina mongolica* in seawater. *Fishery Journal* (*Chinese*) 22, 17-23.

He, Z.H., J.G. Qin, Y. Wang, H. Jiang and Z. Wen, 2001. Biology of Moina mongolica (Moinidae, Cladocera) and perspective as live food for marine fish larvae: review. *Hydrobiologia* 457, 25-37.

Hebert, P.D.N. and C.C. Wilson, 2000. Diversity of the genus *Daphniopsis* in the saline waters of Australia. *Canadian Journal of Zoology* 78, 794-808.

Hecht, T. and S. Appelbaum, 1988. Observations on intraspecific aggression and coeval sibling cannibalism by larva and juvenile *Clarios pwiepinus* (Clariidae: Pisces) under controlled conditions. *Journal of Zoology* 214, 21-44.

Hecht, T. and A.G. Pienaar, 1993. A review of cannibalism and its implications in fish larviculture. *Journal of the World Aquaculture Society* 24, 246-261.

Hino, A. and R. Hirano, 1976. Ecological studies on the mechanism of bisexual reproduction in the rotifer *Brachionus plicatilis*: I. General aspects of bisexual reproduction inducing factors. *Nippon Suisan Gakkasihi* 42, 1093-1099.

Hirayama, K. and H. Funamoto, 1983. Supplementary effect of several nutrients on nutritive deficiency of baker's yeast for population growth of the rotifer *Brachionus plicatilis*. *Nippon Suisan Gakkaishi* 49, 505-510.

Hirayama, K., I. Maruyama and T. Maeda, 1989. Nutritional effect of freshwater *Chlorella* on growth of the rotifer *Brachionus plicatilis*. *Hydrobiologia* 186/187, 39-42.

Holt, G.J., 1993. Feeding larval red drum on microparticulate diets in a closed recirculating water system. *Journal of the World Aquaculture Society* 24, 225-230.

Hontoria, F., J.H. Crowe, L.M. Crowe and F. Amat, 1994. Potential use of liposomes in larviculture as a delivery system through *Artemia* nauplii. *Aquaculture* 127, 255-264.

Huo, J.Z., H.J. Nelis, P. Lavens, P. Sorgeloos and A.P. DeLeenheer, 1996. Determination of vitamin E in aquatic organisms by high-performance liquid chromatography with fluorescence detection. *Analytical Biochemistry* 242, 123-128.

Izquierdo, M.S., T. Arakawa, T. Takeuchi, R. Haroun and T. Watanabe, 1992. Effect of n-3 HUFA levels in Artemia on growth of larval Japanese flounder (*Paralichthys olivaceus*). *Aquaculture* 105, 73-82.

James, C.M., S. Al-Hinty and A.E. Salman, 1989. Growth and n-3 fatty acid and amino acid composition of microalgae under different temperature regimes. *Aquaculture* 77, 337-351.

Jenkins, W.E. and T.I.J. Smith, 1999. Pond nursery production of southern flounder (*Paralichthys lethostigma*) and weaning to commercial diets. *Aquaculture* 176, 173-180.

Johnson, D.W. and I. Katavic, 1984. Mortality, growth and swim bladder stress syndrome of sea bass (*Dicentrarchus labrax*) larvae under varied environmental conditions. *Aquaculture* 38, 67-78.

Jones, D.A., M.S. Kamarudin and L. Le Vay, 1993. The potential for replacement of live feeds in larval culture. *Journal of the World Aquaculture Society* 24, 199-210.

Kanazawa, A., S. Koshio and S.I. Teshima, 1989. Growth and survival of larval red sea bream *Pagrus major* and Japanese flounder *Paralichthys olivaceus* fed microbound diets. *Journal of the World Aquaculture Society* 20, 31-37.

Katavic, I., J. Jug-dujakovic and B. Glamuzina, 1989. Cannibalism as a factor affecting the survival of intensively cultured sea bass (*Dicentrarchus labrax*) fingerlings. *Aquaculture* 77, 135-143.

Kelly, S.P., S.D. Larsen, P.M. Collins and N.Y.S. Woo, 2000. Quantitation of inert feed ingestion in larval silver sea bream (*Sparus sarba*) using auto-fluorescence of alginate-based microparticulate diets. *Fish Physiology and Biochemistry* 22, 109-117.

Kihara, M., S. Ogata, N. Kawano, I. Kubota and R. Yamaguchi, 2002. Lordosis induction in juvenile red sea bream, *Pagrus major*, by high swimming activity. *Aquaculture* 212, 149-158.

Kim, B.G., S. Divakaran, C.L. Brown and A.C. Ostrowski, 2001. Comparative digestive enzyme ontogeny in two marine larval fishes: Pacific threadfin (*Polydactylus sexfilis*) and bluefin trevally (*Caranx melampygus*). *Fish Physiology and Biochemistry* 24, 225-241.

Kogane, T., A. Hagiwara and K. Imaizumi, 1997. Temperature regulation of resting egg formation of the euryhaline rotifer *Brachionus plicatilis* (Kamiura strain). *Hydrobiologia* 358, 167-171.

Kolkovski, S., 2001. Digestive enzymes in fish larvae and juveniles-implications and applications to formulated diets. *Aquaculture* 200, 181-201.

Kolkovski, S., W. Koven and A. Tandler, 1997a. The mode of action of Artemia in enhancing utilization of microdiet by gilthead seabream *Sparus aurata* larvae. *Aquaculture* 155, 193-205.

Kolkovski, S., A. Tandler and M.S. Izquierdo, 1997b. Effects of live food and dietary digestive enzymes on the efficiency of microdiets for seabass (*Dicentrarchus labrax*) larvae. *Aquaculture* 148, 313-322.

Kolkovski, S., A. Tandler, G.W. Kissil and A. Gertler, 1993. The effect of dietary exogenous digestive enzymes on ingestion, assimilation, growth and survival of gilthead seabream (*Sparus-aurata*, Sparidae, linnaeus) larvae. *Fish Physiology and Biochemistry* 12, 203-209.

Koumoundouros, G., P. Divanach and M. Kentouri, 2001. The effect of rearing conditions on development of saddleback syndrome and caudal fin deformities in *Dentex dentex* (L.). *Aquaculture* 200, 285-304.

Koven, W., R. van Anholt, S. Lutzky, I. Ben Atia, O. Nixon, B. Ron and A. Tandler, 2003. The effect of dietary arachidonic acid on growth, survival, and cortisol levels in different-age gilthead seabream larvae (*Sparus auratus*) exposed to handling or daily salinity change. *Aquaculture* 228, 307-320.

Koven, W.M., A. Tandler, G.W. Kissil and D. Sklan, 1992. The importance of n-3 highly unsaturated fatty acids for growth in larval *Sparus aurata* and their effect on survival, lipid composition and size distribution. *Aquaculture* 104, 91-104.

Kraul, S., 1993. Larviculture of the mahimahi *Coryphaena hippurus* in Hawaii, USA. *Journalof the World Aquaculture Society*. 24, 410-421.

Lavens, P. and P. Sorgeloos, 1996. Manual on the production and use of live food for aquaculture, *FAO Tech. Pap.* 361, 295 pp.

Lavens, P. and P. Sorgeloos, 2000. The history, present status and prospects of the availability of *Artemia* cysts for aquaculture. *Aquaculture* 181, 397-403.

Lee, C.-S., 2003. Biotechnological advances in finfish hatchery production: a review. *Aquaculture* 227, 439-458.

Lee, C.-S., N.H. Marcus and P.J. O'Bryen, 2005. Copepods in aquaculture. Blackwell Pub., Ames, Iowa, xiii, 269 pp.

Lee, C.S. and A.C. Ostrowski, 2001. Current status of marine finfish larviculture in the United States. *Aquaculture* 200, 89-109.

Lee, C.S., P.S. Leung and M.S. Su, 1997. Bioeconomic evaluation of different fry production systems for milkfish (*Chanos chanos*). *Aquaculture* 155, 367-376.

Léger, P., D.A. Bengtson, K.L. Simpson and P. Sorgeloos, 1986. The use and nutritional value of *Artemia* as a food source. *Oceanography and Marine Biolology* 24, 521-623.

Lindner, P., A. Eshel, S. Kolkovski, A. Tandler and S. Harpaz, 1995. Proteolysis by juvenile sea bass (*Dicentrarchus labrax*) gastrointestinal enzymes as a method for the evaluation of feed proteins. *Fish Physiology and Biochemistry* 14, 399-407.

Lopez-Alvarado, J., C.J. Langdon, S.-I. Teshima and A. Kanazawa, 1994. Effects of coating and encapsulation of crystalline amino acids on leaching in larval feeds. *Aquaculture* 122, 335-346.

Lubzens, E., A. Tandler and G. Minkoff, 1989. Rotifers as food in aquaculture. *Hydrobiologia* 186/187, 387-400.

Lubzens, E., G. Kolodny, B. Perry, N. Galai, R. Sheshinski and Y. Wax, 1990. Factors affecting survival of rotifers (*Brachionus plicatilis* O.F. Muller) at 4°C. *Aquaculture* 91, 23-47.

Lyon, D.D. and M.R. Fisher, 1998. Temperature and pH-related mortality of red drum larvae in fertilized culture ponds. *The Progresive Fish-Culturist* 60, 227-230.

Madsen, L. and I. Dalsgaard, 1999. Vertebral column deformities in farmed rainbow trout (*Oncorhynchus mykiss*). *Aquaculture* 171, 41-48.

Martinez-Jeronimo, F., M. Elias-Gutierrez and E. Suarez-Morales, 2004. A redescription of *Moina hutchinsoni*, a rare cladoceran (Branchiopoda : Anomopoda) found in remnants of a Mexican saline lake, with notes on its life history. *Journal of Crustacean Biology* 24, 232-245.

Maruyama, I. and K. Hirayama, 1993. The culture of the rotifer Brachionus plicatilis with *Chlorella vulgaris* containing vitamin B_{12} in its cells. *Journal of the World Aquaculture Society* 24, 194-198.

Mayer, C.M. and D.H. Wahl, 1997. The relationship between prey selectivity and growth and survival in a larval fish. *Canadian Journal of Fisheries and Aquatic Sciences* 54, 1504-1512.

McEvoy, L.A., J.C. Navarro, F. Amat and J.R. Sargent, 1997. Application of soya phosphatidylcholine in tuna orbital oil enrichment emulsions for Artemia. *Aquaculture International* 5, 517-526.

Menoyo, D., C.J. Lopez-Bote, A. Diez, A. Obach and J.M. Bautista, 2007. Impact of n - 3 fatty acid chain length and n-3/n-6 ratio in Atlantic salmon (*Salmo salar*) diets. *Aquaculture* 267, 248-259.

Merchie, G., P. Lavens and P. Sorgeloos, 1997. Optimization of dietary vitamin C in fish and crustacean larvae: a review. *Aquaculture* 155, 165-181.

Merchie, G., P. Lavens, P. Dhert, M. Dehasque, H. Nelis, A. De Leenheer and P. Sorgeloos, 1995. Variation of ascorbic acid content in different live food organisms. *Aquaculture* 134, 325-337.

Merchie, G., P. Lavens, P. Dhert, M.G.U. Gomez, H. Nelis, A. DeLeenheer and P. Sorgeloos, 1996. Dietary ascorbic acid requirements during the hatchery production of turbot larvae. *Journal of Fish Biology* 49, 573-583.

Mourente, G., A. Rodriguez, D.R. Tocher and J.R. Sargent, 1993. Effects of dietary docosahexaenoic acid (DHA; 22:6n-3) on lipid and fatty acid compositions and growth in gilthead sea bream (*Sparus aurata* L.) larvae during first feeding. *Aquaculture* 112, 79-98.

Nevejan, N., I. Saez, G. Gajardo and P. Sorgeloos, 2003. Supplementation of EPA and DHA emulsions to a *Dunaliella tertiolecta* diet: effect on growth and lipid composition of scallop larvae, *Argopecten purpuratus* (Lamarck, 1819). *Aquaculture* 217, 613-632.

Nybakken, J.W. and M.D. Bertness, 2005. Marine Biology: an ecological approach. Pearson/Benjamin Cummings, San Francisco, xi, 579 p. pp.

Ohno, A., T. Takahashi and Y. Taki, 1990. Dynamics of exploited populations of the calanoid copepod, *Acartia tsuensis*. *Aquaculture* 84, 27-39.

Øie, G., P. Makridis, K.I. Reitan and Y. Olsen, 1997. Protein and carbon utilization of rotifers (*Brachionus plicatilis*) in first feeding of turbot larvae (*Scophthalmus maximus* L.). *Aquaculture* 153, 103-122.

Oozeki, Y. and K.M. Bailey, 1995. Ontogenic development of digestive enzyme-activities in larval walleye pollock, *Theragra-chalcogramma*. *Marine Biology* 122, 177-186.

Paller, M.H. and W.M. Lewis, 1987. Effects of diet on growth depensation and cannibalism among intensively cultured larval striped bass. *The Progressive Fish-Culturist* 49, 270-275.

Pennak, R.W., 1989. Fresh-water invertebrates of the United States : protozoa to mollusca. Wiley, New York, xvi, 628 p. pp.

Peres, A., C.L. Cahu and J.L.Z. Infante, 1997. Dietary spermine supplementation induces intestinal maturation in sea bass (*Dicentrarchus labrax*) larvae. *Fish Physiology and Biochemistry* 16, 479-485.

Person-Le Ruyet, J., 1989. Early weaning of marine fish larvae onto microdiets: constraints and perspectives. In: *Advances in tropical aquaculture*. *IFREMER Actes de Colloque* 9, 625–642.

Persoone, G. and P. Sorgeloos, 1980. General aspects of the ecology and biogeography of Artemia. In: G. Persoone, P. Sorgeloos, O. Roels and E. Jaspers, Editors, The Brine Shrimps Artemia Ecology, Culturing, Use in Aquaculture vol. 3, Universa Press, Wetteren, Belgium, pp. 3–24.

Qin, J.G. and D.A. Culver, 1992. The survival and growth of larval walleye, *stizostedion vitreum*, and trophic dynamics in fertilized ponds. *Aquaculture* 108, 257-276.

Qin, J.G. and D.A. Culver, 1995. Effect of young-of-the-year walleye (Percidae, *Stizostedion vitreum*) on plankton dynamics and water quality in ponds. *Hydrobiologia* 297, 217-227.

Qin, J.G. and A.W. Fast, 1996. Size and feed dependent cannibalism with juvenile snakehead *Channa striatus*. *Aquaculture* 144, 313-320.

Qin, J.G. and D.A. Culver, 1996. Effect of larval fish and nutrient enrichment on plankton dynamics in experimental ponds. *Hydrobiologia* 321, 109-118.

Qin, J.G. and A.W. Fast, 1997. Food selection and growth of young snakehead Channa striatus. *Journal of Applied Ichthyology* 13, 21-25.

Qin, J.G., S.P. Madon and D.A. Culver, 1995. Effect of larval walleye (*Stizostedion vitreum*) and fertilization on the plankton community - implications for larval fish culture. *Aquaculture* 130, 51-65.

Qin, J.G., A.W. Fast, D. DeAnda and R.P. Weidenbach, 1997. Growth and survival of larval snakehead (*Channa striatus*) fed different diets. *Aquaculture* 148, 105-113.

Reitan, K.I., J.R. Rainuzzo and Y. Olsen, 1994. Influence of lipid composition of live feed on growth, survival and pigmentation of turbot larvae. *Aquaculture International* 2, 33-48.

Ribeiro, L., J.L. Zambonino-Infante, C. Cahu and M.T. Dinis, 1999. Development of digestive enzymes in larvae of *Solea senegalensis*, Kaup 1858. *Aquaculture* 179, 465-473.

Ronnestad, I., S. Helland and O. Lie, 1998. Feeding Artemia to larvae of Atlantic halibut (*Hippoglossus hippoglossus* L.) results in lower larval vitamin A content compared with feeding copepods. *Aquaculture* 165, 159-164.

Sadler, J., P.M. Pankhurst and H.R. King, 2001. High prevalence of skeletal deformity and reduced gill surface area in triploid Atlantic salmon (*Salmo salar* L.). *Aquaculture* 198, 369-386.

Sakamoto, M., D.L. Holland and D.A. Jones, 1982. Modification of the nutritional composition of *Artemia* by incorporation of polyunsaturated fatty acids using micro-encapsulated diets. *Aquaculture* 28, 311-320.

Salhi, M., M.S. Izquierdo, C.M. Hernandez-Cruz, M. Gonzalez and H. Fernandez-Palacios, 1994. Effect of lipid and n-3 HUFA levels in microdiets on growth, survival and fatty acid composition of larval gilthead seabream (*Sparus aurata*). *Aquaculture* 124, 275-282.

Sarasquete, M.C., A. Polo and M. Yufera, 1995. Histology and histochemistry of the development of the digestive system of larval gilthead seabream, *Sparus aurata* L. *Aquaculture* 130, 79-92.

Sargent, J., G. Bell, L. McEvoy, D. Tocher and A. Estevez, 1999a. Recent developments in the essential fatty acid nutrition of fish. *Aquaculture* 177, 191-199.

Sargent, J., L. McEvoy, A. Estevez, G. Bell, M. Bell, J. Henderson and D. Tocher, 1999b. Lipid nutrition of marine fish during early development: current status and future directions. *Aquaculture* 179, 217-229.

Sargent, J.R., L.A. McEvoy and J.G. Bell, 1997. Requirements, presentation and sources of polyunsaturated fatty acids in marine fish larval feeds. *Aquaculture* 155, 117-127.

Schipp, G.R., J.M.P. Bosmans and A.J. Marshall, 1999. A method for hatchery culture of tropical calanoid copepods, *Acartia* spp. *Aquaculture* 174, 81-88.

Scott, J.M., 1981. The vitamin B_{12} requirement of the marine rotifer *Brachionus plicatilis*. *Journal of the Marine Biological Association of the United Kingdom* 61, 983-994.

Sfakianakis, D.G., G. Koumoundouros, L. Anezaki, P. Divanach and M. Kentouri, 2003. Development of a saddleback-like syndrome in reared white seabream *Diplodus sargus* (Linnaeus, 1758). *Aquaculture* 217, 673-676.

Shields, R., 2001. Larviculture of marine finfish in Europe. *Aquaculture* 200, 55-88.

Soliman, A.K., K. Jauncey and R.J. Roberts, 1986. The effect of varying forms of dietary ascorbic acid on the nutrition of juvenile tilapias (*Oreochromis niloticus*). *Aquaculture* 52, 1-10.

Sorgeloos, P., P. Dhert and P. Candreva, 2001. Use of the brine shrimp, *Artemia* spp., in marine fish larviculture. *Aquaculture* 200, 147-159.

Sorgeloos, P., E. Bossuyt, E. Lavina, M. Baeza-Mesa and G. Persoone, 1977. Decapsulation of Artemia cysts: A simple technique for the improvement of the use of brine shrimp in aquaculture. *Aquaculture* 12, 311-315.

Southgate, P.C. and P.S. Lee, 1993. Notes on the use of microbound artificial diets for larval rearing of sea bass (*Lates calcarifer*). *Asian Fisheries Science* 6, 245-247.

Southgate, P.C. and G.J. Partridge, 1998. Development of artificial diets for marine finfish larvae: problems and prospects. In: DeSilva, S.S. (Ed.), Tropical Mariculture. Academic Press, London, pp. 151-169.

Støttrup, J.G. and N.H. Norsker, 1997. Production and use of copepods in marine fish larviculture. *Aquaculture* 155, 231-247.

Støttrup, J.G., K. Richardson, E. Kirkegaard and N.J. Pihl, 1986. The cultivation of Acartia tonsa Dana for use as a live food source for marine fish larvae. *Aquaculture* 52, 87-96.

Stroband, H.W.J. and A.G. Kroon, 1981. The development of the stomach in *Clarias lazera* and the intestinal absorption of protein macromolecules. *Cell Tissue Research* 215, 397-415.

Takeuchi, T., 2001. A review of feed development for early life stages of marine finfish in Japan. *Aquaculture* 200, 203-222.

Tanaka, M., 1971. Studies on the structure and function of the digestive system in teleost larvae: III. Development of the digestive system during postlarval stage. *Japanese Journal Ichthyololgy* 18 164-174.

Tandler, A. and S. Kolkovski, 1992. Rates of ingestion and digestibility as limiting factors in the successful application of microdiets in gilthead seabream *Sparus aurata* larvae. *Israeli Journal of Aquaculture /Bamidgeh* 44, 128-129.

Tolomei, A., C. Burke, B. Crear and J. Carson, 2004. Bacterial decontamination of on-grown *Artemia*. *Aquaculture* 232, 357-371.

Tong, S.Y., C.H. Liu and X.T. Wang, 1988. Appraisement and analysis of nutrient composition for *Moina mongolica* Daddy. *Journal of Dalian Fisheries University* 11, 29-33

Van Anholt, R.D., W.M. Koven, S. Lutzky and S.E. Wendelaar Bonga, 2004. Dietary supplementation with arachidonic acid alters the stress response of gilthead seabream (*Sparus aurata*) larvae. *Aquaculture* 238, 369-383.

Walford, J., T.M. Lim and T.J. Lam, 1991. Replacing live foods with microencapsulated diets in the rearing of seabass (*Lates calcarifer*) larvae: do the larvae ingest and digest protein-membrane microcapsules? *Aquaculture* 92, 225-235.

Watanabe, T. and V. Kiron, 1994. Prospects in larval fish dietetics. *Aquaculture* 124, 223-251.

Watanabe, T., C. Kitajima and S. Fujita, 1983. Nutritional values of live organisms used in Japan for mass propagation of fish: A review. *Aquaculture* 34, 115-143.

Watanabe, T., C. Kitajima, T. Arakawa, K. Fukusho and S. Fujita, 1978. Nutritional quality of rotifer *Brachionus plicatilis* as a living feed from the viewpoint of essential fatty acids for fish. *Nippon Suisan Gakkaishi* 44, 1109-1114.

Watanabe, Y., 1982. Intracellular digestion of horseradish peroxidase by the intestinal cells of teleost larvae and juveniles. *Bulletin of Japanese Society of Scientific Fisheries* 48, 37-42.

Yoshimura, K., A. Hagiwara, T. Yoshimatsu and C. Kitajima, 1996. Culture technology of marine rotifers and the implications for culture of marine fish in Japan. *Marine and Freshwater Research* 47, 217-222.

Yoshimura, K., K. Usuki, T. Yoshimatsu, C. Kitajima and A. Hagiwara, 1997. Recent development of a high density mass culture system for the rotifer *Brachionus rotundiformis* Tschugunoff. *Hydrobiologia* 358, 139-144.

Yufera, M., C. Fernandez-Diaz and E. Pascual, 1995. Feeding rates of gilthead seabream (*Sparus aurata*), larvae on microcapsules. *Aquaculture* 134, 257-268.

Yufera, M., C. Fernandez-Diaz and E. Pascual, 2005. Food microparticles for larval fish prepared by internal gelation. *Aquaculture* 248, 253-262.

Zambonino Infante, J.L. and C. Cahu, 1994. Development and response to a diet change of some digestive enzymes in sea bass (*Dicentrarchus labrax*) larvae. *Fish Physiology and Biochemistry* 12, 399-408.

Zambonino Infante, J.L.Z. and C.L. Cahu, 1999. High dietary lipid levels enhance digestive tract maturation and improve *Dicentrarchus labrax* larval development. *Journal of Nutrition* 129, 1195-1200.

In: Aquaculture Research Trends
Editor: Stephen H. Schwartz, pp.149-184
ISBN 978-1-60456-217-0

Chapter 4

ROLE OF BACTERIA IN THE CHILLED STORAGE AND CRYOPRESERVATION OF SPERM IN AQUATIC ANIMALS: A REVIEW

Subuntith Nimrat[1*] and Verapong Vuthiphandchai[2]
[1]Department of Microbiology, Environmental Science Program,
Faculty of Science, Burapha University, Chonburi 20131, Thailand
[2]Department of Aquatic Science, Faculty of Science, Burapha University,
Chonburi 20131, Thailand

ABSTRACT

Bacterial contamination is considered a major issue for the management of sperm preservation in aquatic animals. Chilled storage and cryopreservation of aquatic animal sperm have been used to meet the growing demand for seeds. Chilled storage of sperm has been effectively used in a number of fish, but in marine invertebrates has been limited to shrimp, lobster, echiura and shellfish. Successful and reproducible cryopreservation has proven to be a reliable method for preserving genetic lines of fish for aquaculture and conservation, but among invertebrates is limited to polycheates, cnidaria, crustaceans (shrimp and crab), shellfish (oysters, hard clams and abalone) and echinoderms (sea urchin and sea cucumber). Few studies have documented changes in bacterial type and number during chilled storage or cryostorage. Access of good quality chilled-stored or cryopreserved sperm is necessary for the management of risk associated with sperm preservation. The presence of bacteria in sperm of aquatic animals can result in deleterious effects on sperm quality and viability if left uncontrolled. Potential pathogens present on the preserved sperm of aquatic animals may affect the health of broodstock and offspring. This paper summarizes the current knowledge pertaining to bacterial contamination of aquatic animal sperm, including bacterial types and sources of contaminants, and gives an overview of the incidence of bacterial contamination, biosecurity and risk assessment and effects of bacterial contamination on offspring quality. These ideas, if taken into account in commercial aquaculture and conservation of

* Corresponding author. Tel. +66-38-745-900 ext. 3120; Fax: +66-38-393-490, E-mail address of corresponding author: subunti@buu.ac.th.

aquatic animals, are almost certain to increase the value added or protection of at least some cultured species. Microbial contamination risk of chilled-stored or cryopreserved sperm after short-term or long-term storage is discussed. Future direction and new avenues of future research on chilled storage and cryopreservation of sperm in aquatic animals are proposed.

1. A Short Overview of Theory on Chilled Stored- and Cryopreserved Sperm

1.1. Sperm Quality

Spermatozoa of aquatic animals are characterized by a wide divergence in their structural organization. In general, spermatozoa are composed of a head, a small midpiece and a tail region. Fish sperm usually do not have an acrosome at the head region (Lahnsteiner and Patzner, 1995) while crustacean and molluscan sperm do. It is generally established that spermatozoa in the testis and seminal plasma have no potential for motility. Initiation of sperm motility occurs after leaving the testis and during movement through the vas deferens in internal fertilization species such as documented in elasmobranches, birds, reptiles and mammals (Alavi and Cosson, 2006), or release into the water in external fertilization such as teleosts (Stoss, 1983; Morisawa, 1985). In viviparous and ovoviviparous species, sperm motility is initiated after discharge into the female genital tract (Stoss, 1983; Billard et al., 1995). Factors regulating initiation of sperm motility in teleosts are changes in the ionic and osmotic environment of the sperm cells (reviewed by Alavi and Cosson, 2006). Composition of seminal plasma and osmolality prevent sperm motility in fish sperm duct. Sperm motility of rainbow trout (*Oncorhynchus mykiss*) is controlled by $K+$ concentration (Morisawa et al., 1983). On the contrary, sperm motility of cyprinids (Morisawa et al., 1983) and marine fishes is regulated by osmolality (Oda and Morisawa, 1993; Krasznai et al., 2003). Spermatozoa that are immotile in solutions isotonic to the seminal plasma become motile when exposed to hypotonic environments in freshwater teleosts (Perchec-Poupard et al., 1997) and hypertonic environments in marine teleosts (Oda and Morisawa, 1993). Therefore, it is critical that teleost sperm is not prematurely activated prior to chilled storage or cryopreservation for subsequent use. In contrast, crustacean (shrimp and crab) spermatozoa are immotile after released from the males into the females.

In addition to milt production, sperm quality is another important aspect that has received relatively little attention, compared with egg quality. Little is known about the determinants of sperm quality in fish, compared to mammals, although interspecific differences have been described (Levanduski and Cloud, 1988; Aas et al., 1991). To date, most studies of sperm quality in fish have focused on commercially important species, salmonids, cyprinids and marine species (Suquet et al., 2000). Sperm quality is a major issue in production aquaculture, for example in terms of fertilization capacity, potency of male spawner and other issues. Sperm quality in fish is based on analysis of a milt sample expressed by abdominal pressure. Different sperm quality parameters, such as sperm motility, sperm concentration, spermatocrit and fertilizing capacity have been used to evaluate the quality of fish sperm (reviewed by Billard et al., 1995). Sperm motility, the most commonly used parameter to evaluate sperm quality, is often described as a percentage of motile sperm and intensity and duration of motility (Chambeyron and Zohar, 1990; Miura et al., 1992). Sperm motility involves complex chemical, electrochemical and mechanochemical processes. Intrinsic processes affecting sperm motility are related to metabolic activity, such as calcium transport, protein phosphorylation and adenosine triphosphate (ATP) or cyclic adenosine 3'5' monophosphate (Benau and Terner, 1980). A relationship between sperm motility and fertilization efficiency has been assumed by several authors (Stoss, 1983; Billard and Cosson, 1992). However, sperm motility is not always a reliable indicator of fertilization efficiency (Terner, 1986; Aas et al., 1991). Sperm concentration, the number of sperm per ml of milt, is highly variable among fish species, ranging from 2×10^{6} to 6.5×10^{10} (reviewed by Leung and Jamieson, 1991). Spermatocrit is a measure of the percentage of the total volume of milt occupied by sperm. The fertilizing capacity is usually considered to be the most conclusive test for sperm quality (Aas et al., 1991). Recent advances in computer-assisted sperm analysis (CASA) technology have provided new means for assessing sperm morphology as indicative of sperm quality in fish (Kime et al., 2001). However, sperm morphology studies in fish have received relatively minor interest, compared to studies in human or domesticated animals (swine, cattle and stallions). Sperm quality parameters in fish commonly show large individual variations due to genetic variability, ageing of spermatozoa, seasonality, urine contamination and collection techniques (Leung and Jamieson, 1991; Rana, 1995; Dreanno et al., 1998). In most fish species, a decline in milt quality throughout the spawning season has been reported (Legendre and Billard, 1980; Piironen, 1985; Munkittrick and Moccia, 1987) due to ageing of spermatozoa (Rana, 1995; Suquet et al., 1998; Babiak et al., 2006a).

1.2. Short-Term Storage of Sperm

Preservation of aquatic animal sperm at low temperature is a useful biotechnology that facilitates hatchery management, and has been reported in a number of species for the purpose of aquaculture or conservation. Chilled storage is one of the most promising techniques employed to preserve sperm of fish or spermatophores of marine invertebrates, for example, striped bass, *Morone saxatilis* (Vuthiphandchai and Zohar, 1999), Atlantic cod, *Gadus morhua* (DeGraaf and Berlinsky, 2004), freshwater shrimp, *Macrobrachium rosenbergii* (Chow, 1982), lobster, *Homarus americanus* (Ishida et al., 1986), black tiger shrimp, *Penaeus monodon* (Nimrat et al., 2005), white shrimp, *Litopenaeus vannamei* (Nimrat et al., 2006), echiura, *Urechis unicinctus* (Kang et al., 2004) and Eastern oyster, *Crassostrea virginica* (Paniagua-Chavez et al., 1998). Asynchrony in gamete production between females and males, and decline in semen quality towards the end of the spawning season allows utilization of low temperature preservation of sperm for artificial propagation. Development of a reliable, simple and efficient protocol for short-term (chilled-storage) or long-term (cryopreserved) storage of sperm is a desirable method for any cultured species. However, the most important factors regulating the success of low temperature preservation include individual male variability, initial sperm quality, urine or feces contamination and storage conditions. In most fish species, contamination of sperm by urine during stripping is unavoidable due to the proximity of sperm duct and ureter. Urine contamination often leads to an alteration of semen quality and is also responsible for the observed variations of milt characteristics. The need for oxygen is also related to the ATP stores, which become depleted in anoxic conditions resulting in loss of motility. The use of antibiotics in semen to prevent microbial growth is considered as a critical aspect for prolonging viability of stored semen (Stoss, 1983, Segovia et al., 2000; DeGraaf and Berlinsky, 2004; Nimrat et al., 2005, Nimrat et al., 2006).

Several studies have been published on preservation of aquatic animal sperm at chilled-stored or sub-zero temperatures. The easiest approach is to store semen undiluted on ice (Paniagua-Chavez et al., 1998). However, the successful storage period is usually short since undiluted sperm under gaseous atmosphere becomes dehydrated due to water evaporation, oxygen depletion and bacterial growth. On the contrary, storage of diluted milt allows for a longer storage period when semen is diluted in suitable extender containing antibiotics. Chilled storage of sperm is performed by diluting spermatozoa into extender prior to storage at 0-4°C for subsequent use in artificial fertilization. However, appropriate diluents are not developed for sperm of many aquatic animal species (Rana, 1995). The suitable composition of extender provides better control of physiochemistry in term of osmolality and prolongs the viability of stored semen during storage (Scott and Baynes, 1980; Stoss, 1983). Successful storage of chilled-stored semen depends on a number of factors, including storage temperature, type of gaseous atmosphere, spermatozoa sedimentation, and replenishment of the storage medium, type of extender, dilution ratio and antibiotics (Scott and Baynes, 1980; Stoss and Refstie, 1983; Jensen and Aiderdice, 1984; Segovia et al., 2000; Babiak et al., 2006b). Short-term storage of sperm is a useful technique that facilitates hatchery management during artificial insemination. Prolonging sperm motility or viability of diluted sperm or spermatophores after post-collection has been reported with different degrees of success in a number of cultured species, i.e. Atlantic halibut, *Hippoglossus hippoglossus* L. for 79 days (Babiak et al., 2006b), Atlantic cod, *Gadus morhua* for 40 days (DeGraaf and Berlinsky et al., 2004), black tiger shrimp, *Penaeus monodon* for 35 days (Nimrat et al., 2005), white shrimp, *Litopenaeus vannamei* for 42 days (Nimrat et al., 2006), lobster, *Homarus americanus* for 289 days (Ishida et al., 1986). This technique does not require expensive equipment, and is easy to perform without specific training.

1.3. Long-Term Storage of Sperm

Cryopreservation of spermatozoa has been used in breeding of domesticated animals and is slowly being applied to aquaculture. Cryopreservation of spermatozoa of aquatic animals requires species-specific protocols, which needs optimization of equilibration period, type and concentration of cryoprotectants, cooling rate, and thawing rate (Lahnsteiner et al., 1996; Chao and Liao, 2001; Dong et al., 2005; Vuthiphandchai et al., 2007). Although cryopreservation protocols give excellent results, they are costly and difficult to apply in the field and not yet widely used. Little success has been achieved thus far in sperm cryopreservation of aquatic animals in large scale production (Christensen and Tiersch, 1997). Due to widespread use of semen banks in aquaculture, simple cryopreservation of aquatic animal sperm in a large scale is required to meet the demand of a reliable cost-effective method for artificial fertilization of the rapidly expanding numbers of lines of aquatic animal species. However, successful and reproducible cryopreservation has proven to be a difficult problem in some studies. The underlying factors responsible for success or failure include sperm collection method, type and concentration of cryoprotectant, cooling and thawing rates (Chao and Liao, 2001; Vuthiphandchai et al., 2007). Cryopreservation of spermatozoa or spermatophores of aquatic animals is usually involves dilution the spermatozoa or spermatophore into an extender containing cryoprotectant before freezing and storage in liquid nitrogen. Highest motilities and/or viability of spermatozoa have been achieved in a number of studies. The major motility and/or viability loss after freezing usually occurs at intermediate temperatures, i.e., -10 and -30°C. Membrane sensitivity associated with cryopreservation has been reported as the main factor causing cellular damage during freezing and thawing (He and Woods, 2004). Change in the membrane stability results in an increase in the permeability of membrane to water and cations, which is related to change in spermatozoa volume (Hammerstedt et al., 1990; Cabrita et al., 2001).

Cryopreservation of fish spermatozoa has been studied in a number of cultured species, especially in cyprinids (Lahnsteiner et al., 2000), salmonids (Stoss, 1983; Lahnsteiner et al., 1996), African catfish, *Clarias gariepinus* (Viveiros et al., 2000), European catfish (Linhart et al., 1993), and channel catfish, *Ictalurus punctatus* (Christensen and Tiersch, 2005). However, cryopreservation of marine invertebrates is currently limited to polycheates, cnidaria, crustaceans (shrimp and crab; Jeyalectumie and Subramoniam, 1989, Gwo, 2000; Vuthiphandchai et al., 2007), shellfish (oyster, hard clam and abalone; Gwo et al., 2002, Dong et al., 2005) and echinoderms (sea urchin and sea cucumber; Asahina and Takahashi, 1978).

2. Bacterial Contamination in the Chilled Stored- and Cryopreserved Sperm

2.1. Occurrence of Bacteria in Aquatic Animals

Recently, the occurrence of bacterial flora in aquatic animals was reported by several authors (Stoss and Refstie, 1983; Jenkins and Tiersch, 1997; Nimrat et al., 2005; Ishida et al., 2006; Nimrat et al., 2006; Nimrat et al., 2007). Different types and number of bacterial load of aquatic animals varies depending on the species of animal, and fluctuates following the change of the environments where they live (e.g. Ringo and Gatetsoupe, 1998; Gatesoupe, 1999; Colakoglu et al., 2006). A wide range of microorganisms present in natural water and aquaculture ponds have been reported (Tiersch and Jenkins, 2003). Most microflora in aquatic animals are transient and the composition of the bacterial flora in aquatic animal organs shows high variation among types of aquatic animals and different time points (e.g. Kvasnikov et al., 1977; Moriarty, 1990; Spanggaard et al., 2000; Molinari et al., 2003; Bucio et al., 2006).

In marine animals, *Vibrio* are the most common genus among the Gram-negative bacteria associated in marine animals such as crustaceans (Vanderzant et al., 1971; Kitting, 1987; Moriarty, 1990; Dempsey and; Lightner, 1993), marine fish (Sakata, 1990) and bivalves (Prieur et al., 1990). Oyster, clams and mussels have been described as natural reservoirs of *Vibrio* (Roberts et al., 1990; Rippey, 1994; Wilson and Moore, 1996; Hoi et al., 1998). *Vibrio* is also a dominant component in the gut microflora of abalone, *Haliotis discus hannai* (Tanaka et al., 2004). Normanno et al. (2006) reported the occurrence of *V. vulnificus* from retail outlets in 47 and 17 out of 600 samplings of *Mytilus galloprovincialis* accounting for 7.83% and 2.83%, respectively.

However, some *Vibrio* species (e.g. *V. harveyi*, *V. parahaemolyticus*, *V. anguillarum* and *V. vulnificus*) have been reported as the causing agents of mortalities in *Penaeus indicus* (Prayitno and Latchford, 1995), *Penaeus monodon* (Lavilla-Potigo et al., 1998), cultured penaeids (Song and Lee, 1993; Lee et al., 1996; Liu et al., 1996a, b), abalone *H. diversicolor supertexta* (Huang et al., 2001; Liu et al., 2001; Cai et al., 2006) and marine fish (Lee et al., 2002; Liu et al., 2003). *V. cholerae* were present in 1.5% of brackish water cultured tropical prawn (Reilly and Twiddy, 1992). Colakoglu et al. (2006) demonstrated that 36.9% of the pathogenic bacterial isolates belonged to the three species of *Vibrio*; *V. alginolyticus*, *V. vulnificus* and *V. parahaemolyticus*. The occurrence frequencies of these three species were as follows; 26.7% for *V. alginolyticus*, 9.4% for *V. vulnificus*, and 0.8% for *V. parahaemolyticus*.

Pseudomonas is the next commonly found Gram-negative bacteria following Vibrios in marine animals (Vanderzant et al., 1971; Dempsey and Kitting, 1987; Moriarty, 1990; Prieur et al., 1990; Sakata, 1990; Lightner, 1993). *Pseudomonas* has also been reported in Brazilian prawn hatcheries causing the "black spot" bacterial necrosis and gill obstruction (Lombardi and Labao, 1991). Sometimes, a combination of Vibrionaceae and *Pseudomonas* has been found in moribund shrimps (Costa et al., 1998). In addition, *Aeromonas* is also found in marine animals. For example, Colakoglu et al. (2006) reported that 29.1% of pathogenic bacteria belonged to the genus *Aeromonas*. However, *Aeromonas hydrophila* species was the only species identified within *Aeromonas* genus among the isolates obtained during the course of this study. Jeyasekaran et al. (2006) also found *Aeromonas* as one of the bacterial flora of Indian white shrimp (*Penaeus indicus*) accounting for 38% of the total bacteria flora.

Besides the commonly found bacteria as described, bacterial flora of Indian white shrimp (*Penaeus indicus*) also consists of a variety types of bacteria such as *Aeromonas*, *Pseudomonas*, *Vibrio*, *Flavobacterium and Serratia* (Jeyasekaran et al., 2006). Moreover, molluscan shellfish are also prone to contamination by fecal pathogens, primarily, *Salmonella*, *Shigella*, *Escherichia coli* and enteric viruses from sewage-polluted waters (Roberts et al., 1990; Rippey, 1994; Wilson and Moore, 1996). Reilly and Twiddy (1992) found *Salmonella* in 16.0% of brackish water cultured tropical prawn. In addition, other Gram-negative rods including *Flavobacterium*, *Pseudomonas* and *Aeromonas*, as well as the Gram-positive bacteria, *Mycobacterium marinum* and the occasional *Micrococcus*, may be involved in bacterial disease syndromes in penaeid shrimp (Costa et al., 1998). As a consequence, several types of pathogenic and nonpathogenic bacteria have been described. Table 1 shows examples of the bacterial genera isolated from two types of shrimps.

In freshwater aquatic animals, *Aeromonas*, *Plesiomonas* and Enterobacteriaceae are dominant bacteria (Sakata, 1990). *Aeromonas* are motile Gram-negative, facultatively anaerobic mesophilic, non-spore forming, rod-shaped bacteria (Roberts et al., 1996; Schubert 1991). Its principal reservoir is the aquatic environment such as freshwater lakes and streams, sewage and brackish water and wastewater systems. *A. hydrophila* is a significant bacterium which constitutes part of the normal flora of fish and other aquatic animals (Popovic et al., 2000) but it is probably the most commonly encountered bacterial pathogen of freshwater fishes. Several species of these motile rods may be responsible for motile aeromonad disease (MAD), also known as motile aeromonad septicemia (MAS).

Table 1. Frequency of the bacterial genera isolated from penaeid shrimps

Identification	*Penaeus stylirostris* (Costa et al., 1998)	*Penaeus indicus* (Jeyasekaran et al., 2006)
Vibrionaceae	29.7 %	57 %
Pseudomonadaceae	11.45 %	28 %
Bacillus	2.41 %	-
Staphylococcus	4.22 %	-
Corynebacterium	0.4 %	-
Alteromonas	1.2 %	-
Flavobacterium	1.61 %	8 %
Micrococcus	0.6 %	-
Acinetobacter	0.8 %	-
Moraxella	1.2 %	-
Cytophaga	0.2 %	-
Flexibacter	0.4 %	-
Serratia	-	7 %

In addition to *A. hydrophila*, bacteria which have been implicated in MAD include *A. sobria*, *A. caviae* and *A. veronii*, these ubiquitous organisms and opportunistic pathogens take advantage of stressed and immunocompromised fish.

Lactic acid bacteria are also found in freshwater and other aquatic animals even though they are not a dominant bacterial group (Ringo and Gatesoupe, 1998). Lactic acid bacteria are Gram-positive, usually nonmotile, nonsporulating bacteria, containing both rods (lactobacilli and carnobacteria) and cocci (streptococci). This group consists of *Streptococcus*, *Leuconostoc*, *Pediococcus*, *Aerococcus*, *Enterococcus*, *Vagococcus*, *Lactobacillus* and *Carnobacterium* (Ringo and Gatesoupe, 1998). Lactobacilli were reported for the first time from the intestines of wild European eel, perch, rudd, ruffe, bleak, silver bream, chub, somnul and farmed African catfish (Kvasnikov et al., 1977; Ringo and Gatesoupe, 1998; Gonzalez et al., 2000; Bucio et al., 2006). In addition, Kvasnikov et al. (1977) found the incidence of lactobacilli and other lactic acid-producing bacteria from the intestines of several types of fish at different stages (larval, fry and fingerling stages) cultured in Ukrania. The presence of other lactic acid bacteria, especially carnobacteria within the intestinal tracts of salmonid species like Arctic charr (*Salvelinus alpinus*), Atlantic salmon (*Salmo salar*), rainbow trout (*Oncorhyncus mykiss*) (Ringo and Gatesoupe, 1998; Ringo et al., 1995; Ringo and Olsen, 1999; Spanggaard et al., 2000) and brown trout (Gonzalez et al., 2000), has also been reported.

Generally, lactic acid bacteria are considered to be nonpathogenic bacteria however some types of lactic acid bacteria (e.g. *Streptococcus*, *Enterococcus*, *Lactobacillus*, *Carnobacterium* and *Lactococcus*) can cause the diseases, for example, in rainbow trout (Hoshina, 1956) and yellowtail (Kusuda et al., 1976). The presence of *Enterococcus* and *Lactococcus garviae*, an emerging zoonotic pathogen that has been isolated from cattle, fish, and humans, were observed in common carp (*Cyprinus carpio*) collected from the Thajin river of Thailand. (Cai et al., 1999). Table 2 summarizes the properties of lactic acid bacteria and some pathogenic lactic acid bacteria in a variety of fish.

Table 2. Properties of lactic acid bacteria and some reports of diseases caused by lactic acid bacteria in cultured fish species

Genus	Pathogenic lactic acid bacteria	Fish species	References
Streptococcus	*Streptococcus*	*Plecoglossus altivelis* *Paralichthys olivaceus* *Fundulus grandis* *Siganus canaliculatus* Atlantic salmon Grey mullet Channel catfish	Kitao et al. (1981) Miyazaki (1982) Rasheed and Plumb (1984) Foo et al. (1985) Karaseva et al. (1991) Lin et al. (1993) Teska and Shotts (1994)
	S. difficile	Rainbow trout *Tilapia*	Eldar et al. (1994) Eldar et al. (1994)
	S. faecalis subsp. *liquefaciens*	Rainbow trout *Ictalurus nebulosus*	Ghittino and Prearo (1992) Teskeredzic et al. (1993)
	S. iniae[a]	Yellowtail Rainbow trout Hybrid tilapia Hybrid striped bass *Siganus fuscescens*	Kusuda (1992) Eldar et al. (1994) Perera et al. (1994) Stoffregen et al. (1996) Sugita (1996)
	S. milleri group	Carp	Austin and Robertson (1993)
	S. parauberis	Turbot	Domenech et al. (1996)
Lactococcus	*L. garlieae*[b]	Yellowtail Rainbow trout	Kusuda et al. (1991) Ghittino et al. (1995)
	L. piscium	Rainbow trout	Williams et al. (1990)
Enterococcus	*Enterococcus*-like	Rainbow trout Turbot	Carson et al. (1993) Toranzo et al. (1994)
	Enterococcus	Turbot	Toranzo et al. (1995)
Vagococcus	*V. salmoninarum*	Salmonids	Schmidtke and Carson (1994)
Lactobacillus	*Lactobacillus*	Pacific salmon Rainbow trout	Earp et al. (1953) Evelyn and McDermot (1961) Ross and Toth (1974)
Carnobacterium	*Carnobacterium*	Channel catfish *Ictalurus nebulosus* Striped bass	Baya et al. (1991) Baya et al. (1991) Baya et al. (1991)
	C. piscicola[c]	Salmonids Rainbow trout	Hiu et al. (1984) Toranzo et al. (1993)

[a] Junior synonym *S. shiloi* (Eldar et al., 1994, 1995).

[b] First described as *Streptococcus* (Kusuda et al., 1976., then as *E. seriolicida* (Kusuda et al., 1991; Domenech et al., 1993; Eldar et al., 1996).

[c] First described as *L. piscicola* (Collins et al., 1987).

As described previously, the associated microorganisms in both marine and freshwater animals may vary with environmental factors (temperature, salinity and pH) in their environments (Lesel, 1990; Hamid et al., 1978; Sakata et al., 1980; Ringo and Strom, 1994; Reilly and Twiddy, 1992). For an example, Kvasnikov et al. (1977) suggested that the changes in the composition of lactobacilli and other lactic acid-producing bacteria were related to the season of the year and life-stage of the fish. With differences in temperature exposure, Bucio et al. (2006) reported that there was a change in the number of lactobacilli species in the intestines of edible fresh water fish during the warm and cold seasons. Higher numbers of bacteria were isolated in the warm than in the cold season. In contrast, lactobacilli were found in the intestines of edible farmed fish reared in a recirculation system even under the warm water conditions. This information confirms that the changes in the environments between natural (river: warm and cold water) and aquaculture environments (a recirculation system) can lead to changes in numbers of the lactobacilli bacteria in freshwater fish in Netherlands

The percentage of isolated species and the percentage of isolates from any particular species vary significantly among three gastrointestinal tract regions. Molinari et al. (2003) reported that nile tilapia (*Oreochromis niloticus*) had higher bacterial numbers in anterior and posterior gut than in stomach of fish cultured in a semi-intensive system. *Aeromonas hydrophila*, *A. veronii*, *Burkholderia cepacia*, *Chromobacterium violaceum*, *Citrobacter freundii*, *Escherichia coli*, *Flavimonas oryzihabitans* and *Plesiomonas shigelloides* were the most frequently isolated Gram-negative bacilli. From these samples, only *P. shigelloides* was present in all three gastrointestinal regions, while *A. hydrophila*, *E. coli*, and *F. oryzihabitans* were isolated only in the stomach. *Citrobacter freundii* and *B. cepacia* were found only in the posterior gut. *Chromobacterium violaceum* was the dominant bacteria isolated from stomach and anterior gut.

Mahmoud et al. (2004) reported that the bacterial flora of common carp (*Cyprinus carpio*) were identified to seven genera and two families: *Acinetobacter*, *Alcaligenes*, *Bacillus*, *Flavobacterium*, *Micrococcus*, *Moraxella*, *Pseudomonas*, Enterobacteriaceae and Vibrionaceae. The predominant bacteria of carp in this study consisted of *Flavobacterium* (37%) and Vibrionaceae (33%) in skin, *Flavobacterium* (33%) in gill and Vibrionaceae (63%) and *Flavobacterium* (37%) in intestine. Kennedy et al. (2006) reported that larval samples of giant freshwater prawn, *Macrobrachium rosenbergii*, had a bacterial load of 10^6 to 10^7 CFU/10 larval and *Vibrio* ranged from 10^2 to 10^4 CFU/10 larvae. Among bacterial flora of commom carp (*Cyprinus carpio*), Gram-negative bacteria were founded to be more dominant than Gram-positive bacteria. The predominant Gram-negative bacteria comprised of *Aeromonas*, *Pseudomonas* and *Vibrio* while the dominant Gram-positive bacteria were *Bacillus* and non-spore formers. Anderson et al. (1989) identified *Alcaligenes* and *Enterobacter* from *M. rosenbergii* larvae with mid-cycle disease. Interestingly, luminescent *Vibrio* has not only caused serious mortalities in marine shrimp (Lavilla-Pitogo et al., 1990; Karunasagar et al., 1994) but also has been reported to cause serious mortalities in *M. rosenbergii* hatcheries (Tonguthai, 1995). Miyamoto et al. (1983) identified 13 genera of bacteria from *M. rosenbergii* larvae, including *Aeromonas*, *Pseudomonas* and *Vibrio* as the dominant flora, while Anderson et al. (1989) who studied the aerobic heterotrophic bacterial flora in *M. rosenbergii* hatcheries in Malaysia have reported *Alcaligenes* and *Vibrio* to be the most frequently encountered genera.

2.2. Type and Number of Bacteria in Sperms of Aquatic Animals

Although the microflora of aquatic animals has been reported by several authors (Ringo and Gatesoupe, 1998; Colakoglu et al., 2006), few studies have monitored the microflora in spermatophores or sperm in aquatic animals (Jenkins and Tiersch, 1997; Nimrat et al., 2006; Nimrat et al., 2007). Martin et al. (1987) reported that bacterial colonization was present on the hardened spermatophore surface of spiny lobster (*Panulirus interruptus*), but was not found to penetrate into deeper layers of spermatophores due to antibacterial activity of phenolic substances inside the spermatophores. Nimrat et al. (2005) reported the bacterial profile and the identity of dominant bacteria in spermatophores of black tiger shrimp (*Penaeus monodon*) after chilled storage with extenders at 2-4°C. At the beginning of the experiment, average values of total bacteria count in spermatophores of black tiger shrimp ranged between $(10.3\pm2.3) \times 10^3$ and $(17\pm4.7) \times 10^3$ CFU/g. The dominant types of those bacteria were *Bacillus*, *Staphylococcus* and *Pseudomonas aeruginosa*. Jeyasekaran et al. (2006) reported *Aeromonas*, *Pseudomonas*, *Vibrio*, *Flavobacterium* and *Serratia* to be predominant in fresh raw shrimp. Those characteristics of some dominant bacteria found in spermatophores of *P. monodon* during chilled storage are shown in Table 3.

Of the bacterial taxa detected in this study, *Bacillus* is a Gram-positive, rod shaped with a large spore in the vegetative cell; colonies were described as white colonies with a rough perimeter. This genus was better adapted to live in soil and water (Gatesoupe, 1999; Green et al., 1999; Moriarty, 1999).

Table 3. Characteristics used to differentiate some dominant bacteria in spermatophores of *P. monodon* (Nimrat et al., 2005)

Characteristics	*Staphylococcus*	*Bacillus*	*Pseudomonas aeruginosa*
Gram reaction	Positive	Positive	Negative
Cell morphology	Cocci (cluster)	Rod	Rod
Blood agar hemolysis	Alpha-hemolysis	Gamma-hemolysis	Gamma- hemolysis
Growth on MacConkey	-	-	+
Catalase test	+	+	+
Oxidase test	-	-	+
Coagulase test	-		
String test			+
TSI (Triple sugar slant/acid, iron) test	Acid slant/acid, H_2S -, gas -	Alkaline slant/alkaline, H_2S -, gas -	Acid H_2S -, gas -
Motility test	-	+	+
Utilization of citrate	-	-	+
LIA (Lysine iron agar) test	+	-	-
Utilization of urea	-	-	-
Growth in 6% NaCl	+	-	-
OF glucose test			+/-

Staphylococcus is Gram-positive, spherical in shape, formed characteristic of yellow colonies with a round prominent shape and smooth perimeter on the surface of culture media.

Pseudomonas species, commonly found bacteria in soil and water, are non-enteric Gram-negative bacilli that are primarily human opportunists (Allen et al., 2004; Palleroni, 2005). This genus can be found in the environment (Sader and Jones, 2005). Additionally, *Pseudomonas* was one of the most frequent shrimp pathogens, representing 7.2% of all bacterial pathogens. Mortality rates of *Pseudomonas* infections are generally high in marine shrimp (Costa et al., 1998). *Pseudomonas aeruginosa* is a leading cause of nosocomial infection for humans (Hanberger et al., 2001; Jones et al., 2002).

2.3. Bacterial Profiles in the Chilled Stored Sperm/Spermatophores of Aquatic Animals

One of the most important parameters representing the quality and the shelf-life of chilled-stored sperm/spermatophores in aquatic animals is the microbiological profile in sperm/spermatophores (Jenkins and Tiersch, 1997; Nimrat et al., 2005; Nimrat et al., 2006). Microbiological profiles and management practices can influence the success of the production (Kennedy et al., 2006). Therefore, documentation of the number and types of bacterial contamination in chilled-stored sperm is required. The presence of bacteria in the sperm mass of preserved spermatophores has also been reported for lobster, *H. americanus* by Ishida et al., 1986, and for the freshwater prawn, *M. rosenbergii* by Chow, 1982. In fish and crab, bacterial load on the sperm mass or spermatophores has been also reported (Stoss and Refstie, 1983; Saad et al., 1988; Jenkins and Tiersch, 1997; Benhalima and Moriyasu, 2001).

Nimrat et al. (2005) studied chilled-stored spermatophores collected from wild-caught black tiger shrimp broodstock and found initial bacterial loads of 10^4 CFU/g, which then increased significantly during 42 days of storage. *Pseudomonas aeruginosa* was prevalent at the beginning of the experiment at $6.7 x 10^3$ CFU/g. In addition, *Bacillus*, *Staphylococcus* and *Pseudomonas aeruginosa* were identified as the predominant bacteria occurring during chilled storage of black tiger shrimp spermatophores. Nimrat et al. (2006) developed a protocol for the chilled storage of white shrimp spermatophores and to evaluate levels of bacterial propagation during such storage. Spermatophores were immersed in four extenders, mineral oil, Ringer's solution, phosphate buffer and 0.85% NaCl, and stored at low temperature (2-4°C) for 35 days. Characteristics of preserved spermatophores changed the least and viable sperm was highest when spermatophores were stored in mineral oil. Spermatophores preserved with mineral oil appeared morphologically normal. *Bacillus circulans*, *Staphylococcus hominis* and *S. lugdunensis*, *S. sciuri*, *S. xylosus* and *Micrococcus* were identified as the predominant bacteria during chilled storage, and total bacterial counts gradually increased during the experiment.

2.3.1. Effect of Extenders on Bacterial Profiles in the Chilled Storage Process

Extender is considered to be an important buffer consisting of essential inorganic compounds and sugars that determine success or failure of sperm chilled storage. Different types of extenders used in the chilled storage of sperm/spermatophores are reported to preserve the morphology, motility and/or viability of sperm/spermatophores (Scott and

Baynes, 1980; Ishida et al., 1986; Vuthiphandchai and Zohar, 1999; Nimrat et al., 2005; Nimrat et al., 2006; Babiak et al., 2006c).

Nimrat et al. (2005) have reported the influence of different types of extenders for chilled storage of black tiger shrimp spermatophores and to evaluate bacterial propagation during spermatophore chilled storage. Spermatophores were selected and preserved using four different extenders, namely mineral oil, Ringer's solution, phosphate buffer and 0.85% sodium chloride, and stored at low temperature (2-4°C) for 42 days. Results showed that there were three dominant types of bacteria present in preserved spermatophores, *Bacillus*, *Staphylococcus* and *Pseudomonas aeruginosa*_(Table 4). The presence of a *Bacillus* was observed prominently in spermatophores preserved with Ringer's solution, compared to other extenders (Figures 1). Highest values of *Bacillus* in spermatophores preserved with Ringer's solution (30.3±8.1) x 10^3 CFU/g) were observed on day 28. The number of *Staphylococcus* on spermatophores preserved with mineral oil was significantly higher ($P<0.05$) than that found for other extender treatments, with values of (219.3±4.9) x 10^3 CFU/g on day 42 (Figures 2). The number of *P. aeruginosa* present on spermatophores preserved with phosphate buffer and 0.85% NaCl was significantly higher ($P<0.05$) than those of mineral oil and Ringer's solution (Figures 3).

Table 4. Characteristics used to differentiate the dominant bacteria (Nimrat et al., 2005)

Characteristics	*Staphylococcus*	*Bacillus*	*Pseudomonas aeruginosa*
Gram reaction	Positive	Positive	Negative
Cell morphology	Cocci (cluster)	Rod	Rod
Blood agar hemolysis	Alpha-hemolysis	Gamma-hemolysis	Gamma- hemolysis
Growth on MacConkey	-	-	+
Catalase test	+	+	+
Oxidase test	-	-	+
Coagulase test	-		
String test			+
TSI (Triple sugar slant/acid, iron) test	Acid slant/acid, H_2S -, gas -	Alkaline slant/alkaline, H_2S -, gas -	Acid H_2S -, gas -
Motility test	-	+	+
Utilization of citrate	-	-	+
LIA (Lysine iron agar) test	+	-	-
Utilization of urea	-	-	-
Growth in 6% NaCl	+	-	-
OF glucose test			+/-

Figure 1 Contamination of *Bacillus* (x 10^3 CFU/g) in black tiger shrimp spermatophores during a 42-d storage period with various extenders. Bars with different letters are significantly different ($P<0.05$) within the same sampling date (Nimrat et al., 2005).

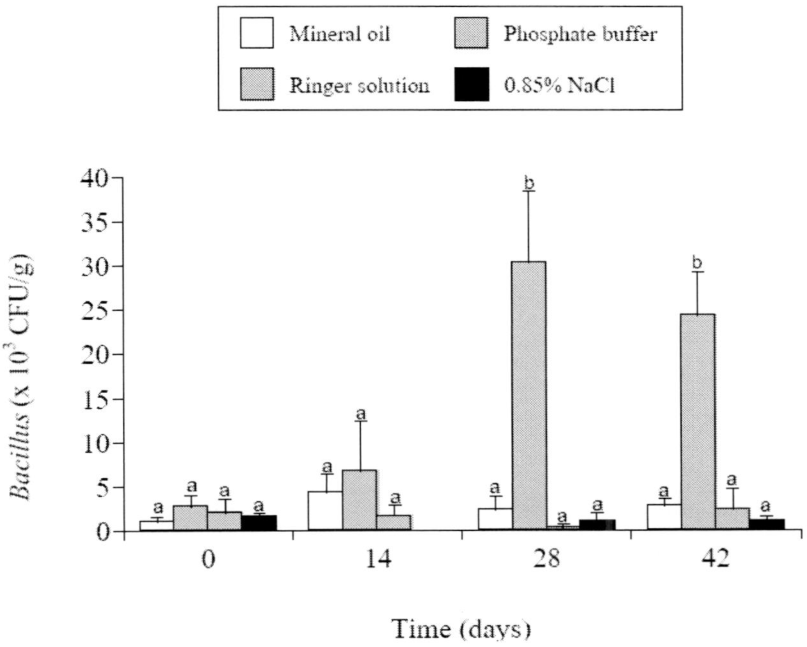

Figure 1. Contamination of *Bacillus* (x 10^3 CFU/g) in black tiger shrimp spermatophores during a 42-d storage period with various extenders. Bars with different letters are significantly different ($P<0.05$) within the same sampling date (Nimrat et al., 2005).

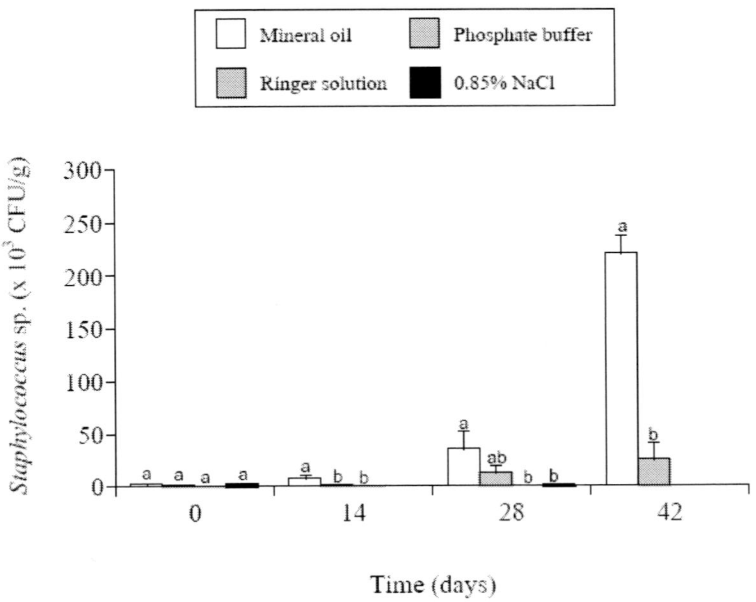

Figure 2. Contamination of *Staphylococcus* (x 10^3 CFU/g) in black tiger shrimp spermatophores during a 42-d storage period with various extenders. Bars with different letters are significantly different ($P<0.05$) within the same sampling date (Nimrat et al., 2005).

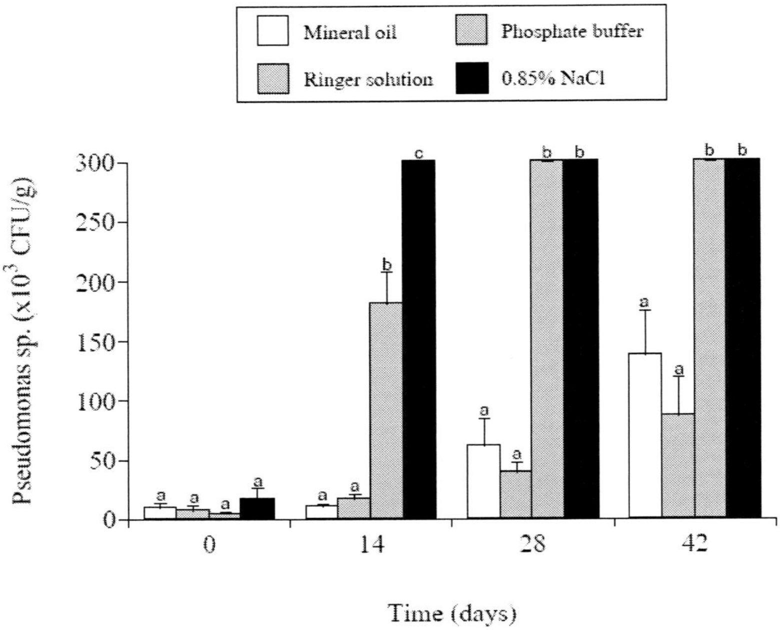

Figure 3. Contamination of *Pseudomonas aeruginosa* (x 10^3 CFU/g) in black tiger shrimp spermatophores during a 42-d storage period with various extenders. Bars with different letters are significantly different (P<0.05) within the same sampling date. Values of 300 x 10^3 CFU/g on the Y-axis represented the number of bacteria over 300 x 10^3 CFU/g (Nimrat et al., 2005).

Figure 3 Contamination of *Pseudomonas aeruginosa* (x 10^3 CFU/g) in black tiger shrimp spermatophores during a 42-d storage period with various extenders. Bars with different letters are significantly different (P<0.05) within the same sampling date. Values of 300 x 10^3 CFU/g on the Y-axis represented the number of bacteria over 300 x 10^3 CFU/g (Nimrat et al., 2005).

Nimrat et al. (2006) showed that the different types of extenders used in chilled storage resulted in occurrence of different numbers and types of bacteria present in preserved spermatophores of white shrimp (*Litopenaeus vannamei*). Table 5 shows the total bacterial counts present in *L. vannamei* spermatophores during a 35-day chilled-stored period with various extenders. At the beginning of the experiment (day 0), no bacteria were detected in any spermatophores. However, total bacterial counts in all treatments had increased slightly by day 14 of the storage period, with means ranging from 31.7±6.5 CFU/g to 50±3.7 CFU/g. After 35 days storage, spermatophores in the mineral oil had the highest level (2050.0±180.6 CFU/g) but this was not significantly higher (P>0.05) than the levels in Ringer's solution and phosphate buffer. Identification of bacteria by various biochemical tests and API kits indicated six dominant types in preserved spermatophores, *Bacillus circulans*, *Staphylococcus hominis*, *S. lugdunensis*, *S. sciuri*, *S. xylosus* and *Micrococcus*. *Bacillus* colonies appeared as white colonies with a rough perimeter, Gram-positive and rod shaped with a large spore in the vegetative cell. *Staphylococcus* and *Micrococcus* formed yellow or white colonies with a round prominent shape and smooth perimeter on the surface of culture media; they were Gram- positive and spherical in shape. *B. circulans* was only observed to occur in spermatophores preserved with mineral oil after 21 days storage. *S. hominis* was only observed on day 14 of storage in all treatments. Number of *S. lugdunensis* present in spermatophores preserved with mineral oil was significantly higher ($P<0.05$) than that of other extenders. However, S. *lugdunensis* was only observed on day 14 of storage. Numbers of *S. sciuri* and *Micrococcus* in spermatophores preserved with Ringer's solution were significantly higher ($P<0.05$) than those found for other extender treatments. Number of *S. xylosus* in spermatophores preserved with 0.85% NaCl was significantly higher ($P<0.05$) than those found in the other three treatments. There were no *Vibrio*, *Pseudomonas* or *Aeromonas* detected during the experiment.

2.3.2. Role of Contaminated Bacteria in Chilled Storage Process

The presence of bacteria in the sperm mass of chilled-stored spermatophores has been reported in some species of aquatic animals such as in freshwater shrimp (Chow, 1982), lobster (Ishida et al., 1986), black tiger shrimp (Nimrat et al., 2005) and white shrimp (Nimrat et al., 2006). Bacteria may cause a negative effect on fertilization rate of preserved spermatophores and on the health of artificially inseminated female broodstock or seed. Assessment of bacterial propagation during chilled storage of spermatophores provides information on the status of bacterial contamination of preserved spermatophores, which can be controlled by antibiotics. Although maintenance of spermatophores at low temperatures of about 2-4°C may aid in retarding but not preventing the growth of bacteria during storage (Ishida et al., 1986). During this storage, an increase in bacterial growth still occurred because sperm mass acted as an enriched food source for those bacteria (Subramoniam, 1993). This phenomenon was confirmed by Nimrat et al. (2005), who preserved *P. monodon* spermatophores at 2-4°C. The presence of bacteria in the sperm mass of preserved spermatophores has also been reported in lobster, *Homarus americanus* (Ishida et al., 1986).

The increase in bacterial population caused the degenerative changes of spermatophores (Gonzalez-Rodriguez et al., 2002; Nimrat et al., 2005; Nimrat et al., 2006). For instance, Chow (1982) reported that the propagation of bacilli inside spermatophores of freshwater shrimp, *Macrobrachium rosenbergii* during chilled storage accelerated the degeneration of matrices, which resulted in the degeneration of sperm (Subramoniam, 1993). Bray et al. (1985) reported that in 60% of cases, the presence of bacteria in the spermatophores of the white shrimp *P. setiferus* resulted in a reduction in the number of viable sperm cells.

Table 5. Total bacterial counts (CFU/g) present in white shrimp spermatophores during a 35-day storage period with various extenders at 2-4°C (Nimrat et al., 2006)

Extenders	Storage period (days)					
	0	7	14	21	28	35
Mineral oil	$0^{a,2}$	$0^{a,2}$	$36.7±4.2^{ab,2}$	$88.3±7.9^{ab,2}$	$2,550.0±574.2^{a,1}$	$2,050.0±180.6^{a,1}$
Ringer's solution	$0^{a,3}$	$0^{a,3}$	$31.7±6.5^{b,3}$	$55.0±10.6^{c3}$	$1,916.7±83.3^{a,1}$	$1,720.0±122.9^{ab,2}$
Phosphate buffer	$0^{a,2}$	$0^{a,2}$	$38.3±7.0^{ab,2}$	$65.0±10.9^{bc2}$	$1,850.0±199.5^{a,1}$	$1,666.7±132.6^{ab,1}$
0.85% NaCl	$0^{a,3}$	$0^{a,3}$	$50.0±3.7^{a,3}$	$116.7±9.9^{a3}$	$1,650.0±143.2^{a,1}$	$1,383.3±86.3^{b,2}$

Values with different letters in a column are significantly different among treatments ($P<0.05$)
Values with different numbers in a row are significantly different among sampling period ($P<0.05$)

Dougherty and Dougherty (1989) also reported development of varying numbers of individual and aggregated yellow-brown to black polymorphic pigmented droplets and degenerative sperm in black spermatophores collected from pond-cultured *L. vannamei*. In addition, Nimrat et al. (2006) indicated that bacterial contamination degenerated the preserved spermatophores of white shrimp (*L. vannamei*)

2.3.3. Effect of Antibiotics on Contaminated Bacteria in Chilled Stored Sperm

Preserved sperm or spermatophore samples of aquatic animals is far from being bacteria-free. Aquatic animal broodstocks are normally raised in the hatcheries or ponds that are not completely sterile. Consequently, freshly collected sperm or spermatophores from a number of aquatic animals commonly contain bacterial contamination. As bacterial contamination is usually difficult to avoid when collecting sperm or spermatophores, preservative antimicrobials have been regarded an essential constituent in sperm extender compositions. Early sperm extender formulae were mostly relied on a combination of penicillin and streptomycin. Presently, more effective broad-spectrum antimicrobials are available for use.

When bacteria contaminate preserved seeds, they degenerate the quality and shelf-life of chilled stored sperm as previously described (Nimrat et al., 2005; Nimrat, 2006). In the case of being contaminated with pathogenic bacteria for aquatic animals, the hazardous pathogens can proliferate and pose a serious threat to the offspring of those aquatic animals when seeds are fertilized and cultured. Those pathogenic bacteria can cause disease outbreaks in the next cultivation. Additionally, when the chilled sperm/spermatophores are contaminated with human pathogenic bacteria, they may serve as a vehicle in the transmission of foodborne infections for human who consume this contaminated aquatic animals (Hansen et al., 1993; Tulsner, 1994). Therefore, effective processes for removal of contaminated bacteria in chilled storage of sperm/spermatophores should be further developed. Antibiotics are considered to be capable of inhibiting or killing the undesired bacteria in contaminated sperm/spermatophores. However, there are few reports regarding the use of antibiotics to control and prevent the growth of bacteria during chilled storage of sperm/spermatophores in aquatic animals e.g. common carp, *Cyprinus carpio* (Saad et al., 1988), Atlantic salmon, *Salmo salar* (Stoss and Refstie, 1983), nile tilapia, *Oreochromis niloticus*, black tiger shrimp, *P. monodon* (Nimrat et al., 2005) and white shrimp, *L. vannamei* (Nimrat et al., 2006).

Nimrat et al. (2005) reported the use of a combination of penicillin and streptomycin at concentrations of 0.1%, 1%, 2% and 3% in chilled-stored spermatophore of black tiger shrimp (*P. monodon*) for controlling the types and number of bacterial contamination in preserved spermatophores. Spermatophores preserved in mineral oil with antibiotic did not appear to have been affected in terms of their morphology or color. However, the apparent viability of sperm was significantly improved after 7 days storage when antibiotic was present. Similar values of sperm viability after 42 days with and without penicillin-streptomycin 0.1%, 1%, 2% and 3% may suggest that additional antibiotic use had no apparent detrimental effect on sperm viability. However, supplementation of penicillin-streptomycin at the concentrations of 0.1%, 1%, 2% and 3% can efficiently reduce bacterial contamination in preserved spermatophores since bacterial contamination was greatly reduced after 14 days before declining to an undetectable level after 42 days of chilled storage. There was no significant difference ($P>0.05$) in total bacteria counts of spermatophores preserved with four levels of penicillin-streptomycin (Figures 4). Penicillin-streptomycin at 0.1% to 3% strongly inhibited bacteria growth during chilled storage of *P. monodon* spermatophores.

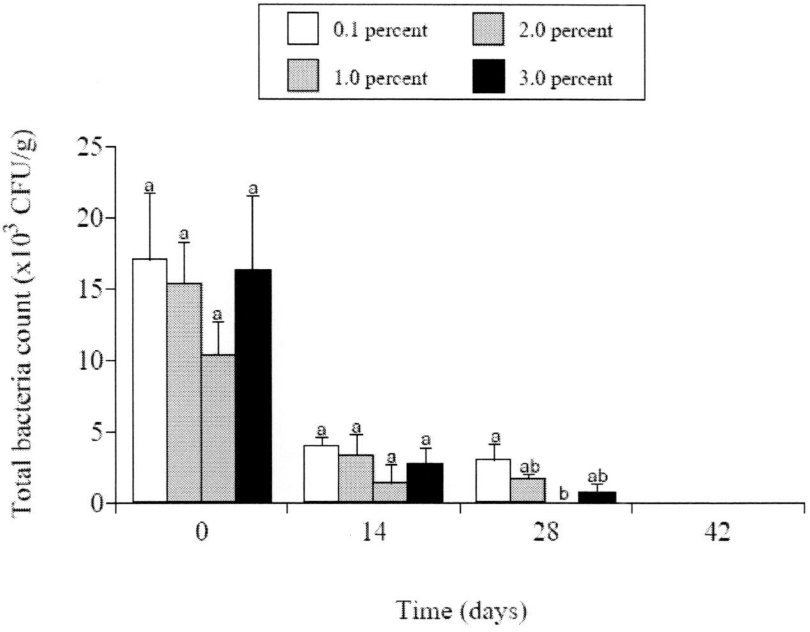

Figure 4.Total bacterial count (x10^3 CFU/g) present in black tiger shrimp spermatophores during a 42-d storage period with mineral oil supplemented with penicillin-streptomycin at 0.1%, 1%, 2% and 3%. Bars with different letters are significantly different ($P<0.05$) within the same sampling date (Nimrat et al., 2005).

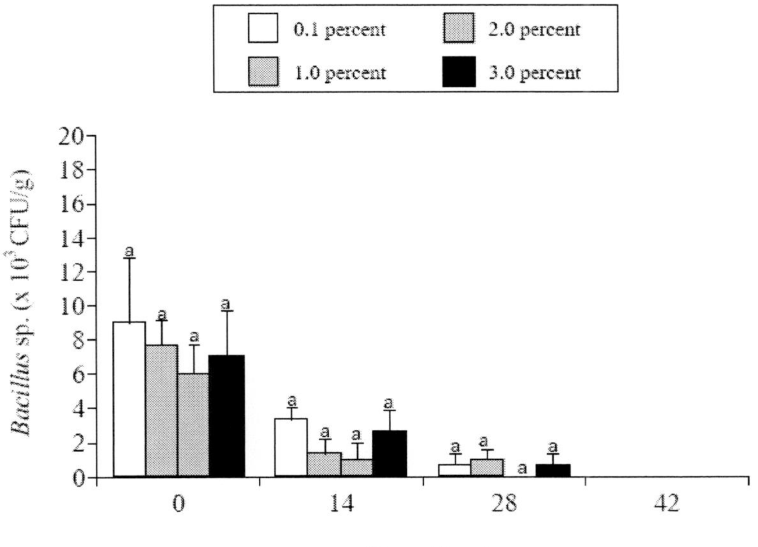

Figure 5. Contamination of *Bacillus* (x 10^3 CFU/g) in black tiger shrimp spermatophores during a 42-d storage period with mineral oil supplemented with penicillin-streptomycin at 0.1%, 1%, 2% and 3%. Bar with similar letter is not significantly different ($P>0.05$) within the same sampling date (Nimrat et al., 2005).

Three predominant bacteria, *Bacillus*, *Staphylococcus* and *Ps. aeruginosa*, were also present in preserved spermatophores containing varying levels of penicillin-streptomycin during the storage period. The number of *Bacillus*, *Staphylococcus* and *Ps. aeruginosa* found among treatments at the beginning of experiment ranged from (6±1.7) x 10^3 to (9±3.8) x 10^3 CFU/g, (1±0.6) x 10^3 to (4.3±1.9) x 10^3 CFU/g, and (3.3±0.9) x 10^3 to (6.7±3.4) x 10^3 CFU/g, respectively, and declined to an undetectable level by day 42 (Figures 5-7). These findings indicate the addition of all four concentrations of penicillin-streptomycin efficiently controlled bacteria growth during the chilled storage of spermatophores. Taking into account cost effectiveness and less concentration of antibiotics utilized, 0.1% penicillin-streptomycin should be suitable for bacterial control in chilled storage of *P. monodon* spermatophores.

Nimrat et al. (2006) have demonstrated the effect of antibiotic concentrations (penicillin-streptomycin) on spermatophores of white shrimp (*L. vannamei*) by chilled storage. Spermatophores were preserved in only mineral oil with or without 0.1% penicillin-streptomycin, and evaluated for changes in external morphology of spermatophores, sperm viability and total bacteria count every week during a 35-day experimental period. Percentages of viable sperm (69.5±3.9%) were significantly higher ($P<0.05$) among spermatophores preserved in mineral oil with 0.1% antibiotic compared with those preserved only in mineral oil (57.7±3.4%) over 35-days. The number of total bacteria in the treatment with mineral oil ranged between 28.3±4.8 and 2,416.7±299.4 CFU/g, but in mineral oil containing antibiotic bacteria were undetectable as shown in Table 6.

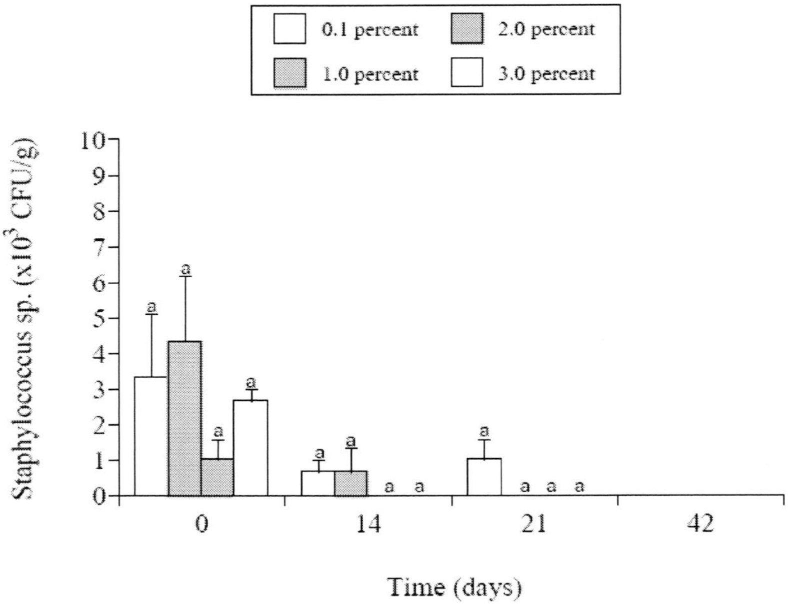

Figure 6. Contamination of *Staphylococcus* (x 10^3 CFU/g) in black tiger shrimp spermatophores during a 42-d storage period with mineral oil supplemented with penicillin-streptomycin at 0.1%, 1%, 2% and 3%. Bar with similar letter is not significantly different ($P > 0.05$) within the same sampling date (Nimrat et al., 2005).

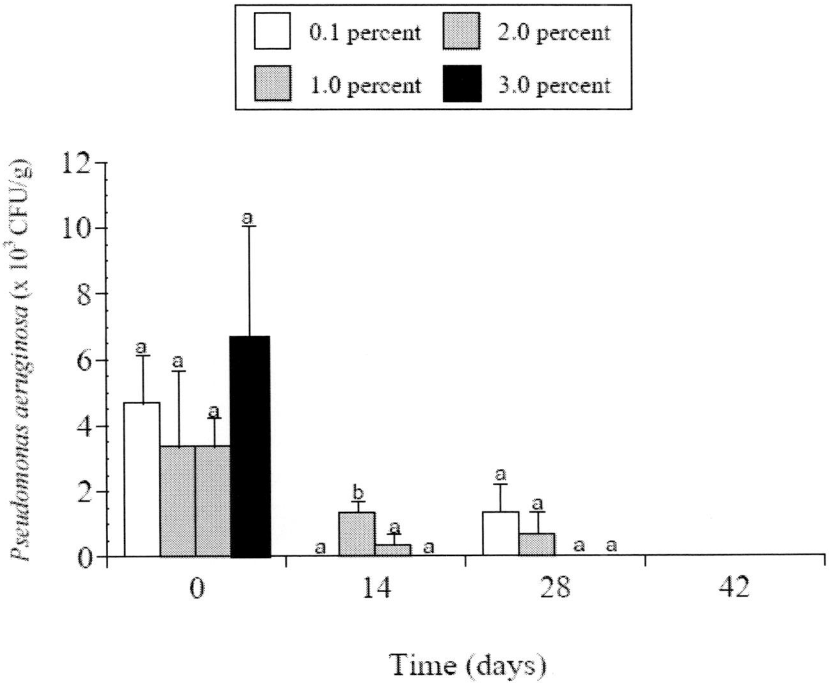

Figure 7. Contamination of *Pseudomonas aeruginosa* (x 10^3 CFU/g) in black tiger shrimp spermatophores during a 42-d storage period with mineral oil supplemented with penicillin-streptomycin at 0.1%, 1%, 2% and 3%. Bar with similar letter is not significantly different (P>0.05) within the same sampling date (Nimrat et al., 2005).

2.4. Bacterial Profiles in Cryopreserved Sperm/Spermatophores of Aquatic Animals

2.4.1. Change in Bacterial Flora of Cryopreserved Sperm/Spermatophores During Storage in Liquid Nitrogen

Recently, investigation in spermatophore cryopreservation of invertebrates has been successful in some species. However, limited work has focused on the effects of bacterial contamination on cryopreserved spermatophores.

Nimrat et al. (2007) reported on the bacterial flora of black tiger shrimp (*P. monodon*) spermatophores stored in liquid nitrogen for 210 days and found that total heterotrophic bacteria in fresh and cryopreserved spermatophores held in liquid nitrogen for 1 h were not significantly different (P>0.05), with averages of 46.3 ± 2.9 x 10^4 CFU/g and 44.3 ± 3.3 x 10^4 CFU/g, respectively (Table 7). Number of total heterotrophic bacteria decreased significantly (P<0.05) to 2.1 ± 0.4 x 10^4 CFU/g after storage for 30 days, and 0.05 ± 0.02 x 10^4 CFU/g at the end of experiment.

Table 6. Total bacterial counts (CFU/g) present in white shrimp spermatophores during a 35-day storage period in mineral oil with and without antibiotic (0.1% penicillin-streptomycin) supplementation at 2-4°C (Nimrat et al., 2006)

Extenders	Storage period (days)					
	0	7	14	21	28	35
Without antibiotic	$0^{a,2}$	$0^{a,2}$	$28.3{\pm}4.8^{a,2}$	$95.0{\pm}5.6^{a,2}$	$2{,}300.0{\pm}432.0^{a,1}$	$2{,}416.7{\pm}299.4^{a,1}$
With antibiotic	$0^{a,1}$	$0^{a,1}$	$0^{b,1}$	$0^{b,1}$	$0^{b,1}$	$0^{b,1}$

Values with different letters in a column are significantly different between treatments ($P<0.05$)
Values with different numbers in a row are significantly different among sampling period ($P<0.05$)

Total *Vibrio* count in the fresh spermatophores and cryopreserved spermatophores immersed in liquid nitrogen for 1 h were 300 x 10^3 CFU/g (Table 7). The number of *Vibrio* declined significantly ($P<0.05$) to 18.7 ± 4.1 x 10^3 CFU/g and 0.9 ± 0.3 x 10^3 CFU/g after 30 and 90 days storage, respectively. *Vibrio* were not detected after 120 days storage. Total *Pseudomonas* count of the control were 300 x 10^3 CFU/g, significantly higher ($P<0.05$) than in cryopreserved spermatophores in liquid nitrogen for 1 h (111.7 ± 5.9 x 10^3 CFU/g) (Table 7). Cryopreserved spermatophores held in liquid nitrogen between 30-210 days had low levels of *Pseudomonas*, ranging between 0.4 ± 0.1 CFU/g and 1.2 ± 0.3 CFU/g. However, *Ps. aeruginosa* was not present in cryopreserved spermatophores after 90 days of storage. Potential or opportunistic bacterial pathogens from fresh and cryopreserved spermatophores for 1 h were represented by 23 species (Table 8). *Staphylococcus* nonaureus, *Micrococcus*, *Proteus*, *Ochrobactrum anthropi*, *Corynebacterium*, *Bacillus licheniformis* and *Pseudomonas* were present in cryopreserved spermatophores throughout the full 210 days of storage. *Escherichia fergusonii*, *Brevibacterium*, *Rhodococcus* and *Bacillus megaterium* were present in cryopreservaed spermatophores within 180 days of storage. *Proteus mirabilis*, *Enterobacter*, *Vibrio*, *Kocuria varians*, *Bacillus cereus*, *B. amyloliquefaciens*, *B. circulans*, *Ps. aeruginosa* and *Burkholderia cepacia* were not detected in cryopreserved spermatophores when stored longer than 90 or 120 days.

Based on our findings in table 7 and table 8, a reduction in bacteria of cryopreserved spermatophores over time may be associated with the chilling sensitivity of bacteria (Gram and Dalgaard, 2002). Contrary to total heterotropic bacteria and total *Vibrio*, a significant decline in total *Pseudomonas* within 1 h of cryostorage indicated that total *Pseudomonas* was more sensitive to freezing than total heterotrophic bacteria and total *Vibrio*. The absence of total *Vibrio* after 120 days of storage, compared to the other taxa of bacteria, suggests the former were less tolerant to ultra cold storage. A decline in the numbers of the three bacterial groups to an average level less than 0.05 x 10^4 CFU/g or undetectable level demonstrated the reduced risk of microbial contamination of cryopreserved spermatophores during long-term cold storage in liquid nitrogen. This is of great concern to breeding programs and disease transmission by animal and human health authorities (Bielanski, 2005a; 2005b). Therefore, cryopreservation of *P. monodon* spermatophores resulted in a significant reduction of contaminated bacteria. This study reported on the first assessment of microbial contamination in cryopreserved spermatophores of *P. monodon* and showed that cryopreservation of *P. monodon* spermatophores can eliminate contamination by *Vibrio* and *Ps. aeruginosa*, both pathogenic bacteria for black tiger shrimp. There were several theories as to why microorganisms are killed or at least injured during the freezing process (Geiges, 1996). It has been suggested that microorganisms suffer damage to cell walls and membranes during freezing from extra- and intracellular ice formation, electrolyte imbalance resulting from dehydration and increased solute concentration, and shrinkage of cells resulting from osmosis (Baust, 1963).

A similar pattern of change in bacterial counts with cryopreservation was reported for bovine semen by other investigations. Mazurova et al. (1975) demonstrated a significant decline in the number of contaminated bacteria in bull sperm immediately after freezing and after three months of sperm storage in liquid nitrogen. Research carried out by several authors (Le et al., 1997; Bielanski and Stewart, 1996; Bielanski et al., 2003) showed that *Pseudomonas*, *Streptococcus*, *Staphylococcus* and *Stenotrophomonas maltophilia*, which were the most common contaminant bacteria of bovine semen, have suppression effects on fertilization and embryonic development. Bielanski et al. (2003) found *Staphylococcus sciuri*, *Proteus vulgaris*, *Escherichia coli*, *Corynebacterium*, *Corynebacterium xerosis* and *Bacillus sphaericus* in cryopreserved bull semen stored in liquid nitrogen for 6-35 years, while *Ochrobactrum anthropi* was isolated from cryopreserved bull embryo stored for 15 years. Miyamoto-Shinohara et al. (2000) reported different abilities of the bacteria to tolerance freeze-drying conditions during 16 years of storage. It is interesting to further investigate the effects of antibiotic supplementation on cryopreservation of aquatic animal sperm/spermatophores. Can addition of antibiotics control bacterial growth or inhibit the occurence of microbial risk of potential bacteria?. Nimrat et al. (2005) reported the addition of an antibiotic in sperm extender for chilled storage of black tiger shrimp spermatophores at 2-4°C resulted in successful control of potential bacteria, particularly, *Ps. aeruginosa*, without having a negative effect on sperm viability and fertilization capacity.

Table 7. Summary of total heterotrophic bacteria, total Vibrio and total Pseudomonas (Nimrat et al., 2007)

Time (days)	Type of bacteria (CFU/g)		
	Total heterotrophic bacteria x 10^4	Total *Vibrio* x 10^3	Total *Pseudomonas* x 10^3
Control	46.3 ± 2.9^a	300 ± 0^a	300 ± 0^a
0	44.3 ± 3.3^a	300 ± 0^a	111.7 ± 5.9^b
30	2.1 ± 0.4^b	18.7 ± 4.1^b	1.2 ± 0.3^c
60	1.9 ± 0.2^b	6.2 ± 3.1^c	1.1 ± 0.1^c
90	1.7 ± 0.1^b	0.9 ± 0.3^d	0.8 ± 0.2^c
120	1.2 ± 0.2^b	0^c	0.7 ± 0.2^c
150	0.08 ± 0.02^b	0^c	0.7 ± 0.1^c
180	0.07 ± 0.02^b	0^c	0.6 ± 0.1^c
210	0.05 ± 0.02^b	0^c	0.4 ± 0.1^c

Note: Different superscript letters on the column indicated significant difference ($P<0.05$) in the number of bacteria during cryostorage.

Table 8. Summary of type of bacteria found in cryopreserved spermatophores during the experiment (Nimrat et al., 2007)

Type of bacteria	Time (days)								
	Control	1h	30	60	90	120	150	180	210
1. Gram-positive cocci									
1.1 *Staphylococcus* nonaureus	+	+	+	+	+	+	+	+	+
1.2 *Micrococcus*	+	+	+	+	+	+	+	+	+
2. Enterobacteriaceae									
2.1 *Escherichia fergusonii*	+	+	+	+	+	+	+	+	-
2.2 *E. coli*	+	+	+	+	-	-	-	-	-
2.3 *Proteus*	+	+	+	+	+	+	+	+	+
2.4 *Proteus mirabilis*	+	+	+	+	-	-	-	-	-
2.5 *Alcatigenes xylosoxidanes*	+	+	+	+	+	+	+	-	-
2.6 *Enterobacter*	+	+	+	+	-	-	-	-	-
3. *Vibrio*	+	+	+	+	+	-	-	-	-
4. *Brevibacterium*	+	+	+	+	+	+	+	+	-
5. Other gram-negative rod									
5.1 *Ochrobactrum anthropi*	+	+	+	+	+	+	+	+	+
5.2 *Kocuria varians*	+	+	+	+	-	-	-	-	-
6. Gram-positive rod									
6.1 *Corynebacterium*	+	+	+	+	+	+	+	+	+
6.2 *C. aquaticum*	+	+	+	+	+	+	+	-	-
6.3 *Rhodococcus*	+	+	+	+	+	+	+	+	-
6.4 *Bacillus cereus*	+	+	+	+	+	-	-	-	-
6.5 *B. amyloliquefaciens*	+	+	+	+	+	-	-	-	-
6.6 *B. circulans*	+	+	+	+	+	-	-	-	-
6.7 *B. megaterium*	+	+	+	+	+	+	+	+	-
6.8 *B. licheniformis*	+	+	+	+	+	+	+	+	+
7. Pseudomonaceae									
7.1 *Pseudomonas*	+	+	+	+	+	+	+	+	+
7.2 *Ps. aeruginosa*	+	+	+	+	-	-	-	-	-
7.3 *Burkholderia cepacia*	+	+	+	+	+	-	-	-	-

Note: Control refers to freshly collected spermatophores without freezing, while 1 h refers to cryopreserved spermatophores stored in liquid nitrogen for 1 h

3. Biosecurity and Risk Assessment of Cryopreserved Sperm

The safe storage of sperm following cryopreservation is of concern to animals and human health authorities (Clarke, 1999; Bielanski et al., 2003; Tiersch and Jenkins, 2003; Levy et al., 2004). Generally, it is commonly assumed that liquid nitrogen is sterile and cryopreserved sperm kept in liquid nitrogen remain viable and unchanged during long-term storage. Indeed when manufactured liquid nitrogen usually has a very low particulate, water and microbial content (Morris, 2005). Use of liquid nitrogen for long-term storage of cells has been considered as a realistic choice for preserving cell viability. Effective long-term cryogenic storage of cells and tissues requires storage of samples at -196°C. Cryopreservation, storage and transport of cryopreserved sperm without bacterial contamination are desirable to ensure the disease-free status of donor stock. Sources of contamination during storage in liquid nitrogen may also be derived from different causes. Ice sediment can accumulate within liquid nitrogen tank during long-term storage and has been implicated as a source of contamination of cryopreserved samples (Piasecka-Serafin, 1972). Contamination of liquid nitrogen can be occurred by entrapping of air borne microbes into ice crystal that may form in the air above the liquid nitrogen (Morris, 2005). Opening the lid of liquid nitrogen tank generates significant turbulence and will mix nitrogen gas with external air. The microbial load then easily falls into the liquid nitrogen. Removal of storage components into the air or routine sampling of cryopreserved samples may result in further contamination from the air. In fact, bacteria responsible for contamination initially gain access to cryopreserved sperm from sample collection process. Other sources that have been identified include environment, personnel hygiene, water used for extender and cryoprotectant preparation, and cross-contamination in liquid nitrogen tanks. With the understanding that bacterial contamination is unavoidable when collecting aquatic animal sperm, freshly collected sperm are easily contaminated with bacteria present on the body surface (Harper and Talbot, 1984). The presence of bacteria in intact sperm has also been reported (Chow, 1982; Ishida et al., 1986). Since there are numerous steps associated with freezing and thawing of cryopreserved sperm, bacterial contamination may easily occur if proper care has not been considered during the cryopreservation process. Storage of cryopreserved samples in the vapour phase of liquid nitrogen has been proposed to prevent microbial contamination because there is no direct pathway for cross contamination. This could be achieved by storage in a system that does not require immersion of the materials within the liquid phase, although concern has been expressed that this could compromise cell viability (Tedder et al., 1995). Nevertheless, the presence of airborne ice crystal in the vapour phase of liquid nitogen may still pose a risk (Fountain et al., 1997).

Microbiological cross-contamination in sperm banks has raised concern, in many sperm bank managers who have performed several approaches aimed at reducing the perceived risks. Contamination of cryopreserved sperm may also occur during storage of cells in liquid nitrogen, which is undesirable. In mammal, cross-contamination during storage of biological material in liquid nitrogen has been reported (Tedder et al., 1995; Fountain et al., 1997; Bielanski et al., 2003). A heavy contamination by *Aspergillus* was observed in a liquid nitrogen tank containing haematopoietic stem cells (Foutain et al., 1997). Hawkins et al. (1996) reported the cross-contamination of hepatitis B virus in liquid nitrogen using nucleotide sequence analysis during bone marrow storage. Viruses have also been observed to survive direct exposure to liquid nitrogen, such as vesicular stomatitis virus (Schafer et al., 1976) and herpes simplex virus, adenovirus (Jones and Darville, 1989). As a result, cross-contamination during sperm cryopreservation in humans has been taken seriously to prevent infection in assisted reproduction technology. On the contrary, none of any studies in cryopreservation of aquatic animal sperm has been reported the cross-contamination of bacteria, viruses or fungi. Although we were unable to demonstrate contamination of liquid nitrogen due to the lack of liquid nitrogen sampling for bacterial assessment cross contamination in liquid nitrogen in our studies dealing with cryopreservation of *P. monodon* spermatophores might not have occurred or was minimal since the types of bacteria identified in cryopreserved spermatophores during long-term storage was unchanged (Nimrat et al., 2007; Vuthiphandchai et al., 2007). Inadequate or faulty sealing of cryovials or straws could adsorb contaminated liquid nitrogen, resulting in cross-infection during subsequent use of thawed samples (Hawkins et al., 1996; Russel et al., 1997). Clarke (1999) reported that cryovials with or without sealed O-rings from some manufacturers adsorbed liquid nitrogen during 3 h immersion in liquid nitrogen, whereas liquid nitrogen condensation was not adsorbed inside the cryovials stored in the liquid nitrogen phase for 16-24 h. The seepage of liquid nitrogen into submersed cryovials or straws occurs due to formation of a vacuum caused by the condensation of the gaseous nitrogen. The sealing of cryovials could prevent cross-contamination of bacteria in frozen samples to liquid nitrogen due to the use of sealed-O-rings. Chen et al. (2006) The use of sterile techniques was effective in the control or minimization of microbial contamination, although sperm cryopreservation protocol in fish in many studies has not involved aseptic techniques (Stoss, 1983; Lahnsteiner et al., 2000; Chao and Liao, 2001; DeGraaf and Berlinsky, 2004). Bielanski (1997) demonstrated all sealed samples of embryos stored in contaminated liquid nitrogen tested negative for the presence of BVDV and BHV-1 viruses. Therefore, cryovials or straws that were obviously cracked or have lost their seal at the end should not be thawed for subsequent use to prevent disease transmission. Cutting the thawed straws with sterile scissors or a scalpel blade or using male cap with cryovials for subsequent use are recommended for good sanitation practice to prevent contamination. Use of separate liquid nitrogen containers and high security straws for long-term storage of frozen samples is recommended to avoid cross contamination with other samples.

4. Concluding Remarks and Recommendations

Some aspects of the role of bacteria in chilled storage and cryopreservation of sperm in aquatic animals has been thoroughly studied and are well understood, in particular, development of protocols for chilled storage and cryopreservation. However, specific protocols are required to reduce the damage induced by preservation techniques for each species. The possibility of initial bacterial infection in sperm of aquatic animals is needed to take into consideration for preservation of aquatic animal sperm in addition with other sources of contamination. The control of the absence of bacteria in sperm before sperm chilled storage or cryopreservation is recommended for management of sperm preservation in aquatic animals and microbial risk associated with sperm preservation. While the benefits of chilled storage of aquatic animal reproductive cell are well established for fish species, crustacean spermatophores have received less attention. Microbial contamination inside cryopreserved sperm of aquatic animals where freezing may interact is only partially elucidated in black tiger shrimp. Storage of cryopreserved spermatophores of *P. monodon* resulted in a significant decline of contamination bacteria and low microbial risk of pathogenic bacteria. Published studies examining bacterial contamination of extended sperm in aquatic animal are even sparser. Study on cross-contamination during development of sperm cryopreservation protocols of aquatic animals are areas of future research. Futhermore, research must be undertaken to examine the effect of microbial contamination on the health of broodstock and larvae after undertaking artificial insemination with cryopreserved sperm. In the perspective of biosecurity and risk assessment of cryopreserved sperm, safe collection and processing of sperm needs to be taken in consideration especially for acceptance in international trade. The effectiveness of the antibiotics used for chilled storage and cryopreservation of aquatic animal sperm is needed for identification of appropriate antibiotic type and optimal dose required for artificial insemination since the use of anitibiotic with preserved sperm has not been systematically studied in cultured species of aquatic animals. Preservation of aquatic animal sperm should in future be handled in a way which minimizes the chance of inadvertent transfer of microorganisms to stored sample or offspring.

ACKNOWLEDGEMENTS

We thank our colleagues (Siriporn Kodcharat, Aritsara Keatsaksit, Sunisa Suksawat and Prasert Siriboonlamoom) in the Departments of Aquatic Science and Microbiology, Faculty of Science who were involved in our studies that form the basis of the present paper. We also thank Dr. Kashane Chalermwat for comments on the manuscript.

REFERENCE

Aas, G.H., Refstie, T. and Gjerde, B. 1991. Evaluation of milt quality of Atlantic salmon. *Aquaculture* 95: 125-132.

Alavi, S.M.H. and Cosson, J. 2006. Sperm motility in fishes: (II) Effects of ions and osmolality: A review. *Cell Biology Internationa*l 30: 1-14.

Allen, M.A., Edberg, S.E. and Reasoner, D.J. 2004. Heterotrophic plate count bacteria-what is their significance in drinking water?. *International Journal of Food Microbiology* 92: 265-274.

Anderson, I.G., Shamsudin, M.N. and Nash, G. 1989. A preliminary study on the aerobic heterotrophic bacterial flora in giant freshwater prawn, *Macrobrachium rosenbergii*, hatcheries in Malaysia. *Aquaculture* 81: 213-223.

Asahina, E. and Takahashi, T. 1978. Freezing tolerance in embryos and spermatozoa of the sea urchin. *Cryobiology* 5:122-127.

Austin, B. and Robertson, P.A.W. 1993. Recovery of *Streptococcus milleri* from ulcerated koi carp *Cyprinus carpio* in the U.K. Bulletin of the European Association for Fish Pathologists 13: 207-209.

Babiak, I., Ottesen, O., Rudolfsen, G. and Johnsen, S. 2006a. Quantitative characteristics of Atlantic halibut, *Hippoglossus hippoglossus* L., semen throughout the reproductive season. *Theriogenology* 65:1587–1604.

Babiak, I., Ottesen, O., Rudolfsen, G. and Johnsen, S. 2006b. Chilled storage of semen from Atlantic halibut, *Hippoglossus hippoglossus* L. I: Optimizing the protocol. *Theriogenology* 66: 2025–2035.

Babiak, I., Ottesen, O., Rudolfsen, G. and Johnsen, S. 2006c. Chilled storage of semen from Atlantic halibut, *Hippoglossus hippoglossus* L. II: Effect of spermiation advancement, catheterization of semen, and production-scale application. *Theriogenology* 66: 2036–2046.

Baust, J.G. 1963. Mechanisms of cryoprotection in freezing tolerant animal systems. *Cryobiology* 10: 197-205.

Baya, A.M., Toranzo, A.E., Lupiani, B., Li, T., Roberson, B.S. and Hetrick, F.M. 1991.

Biochemical and serological characterization of *Carnobacterium* spp. isolated from farmed and natural populations of stripped bass and catfish. *Applied and Environmental Microbiology* 57: 3114-3120.

Benau, D. and Terner, C. 1980. Initiation, prolongation, and reactivation of the motility of salmonid spermatozoa. *Gamete Research* 3: 247-257.

Benhalima, K. and Moriyasu, M. 2001. Prevalence of bacteria in the spermathecae of female snow crab, *Chionoecetes opilio* (Brachyura: Majidae). *Hydrobiologia* 449: 261-266.

Bielanski, A. 1997. A review on disease transmission studies in relationship to production of embryos by in vitro fertilization and to related new reproductive technologies. *Biotechnology Advances* 15: 633-656.

Bielanski, A. 2005a. Experimental microbial contamination and disinfection of dry (vapour) shipper dewars designed for short-term storage and transportation of cryopreserved germplasm and other biological specimens. *Theriogenology* 63: 1946-1957.

Bielanski, A. 2005b. Non-transmission of bacterial and viral microbes to embryos and semen stored in the vapour phase of liquid nitrogen in dry shippers. Cryobiology 50: 206-210.

Bielanski, A. and Stewart, B. 1996. Ubiquitous microbes isolated from in vitro fertilization (IVF) system. *Theriogenology* 45: 269.

Bielanski, A., Bergeron, H., Lau, P.C.K. and Devenish, J. 2003. Microbial contamination of embryos and semen during long term banking in liquid nitrogen. *Cryobiology* 46: 146-152.

Billard, R. and Cosson, M.P. 1992. Some problems related to the assessment of sperm motility in freshwater fish. *Journal of Experimental Zoology* 261: 122-131.

Billard, R., Cosson, J., Crim, L.W. and Suquet, M. 1995. Sperm physiology and quality. In N.R. Bromage and R.J. Roberts (Eds.), Broodstock management and egg and larval quality (pp. 25-52). London: Blackwell Science.

Bray, W.A., Leung-Trujillo, J.R., Lawrence, A.L. and Roberts, S.M. 1985. Preliminary investigation of the effects of temperature, bacterial inoculation, and EDTA on sperm quality in captive *Penaeus setiferus*. *Journal of the World Mariculture Society* 16: 250–257.

Bucio, A., Hartemink, R., Schrama, J.W., Verreth, J. and Rombouts, F.M. 2006. Presence of lactobacilli in the intestinal content of freshwater fish from a river and from a farm with a recirculation system. *Food Microbiology* 23: 476-482.

Cabrita, E., Anel, L. and Herraez, M.P. 2001. Effect of external cryoprotectants as membrane stabilizers on cryopreserved rainbow trout sperm. *Theriogenology* 56: 623–635.

Cai, J., Han, Y. and Wang, Z. 2006. Isolation of *Vibrio parahaemolyticus* from abalone (*Haliotis diversicolor supertexta* L.) postlarvae associated with mass mortalities. *Aquaculture* 257: 161-166.

Cai, Y., Suyanandana, P., Saman, P. and Benno, Y., 1999. Classification and characterization of lactic acid bacteria isolated from the intestine of common carp and fresh water prawns. *Journal of General and Applied Microbiology* 45:177-184.

Carson, J., Gudkovs, N. and Austin, B. 1993. Characteristics of an *Enterococcus*-like bacterium from Australia and South Africa, pathogenic for rainbow trout, *Oncorhynchus mykiss* Walbaum. *Journal of Fish Diseases* 16: 381-388.

Chambeyron, F. and Zohar, Y. 1990. A diluent for sperm cryopreservation of gilthead seabream, *Sparus aurata*. *Aquaculture* 90: 345-352.

Chao, N.H. and Liao, I.C. 2001. Cryopreservation of finfish and shellfish gametes and embryos. *Aquaculture* 197: 161-189.

Chow, S. 1982. Artificial insemination using preserved spermatophores in the palaemonid shrimp *Macrobrachium rosenbergii*. Bulletin of the Japanese Society of Scientific Fisheries 48: 1693-1695.

Christensen, J.M. and Tiersch, T.R. 1997. Cryopreservation of channel catfish spermatozoa: effect of cryoprotectant, straw size, and formulation of extender. *Theriogenology* 47: 639-645.

Christensen, J.M. and Tiersch, T.R. 2005. Cryopreservation of channel catfish sperm: effects of cryoprotectant exposure time, cooling rate, thawing conditions, and male-to-male variation. *Theriogenology* 63: 2103–2112.

Clarke, G.N. 1999. Sperm cryopreservation: is there a significant risk of cross-contamination? Human Reproduction 14: 2941-2943.

Colakoglu, F.A., Sarmasik, A. and Koseoglu, B. 2006. Occurrence of *Vibrio* spp. And *Aeromonas* spp. in shellfish harvested off Dardanelles cost of Turkey. Food Control 17: 648-652.

Collins, M.D., Farrow, A.E., Phillips, B.A., Ferusu, S. and Jones, D. 1987. Classification of *Lactobacillus divergens*, *Lactobacillus piscicola* and some catalase-negative sporogenous, rod-shaped bacteria from poultry in a new genus, *Carnobacterium*. *International Journal of Systematic Bacteriology* 37: 310-316.

Costa, R., Mermoud, I., Koblavi, S., Morlet, B., Haffner, P., Berthe, F., Legroumellec, M. and Grimont, P. 1998. Isolation and characterization of bacteria associated with a *Penaeus stylirostris* disease (Syndrome 93) in New Caledonia. *Aquaculture* 164: 297-309.

DeGraaf, J.D. and Berlinsky, D.L. 2004. Cryogenic and refrigerated storage of Atlantic cod (*Gadus morhua*) and haddock (*Melanogrammus aeglefinus*) spermatozoa. *Aquaculture* 234: 527-540.

Dempsey, A.C. and Kitting, C.L. 1987. Characteristics of bacteria isolated from penaeid shrimp. Crustaceana 52: 90-94.

Dong, Q., Huang, C., Eudeline, B. and Tiersch, T.R. 2005. Systematic factor optimization for cryopreservation of shipped sperm samples of diploid Pacific oysters, *Crassostrea gigas*. Cryobiology 51: 176-197.

Dougherty, W.J. and Dougherty, M.M. 1989. Electron microscopical and histochemical observations on melanized sperm and spermatophores of pond-cultured shrimp, *Penaeus vannamei*. *Journal of Invertebrate Pathology* 54: 331-343.

Dreanno, C., Suquet, M., Desbruyeres, E., Cosson, J., Delliou, H.L. and Billard, R. 1998. Effect of urine on semen quality in turbot (*Psetta maxima*). Aquaculture 169: 247-262.

Earp, B.J., Ellis, C.H. and Ordal, E.J. 1953. Kidney disease in young salmon. State Washington Department of Fisheries, Special Report No. 1.

Eldar, A., Bejerano, Y. and Bercovier, H. 1994. *Streptococcus shiloi* and *Streptococcus difficile*: two new streptococcal species causing a meningoencephalitis in fish. *Current Microbiology* 28: 139-143.

Eldar, A., Frelier, P.F., Assenta, L., Varner, P.W., Lawhon, S. and Bercovier, H. 1995. *Streptococcus shiloi*, the name for an agent causing septicemic infection in fish, is a junior synonym of *Streptococcus iniae*. *International Journal of Systematic Bacteriology* 45: 840-842.

Eldar, A., Ghittino, C., Asanta, L., Bozzetta, E., Goria, M., Prearo, M. and Bercovier, H. 1996. *Enterococcus seriolicida* is a junior synonym of *Lactococcus garlieae*, a causative agent of septicemia and meningoencephalitis in fish. *Current Microbiology* 32: 85-88.

Evelyn, T.P.T. and McDermot, L.A. 1961. Bacteriological studies of fresh-water fish: I. Isolation of aerobic bacteria from several species of Ontario fish. *Canadian Journal of Microbiology* 7: 375-382.

Foo, J.T.W., Ho, B. and Lam, T.J. 1985. Mass mortality in *Siganus canaliculatus* due to streptococcal infection. *Aquaculture* 49: 185-195.

Fountain, D., Ralston, M., Higgins, N., Gorlin, J.B., Uhl, L., Wheeler, C., Antin, J.H., Churchill, W.H. and Benjamin, R.J., 1997. Liquid nitrogen freezers: a potential source of microbial contamination of hematopoietic stem cell components. *Transfusion* 37: 585-591.

Gatesoupe, F.J. 1999. The use of probiotics in aquaculture: review. *Aquaculture* 180: 147-165.

Geiges, O. 1996. Microbial processes in frozen food. *Advances in Space Research* 18: 109-118.

Ghittino, C., Prearo, M., Bozzetta, E. and Eldar, A. 1995. Pathogenetical characterization of the causative agent of fish streptococcosis in Italy and vaccination trials in rainbow trout. Bollettino Societa Italiana di Patalogia Ittica 7: 2-12.

Ghittino, P. and Prearo, M. 1992. Report of streptococcosis in rainbow trout *Oncorhynchus mykiss* in Italy: preliminary note. Bollettino Societa Italiana di Patalogia Ittica 8: 4-9.

Gonzalez, C.J., Encinas, J.P., Garcia-Lopez, M.L. and Otero, A. 2000. Characterization and identification of lactic acid bacteria from fresh water fishes. *Food Microbiology* 17: 383-391.

Gonzalez-Rodriguez, M.N., Sanz, J.J., Santos, J.A., Otero, A. and Garcia- Lopez, M.L. 2002. Numbers and types of microorganisms in vacuum packed cold-smoked freshwater fish at the retail level. *International Journal of Food Microbiology* 77: 161-168.

Gram, L. and Dalgaard, P. 2002. Fish spoilage bacteria - problems and solutions. *Current Opinion in Biotechnology* 13: 262-266.

Green, D., Wakeley, P.R., Page, A., Barnes, A., Baccigalupi, L., Ricca, E. and Cuttingi, S.M. 1999. Characterization of two *Bacillus* probiotics. *Applied and Environmental Microbiology* 65: 4288-4291.

Gwo, J.C. 2000. Cryopreservation of aquatic invertebrate semen: a review. *Aquaculture Research* 31: 259-271.

Gwo, J.C., Chen, C.W. and Cheng, H.Y. 2002. Semen cryopreservation of small abalone (*Haliotis diversicolor supertexa*). *Theriogenology* 58: 1563-1578.

Hamid, A., Sakata, T. and Kakimoto, D. 1978. Microflora in the alimentary tract of grey mullet: 2. A comparison of the mullet intestinal microflora in fresh and sea water. Bulletin of the Japanese Society of Scientific Fisheries 44: 53-57.

Hammerstedt, R.H., Graham, J.K. and Nolan, J.P. 1990. Cryopreservation of mammalian sperm: what we ask them to survive. *Journal of Andrology* 11: 73-88.

Hanberger, H., Diekema, D., Fluit, A., Jones, R., Struelens, M., Spencer, R. and Wolff, M. 2001. Surveillance of antibiotic resistance in European ICUs. *Journal of Hospital Infection* 48: 161-176.

Hansen, W., Freney, J., Benyegoub, H., Letouzey, M.N., Gigi, J. and Wauters, G. 1993. Severe human infections caused by *Vibrio metschnikovii*. *Journal of Clinical Microbiology* 31: 2529-2530.

Harper, R. and P. Talbot. 1984. Analysis of the epibiotic bacteria of lobster (*Homarus*) eggs and their influence on the loss of eggs from the pleopods. *Aquaculture* 36: 9-26.

Hawkins, A.E., Zuckerman, M.A., Briggs, M., Gilson, R.J.C., Goldstone, A.H., Brink, N.S. and Tedder, R.S., 1996. Hepatitis B nucleotide sequence analysis: linking an outbreak of acute hepatitis B to contamination of a cryopreservation tank. *Journal of Virological Methods* 60: 81-88.

He, S. and Woods, L.C. 2004. Effects of dimethyl sulfoxide and glycine on cryopreservation induced damage of plasma membranes and mitochondria to striped bass (*Morone saxatilis*) sperm. *Cryobiology* 48: 254-262.

Hiu, S.F., Holt, R.A., Sriringanathan, N., Seidler, R.J. and Fryer, J.L. 1984. *Latcobacillus piscicola*, a new species from salmonid fish. International *Journal of Systematic Bacteriology* 34: 393-400.

Hoi, L., Larsen, J.L., Dalsgaard, I. and Dalsgaard, A. 1998. Occurrence of *Vibrio vulnificus* biotypes in Danish marine environments. *Applied and Environmental Microbiology* 64(1): 7-13.

Hoshina, T. 1956. An epidemic disease affecting rainbow trout in Japan. Journal of the Tokyo University of Fisheries 42: 35-46.

Huang, C.Y., Liu, P.C. and Lee, K.K. 2001. Withering syndrome of the small abalone, *Haliotis diversicolor supertexta*, is caused by *Vibrio parahaemolyticus* and associated with thermal induction. Zeitschrift für Naturforschung 56: 898-901.

Ishida, T., Talbot, P. and Kooda-Cisco, M.. 1986. Technique for the long-term storage of lobster (*Homarus*) spermatophores. *Gamete Research* 14: 183-195.

Jenkins, J.A. and Tiersch, T.R. 1997. A preliminary bacteriological study of refrigerated channel catfish sperm. *Journal of the World Aquaculture Society* 28: 282-288.

Jensen, J.O.T. and Aiderdice, D.F. 1984. Effect of temperature on short-term storage of eggs and sperm of chum salmon (*Oncorhynchus keta*). *Aquaculture* 37: 251-265.

Jeyalectumie, C. and Subramoniam, T. 1989. Cryopreservation of spermatophores and seminal plasma of the edible crab *Scylla serrata*. Biological Bulletin 177: 247-253.

Jeyasekaran, G., Ganesan, P., Anandaraj, R., Shakila, R.J. and Sukumar, D. 2006. Quantitative and qualitative studies on the bacteriological quality of Indian white shrimp (*Penaeus indicus*) stored in dry ice. *Food Microbiology* 23: 526-533.

Jones, R.N., Kirby, J.T., Beach, M.L., Biedenbach, D.J. and Pfaller, M.A. 2002. Geographic variations in activity of broad-spectrum β-lactams against *Pseudomonas aeruginosa*: Summary of the worldwide SENTRY Antimicrobial Surveillance Program. *Diagnostic Microbiology and Infectious Disease* 43: 239-243.

Jones, S.K. and Darville, J.M. 1989. Transmission of virus particles by cryotherapy and multi-use caustic pencils: a problem to dermatologists?. *British Journal of Dermatology* 121: 481-486.

Kang, K.H., Shao, M.Y., Kho, K.H. and Zhang, Z.F. 2004. Short-term storage and cryopreservation of *Urechis unicinctus* (Echiura: Urechidae) sperm. *Aquaculture Research* 35(13): 1195-1201.

Karaseva, T.A., Serdyuk, A.V. and Longinova, G.A. 1991. Experience in treatment of *Streptococcus* sp. infection in Atlantic salmon *Salmo salar* L. Int. Council for the Exploration of the Sea. Mariculture Committee F6, 1991, pp. 11.

Karunasagar, I., Pai, R., Malathi, G.R. and Karunasagar, I. 1994. Mass mortality of *Penaeus monodon* larvae due to antibiotic resistant *Vibrio harveyi* infection. *Aquaculture* 128: 203-209.

Kennedy, B., Venugopal, M.N., Karunasagar, I. and Karunasagar, I. 2006. Bacterial flora associated with the giant freshwater prawn *Macrobrachium rosenbergii*, in the hatchery system. *Aquaculture* 261: 1156-1167.

Kime, D.E., Van Look, K.J.W., McAllister, B.G., Huyskens, G., Rurangwa, E. and. Ollevier, F. 2001. Computer-assisted sperm analysis CASA as a tool for monitoring sperm quality in fish. *Comparative Biochemistry and Physiology Part* C 130: 425-433.

Kitao, T., Aoki, T. and Sakoh, R. 1981. Epizootic caused by β-haemoltytic *Streptococcus* species in cultured freshwater fish. *Fish Pathology* (Tokyo) 15: 301-307.

Krasznai, Z., Morisawa, M., Krasznai, Z.T., Morisawa, S., Inaba, K. and Bazsane, Z.K. 2003. Gadolinium, a mechano-sensitive channel blocker, inhibits osmosis-initiated motility of sea- and freshwater fish sperm, but does not affect human or ascidian sperm motility. *Cell Motility and the Cytoskeleton* 55: 232-243.

Kusuda, R. 1992. Bacterial fish diseases in mariculture in Japan with special emphasis on streptococciosis. Bamidgeh 44: 140.

Kusuda, R., Kawai, K., Salati, F., Banner, C.R. and Fryer, J.L. 1991. *Enterococcus seriolicida* sp. *nov.*, a fish pathogen. *International Journal of Systematic Bacteriology* 41: 406-409.

Kusuda, R., Kawai, K., Toyoshima, T. and Komatsu, I. 1976. A new pathogenic bacterium belonging to the genus *Streptococcus*, isolated from epizootic of cultured yellowtail. Bulletin of the Japanese Society of Scientific Fisheries 42: 1345-1352.

Kvasnikov, E.I., Kovalenko, N.K. and Materinskaya, L.G. 1977. Lactic acid bacteria of freshwater fish. Mikrobiologiya 46(4): 619-624 (in English).

Lahnsteiner, F. and Patzner, R.A. 1995. Fine structure of spermatozoa of two marine teleost fishes, the red mullet, *Mullus barbatus* (Mullidae) and the white sea bream, *Diplodus sargus* (Sparidae). *Journal of submicroscopic cytology and pathology* 27: 259-266.

Lahnsteiner, F., Berger, B., Horvath, A., Urbanyi, B. and Weismann, T. 2000. Cryopreservation of spermatozoa in cyprinid fishes. Theriogenology 54: 1477-1496.

Lahnsteiner, F., Patzner, R., and Wiesmann, T. 1996. Semen cryopreservation of salmonid fish: Influence of handling parameters on the post-thaw fertilization rate. *Aquaculture* Research 27(9): 659-671.

Lavilla-Pitogo, C.R., Baticados, M.C.L., Cruz-Lacierda, E.R. and de la Pena, L.D. 1990. Occurrence of luminous bacterial disease of *Penaeus monodon* larvae in the Philippines. *Aquaculture* 91: 1-13.

Le, K.W., Kim, I.H., Son, D.S., Lee, H.J., Lee, D.W., Seo, G.H., Choi, S.H., Yang, B.C., Kim, K.N. and Jung, S.C. 1997. Identification of bacteria derived from frozen bovine semen that resulted in contamination during in vitro fertilization. *Theriogenology* 47: 375.

Lee, K.K., Liu, P.C. and Chuang, W.H. 2002. Pathogenesis of gastroenteritis caused by *Vibrio carchariae* in cultured marine fish. *Marine Biotechnology* 4: 267-277.

Lee, K.K., Yu, S.R., Yang, T.I., Liu, P.C. and Cheng, F.R. 1996. Isolation and characterization of *Vibrio alginolyticus* isolated from diseased kuruma prawn, *Penaeus japonicus*. Letters in Applied Microbiology 22: 111-114.

Legendre, M. and Billard, R. 1980. Cryopreservation of rainbow trout sperm by deep-freezing. Reproduction, Nutrition and Development 20: 1859-1868.

Lesel, R. 1990. Thermal effect on bacterial flora in the gut of rainbow trout and African catfish. In R. Lesel (Ed.), Microbiology in Poecilotherms (pp. 33-38). Amsterdam: Elsevier.

Leung, L.K.P. and Jamieson, B.G.M. 1991. Live preservation of fish gametes. In B.G.M. Jamieson (Ed.), Fish evolution and systematics: Evidence from spermatozoa (pp.245-295). Cambridge: Cambridge University Press.

Levanduski, M.J. and Cloud, J.G. 1988. Rainbow trout (*Salmo gairdneri*) semen: effect of non-motile sperm on fertility. *Aquaculture* 75: 171-179.

Levy, R., Grattard, F., Maubon, I., Ros, A. and Pozzetto, B. 2004. Bacterial risk and sperm cryopreservation. *Andrologia* 36: 282-285.

Lightner, D.V. 1993. Diseases of cultured penaeid shrimp. In J.P. McVey (Ed.), Handbook of Mariculture, vol. 1, Crustacean Aquaculture (pp. 393-486). Boca Raton: CRC Press.

Linhart, O., Billard, R. and Proteau, J.P. 1993. Cryopreservation of European catfish (*Silurus glanis* L.) spermatozoa. *Aquaculture* 115: 347-359.

Liu, P.C., Chen, Y.C. and Lee, K.K. 2001. Pathogenicity of *Vibrio alginolyticus* isolated from diseased small abalone, *Haliotis diversicolor supertexta*. *Microbios* 104: 71-77.

Liu, P.C., Chuang, W.H. and Lee, K.K. 2003. Infectious gastroenteritis caused by *Vibrio harveyi* (*V. carchariae*) in cultured red drum, *Sciaenops ocellatus*. *Journal of Applied Ichthyology* 19: 59-61.

Liu, P.C., Lee, K.K. and Chen, S.N. 1996a. Pathogenicity of different isolates of *Vibrio harveyi* in tiger prawn, *Penaeus monodon*. Letters in Applied Microbiology 22: 413-416.

Liu, P.C., Lee, K.K., Yii, K.C., Kou, G.H. and Chen, S.N. 1996b. Isolation of *Vibrio harveyi* from diseased kuruma prawns *Penaeus japonicus*. *Current Microbiology* 33: 129-132.

Lombardi, J.V. and Laboa, V.I. 1991. Disease and conditioning factors of mortality in larval culture of prawn of the genus Macrobrachium. In Proceeding of the 3rd Brazilian symposium of shrimp culture. João Pessoa.

Mahmoud, B.S.M., Yamazaki, K., Miyashita, K., Il-Shik, S., Dong-Suk, C. and Suzuki, T. 2004. Bacterial microflora of carp (*Cyprinus carpio*) and its shelf-life extension by essential oil compounds. *Food Microbiology* 21: 657-666.

Martin, G.G., Herzig, C. and Narimatsu, G. 1987. Fine structure and histochemistry of the freshly extruded and hardened spermatophore of the spiny lobster *Panulirus interruptus*. *Journal of Morphology* 192: 237-246.

Mazurova, J., Rysanek, M. and Forejtek, M. 1975. Effect of long-term freezing preservation on the level of bacterial contamination of the sperm. Veterinary Medicine (Praha) 20: 147-152.

Miura, T., Yamauchi, K., Takahashi, H. and Nagahama, Y. 1992. The role of hormones in the acquisition of sperm motility in salmonid fish. *Journal of Experimental Zoology* 261: 359-363.

Miyamoto, G., Prock, J., Nakamura, R., Nakagawa, L., Shimojo, R., Sato, V. and Akita, G., 1983. A preliminary microbiological and water quality survey of two Hawaiian prawn (*Macrobrachium rosenbergii*) hatcheries. In G.L. Rogers, R. Day and A. Lim (Eds.),

Proc. First Int. Conf. Warm Water Aquaculture-Crustacea. Utah: Brigham Young University.

Miyamoto-Shinohara, Y., Imaizumi, T., Sukenobe, J., Murakami, Y., Kawamura, S. and Komatsu, Y. 2000. Survival rate of microbes after freeze-drying and long-term storage: Brief communication. *Cryobiology* 41: 251-255.

Miyazaki, T. 1982. Pathological study on streptococciosis histopathology of infected fishes. Fish Pathology (Tokyo) 17: 39-47.

Molinari, L.M., Scoaris, D.D.O., Pedroso, R.B., Bittencourt, N.D.L.R., Nakamura, C.V., Ueda-Nakamura, T., Filho, B.A.D.A. and Filho, B.P.D. 2003. Bacterial microflora in the gastrointestinal tract of Nile tilapia, *Oreochromis niloticus,* cultured in a semi-intensive system. Acta Scientiarum.Biological Sciences 25: 267-271.

Moriarty, D.J.W. 1990. Interactions of microorganisms and aquatic animals, particularly the nutritional role of the gut flora. In R. Lesel (Ed.), Microbiology in Poecilotherms (pp. 217-222). Amsterdam: Elsevier.

Moriarty, D.J.W. 1999. Disease control in shrimp aquaculture with probiotic bacteria. In Proceedings of the 8th International Symposium on Microbial Ecology (pp 237- 243). Halifax: Atlantic Canada Society for Microbial Ecology.

Morisawa, M. 1985. Initiation mechanism of sperm motility at spawning in teleosts. Zoological Science 2: 605-615.

Morisawa, M., Suzuki, K., Shimizu, H., Morisawa, S. and Yasuda, K. 1983. Effects of osmolality and potassium on motility of spermatozoa from freshwater cyprinid fishes. *Journal of Experimental Biology* 107:95-103.

Morris, G.J. 2005. The origin, ultrastructure, and microbiology of the sediment accumulating in liquid nitrogen storage vessels. *Cryobiology* 50: 231–238

Munkittrick, K.R. and Moccia, R.D. 1987. Seasonal changes in the quality of rainbow trout (*Salmo gairdneri*) semen: effect of a delay in stripping on spermatocrit, motility, volume and seminal plasma constituents. *Aquaculture* 64: 147-156.

Nimrat, S., Bart, A.N., Keatsaksit, A. and Vuthiphandchai, V. 2007. Microbial flora of spermatophores from black tiger shrimp (*Penaeus monodon*) declines over long-term cryostorage. *Aquaculture* (In press).

Nimrat, S., Sangnawakij, T. and Vuthiphandchai, V. 2005. Preservation of black tiger shrimp (*Penaeus monodon*) spermatophores by chilled storage. *Journal of the World Aquaculture* Society 36: 76-86.

Nimrat, S., Siriboonlamom, S., Zhang, S., Xu, Y. and Vuthiphandchai, V. 2006. Chilled storage of white shrimp (*Litopenaeus vannamei*) spermatophores. *Aquaculture* 261: 944-951.

Normanno, G., Parisi, A., Addante, N., Quaglia, N.C., Dambrosio, A., Montagna, C. and Chiocco, D. 2006. *Vibrio parahaemolyticus*, *Vibrio vulnificus* and microorganisms of fecal origin in mussels (*Mytilus galloprovincialis*) sold in the Puglia region (Italy): Short communication. *International Journal of Food Microbiology* 106: 219-222.

Oda, S. and Morisawa, M. 1993. Rises of intracellular Ca^{2+} and pH mediate the initiation of sperm motility by hyperosmolality in marine teleosts. Cell Motility and the Cytoskeleton 25:171-178.

Palleroni, N.J. 2005. Genus I. *Pseudomonas*. In D.J. Brenner, N.R. Krieg and J.T. Staley (Eds.), Bergey's manual of systematic bacteriology, vol. 2, part B (pp. 340-344). New York: Spinger.

Paniagua-Chavez, C.G., Buchanan, J.T. and Tiersch, T.R. 1998. Effect of extender solutions and dilution on motility and fertilizing ability of Eastern oyster sperm. Journal of Shellfish Research 17: 231-237.

Perchec-Poupard, G., Gatti, J.L., Cosson, J., Jeulin, C., Fierville, F. and Billard, R. 1997. Effects of extracellular environment on the osmotic signal transduction involved in activation of motility of carp spermatozoa. *Journal of Reproduction and Fertility* 110: 315-327.

Perera, R.P., Johnson, S.K., Collins, M.D. and Lewis, D.H. 1994. *Streptococcus iniae* associated with mortality of *Tilapia nilotica* x *T. aurea* hybrids. *Journal of Aquatic Animal Health* 6: 335-340.

Piasecka-Serafin, M. 1972. The effect of sediment accumulation in containers under experimental conditions on the infection of semen stored directly in liquid nitrogen (-196°C). Bull Acad Pol Sci Biol 20(4): 263-267. (Bulletin de l Academie Polonaise des Sciences. Serie des Sciences Biologiques) Piironen, J. 1985. Variation in the properties of milt from the Finnish landlocked salmon (*Salmo salar* m. sebago Girard) during a spawning season. *Aquaculture* 48: 337-350.

Popovic, N.T., Teskeredzic, E., Strunjak-Perovic, I. and Coz-Rakovac, R. 2000. *Aeromonas hydrophila* isolated from wild freshwater fish in Croatia. Veterinary Research Communications 24: 371-377.

Prayitno, S,B. and Latchford, J.W. 1995. Experimental infections of crustaceans with luminous bacteria related *Photobacterium* and *Vibrio* spp. effect of salinity and pH on infectiosity. Aquaculture 132: 105-112.

Prieur, D., Mevel, G., Nicolas, J.L., Plusquellec, A. and Vigneulle, M. 1990. Interactions between bivalve molluscs and bacteria in the marine environment. Oceanography and Marine Biology: An Annual Review 28: 277-352.

Rana, K. 1995. Preservation of gametes. In N.R. Bromage and R.J. Roberts (Eds.), Broodstock management and egg and larval quality (pp. 53-75). Cambridge: Cambridge University Press.

Rasheed, V.C. and Plumb, J. 1984. Pathogenicity of a non-haemolytic group B *Streptococcus* sp. in Gulf killifish (*Fundulus grandis* Baird and Girard) . *Aquaculture* 37: 97-105.

Reilly, P.J. and Twiddy, D.R. 1992. *Salmonella* and *Vibrio cholerae* in brackish water tropical prawns. *International Journal of Food Microbiology* 16: 293-301.

Ringo, E. and Gatesoupe, F.J. 1998. Lactic acid bacteria in fish. Aquaculture 160, 177-203.

Ringo, E. and Olsen, R.E. 1999. The effect of diet on aerobic bacteria flora associated with intestine of Arctic charr (*Salvelinus alpinus* L.). *Journal of Applied Microbiology* 86: 22-28.

Ringo, E. and Strom, E. 1994. Microflora of Arctic charr, *Salvelinus alpinus* L.: gastrointestinal microflora of free-living fish and effect of diet and salinity on intestinal microflora. *Aquaculture and Fisheries Management* 25: 623-629.

Ringo, E., Strom, E. and Tabachek, J.A. 1995. Intestinal microflora of salmonids: a review. *Aquaculture Research* 26: 773-789.

Rippey, S.R. 1994. Infectious diseases associated with molluscan shellfish consumption. Clinical Microbiological Reviews 7: 419-425.

Roberts, D., Hooper, W. and Greenwood, M. 1990. Practical food microbiology. London: Public Health Laboratory Service, pp. 51-55.

Roberts, T.A., Baird-Parker A.C. and Tompkin, R.B. 1996. *Aeromonas*, *Microorganisms* in foods, characteristics of microbial pathogens ICMSF, vol. 5. London: Blakie Academic and Professional, pp. 5-19.

Ross, A.J. and Toth, R.J. 1974. *Lactobacillus*-a new fish pathogen? Progressive Fish-Culturist 36: 191.

Russel, P.H., Lyaruu, V.H., Millar, J.D., Curry, M.R. and Watson, P.E. 1997. The potential transmission of infectious agents by semen packing during storage for artificial insemination. *Animal Reproduction Science* 47: 337-342.

Saad, A., Billard, R., Theron, M.C. and Hollebecq, M.G. 1988. Short-term preservation of carp *Cyprinus carpio* semen. *Aquaculture* 71: 133-150.

Sader, H.S. and Jones, R.N. 2005. Antimicrobial susceptibility of uncommonly isolated non-enteric Gram-negative bacilli. *International Journal of Antimicrobial Agents* 25: 95-109.

Sakata, T. 1990. Microflora in the digestive tract of fish and shell-fish. In R. Lesel (Ed.), Microbiology in Poecilotherms (pp. 171-176). Amsterdam: Elsevier.

Sakata, T., Okabayashi, J. and Kakimoto, D. 1980. Variations in the intestinal microflora of *Tilapia* reared in fresh and sea water. Bulletin of the Japanese Society of Scientific Fisheries 46: 313-317.

Schafer, T.W., Everett, J. and Silver, G.H. 1976. Biohazard potential: recovery of infectious virus from liquid nitrogen of a virus repository. *Health Laboratory Science* 13: 23-24.

Schmidtke, L.M. and Carson, J. 1994. Characteristics of *Vagococcus salmoninarum* isolated from diseased salmonid fish. *Journal of Applied Bacteriology* 77, 229-236.

Schubert, R.H.W. 1991. *Aeromonas* and their significance as potential pathogens in water. *Journal of Applied Bacteriology* 70: 131S-135S.

Scott, A.P. and Baynes, S.M. 1980. A review of the biology, handling, and storage of salmonid spermatozoa. *Journal of Fish Biology* 17: 707-739.

Segovia, M., Jenkins, J.A., Paniagua-Chavez, C. and Tiersch, T.R. 2000. Flow cytometric evaluation of anitibiotic effects on viability and mitochondrial function of refrigerated spermatozoa of nile tilapia. *Theriogenology* 53:1489-1499.

Song, Y.L. and Lee, S.P. 1993. Characterization of ecological implication of luminous *Vibrio harveyi* isolated from tiger shrimp (*Penaeus monodon*). Bulletin of the Institute of Zoology Acedemia Sinica 32: 217-220.

Spanggaard, B., Huber, I., Nielson, J., Nielson, T., Appel, K.F. and Gram, L. 2000. The microflora of rainbow trout intestine: a comparison of traditional and molecular identification. Aquaculture 182: 1-15.

Stoffregen, D.A., Backman, S.C., Perham, R.E., Bowser, P.R. and Babish, J.G. 1996. Initial disease report of *Streptococcus iniae* infection in hybrid striped sunshine bass and successful therapeutic intervention with the fluoroquinolone antibacterial enrofloxacin. *Journal of the World Aquaculture Society* 27: 420-434.

Stoss, J. 1983. Fish gamete preservation and spermatozoan physiology. In W.S. Hoar, D.J. Randall and E.M. Donalson (Eds.), Fish Physiology, vol.9, part B (pp. 305-350). Orlando: Academic Press.

Stoss, J. and Refstie, T. 1983. Short-term storage and cryopreservation of milt from Atlantic salmon and sea trout. *Aquaculture* 30: 229-236.

Subramoniam, T. 1993. Spermatophores and sperm transfer in marine crustaceans. In J.H.S. Blaxter and A.J. Southward (Eds.), Advances in marine biology, vol. 29 (pp. 129-214). San Diego: Academic Press.

Sugita, A. 1996. A case of streptococcicosis in dusky spinefoot. Fish Pathology (Tokyo) 31: 47-48.

Suquet, M., Dreanno, C., Dorange, G., Normant, Y., Quemener, L., Gaignon, J.L. and Billard, R. 1998. The ageing phenomenon of turbot spermatozoa: effects on morphology, motility and concentration, intracellular ATP content, fertilization, and storage capacities. *Journal of Fish Biology* 52: 31-41.

Suquet, M., Dreanno, C., Fauvel, C., Cosson, J. and Billard, R. 2000. Cryopreservation of sperm in marine fish. *Aquaculture Research* 31: 231-243

Tanaka, R., Ootsubo, M., Sawabe, T., Ezura, Y. and Tajima, K. 2004. Biodiversity and in situ abundance of gut microflora of abalone (*Haliotis discus hannai*) determined by culture-independent techniques. *Aquaculture* 241: 453-463.

Tedder, R.S., Zuckerman, M.A., Goldstone. A.H., Hawkins, A.F., Fielding, A., Briggs, E.M., Irwin, D., Blair, D., Gorman, A.M., Patterson, K.G., Linch, D.C., Heptonstall, J. and Brink, N.S. 1995. Hepatitis B transmission from contaminated cryopreservation tank. *Lancet* 346: 137–140.

Terner, C. 1986. Evaluation of salmonid sperm motility for cryopreservation. *The Progressive Fish-Culturist* 48: 230-232.

Teska, J.D. and Shotts, E.B. 1994. Non-haemolytic group B streptococci from humans, fish and frogs. Biomedical Letters: 195-201.

Teskeredzic, E., Grahek, D., Malnar, L., Teskeredzic, Z. and Hacmanjek, M. 1993. Bacterial disease of the brown bullhead (*Amiurus nebulosus* L.). *Ribarstvo* 48: 5-11.

Tiersch, T.R. and Jenkins, J.A. 2003. Biosecurity and regulatory considerations for transfer of cryopreserved sperm and early life stages of aquatic species. In C.S. Lee and P.J O'Bryen (Eds.), Biosecurity in aquaculture production systems: Exclusion of pathogens and other undesirables (pp. 171-198). Baton Rouge: The World Aquaculture Society.

Tonguthai, K. 1995. Diseases of the freshwater prawn, *Macrobrachium rosenbergii*. The Aquatic Animal Health Research Institute Newsletter 4(2): 1-4.

Toranzo, A.E., Cutrin, J.M., Nunez, S., Romalde, J.L. and Barja, J.L. 1995. Antigenic characterization of *Enterococcus* strains pathogenic for turbot and their relationship with other gram-positive bacteria. *Disease of Aquatic Organisms* 21: 187-191.

Toranzo, A.E., Devesa, S., Heinen, P., Riaza, A., Nunez, S., Figueras, A. and Barja, J.L. 1994. Streptococcosis in cultured trout caused by an *Enterococcus*-like bacterium. Bulletin of the European Association for Fish Pathologists 14: 19-23.

Toranzo, A.E., Romalde, J.L., Nunez, S., Figueras, A. and Barja, J.L. 1993. An epizootic in farmed, marked-size rainbow trout in Spain caused by a strain of *Carnobacterium piscicola* of unusual virulence. *Disease of Aquatic Organisms* 17: 87-99.

Tulsner, M. 1994. Fischverarbeitung I. Hamburg: Behr's Verlag, p. 301.

Vanderzant, C., Nickelson, R. and Judkins, P.W. 1971. Microbial flora of pond-reared brown shrimp (*Penaeus aztecus*). Applied Microbiology 21: 916-921.

Viveiros, A.T.M., So, N. and Komen, J. 2000. Sperm cryopreservation of African catfish (*Clarias gariepinus*): cryoprotectants, freezing rates and sperm: egg dilution ratio. Theriogenology 54, 1305-1308.

Vuthiphandchai, V. and Zohar, Y. 1999. Age-related sperm quality of captive striped bass, *Morone saxatilis*. *Journal of the World Aquaculture Society* 30: 65-72.

Vuthiphandchai, V., Nimrat, S., Kotcharat, S. and Bart, A.N. 2007. Development of a cryopreservation protocol for long-term storage of black tiger shrimp (*Penaeus monodon*) spermatophores. *Theriogenology* 68: 1192-1199.

Williams, A.M., Fryer, J.L. and Collins, M.D. 1990. *Lactococcus piscium* sp. *no*v. A new *Lactococcus* species from salmonid fish. FEMS Microbiology Letters 68: 109-114.

Wilson, I.G. and Moore, J.E. 1996. Presence of *Salmonella* spp. and *Campylobacter* spp. in shellfish. *Epidemiology and Infection* 116: 147-153.

In: Aquaculture Research Trends
Editor: Stephen H. Schwartz, pp.185-212

ISBN 978-1-60456-217-0

Chapter 5

GROWTH, GAMETOGENIC CYCLE AND SEASONAL CHANGES IN BIOCHEMICAL COMPOSITION OF THE PACIFIC OYSTER *CRASSOSTREA GIGAS* IN RELATION TO THE ENVIRONMENTAL CONDITIONS OF THE BIZERT LAGOON, TUNISIA

Salwa Dridi [1,2,*] ***Mohamed Salah Romdhane*** [b] ***and M'hamed Elcafsi*** [1]

1 Faculté des Sciences de Tunis/ Département de Biologie,
Unité de Physiologie et d' Ecophysiologie des Organismes Aquatiques, Campus Universitaire, 2092, El Manar II, Tunis, Tunisia

2Institut National Agronomique de Tunisie/, Département des Sciences de la Production Animale et de la Pêche, Unité d'écosystèmes et ressources aquatiques, 43 Av. Charles Nicole, 1082 Tunis, Tunisia

ABSTRACT

The relationship between growth, gametogenic cycle and the seasonal variations in biochemical composition in the gonad-digestive gland of the bivalve *Crassostrea gigas* was studied from January 2002 until January 2003. Changes in the shell growth and the body weight are associated with metabolite accumulation in the complex gonad – digestive gland, storage utilization of reserves, gametogenic events and environmental parameters, especially food availability. This study reports the first evidence of reproduction of *Crassostrea gigas* in the Bizert lagoon of Tunisia. Initiation of gametogenesis took place in March when water temperature reached 14 °C. Ripe oysters were observed in May at a water temperature of 20 °C. Spawning began in June and reached a peak in September when water temperature ranged between 23 °C and 27 °C. During the latter period, oysters presented gonads with ruptured follicles and residual gametes. The study found that *Crassostrea gigas* showed a clear cycle of energy storage and utilization in the Bizert lagoon. Total biochemical composition was about 6.67 % ± 1.34 of dry weight in lipids, 6.50 % ± 3.01 of dry weight in glycogen and 14.34 % ± 3.60

* Corresponding Author: salwadridi@yahoo.fr / Phone : + 216 99558004; Fax: +216 885480.

of dry weight in protein and. Lipid and glycogen concentrations in the gonad – digestive gland were inversely related: maximum concentrations of glycogen occurred between December and February and corresponded with minimum concentrations of lipids. High glycogen contents recorded in winter supported the processes of gametogenesis and then decreased progressively to attain low values at maturity and during summer corresponded to the first emission of gametes (June) and the beginning of total spawning, however lipids are accumulated in gonads during the period of maximum ripeness (spring) and decrease in summer. During autumn, the increase of total fatty acids values appeared to be related to the available food in association with an increase in the chlorophyll *a* concentrations. Protein contents show many variations throughout the year. High protein levels occurred in May and in July which corresponded to oocyte maturation and then decreased in late summer (beginning of total spawning). At the expense of the available food in autumn, protein concentrations increase again to assure recovery of the tissue weight of the bivalve, which induced an increase in the condition indices values. The process of growth was analyzed during the two phases of the gametogenic cycle. Correlation between the different parameters of growth models noted during sexual repose is greater than those determined during the gonadal activity, including ripeness and spawning. The gametogenic cycle influence clearly the lengthwise and weight modals of growth by affecting the bivalve with physiologic troubles traduced by win or loose of the organic matter. During the sexual repose, fatty acids are accumulated in the *C. gigas* tissues (total SFA, MUFA and PUFA (n-3) show the highest values (respectively 26.38 % ± 1.78; 15.84 % ± 0.56; 46.94 % ± 0.61 of the total fatty acids). The glycogen levels are also attempt the highest during the sexual repose (11,52 % ± 4,64 of the tissue dry weight). These biochemical components of the *C. gigas* tissues contribute in the increases of shell length, width and thickness respectively (2,04 ± 0,18 cm; 0,58 ± 0,08 cm and 0,90 ± 0,06 cm from January 02 to February 02 and 1,25 ± 0,33 cm; 0,65 ± 0,11 cm and 0,28 ± 0,07 cm from November 02 to January 03) during this phase of the gametogenic cycle. However during the gametogenic activity (March 02 – September 02) total SFA, MUFA and PUFA (n-3) show the lowest values (respectively 23.13 % ± 1.38; 12.35 % ± 0.55; 38.15 % ± 0.84 of the total fatty acids) also the glycogen levels decrease and attempt the minimum in September (2.09 % ± 1.98 of the tissue dry weight) which increase the shell size by only 1,12 ± 0,08 cm, 0,64 ± 0,80 cm and 0,57 ± 0,12 cm in length, width and thickness. Protein levels are comprised between 9,5 % ± 2,19 and 21,37 % ± 1,50 of the total tissues dry weight. At the expense of the available food in autumn, protein concentrations increase to assure recovery of the tissue weight of the bivalve. This metabolite has not any clear correlations with the shell growth but contribute in supporting the bivalve essentially during summer when lipids and glycogen reserves are depleted.

Keywords: *Bivalve, Body and shell growth, chlorophyll a, Energy, Fatty acids, Gametogenic cycle, Glycogen, Protein.*

1. INTRODUCTION

Oyster aquaculture industry is one of the oldest aquaculture activities worldwide. The evolution of this industry began in the 20th century after the establishment of reliable technologies,(Lubet, 1991). In 2000, world production of aquaculture mollusc has attained 10,73 millians of tones representing 30,15 % of the aquacultural production in all sectors, which attained 35,58 millians of tones. The bivalve industry mainly concerned USA, China,

Japan, Corea, France, Spain and the Netherlands. (FAO, 2002). The major production was represented by oysters (37,37% of the total production in mollusks, 98,32% of which were *Crassostrea gigas*, 24,6% clams, 12,28% mussels and 10,75% scallops Saint Jaques (FAO, 2002).

Studies on the seasonal variations in the biochemical composition of marine bivalves have been numerous (Taylor and Venn, 1979; Mann, 1979; Li et al., 2000; Dridi et al., 2007).

Some workers have pointed out that the amount of energy available to any organism is generally limited in some way and that a trade-off between reproductive effort and continued growth exists (Fisher 1958, Calow 1983). In this context, Gabbott (1983), specify that seasonal metabolic activities in molluscs result from complex interactions among food availability, environmental conditions, growth and reproduction. In general when food is abundant, reserves accumulate prior to gametogenesis in the form of glycogen, lipid and protein substrates, and subsequently are utilized in the production of gametes when metabolic demand is high (Mathieu and Lubet, 1993). The specifics of which substrates are important and how the timing of their consumption varies between species as well as among populations of the same species (Giese, 1969; Bayne, 1976; Sastry, 1979 and Barber and Blake, 1981).

Although the Pacific oyster *Crassostrea gigas* is one of the most commercially important bivalves in the world, characterized by the most suitable organisms for analyzing growth relationships because the individual growth pattern is faithfully recorded in their incrementally growing shells (Lutz and Rhoads 1980; Jones 1983). The performance of *Crassostrea gigas* (growth and annual gametogenic cycle) has previously been the subject of many experimental studies. (Perdue 1984; Maurer and Borel, 1986; Berthomé et al. 1986; Dinamani 1987; Maurer 1989; Haure and Baud 1990; Ruiz et al. 1992; Steele and Mulcahy 1999; Berthelin et al. 2000; Chavez Villalba et al. 2001, Ren et al. 2003 and Dridi et al. 2006, 2007).

In Tunisia, the oyster aquaculture industry is relatively a recent activity. The cupped oyster *C. gigas* has been farmed for the first time in the Bizert lagoon in 1972. Spats were imported directly from Japan, and then from France (Medhioub, 1993). Continued oyster production in Tunisia since then remains dependent of the importation of juveniles from France. To be able to improve the methods of cultivating this bivalve, detailed knowledge of it's ecobiology in our country is fundamental to avoid us the foreign spat importation, which represented until now an economic handicap to this industry.

Considering these reasons, an investigation of seasonal changes in the biochemical composition in relation to growth, gametogenic cycle and environmental conditions in *C. gigas* of the Bizert lagoon, was aimed at in this study, to determine periods of low energetic condition which increase the fragility of oysters in confronting the different environmental stresses.

2. Materials and Methods

2.1. Study Area and Sampling

Bizert lagoon is located on the north of Tunisia, between 37° 08' and 37° 14' N and 9° 46' and 9° 58' E (Figure 1).

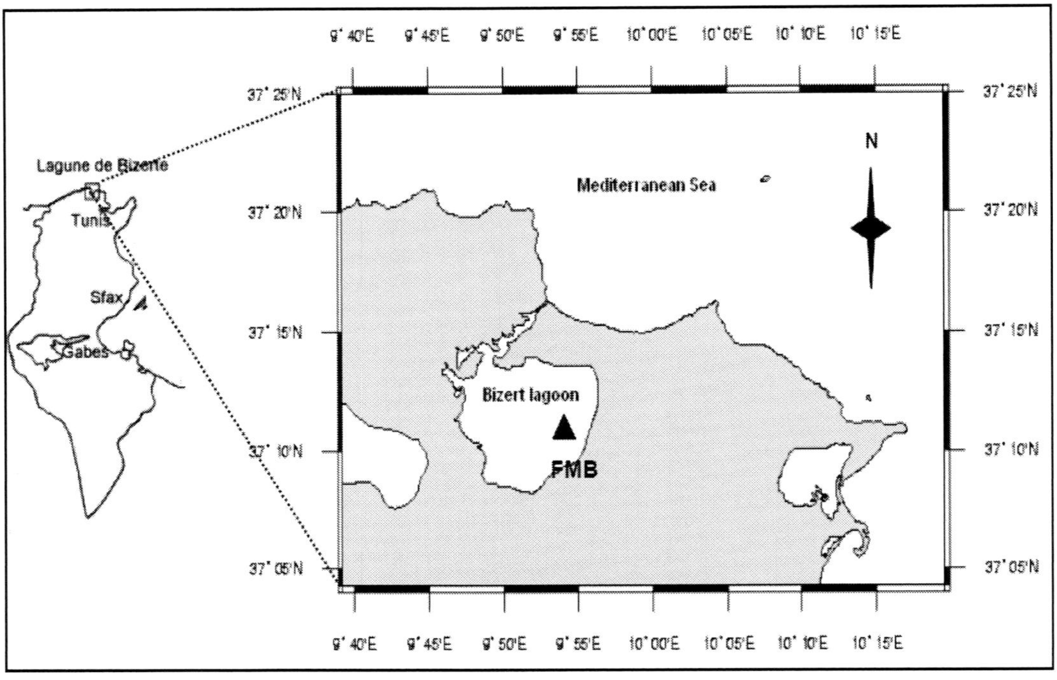

Figure 1. Map of the oyster parc (FMB) in the Bizert lagoon (▲)

Samples of adult oysters grown in suspended culture were collected from the oyster farm FMB (Ferme Marine de Bizerte) situated at 37° 09 38' N and 9° 53.55' E in the Bizert lagoon at monthly intervals from February 2002 to January 2003. Oysters was 12 months old, measure 6.69 cm ± 0.75 SE (maximum anterior posterior difference) and weighed 30.94 g ± 10.82 SE at the start of the trial.

2.2. Environmental Parameter Recording

At the sampling site, a temperature and salinity measurements (with respectively thermometer at 0,01 °C accuracy and salinometer type WTW.LF 191) were measured every month at the point of collection. Chlorophyll *a* measurements were also performed every month at the site of culture and were evaluated using fluorimetric methods (Aminot and Chaussepied, 1983). The average surface water temperature in the area of oyster collection fluctuated between a minimum of 10.9 °C and a maximum of 28° C during the period of study. Highest monthly mean values occurred in August and lowest temperature in January. The annual mean temperature was 19.7°C ± 5.8°C. Salinity varied between 38 ‰ and 33 ‰ during the period of study. Salinity's values are highest after summer evaporation caused by the temperature's increase, but these values decreases during the rainy season. The chlorophyll *a* level displayed marked seasonal changes. Peaks in chlorophyll *a* were observed in May (2,4 µg/l) and November (1,2 µg/l). (Figure 2)

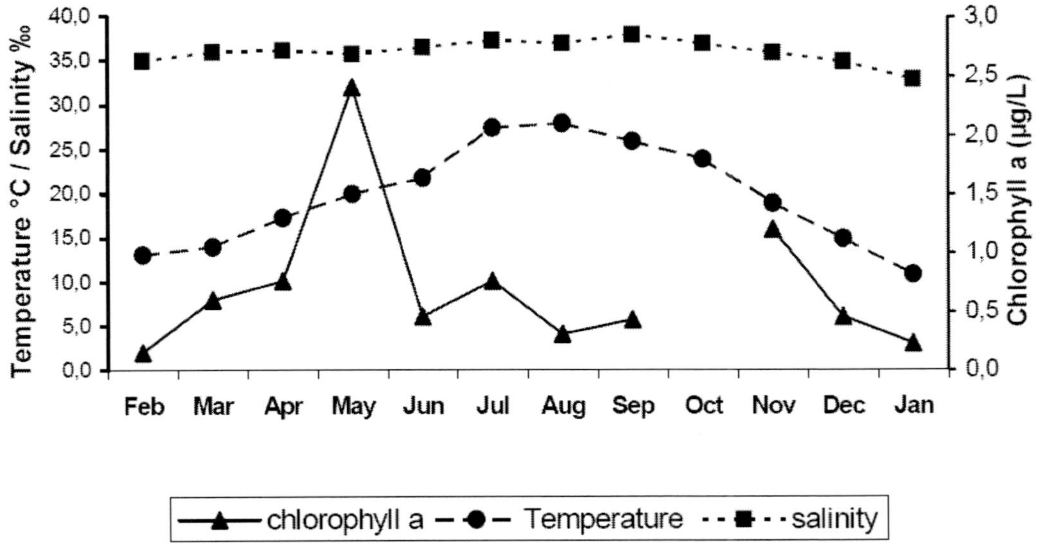

Figure 2. Monthly variation of temperature, salinity and chlorophyll "*a*" from February 2002 to January 2003.

2.3. Histological Techniques

From February 2002 to January 2003 monthly samples of between 10 and 15 oysters were collected.

The tissue, of each oyster was removed from the shell, weighted and sectioned. One transverse cut was made from the gills and palps, with a second transverse cut made a few millimeters below the first, (Steele et al. 1999). The resulting segment of body tissue was fixed in Bouin's solution, and was dehydrated through an increasing ethanol concentration series. Dehydrated samples were cleared and embedded in paraffin following a standardized procedure (Lango-Reynoso et al. 2000). Sections 6 µm thick were cut mounted on glass slides and stained with a solution of Groat's Hematoxylin and Eosin (Gabe, 1968).

Each histological section of gonadal tissue was examined in detail under a microscope to assess the sex and stage of gonadal development.

Sex and gametogenic stage was done according to the reproductive scale reported by Lubet (1959): Stage 0; undifferentiated gonads, characterized by empty gonads (sexual repose). There is no presence of follicles peripherally to the digestive gland, making sex determination difficult. Stage I; Gonad early – development, known by the decrease of the interfollicular connective tissue after gonia multiplication and expansion of the follicles, which contain oogonia or spermatogonia and primary oocytes or spermatocytes. Stage II; Gonad late – development. An increase of the oocyte diameter was observed. Secondary oocytes and spermatocytes predominate in the follicles. Some spermatozoa could be observed. Stage III A; Intense gametogenic activity, characterized by disappearance of the interfollicular connective tissue. Male oyster gonads with spermatids oriented toward the center of the follicle lumen, some times with a narrow band stage on the wall. The females exhibit a wide spectrum of oocyte sizes in all gametogenic stages including some free oocytes: Adhering oocyte and peduncular oocyte. Stage III B; Gonad maturation, during which the male oyster gonads present homogenous aspect after disappearance of the spermatids running toward the center of the follicle. The female gonads present many mature oocytes with a polygonal shape that appeared to be free within the follicular lumen. Stage III D; Spent gonads, characterized by the presence of ruptured follicles and residual gametes in the gonads of oysters. Connective tissues reappear between follicles.

2.4. Biometric Measurements and Condition Indice

During this study, four hundred and eighty seven individuals of *C. gigas* were collected for length (L), thickness (T), width (W), total wet weight (W wt) , flesh wet weight (Wwf), flesh dry weight (Wdf) and valves dry weight (Wdv) measurements. The linear measurements were made using a slide caliper (accuracy ± 0,001 m), however, weights were recorded using 0.001g electronic precision balance. For the dry weights measurements, the valves of each individual were opened carefully and all the flesh removed from the shells. Before weighing, excess moisture was removed from all parts of the animal using absorbent paper. After recording the total wet weight, shell measurements, the tissues were oven-dried for 72 h at 60°C, and the dry weight is determined.

Condition index of the oysters was calculated as the ratio of the dry weight of the soft parts/ dry weight of shell X 100 (Walne and Mann, 1975).

$$IC = X\ 100 \quad \frac{\text{Dry weight of the soft parts}}{\text{Dry weight of shell}}$$

2.5. Biochemical Analysis

According to Lubet, (1991), the reserve tissue of oystreidae is well individualized. It's an connective tissue with vesicles cells containing glycogen and lipid droplets which constitute a nappy solder to the mantle and enveloping the visceral mass (gonad - digestive gland). For that reason, the total lipids, glycogen and protein were determined in gonad-digestive gland of each animal. Oysters destiny for the biochemical analysis were sampling every month and measure between 7,00 and 9,00 cm (maximum anterior posterior difference). Extractions were made on each oyster separately.

The total lipid extraction was performed according to Folch et al., (1957) method (chloroform/ methanol; 2:1, vol/vol), as modified by Bligh and Dyer (1959). Fatty acid methyl esters were obtained according to Metcalfe et al. (1966).

Fatty acid components of lipids were analyzed on a Hewlett Packard 4890 D gas. Liquid chromatograph equipped with capillary column (30 m x 0,25 mm ID, 0,25µm film) maintained between 150°C and 210° C for all the time of analysis. Fatty acid methyl esters were injected in a split/splitless capillary injector (ratio 100:1) maintained at 230°C and detected by a flame ionization detector maintained at 250°C. The identities of the different fatty acids were obtained through comparison with the retention times of standard fatty acid methyl esters. Peaks comprising less than 0.2 % of the total area were not considered. The following shorthand notation was used (C:X (n-y)), where C is the number of carbon atoms, X is the number of double bonds and y is the position of the first double bond counted from the terminal methyl group. Each data point was the mean of six replicates. Total lipid results were expressed as percentage of total gonad-digestive gland dry weight.

Glycogen extraction was based on the enzymatic method of Dubois et al., (1956) using glucose solution (100mg/l). Each data point was the mean of ten replicates.

Protein determination was based on the bovine serum albumine used as standard according to Lowry et al., (1951) method. Each data point was the mean of ten replicates.

2.6. Data Analysis

Data obtained from shell measurements were used to determine morphometric relationships among shell growth axis of length, thickness and width. For isometric relationships, we used linear regression tests to obtain descriptive equations, and the goodness of fit was described using the correlation coefficient (r). The allometric growth relationships of length, thickness and width with the Wwt were estimated using potential regression and the goodness of fit were described using the correlation coefficient (Sokal and Rohlf, 1969).

The t-test (Student's t) and the multiple comparison method of Duncan at $P< 0.05$ were used for comparison of the linear regression slopes. Data of the biochemical composition were analyzed for significant differences of means, ANOVA, and inspected by Duncan test at the level of ($p <0.05$). Statistical analyses were carried out using the software-program STATISTICA 6.0.

3. Results

3.1. Gametogenesis

The reproductive cycle of C. gigas was studied in the Bizert lagoon from February 2002 until January 2003. Sex and gametogenetic stage was done according to the reproductive scale reported by Lubet (1959).

Stage 0. Undifferentiated Gonads

This stage was observed in February, September, November, December 2002 and January 2003 and was characterized by considerable developed connective tissue full of energy reserves with little follicular material. During this period gonads were empty (sexual repose). There was no presence of follicles peripherally to the digestive gland, making sex determination difficult.

Stage I. Gonad Early - Development

This stage was established in March and April 2002. The interfollicular connective tissue decreased after gonia multiplication and expansion of the follicles, which contained oogonia or spermatogonia and primary oocytes or spermatocytes.

Stage II. Gonad Late - Development

This stage was detected in April 2002. An increase of the oocyte diameter was observed. Secondary oocytes and spermatocytes predominate in the follicles. Some spermatozoa could be observed.

Stage III A. Intense gametogenic activity

In May 2002, the interfollicular connective tissue disappeared. Male oyster gonads presented spermatids oriented toward the center of the follicle lumen, some times with a narrow band stage on the wall.

The females exhibited a wide spectrum of oocyte sizes in all gametogenic stages including some free oocytes.

1. Adhering oocyte: Oocytes were adhered to the follicle wall on a large surface. Their cytoplasm showed a darker coloration as it got richer in vitellus. Cells contained a distinct nucleus and nucleolus.
2. Peduncular oocyte: Adhesion of oocytes to the follicle wall was limited to a peduncle. Then cells became longer because of the vitell accumulation and the migration of the nucleus to the free extremity of the cells.

Stage III B. Gonad maturation

This stage was observed in May, July and August 2002.

The male oyster gonads presented homogenous aspect after disappearance of the spermatids rumming toward the center of the follicle. The female gonads presented many mature oocytes with a polygonal shape that appeared to be free within the follicular lumen and able to expelle.

Stage III D. Spent gonads

This stage occured in June and September 2002.

Oysters presented ruptured follicles and residual gametes. Connective tissues reappeared between follicles.

The percentage distribution of different gametogenetic stages observed during the period of study is illustrated in Figure (3).

Gametogenesis began in March or April, after a period of a sexual inactivity during the months of November, December and January. In February, more then 35 % of oysters were at stage I of gametogenic development and in March, 25 % were at stage II. Gonads become ripe from May onwards. Although some individuals were at stage III B in May, the spawning activity (stage III D) was observed in June (75 %), Jully (5 %), August (20 %) and September (100 %). Between November, December and January most of the oysters returned to the resting phase. We could therefore distinguish tow phases in the gonadal cycle of *C. gigas* in the Bizert lagoon: a resting phase from November to January and gametogenesis during the rest of the year, including ripeness and spawning in the summer.

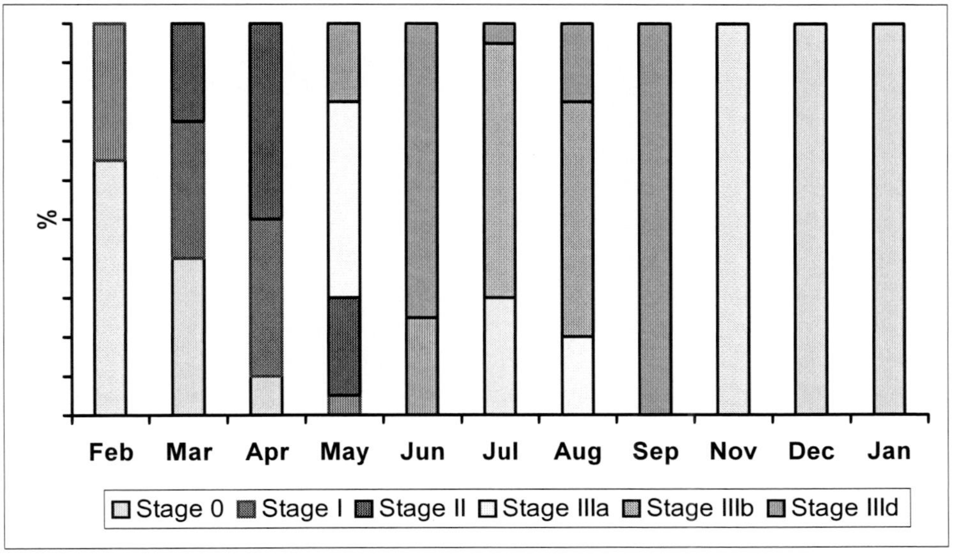

Figure 3. Seasonal distribution of *Crassostrea gigas* at different stages of gonad development, according to Lubet (1959).

3.2. Condition Indices

According to Orban et al. (2006), the commercial quality and physiological state of bivalve molluscs are adequately described by the Condition Index, a parameter of economic relevance reflecting the ecophysiological conditions and the health of animals.

The monthly pattern of CI in *C. gigas* varied between 2.31 ± 0.73 and 4.96 ± 1.16 (Figure 4). During gametogenic development, the CI remained high values (4.31 ± 0.90) and then decreased significantly ($p<0.05$) to 2.31 ± 0.73 when the animals were ripe (June). The values of CI recovered between July and August, and then remains a significative decrease ($p<0.05$) in September, after massive spawning at the end of the summer. At sexual inactivity, CI levels increased, reaching maximum values essentially in winter (4.97 ± 1.16).

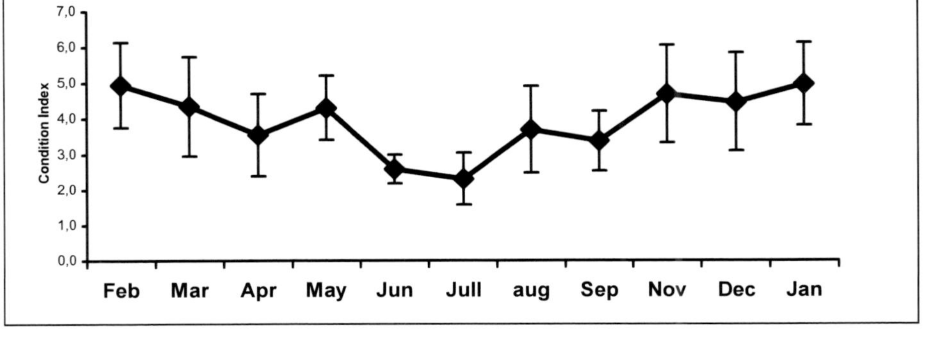

Figure 4. Evolution of the condition index (CI) of Crassostrea gigas from February 2002 to January 2003.

3.3. Biometric Measurements

Mean biometric data of *C. gigas* in the Bizert lagoon from January 2002 to January 2003 (shell length, width and thickness (cm) and total wet weight (g)) varied throughout the year (Figure 5). Differences among average growth / month rates per phase of the gametogenic cycle were observed for length during the sexual repose (January 02 – February 02) 2,04 ± 0,18 cm, gametogenic activity (March 02 – September 02) 1,12 ± 0,08 cm and sexual repose again (November 02 – January 03) 1,12 ± 0,33 cm. For width during sexual repose (January 02 – February 02) 0,58 ± 0,08 cm, gametogenic activity (March 02 – September 02) 0,64 ± 0,80 cm and sexual repose again (November 02 – January 03) 0,65 ± 0,11 cm. For thickness during sexual repose (January 02 – February 02) 0,90 ± 0,06 cm, gametogenic activity (March 02 – September 02) 0,57 ± 0,12 cm and sexual repose again (November 02 – January 03) 0,28 ± 0,07 cm. Variations in the total wet weight are dependent of win or loose of the organic matter in relation with the gametogenic cycle. We observe two depressions of the total wet weight levels in June and in September in relation with spawning.

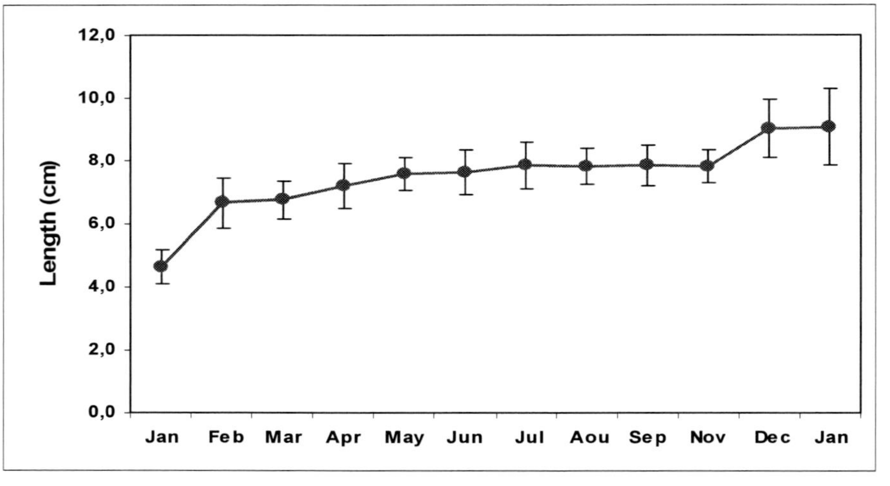

(a)

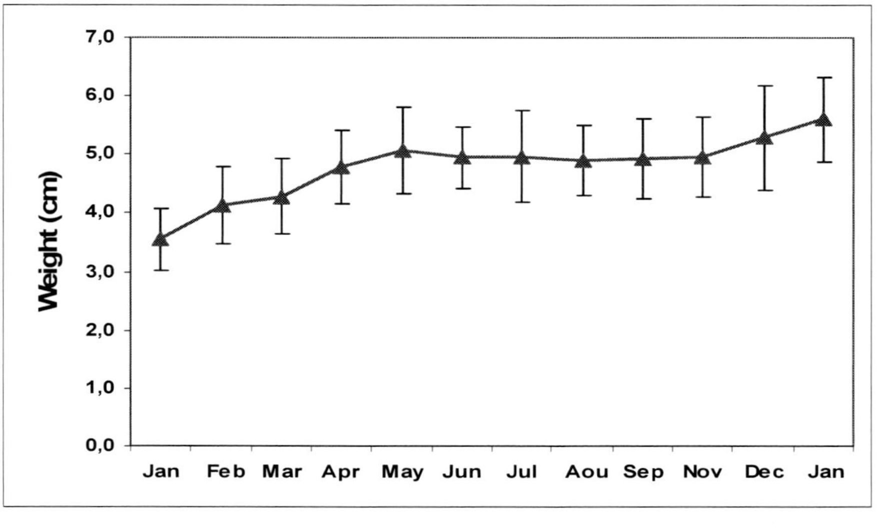

(b)

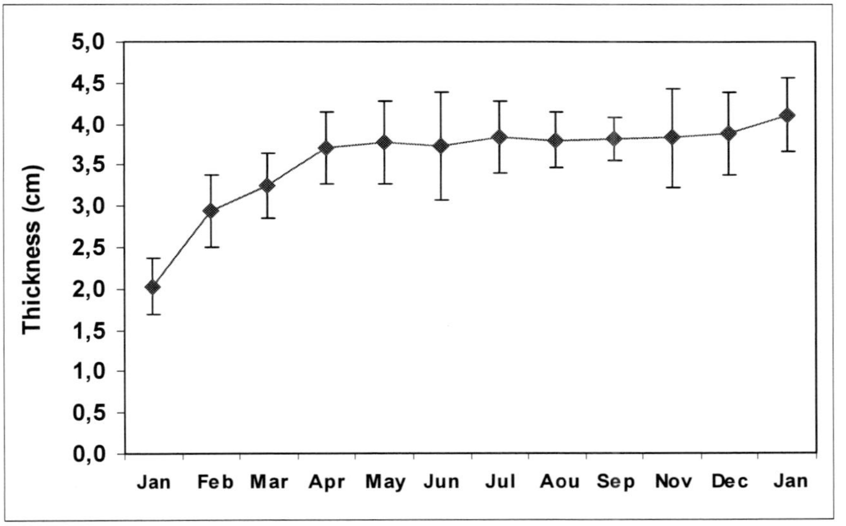

(c)

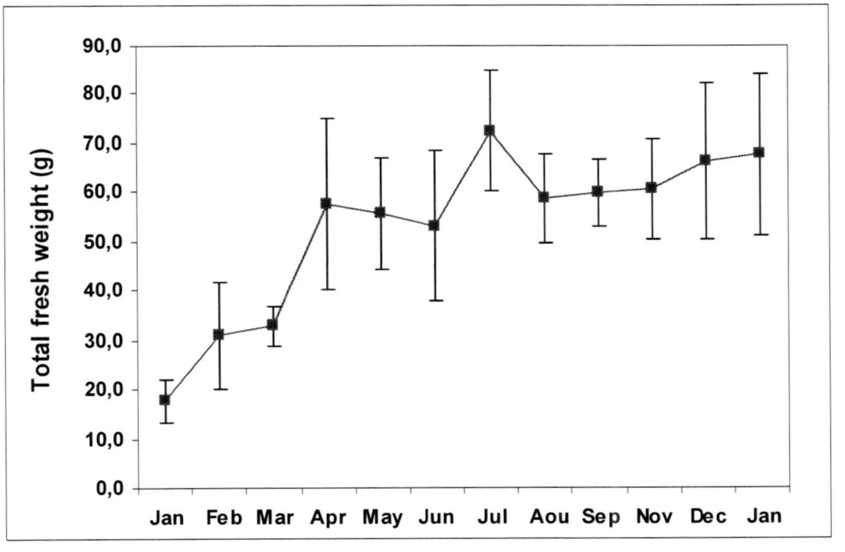

(d)

Figure 5 (a, b, c, d): Mean biometric data evolution of *Crassostrea gigas* in the Bizert lagoon from January 2002 to January 2003 (shell length, width and thickness (cm) and total wet weight (g)).

The comparison of length, thickness and width between them was carried out using linear regression of logarithmically transformed data of the form where shown in table. 1:

1- Log (L) = Log a + b Log (W)
2- Log (T) = Log a + b Log (L)
3- Log (T) = Log a + b Log (W)
b values calculated were compared to 1:
b > 1: Allometric relationship increased
b < 1: Allometric relationship diminished

Table 1. Allometric relationship between lengthwise growths

Period	Regression equations	N	r	t
January 02 – January 03	Log W = 0.550 Log L + 0.426	487	0.58	**
	Log T = 0.919 Log L - 0.601	487	0.75	**
	Log T = 0.744 Log W + 0.075	487	0.57	**

N: number of individuals
r: Correlation coefficients of the regression equation.
**: Student's test significant ($t > 1.96$)

Results of growth in this study are related to gametogenic activities determined by Dridi et al., (2006). The gametogenic cycle comprised two phases: gametogenesis including ripeness and spawning (March- September) and sexual repose during the rest of the year.

The isometric growth patterns among the different axes were: length and width (Log W = 0.555 Log L + 0.400; r = 0,64 ($p < 0.05$)), length and thickness (Log T = 0.933 Log L - 0.686; r = 0,82 ($p < 0.05$)) and thickness and width (Log T = 0.852 Log W - 0.172; r = 0,64 ($p < 0.05$)) during the sexual repose. However during the gametogenic activity the allometric relationship between length and width (Log W = 0.326 Log L + 0.888; r = 0,22 ($p < 0.05$)), length and thickness (Log T = 0.133 Log L + 1.027; r = 0,1) and thickness and width (Log T = 1.091 + 0.132 Log W; r = 0,14 ($p < 0.05$)).

The relationships: Length (L) considered as size reference with total wet weight, wet flesh weight, dry flesh weight and valves dry weight, are shown in (table 2) and described by the following equations:

1- Log (L) = Log a + b Log (Wt)
2- Log (Wwf) = Log a + b Log (L)
Log (Wdf) = Log a + b Log (L)
Log (Wdv) = Log a + b Log (L)
b values calculated were compared to 3:
b > 3: Allometric relationship increased
b < 3: Allometric relationship diminished

Table 2. Allometric relationship between Length and the different measurements of weight

Period	Regression equations	N	r	t
January 02 – January 03	Log Wwt = 2.025 Log L - 0.154	487	0.83	**
	Log Wwf = 1.833 Log L - 2.135	330	0.52	**
	Log Wdf = 1.661 Log L - 3.190	330	0.44	**
	Log Wdv = 1.584 Log L + 0.268	330	0.56	**

The allometric relationship between length and the parameters of weight were: length and total fresh weight (Log Wwt = 1,967 Log L - 0,128; r = 0,88 ($p < 0.05$)); length and the fresh weight of the flesh (Log Wwf = 2.456 Log L - 3.260; r = 0,73 ($p < 0.05$)), length and the dry weight of the flesh (Log Wdf = 2.306 Log L - 4.384, r = 0,65 ($p < 0.05$)) and length and the dry weight of the valves (Log Wdv = 2.048 Log L - 0.739, r = 0,69 ($p < 0.05$)) during the sexual repose. However during the gametogenic activity, the allometric relationship between length and the parameters of weight were: length and total fresh weight (Log Wwt = 1.390 Log L + 1. 185; r = 0,47 ($p < 0.05$)), length and the fresh weight of the flesh (Log Wwf = 0.713 Log L + 0.029; r = 0,21 ($p < 0.05$)), length and the dry weight of the flesh (Log Wdf = 0.575 Log L -1.078; r =0,15 ($p < 0.05$)), and length and the dry weight of the valves (Log Wdv = 1.166 Log L + 1.140; r = 0,42 ($p < 0.05$)).

3.4. Biochemical Analysis

The changes in biochemical composition of gonad –visceral mass of *C. gigas*, throughout the year, are shown in Figure a, b, c, d.

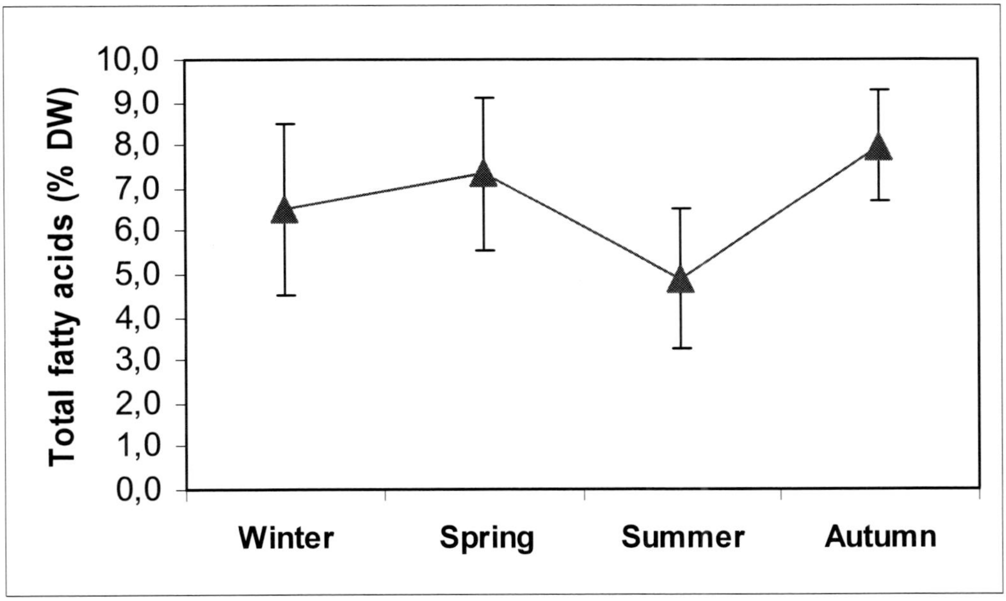

(a)

Figure 6 a. Seasonal variations in total lipid content (mg/ g dry weight) of the gonad-visceral mass of *Crassostrea gigas* in the Bizert lagoon.

(b)

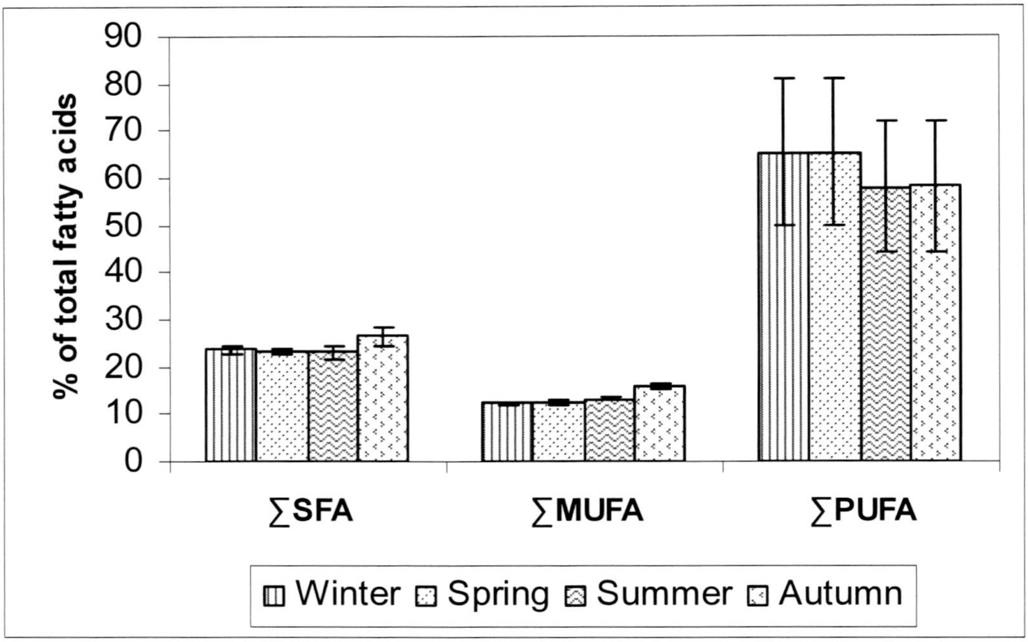

Figure 6 b. Seasonal variations in PUFA, MUFA and SFA (% of total fatty acids), of the gonad-visceral mass of *Crassostrea gigas* in the Bizert lagoon.

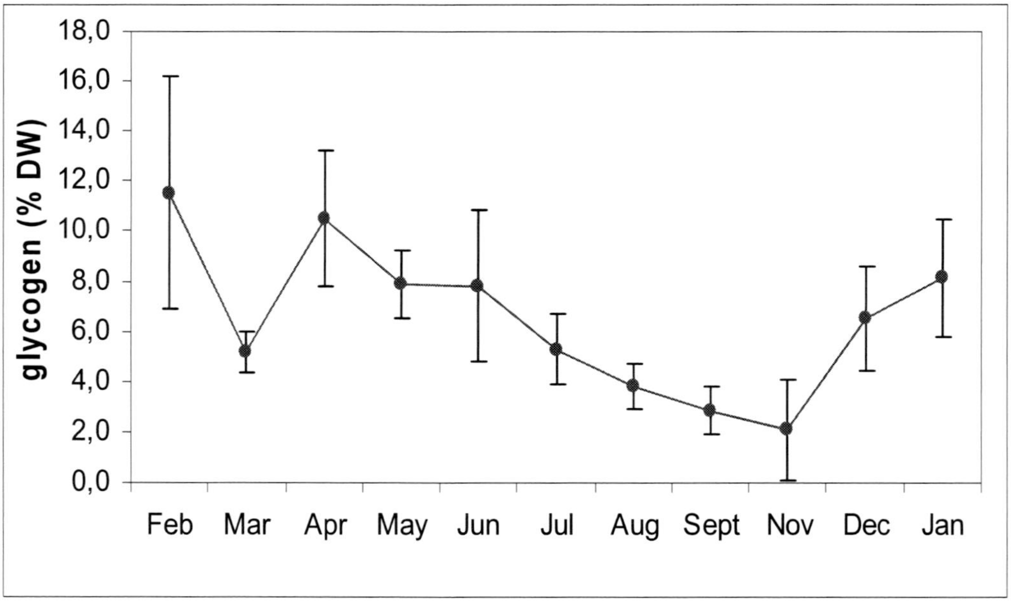

(c)

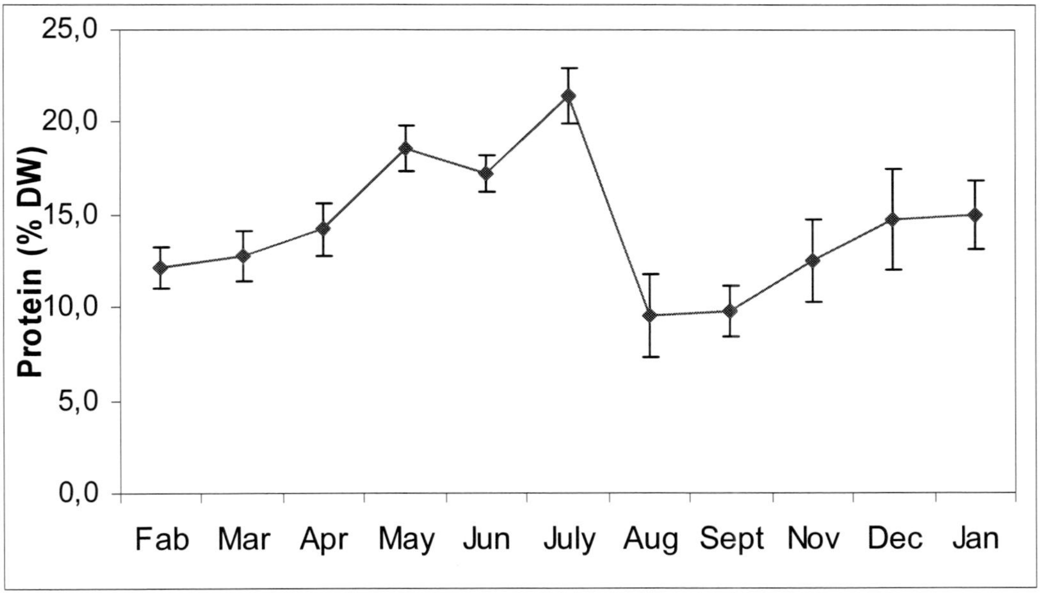

(d)

Figure 6 c, d. Seasonal variations in glycogen (c) and protein (d) contents (mg/ g dry weight) of the gonad-visceral mass of *Crassostrea gigas* in the Bizert lagoon.

The total fatty acids compositions from the gonad-visceral mass were shown in Table 3.

In the present study seasonal variation of total fatty acid (% DW) indicated that highest content of total fatty acids are noted during spring (7.33 ± 1.78) which corresponded to sexual maturity and also during autumn (7.97 ± 1.31) $p<0.05$, which corresponded to the increase of nutrient availability demonstrated by the Chlorophyll *a* level elevation in November (1,2 µg/l). However a significative decrease ($p<0.05$) of total fatty acid occur during summer (4.87 ± 1.60) which corresponded to the first partially spent (June) and began of total spawning (August) (Figure 6 a).

During the sexual repose, fatty acids are accumulated in the *C. gigas* tissues (total SFA, MUFA and PUFA (n-3) show the highest values (respectively 26.38 % ± 1.78; 15.84 % ± 0.56; 46.94 % ± 0.61 of the total fatty acids). SFA is dominated by C16:0 (20,80 % ± 3,95 of total fatty acids). MUFA is dominated by C18:1 (*n*-7) (5.26 % ± 1.20 of total fatty acids) and PUFA (*n*-3) is dominated by C18:3 (*n*-3) (4,38 % ± 0,95); C18:4 (*n*-3) (7,88 % ± 1,42); C20:5 (*n*-3) (12,15 % ± 1,71) and C22:6 (*n*-3) (21,22 % ± 1,10). However during the gametogenic activity (March 02 – September 02) total SFA, MUFA and PUFA (*n*-3) show the lowest values (respectively 23.13 % ± 1.38; 12.35 % ± 0.55; 38.15 % ± 0.84 of the total fatty acids).

During the year of study, levels of main fatty acids in the total lipid fraction of the gonad-digestive gland (% of total fatty acids), remain more or less constant except for the (*n*-3) polyunsaturated fatty acids, especially 22:6 (*n*-3) which remained a significative decrease ($p<0.05$) from spring to summer, and then increase significantly ($p<0.05$) from summer to autumn, due to the elevation of chlorophyll *a* levels registered in November.

Table 3. Seasonal changes in the total fatty acid composition of gonad – digestive gland of *Crassostrea gigas* in the Bizert lagoon from February 2002 to January 2003

Fatty acids	Winter	Spring	Summer	Autumn
	Mean SD	Mean SD	Mean SD	Mean SD
14 : 0	2.55 ± 1.87	1.15 ± 1.28	1.22 ± 0.64	2.46 ± 0.27
15 : 0	0.42 ± 0.21	0.28 ± 0.14	0.51 ± 0.30	0.72 ± 0.11
16 : 0	19.77 ± 1.41	20.84 ± 1.17	20.05 ± 3.29	20.80 ± 3.95
18 : 0	0.91 ± 0.13	0.85 ± 0.05	1.33 ± 0.71	2.32 ± 1.18
∑SFA	23.66 ± 0.86	23.14 ± 0.65	23.13 ± 1.38	26.38 ± 1.78
16:1(n-9)	0.21 ± 0.19	0.33 ± 0.30	0.45 ± 0.33	0.49 ± 0.15
16:1(n-7)	1.83 ± 0.74	1.57 ± 1.75	1.17 ± 0.97	1.67 ± 0.78
18:1(n-9)	1.78 ± 0.89	1.57 ± 0.36	2.23 ± 0.84	2.26 ± 1.81
18:1(n-7)	3.47 ± 0.44	3.69 ± 0.71	3.72 ± 1.06	5.26 ± 1.20
20:1(n-9)	1.91 ± 0.12	1.94 ± 0.76	1.88 ± 0.56	2.66 ± 0.54
20:1(n-7)	2.01 ± 0.21	2.09 ± 0.25	2.93 ± 0.96	3.00 ± 0.94
20:1(n-11)	1.19 ± 0.17	1.14 ± 0.12	0.61 ± 0.51	0.48 ± 0.31
∑MUFA	12.40 ± 0.31	12.35 ± 0.55	12.93 ± 0.26	15.84 ± 0.56
18:2(n-6)	10.64 ± 0.96	11.93 ± 3.52	10.01 ± 3.50	7.47 ± 2.96
18:3(n-6)	0.17 ± 0.06	0.16 ± 0.07	0.14 ± 0.09	1.14 ± 0.55
20:2(n-6)	0.39 ± 0.91	0.26 ± 0.04	0.22 ± 0.05	0.80 ± 0.46
20:3(n-6)	0.18 ± 0.18	0.13 ± 0.12	0.20 ± 0.12	0.12 ± 0.12
20:4(n-6)	1.32 ± 0.16	1.27 ± 0.39	1.98 ± 0.56	2.22 ± 0.10
22:5(n-6)	0.51 ± 0.04	0.47 ± 0.15	0.53 ± 0.30	0.82 ± 0.10
22:4(n-6)	1.64 ± 0.24	2.26 ± 0.75	2.43 ± 0.55	1.96 ± 0.78
∑PUFA (n-6)	14.86 ± 0.32	16.50 ± 0.27	15.53 ± 1.23	14.55 ± 0.28
18:3(n-3)	4.38 ± 0.95	4.00 ± 0.82	3.39 ± 0.89	3.12 ± 1.11
18:4(n-3)	7.88 ± 1.42	5.89 ± 1.33	4.18 ± 1.57	3.27 ± 0.52
20:4(n-3)	1.04 ± 0.25	0.73 ± 0.06	0.62 ± 0.22	0.73 ± 0.24
20:5(n-3)	12.15 ± 1.71	11.00 ± 1.49	10.68 ± 1.18	11.43 ± 1.61
22:5(n-3)	0.74 ± 0.18	0.81 ± 0.26	1.19 ± 0.23	0.77 ± 0.25
22:6(n-3)	20.40 ± 0.57	21.94 ± 1.73	17.73 ± 2.48	21.22 ± 1.10
21:5(n-3)	0.32 ± 0.28	0.24 ± 0.26	0.33 ± 0.29	0.69 ± 0.43
∑PUFA (n-3)	46.94 ± 0.61	44.63 ± 0.67	38.15 ± 0.84	41.25 ± 0.52
22 : 2 i	0.15 ± 0.13	0.67 ± 0.33	0.80 ± 0.14	0.05 ± 0.05
22 : 2 j	0.60 ± 0.52	0.38 ± 0.45	0.65 ± 0.55	0.07 ± 0.06
18 : 2 (n-4)	2.66 ± 0.33	2.95 ± 1.08	2.78 ± 0.98	2.16 ± 0.16
∑PUFA	65.22 ± 15.49	65.15 ± 15.55	57.93 ± 13.69	58.09 ± 13.93
(n-3) / (n-6)	3.15	2.70	2.45	2.83

In the total fatty acid fraction we found two C22 dienoic acids, 22:2i and 22:2j, which we considered to be 22:2Δ7,13 and 22:2Δ7,15, respectively (Klingensmith, 1982). The (*n*-3) / (*n*-6) ratio in the total fatty acid fraction remained high levels in winter and low values occurs during summer (Figure 6 b). Seasonal fluctuation of the glycogen contents (% DW) stored in gonad-digestive gland shows progressive significative decrease from April (10.49 ± 2.68)

$p<0.05$ to September (2.09 ± 1.98). Thereafter it increases significantly ($p<0.05$) during winter especially in February (11.52 ± 4.65) which follow period of favorable environmental conditions (food availability) and sexual repose (Figure c).

High values of proteins (% DW) are noted during May (18.52 ± 1.23) $p<0.05$ and July (21.37 ± 1.50) ($p<0.05$) which corresponded to gametogenic maturity and then successed by significative low values recorded during August (9.5 ± 219) $p<0.05$ coincided with lost of organic matter due to the beging of total spawning (Figure 6 d).

4. DISCUSSION

Several studies on the sexual maturation of bivalves have shown that the reproductive activity was influenced by environmental factors like temperature, food availability (Robinson, 1992; Mathieu, 1994) and pollutants (Paez- Osuna *et al.* 1995; Morcillo and Porte, 2000). Manipulating physical (temperature) and nutritional (phytoplankton supply) factors may affect gonadal development, either by accelerating gametogenesis or by retarding gonadal maturation (Gallager and Mann, 1986).

We were able to observe gametogenesis of *C. gigas* in the Bizert lagoon since March 2002. Spawning began in June and September 2002 after which sexual repose took place. During this late gametogenic stage, the interfollicular connective tissue occupied the majority of the gonadal tissue. This tissue represents the energy pool, which accumulates metabolites during the winter and before initiation of gametogenesis.

When gametogenetic activity started, gonia multiplication took place until April. During May, we observed mature gametes, spermatozoa oriented with their tails toward the follicle lumen and ova with a polygonal form totally filling the follicle. The maturation of oyster germ cells is a function of temperature as suggested by Muranaka and Lannan (1984).

This phenomenon was also observed until July and August indicating gonadic restoration. Lubet (1991), revealed in *C. gigas* from Arcachon (France) that gonad restoration in July/August preceded spawning in June/onset July and was followed by others at the end of August onset September. However, males presented in the same period (May) spermatozoa with their tails oriented toward the follicle lumen ready to expelle.

In the Bizert lagoon, gametes were mature when water temperature reached 20°C (May).

Temperature is considered to be responsible for gametogenesis initiation, but it also plays a major role in spawning. Lubet (1976), showed that thermic shock had an impact on switching gamete release.

In the Bizert lagoon there is no tide, so spawning takes place thanks to other environmental factors like water warming in the summer (Lubet, 1959). During the sexual cycle of *C.gigas* in the Bizert lagoon, release gametes occur also under the influence of the high temperatures of June and September (23°C and 27°C respectively). During this period, oysters have gonads with ruptured follicles and residual gametes. Thus there is a short period of inactivity, from November to February when reproductive activity is minimal, following by rapid development and proliferation of the gonad, accompanied by rapid increase in tissue weight- especially gonad-visceral mass- during the spring corresponded with available food in association with an increase in the chlorophyll *a* concentration.

As in other bivalve species, storage metabolism in oysters is closely associated with reproductive events and represents an important biochemical pathway (Berthelin et al. 2000). Elsewhere, Mackie and Ansell (1993) reported that environmental factors apparently play a dominant role in determining events in the storage cycles.

The dry body condition index of *C. gigas* in the Bizert lagoon reflects its reproductive activity during most of the year. At the start of the spawning period (August) there was a rapid decrease in the tissue weights and condition index. At the beginning of autumn and in winter, the abundance of food available allowed recovery of tissue weight, and was accompanied by an increase in the values of both condition indices and the gonadic weight and reached a maximum on winter which corresponds to the sexual repose during which reserves are accumulated in oysters tissues and are considered to be used principally in gametogenesis and lost in spawning. According to (Falk-Petersen et al. 1986; Gallager et al. 1986; Helm et al. 1973; Tocher et al. 1988; Whyte JNC et al. 1990), the weight increase brought about an accumulation, or enrichment, of lipids in absolute amounts, corresponding with the female gametogenesis process.

Analysis of the growth modals of *C. gigas* gives the opportunity to understand the evolution of the morphological characters of the different body measurements of this cupped oyster during it's biological cycle in the Bizert lagoon.

Temperature and microalgae make up affected the survival and growth in all ontogenic stages in the Pacific oyster (Bougrier et al., 1995; Friedman et al., 1998; Masson et al., 1998 and Bayne 2002).

In this study, we remarked better correlation between lengthwise and weight parameters of oysters during the period from November to February when temperature values ranged from 19°C to 13.1°C however the oysters growth is affected during the rest of the year (from March to September) when temperature values starts at 14°C, attempts 28°C in August and decreased to 26°C in September. According to Friedman et al., 1998, *Crassostrea gigas* growth is rapid at 24°C and ceases around 30°C and Shpigel et al., 1992 found evidence that Pacific oysters either diploid or triploids, were metabolically stressed at 30°C.

According to Garton and Johonson, 2000, factors other than water temperature, such as food quantity and quality were responsible for accelerating the shell growth rate of mussels. These deductions are in accord with Chew et al., (1989) reports, who establish that the growth of clams is affected by temperature, salinity, exposure regimes and food availability (chlorophyll *a* concentration). According to Wallace and Reinsnes (1985); Wilson, (1987) and Lodeiros et al., (1998), bivalve shells are made from calcium carbonate. The Japanese pearl oyster *Pinctada fucuta martensii* can assimilate calcium from sea water with the mantle or gill, and the uptake efficiency is enhanced by the ingestion of plankton food (Watabe et al., 1957 and Kuwatani, 1964). Thus, the shell growth has an important relationship with quantity of food.

In the Bizert lagoon, water is characterized by two elevations of chlorophyll *a* concentrations. The first is noted in autumn (November) when chlorophyll *a* level is about 1,2 μg/l and the second is in spring (May) when the chlorophyll *a* concentration attempt 2,4 μg/l. Hence we conclude that the increase of shell size from November to January is due to the autumnal chlorophyll *a* elevation. It has been shown that the shell growths of oysters (*Ostrea edulis*) and scallops (*Pecten maximus*) are affected by food quality, defined as organic content of the suspended solids, and current speed (Wilson 1987), and that of mussel *Mytilus edulis* (Linnaeus) is depressed by low salinity (Almada-Villela 1984). But in the Bizert lagoon we didn't observe any increase in the shell size around May, although the chlorophyll *a* levels are maximal in spring, in fact, this period of the year corresponds to the sexual maturation of *C. gigas* in the Bizert lagoon. According to Wallace and Reinsnes (1985), gonad tissue maturation in bivalves has been shown to depress shell growth. The relationship between growth and reproduction, from the viewpoint of physiological energetics was discussed theoretically by some ecologists (Pianka and Parker, 1975; Sebens, 1979). Other workers have empirically examined the amount of energy available for growth and reproduction at given age for many bivalves (Rodhouse, 1978 for *Ostrea edulis* and Bayne et al. 1983 for *Mytillus edulis*). These authors confirmed that the energy available for growth and reproduction varies with body size and depends on the difference between energy intake and metabolic cost.

In addition, some workers have suggested that the length of the interval of shell growth is closely related to sexual maturity patterns and that the inverse relationship between the length of the pre-reproductive stage and temperature exists in many mollusks (as the Atlantic squid *Sthenoteuthis pteropus* (Zuev et al., 1979) and the pteriopod *Clione limacine* (Lalli and Conover, 1973).

In this study correlation index values (r) varied during the different phases of gametogenic cycle of *C. gigas* determined using the histological techniques. These results establish that correlations between the different parameters of lengthwise growth modals are closer during sexual repose (between 0,64 ($p < 0.05$) and 0.82 ($p < 0.05$)) then those determined during the gonadal activity (between 0.1 and 0.22 ($p < 0.05$)). Also Correlation index values (r) between the different parameters of growth varied all over the year. During sexual repose, correlation between the measurements of weight and length (L) are between 0,65 ($p < 0.05$) and 0,88 ($p < 0.05$), however (r) values calculated during gonadal activity are relatively low (between 0,15 ($p < 0.05$) and 0.47 ($p < 0.05$)).

These observations allow us to conclude that the gametogenic cycle influence clearly the lengthwise and weight modals of growth. These two growth modals are affected by the physiologic troubles of *C. gigas* during the gametogenic activity traduced by win or loose of the organic matter during this period of the year.

During the sexual repose, the biochemical component are accumulated in the *C. gigas* tissues and appears to correlate positively with the amount of energy available for shell growth during this phase of the gametogenic cycle which corresponds to the autumnal elevation of chlorophyll *a* levels in the Bizert lagoon.

According to Falk-Petersen et al. (1986); Gallager and Mann, (1986); Helm et al. (1973); Tocher et al. 1988; Whyte JNC et al. 1990), the weight increase brought about an accumulation, or enrichment, of lipids in absolute amounts, corresponding with the female gametogenesis process.

A prevalence of total polyunsaturated fatty acids (PUFA, 57.93 - 65.22 % of total fatty acids) over the saturated (SFA, 23.13- 26.38 % of total fatty acids) and monounsaturated ones (MUFA, 12.35- 15.84 % of total fatty acids) was established throughout the period of study, as was also observed in oysters, mussels and cuttlefish from different Mediterranean sea sites (Orban et al., 2002, 2004; Ozyurt, et al. 2006). No significant seasonal changes of total SFA were observed during the year, while total PUFA, MUFA and n-3 PUFA varied significantly ($p<0.05$).

The polyunsaturated fatty acids were the most important fatty acid type especially the (*n*-3) series dominated by eicosapentaenoïc (C20:5 (*n*-3), 10.68-12.15 % of total fatty acids) and docosahexaenoïc acids (C22:6 (*n*-3), 17.33-21.94 % of total fatty acids), followed by total saturated fatty acids, principally the palmitic (C16:0, 19.77-20.84 % of total fatty acids). The monounsaturated fatty acids were mainly represented by the palmitoleïc C16:1(*n*-7), the oleic C18:1(*n*-7) and the gadoleïc C20:1(*n*-7).

The maximal levels of C16:0 are noted in spring (20,84 % ± 1,17 of total fatty acids) and in autumn (20,80 % ± 3,95) which corresponds to the maximal production of chlorophylle *a* of spring and autumn in the Bizert lagoon. According to Thompson et al. (1996), the proportion of C16:0 in the larval diets was positively correlated with higher survival and growth rates of oyster larvae.

According to Von Elert (2004), tissue concentrations of C18:3 (*n*-3) and C18:4 (*n*-3) are considerably higher in *Daphnia galeata*, which indicates higher assimilation of these C18 PUFAs in the fast growing of this crustacean. Results found in the present work confirm this deductions by showing high levels of C18:3 (*n*-3) and C18:4 (*n*-3) in winter which corresponds to better growth index with respectively values (4,38 % ± 0,95 and 7,88 % ± 1,42 of total fatty acids).

Polyunsatured fatty acids followed a seasonal cycle related to the gametogenic cycle, maximal contents of C22:6(n-3) (21.94 ± 1.73) corresponded with oocyte maturation (spring) and minimal contents (17.73 ± 2.48) with the prespawned period. High levels of 22:6 (*n*-3) plays a major role at structural and functional level of cell membranes involved in oogenesis and embryogenesis (Soudant et al. 1996; Soudant et al. 1999). The specific role of 20:5 (*n*-3) in previous studies has been related to energetic functions. Variations in PUFA appeared to be also related to the available food, especially in autumn, were recovery of PUFA levels was observed in association with the increase in the chlorophyll *a* concentration in seawater. As also deduced by (Ojea et al. 2004), high percentage of PUFA during autumn is explained by the abundance of phytoplankton traduced with an increase in the concentration of chlorophyll *a*. Previous reports have shown an inverse relationship between temperature and the amount of polyunsaturated fatty acids in tissue lipids of fish and invertebrates (Chu et al. 1991; Hazel et al. 1979; Holland et al. 1978; Piretti et al. 1988 and Sargent et al. 1976). This is true in the present study in which PUFA followed a seasonal cycle contrary to temperature. High PUFA levels are registered in winter when temperature levels are minimal, however low PUFA levels occur in summer when temperature levels are maximal. Bell et al. (1986) pointed out that a decrease in temperature values produces generally, an increase in the PUFA content in order to maintain the fluidity of the membranes.

According to Webb and Chu, (1983); Whyte et al., (1990) and Caeras et al., (2002), polyunsatured fatty acids (PUFA) with 20 and 22 carbon atoms and more than three double bonds are essential for survival and growth of molluscs. In the same context, Langdon and Waldock, (1981) and Enright et al. (1986) correlated diets rich in 20:5 and 22:6 with high growth rates of juvenile *Crassostrea gigas* and *Ostrea edulis*. Results found on *C. gigas* in this study are in accord with these deductions. Maximal level of C20:5 (*n*-3) is registered during winter (12,15 ± 1,71 % of total fatty acids) which coincide with sexual repose and the increase of (r) levels. But these deductions are contradictory with those found by Thompson et al., (1996) who demonstrate that the proportion of 20:5 (*n*-3) in the larval diet was negatively correlated with the growth rates of the *C. gigas* larvae. The C22:6 (n-3) level is maximal in spring (21,94 % ± 1,73) and corresponds to the maximal concentration of chlorophyll *a* in the culture site. This fatty acid decrease significantly ($p<0.05$) in summer and increase again significantly ($p<0.05$) in autumn (21,22 % ± 1,10 of total fatty acids) during which oysters are in sexual repose inducing better growth of the bivalve and increase in (r) levels and shell length, width and thickness respectively (2,04 ± 0,18 cm; 0,58 ± 0,08 cm and 0,90 ± 0,06 cm from January 02 to February 02 and 1,25 ± 0,33 cm; 0,65 ± 0,11 cm and 0,28 ± 0,07 cm from November 02 to January 03). During this phase of the gametogenic cycle we remarked that total SFA, MUFA and PUFA (n-3) show the highest values (respectively 26.38 % ± 1.78; 15.84 % ± 0.56; 46.94 % ± 0.61 of the total fatty acids). However during the gametogenic activity (March 02 – September 02) total SFA, MUFA and PUFA (n-3) show the lowest values (respectively 23.13 % ± 1.38; 12.35 % ± 0.55; 38.15 % ± 0.84 of the total fatty acids) which increase the shell size by only 1,12 ± 0,08 cm, 0,64 ± 0,80 cm and 0,57 ± 0,12 cm in respectively length, width and thickness. These results allow us to conclude that accumulation of fatty acids in the gonad – digestive gland of oysters contribute in improving growth rates of oyster cultured in the Bizert lagoon.

The *n*-3/ *n*-6 PUFA ratio values (2,45 – 3,15) also confirm the important role of fatty acids in the process of growth. This ratio is maximal in winter (3,15) corresponding to the sexual inactivity and the increase of the correlation coefficient (r) values. According to Sargent et al., (1993), the necessity to maintain a high ratio of (*n*-3)/(*n*-6) in marine organisms has been found to be important for survival and growth.

Glycogen plays a central role in energetic and metabolic supply of gametogenesis in many bivalves (Gabbott, 1975; Bayne et al., 1982; Ruiz et al., 1992; Mathieu and Lubet, 1993). The seasonal changes of the glycogen stored in gonad-digestive gland of the Bizert lagoon oysters is inversely related to the pattern of gonadal tubule development with low values of glycogen in September (2.09 % ± 1.98 of the dry weight) at the spawning time and remain progressive elevation to attempt the highest values in February corresponding to sexual repose and in April while gonad is restricted to reduce tubules constituted by protogonia and few gonia initiating a new gametogenic cycle. Glycogen becomes the most important energy reserve of oysters once this species approach a size of 50 mm, and this shift appeared to be associated with preparation for gametogenesis (Garcia-Esquivel et al., 2000). According to Deslous-Paoli and Heral, (1988), glycogen can be simultaneously an energy source for growth and stored in specific cells as energetic reserve during the vitellogenesis process.

These deductions allow us to conclude that the progressive elevation of glycogen in *C. gigas* tissues, which attempt it's maximum in February, is one of the most important factors that contribute in acceleration the shell growth of *C. gigas* during winter (at sexual repose) in the Bizert lagoon.

Proteins are the main biochemical fraction of bivalve eggs (Holland, 1978) and larvae (i.e. Whyte et al., 1987, 1990; His and Maurer, 1988), yet the importance of this substrate as an energy source during and after metamorphosis has remained controversial (Holland and Spencer, 1973; Bartlett, 1979; Rodriguez et al., 1990). According to (Gabott and Bayne, 1973; Barber and Blake, 1981; Ruiz et al. 1992), proteins serve as an energy reserve supporting the end of gametogenesis. In this study proteins present high values in May and July which corresponds to the period of maximum ripeness (respectively 18.52 ± 1.23 and 21.37 ± 1.50). Protein changes in gonad-visceral mass is closely related to gonad maturation and may therefore reflect protein accumulation in oocytes, in addition to protein contribution to energetic maintenance in periods of reduced glycogen levels, (Berthelin et al. 2000).

Previous biochemical studies have been mainly focused on the role of lipids during and after metamorphosis (Gallager and Mann, 1986; Wilkfors et al., 1996; Caers et al., 1998).

However, recent studies have shown that proteins and lipids were equally important as energy sources during metamorphosis of *Ostrea edulis* (Rodriguez et al., 1990) and *Ostrea chilensis* (Videla et al., 1998). This support the idea that the protein levels in the gonad – digestive gland of *C. gigas* of the Bizert lagoon have not any clear correlations with the shell growth but contribute in supporting the bivalve during the sexual maturity (May and July) when glycogen reserves are depleted.

CONCLUSION:

Analysis of the growth modals on *C. gigas* in this study, give us the opportunity to understand that modals of growth (allometric relationships and mean biometric data) are influenced by the gametogenic cycle and biochemical composition of the tissues. Lipids, glycogen and proteins of the gonad-visceral mass of *C. gigas* are dependent upon the stages of the gametogenic cycle, temperature and chlorophyll *a* concentration in the watersea, played an important role as the gametogenesis and growth fuel and to support the basic metabolic maintenance cost in periods of food storage.

ACKNOWLEDGEMENTS

This research was carried out thanks to Dr Slim TRITAR from FMB (Ferme marine des bivalves) for his support.

REFERENCES:

Almada-Villela, P.C., (1984). The effect of reduced salinity on the shell growth of small Mytillus edulis. Journal of Marine Biology Assosiation of United Kingdom 64. 171-182.

Aminot, A., et M. Chaussepied., (1983). Manuel des analyses chimiques en milieu marin. Centre national pour l'exploitation des océans. CENEXO: 395 pp.

Andersen and Nass 1993;

Barber, B. J., Blake, N.J., (1981). Energy storage and utilisation in relation to gametogenesis in *Argopecten irradians concentricus* (Say). *J. Exp. Mar. Biol. Ecol.* 45, 105-116.

Bartlett, B.R., (1979). Biochemical changes in the Pacific oyster, *Crassostrea gigas* Thunberg, 1795 during larval development and metamorphosis. *Proc. Natl. Shellfish. Assoc.* 69, 202 abstract.

Bayne, B.L., (1976). Aspects of reproduction in bivalve molluscs. In: Vieley, M.L. (Ed.), Estuarine processes. Academic press, New York, pp. 432-448.

Bayne, B.L., Bubel, A., Gabbott, P.A., Livingstone, D.R., Lowe, D.M. Moore, M.N., (1982). Glycogen utilisation and gametogenesis in : *Mytillus edulis* (L.). *Mar. Biol. Lett.* 3, 89-105.

Bayne, B.L., Salkeld, P.N., Worrall, C.M., (1983). Reproductive effort and value in fifferent populations of the marine mussel, *Mytillus edulis* L. *Oecologia* 59, 18-26.

Bayne, B.L., (2002). A physiological comparison between Pacific oysters *Crassostrea gigas* and Sydney Rock oysters *Saccostrea glomerata*: food, feeding and growth ina shared estuarine habitat. *Marine Ecology Progress Series* 232, 163 - 178.

Bell, M.V., Henderson, R.J., Sargent, J.R., (1986). The role of polyunsatured fatty acids in fishes. *Comp. Biochem. Physiol. Part B* 83, 711-719.

Berthelin, C., Kellner, K., Mathieu, M., (2000). Storage metabolism in the Pacific oyster (*Crassostrea gigas*) in relation to summer mortalities and reproductive cycle (West Coast of France). *Comp. Biochem. And phsiol. Part B.* 125, 359-369.

Berthomé, J.P., J. Prou and A. Bodoy, (1986). Performances de croissance de l'huître creuse *Crassostrea gigas*, dans le bassin de Marenne –Oléron entre 1979 et 1982. *Haliotis* 15, 183-192.

Bligh, E.G. and W.J. Dyer., (1959). A rapid method of total lipid extraction and purification. *Can. J. Biochem.* 37, 911-917.

Bougrier, S., Geairon, P., Deslous-Paoli, J.M., Bacher, C. and Jonquieres, G. (1995). Allometric relationships and e¡ects of temperature on clearance and oxygen consumption rates of *Crassostrea gigas* (Thunberg). *Aquaculture* 134, 143-154.

Careas, M., S.D. Utting, P. Coutteau, P.F. Millican, and P. Sorgeloos., (2002). Impact of the supplementation of a docosahexanoic acid-rich emulsion on the reproductive output of oyster broodstock, *Crassostrea gigas*. Marine biology, 140, 1157-1166.

Caers, M., Coutteau, P., Lombeida, P., Sorgeloos, P., (1998). The effect of lipid supplementation on growth and fatty acid composition of *Tapes phillippinarum* spat. *Aquaculture* 162, 287–299.

Calow, P. (1983). Life-cycle patterns and evolution. In Russell-Hunter.W.D.(ED)The Mollusca. 6. Ecology. Academic Press. New York. P. 649-678.

Chavez-Villalba, J., Pommier, J., Andrimisez, J., Povreau, S., Barret, J., Cochard, J.C., and M. Le Pennec., (2001). Broodstock conditionning of the oyster *Crassostrea gigas* : origin and temperature effect. *Aquaculture.* 214, 115-130.

Chew, KK. (1989). Manila clam biology and fishery development in Western North America. In: Manzi JJ, Castagna M (eds). *Clam Mariculture in North America.* Elsevier, Amsterdam. 243–261.

Chu, F.L.E., Greaves, J., (1991). Metabolism of palmitic, linoleic and linolenic acids in adult oysters, *Crassostrea virginica*. *Mar. Biol.* Vol. 110, 229-236.

Deslous-Paoli, J.M., Héral, M., (1988). Biochemical composition and energy value of *Crassostrea gigas* (Thunberg) cultured in the bay of Marennes-Oléron. *Aquat. Living. Resour.* 1, 239-249.

Dinamani, P. (1987). Gametogenic patterns in populations of Pacific oyster, *Crassostrea gigas*, in Northland, New Zealand. Aquaculture, 64, 65-76.

Dridi, S., Romdhane M.S., El Cafsi M. (2006). Evidence of *Crassostrea gigas* reproduction in the Bizert lagoon, Tunisia. *Journal of Biological Research*. Vol 5, 35-45.

Dridi S., Romdhane M.S., El Cafsi M., (2007). Seasonal variation in weight and biochemical composition of the Pacific oyster, *Crassostrea gigas* in relation to gametogenic cycle and environmental conditions of the Bizert lagoon, Tunisia. *Aquaculture*, 263; 238 – 248.

Dubois M., Gilles, K.A., Hamilton, J.K., Rebers P.A., Smith, F., (1956), Colorimetric method for determination of sugars and related substances. *Anal.Chem.* 28, 350-356.

Enright, C.T., G.F. Newkirk., J.S. Craigie, and J.D. Castell., (1986). Growth of juvenile *Ostrea edulis* L. fed *Chaetoceros gracilis* Schutt of varied chemical composition. *J. Exp. Mar. Biol. Ecol.,* 96: 15-26.

Falk-Petersen, S, Falk-Petersen, IB, Sargent, JR, Haug T., (1986). Lipid class and fatty acid composition of eggs from the Atlantic halibut (*Hippoglossus hippoglossus*). *Aquaculture.* 52, 207-11.

FAO, (2002). *Fishery statistics capture production* .Vol 90/1. FAO Statistics series N 166. 617p.

Fisher, R.A. (1958). The genetical theory of natural selection. 2nd edn. Dover. New york.

Folch J., M. Lees, and G.H. Sloane Stanley. (1957). A simple method for the isolation and purification of total lipids from animal tissues. *J. Biol. Chem.* 266, 497-509.

Gabbott, P.A., Bayne, B.L. (1973). Biochemical effects of temperature and nutritive stress on *Mytilus edulis* (L.). *J. Mar. Biol. Assoc. UK.* Vol. 53, 269-286.

Gabbott, P.A., (1975). Storage cycle in marine bivalve molluscs: a hypothesis concerning the relationship between glycogen metabolism and gametogenesis. In: Barnes, H. (Ed.), Proceedings of the 9th European Marine Biology Symposium. Aberdeen University Press, pp.191-211.

Gabbott, P.A., (1983). Developmental and seasonal metabolic activities in marine molluscs. In, The Mollusca. Vol (2). Environmental biochemistry and physiology, edited by P. W. Hochachka, Academic Press, New York: 165-217.

Gallager, S.M., Mann, R., (1986). Growth and survival of larvae of *Mercenaria mercenaria* (L.) and *Crassostrea virginica* (Gmelin) relative to broodstock conditioning and lipid content of eggs. *Aquaculture.* 56, 105-121.

Garcia-Esquivel, Z., Gonzalez-Gomez, M.A., Gomez-Togo, D., Galindo-Bect, M.S., Hernandez-Ayon, M., (2000). Microgeographic differences in growth, mortality and biochemical composition of cultured Pacific oysters *Crassostrea gigas* from San Quintin bay, Northwest Mexico. *J. Shellfish Res.* 19, 789–798.

Gabe M., 1968. *Techniques histologiques*. Ed. Masson et Cie. 1113 p.

Garton, D.W. and Johonson, L., (2000). Variation in growth rates of the zebra mussels, Dreissena polymorpha, within Lake Wawasee. *Freshwater Biology.* 45, 443 – 451.

Giese, A.C., (1969). A new approach to the biochemical composition of the mollusc body. *Oceanogr. Mar. Biol.* 7, 175-229.

Haure, J. and J.P. Baud, (1990). Croissance, engraissement et mortalité de *Crassostrea gigas* en baie de Bourgneuf. Comparaison de quatre secteurs ostréicoles. IFREMER/ RIDRV 90.11,25 pp.

Hazel, J.R., (1979). Influence of thermal acclimatation on membrane lipid composition of rainbow trout liver. *Am. J. Physiol.* 236: 91-101.

Helm, M.M, Holland, D.H. Stephenson, R.R., (1973). The effect of supplementary algal feeding of a hatchery breeding stock of *Osrea edulis* L. on larval vigour. *J Mar Biol Ass* UK 53, 673-84.

His, E., Maurer, D., (1988). Shell growth and gross biochemical composition of oyster larvae *Crassostrea gigas* in the field. *Aquaculture* 69, 185–194.

Holland, D.L. (1978). Lipid reserves and energy metabolism in the larvae of benthic marine invertebrates. In: Malins, D.C.; Sargent, J.R. (eds). Biochemical and biophysical perspectives in marine biology. Vol. 4. New York: *Academic Press*, 85-123.

Holland, D.L., Spencer, B.E., (1973). Biochemical changes in fed and starved oysters, *Ostrea edulis* L. during larval development, metamorphosis and early spat growth. *J. Mar. Biol. Assoc.* U. K. 53, 287–298.

Jones, D.S. (1983). Sclerochronology: reading the record of the molluscan shell. *Am. Scient.* 72, 384 –391.

Klingensmith, J.S., (1982). Distribution of methylene and non methyleneinterrupted dienoic fatty acids in polar lipids and triacylglycerols of selected tissue of the hardshell clam (Mercanaria mercenaria). *Lipids* 17, 976–981.

Kuwatani, Y. (1964). Rearing of the pearl oyster, Pinctada martensil (Dunker), on the diet of Skeletonema costatum. Bulletin of Japanese Society of Scientific Fisheries 30. 104 – 113. (in Japanese, with English abstract).

Lalli, C.M., Conover, R.J., (1973). Reproduction and developpement of Pedocline doliiformis, and a comparison with Clione limaclina (Opisthobranchia: Gymnosomata) MAR. BIOL. 19, 13-22.

Langdon, C.J., Waldock, M.J., (1981). The effect of algal and artificial diets on the growth and fatty acid composition of *Crassostrea gigas* spat. *J. Mar. Biol. Assoc.* U. K. 61, 431–448.

Lango-Reynoso F, Chavez-Villalba J, Cochard J C, Le Pennec M, (2000). Oocyte size, a mean to evaluate the gametogenic development of the pacific oyster, *Crassostrea gigas* (Thumberg, 1793). *Aquaculture* 190: 183-199.

Li, Q., Osada, M., Mori, K., (2000). Seasonal biochemical variations in Pacific oyster gonadal tissue during sexual maturation. *Fisheries Science*. 66, 502 – 508.

Lodeiros, C.J., Renjel, J.J., Freites, L., Morales, F. and Himmelman, J.H. (1998). Growth and survival of the tropical scallop Lyropecten (Nodipecten) nodosus maintained in suspended culture at three depths. *Aquaculture* 165, 41-50.

Lowry, O.H., Roseborouch, N.I., Farrand, A.L., and Randall, R.J., (1951). Protein measurement with the folin phenol reagent. *J. Biol.Chem.*, 263-275.

Lubet P, (1959). Recherches sur le cycle sexuel et l'emission des gamètes chez les mytilidés et les péctinidés (mollusques bivalves). *Revue des travaux de l'institut des pêches maritimes,* 23: 387-542.

Lubet, P., (1991). Bases biologiques de la culture des mollusques. In. (G. Barnabé). Bases biologiques et écologiques de l'aquaculture. 99-210.

Lutz, R.A., Rhoads, D.C., (1980). Growth patterns within the molluscan shell. In: Roads, D. C., Lutz R. A., (eds) Skeletal growth of aquatic organisms. Plenum Press. New York, p. 203-254.

Mackie, L.A., Ansell, A.D., (1993). Differences in reproductive ecology in natural and transplanted populations of *Pecten maximus*: evidence for the existence of separate stocks. *J. Exp. Mar. Biol. Ecol.* 169, 57-75.

Mann, R., (1979). Some biochemical and physiological aspects of growth and gametogenesis in *Crassostrea gigas* and *Ostrea edulis* grownat sustained elevated temperatures. Journal of the Marine Biology Association of the United Kingdom 59, 95-110.

Mason, C.J., Reid, D.D. and Nell, J.A., (1998). Growth characteristics of Sydney rock oysters Saccostrea comercialis in relation to size and temperature. *Journal of ExperimentalMarine Biology and Ecology* 227,155-168.

Mathieu, M., Lubet, P., (1993). Strorage tissue metabolism and reproduction in marine bivalves- abref review. *Inv. Repr. Dev.* 23(2-3), 123-129.

Mathieu M, (1994). Endocrine control of carbohydrate metabolism in mollusc. Perspect. Comp. Endo. In; Davey, K.G., Peter, R.E., Tobe, S.S. (Eds.), Perspectives in Comparative Endocrinology. National Research Council of Canada, Ottawa, pp. 471-474.

Maurer, D., Borel, M. (1986). Croissance, engraissement et cycle sexuel de *Crassostrea gigas* dans la bassin d'Arcachon : comparaison des huîtres âgées de 1 à 2 ans. Haliotis, 15, 125-134.

Maurer, D., (1989). Approche des relations entre la croissance de l'huître Crassostrea gigas et le milieu dans le bassin d'Arcachon. IFREMER/ RIDRV 89. pp 34,33.

Medhioub, M.N., (1993). Aperçu sur la conchyliculture en Tunisie point de situation et réflexions. Revue de l'I.N.A.T. Vol. 8(2), 19-26.

Metcalfe, D., A. Schmitz, and Pelka, J.R., (1966). Rapid preparation of fatty acid seters from lipids for gas chromatographic analysis. *Anal. Chem.* 38, 524-535.

Morcillo Y, Porte C, (2000). Evidence of endocrine distribution in clam- *Ruditapes decussata* – transplanted to a tributyltin- polluted environment. *Environmental Pollution*. 107 : 47-52.

Muranka, M.S. and Lannan, J.E., (1984). Broodstock management of *Crassostrea gigas*: environmental influences on broodstock conditioning. *Aquaculture*. Vol. 39, 217-228.

Ojea, J., A.J. Pazos, D. Martinez, S. Novoa, J.L. Sanchez, M. Abad, (2004). Seasonal variation in weight and biochemical composition of the tissues of *Ruditapes decussatus* in relation to the gametogenic cycle. *Aquaculture,* Vol. 238, pp. 451-468.

Orban, E., Di Lena, G., Masci, M., Nevigato, T., Casini, I., and Caproni, R. (2004). Growth, nutritional quality and safety of oysters (*Crassostrea gigas*) cultured in the lagoon of Venice (Italy). *Journal of the Science of Food and Agriculture*, 84, 1929-1938.

Orban, E., Di Lena, G., Nevigato, T., Casini, I., Marzetti, A., and Caproni, R. (2002). Seasonal changes in meat content, condition index and chemical composition of mussels (*Mytilus galloprovincialis*) cultured in two different Italian sites. *Food Chemistry*. 77, 57-65.

Ozyurt, G., Duysak, O., Akamca, E., and Tureli, C., (2006). Seasonal changes of fatty acids of cuttlefish *Sepia officinalis* L. (Mollusca: Cephalopoda) in the north eastern Mediterranean sea. *Food Chemistry*, 95, 382-385.

Paez-Osuna, F., Frias-Espercueta, M.G., Osuna-Lopez, J.I., (1995). Trace metal concentration in relation to season and gonadal maturation in the oyster *Crassostrea iridescens*. *Marine Environmental Research*. 40: 19-31.

Perdue, J., Erickson, G., (1984). A comparaison of the gametogenic cycle between the Paciphic oyster *Crassostrea gigas* in Washington state. *Aquaculture* 37, 231-237.

Pianka, E.R., Parker, W.S., (1975). Age specific reproductive tactics. Am. Nat. 109, 453-464.

Piretti, M.V., Zuppa, F., Paglinca, G., Taioli, F., (1988). Investigation of the seasonal variations of fatty acid. Constituents in selected tissue of the bivalve mollusc, Scapharca inaequivalvis (Bruguiere). *Comp. Biochem. Physiol.* 89 B, 183-187.

Ren, J.S., Marsden, I.D., Ross, A.H., Schiel, D.R., (2003). Seasonal variation in the reproductive activity and biochemical composition of the Pacific oyster (*Crassostrea gigas*) from the Malborough Sounds, New Zealand. *Journal of Marine Fresh Research*, 37, 171-182.

Robinson, A., (1992). Dietary supplements for reproductive conditioning of *Crassostrea gigas* Kumamoto (Thumberg): I. Effects on gonadal development, quality of ova and larvae through metamorphosis. *Journal of Shellfish Research*. 11: 437-441.

Rodhouse, P.G., (1978). Energy transformations by the oyster *Ostrea edulis* L. in a temperate estuary. J. exp. Mar.Biol. Ecol. 34, 1-22.

Rodriguez, J.L., Sedano, F.J., Garcýa-Martin, L.O., Perez-Camacho, A., Sanchez, J.L., (1990). Energy metabolism of newly settled *Ostrea edulis* spat during metamorphosis. *Mar. Biol.* 106, 109–111.

Ruiz, C., Abad, M., Senado, F., Garcia-Martin, L. O., Sanchez Lopez, J.L., (1992). Influence of seasonal environmental changes on the gamete production and biochemical composition of *Crassostrea gigas* (Thunberg) in suspended culture in EL Grove, Galicia, Spain. *J. Exp. Mar. Biol. Ecol.* 155, 249-262.

Sargent, J.R., Bell, J.G., Bell, M.V., Henderson, R.J., Tocher, D.R., (1993). The metabolism phospholipids and polyunsatured fatty acids in fish. In: Lahlou, B.,Vitiello, P. (Eds.), *Aquaculture: Fundamental and Applied Research*, vol. 43. Coastal and Estuarine StudiesAmerican Geophysical Union, Washington, DC, pp. 103–124.

Sastry, A.N., (1979). Pelecypoda (excl. Ostreidae). In Reproduction of marine invertebrates, Ed. A. C. Giese et al. New York: Academic Press 5, 113-292.

Sebens, K.P., (1979). The energetics of asexual reproduction and colony formation in benthic marine invertebrates. *Am. Zool.* 19, 683-697.

Shpigel, M., Barber, B.J. and Mann, R., (1992). Effects of elevated temperature on growth, gametogenesis, physiology and biochemical composition in diploid and triploid Pacific oyster, *Crassostrea gigas* Thunberg. *Journal of Experimental Marine Biology and Ecology* 161, 15-25.

Sokal, R. and Rohlf, J. (1969). Biometry.W.H. Freeman and Company, San Francisco, CA, USA.

Soudant, P., Moal, J., Marty, Y, Samain, J.F., (1996). Effect of food fatty acid quality on fatty acid composition of class polar lipids in female gonads of *Pecten maximus. J. Exp. Mar. Biol. Ecol.* Vol. 205, 149-63.

Soudant, P., K. Van Ryckeghem, Y., Marty, J. Moal, J.F. Samain, P. Sorgeloos, (1999). Comparison of the lipid class and fatty acid composition between a reproductive cycle in nature and a standard hatchery conditioning of the Pacific Oyster *Crassostrea gigas. Comp. Biochem. Physiol. Part B.* Vol. 123, 209-222.

Steele, S. and Mulcahy, M.F., (1999). Gametogenesis of the oyster *Crassostrea gigas* in southern Ireland. *J. Mar. Biol. Ass.* U.K. 79, 673-686.

Taylor, A.C. and Venn,T.J., (1979). Seasonal variation in weight and biochemical composition of the tissues of the queen scallop *Clamys opercularis* from the Clyde Sea Area. *J. Mar. Biol. Ass. UK* 1979; 59, 605 – 621.

Tompson, P.A. ; M.X. Guo and P.J. Harrison., (1996). Nutritiona value of diets that vary in fatty acid composition for larval Pacific oysters (*Crassostrea gigas*). *Aquaculture*, 143: 379-391.

Tocher, D.R., Harvie, D.G., (1988). Fatty acid composition of the major phosphoglycerides from fish natural tissues; (n-3) and (n-6) polyunsatured fatty acids in rainbow trout (*Salmo gairdneri*) and cod (*Gadus morhua*) brains and retinas. *Fish Physiol Biochem*; 5, 229-239.

Videla, J.A., Chaparro, O.R., Thompson, R.J., (1998). Role of biochemical energy reserves in the metamorphosis and early development of the oyster *Ostrea chilensis*. *Mar. Biol.* 132, 635–640.

Von Elert, E., (2004). Food quality constraints in Daphnia: interspecific differences in the response to the absence of a long chain polyunsaturated fatty acid in the food source. *Hydrobiologia*. 526, 187 – 196.

Wallace, J.C. and Reinsnes, T.G., (1985). The significance of various environmental parameters of growth of the Iceland scallop, Clamys islandica (Pectinidae), in hanging culture. *Aquaculture* 44. 229 – 242.

Walne, P.R, and Mann, R., (1975). Growth and biochemical composition in *Ostrea edulis* and *Crassostrea gigas*. In Barnes, H. (Ed), *Proceedings of the 9th European Marine Biology Symposium*. Aberdeen University Press, pp. 587-607.

Watabe, N., Yoshii, G. and Okada, Y., (1957). Studies of Shell formation in young pearl oyster, Pinctada martensii (Dunker), using Ca^{45}: comparison of the uptake of Ca^{45} from sea water and food. Bulletin of japanese society of Scientific Fisheries 23, 139 – 143. (in japanese with English abstract).

Webb, K.L. and F.L. Chu., (1983). Phytoplankton as a food source for bivalve larvae. In: Pruder, G. D., Langdon C., Conklin, D. (Eds), Proceedings of the second International Conference on Aquaculture Nutrition Biochemival and Physiological Approches to Shellfish Nutrition. Louisiana State University Press, baton Rouge, LA: 272-291.

Whyte, J.N.C., N. Bourne, and C.A. Hodgson., (1990). Nutritional condition of rock scallop, *Crassadoma gigantean* (Gray), Larvae fed mixed algal diets. *Aquaculture* 86, 25-40.

Whyte, J.N.C., Bourne, N., Hodgson, C.A., (1987). Assessment of biochemical composition and energy reserves in larvae of the scallop *Patinopecten yessoensis*. *J. Exp. Mar. Biol. Ecol.* 113, 113–124.

Wilkfors, G.H., Patterson, G.W., Ghosh, P., Lewin, R.A., Smith, B.C., Alix, J.H., (1996). Growth of post-set oyster, *Crassostrea Õirginica*, on high-lipid strains of algal flagellates *Tetraselmis* spp. *Aquaculture* 143, 411–419.

Wilson, J.H., (1987). Environmental parameter controlling growth of *Ostrea edulis* L. and *Pecten maximus* L. in suspended culture. *Aquaculture* 64, 119 – 131.

Zuev, G.V., Nigmatullin, C.M., Nikulsky, V.N., (1979). Growth and life span of *Sthenoteuthis petropus* in the east- central Atlantic. *Zool. Zh.* 58, 1632-1641.

In: Aquaculture Research Trends
Editor: Stephen H. Schwartz, pp.213-235

ISBN 978-1-60456-217-0

Chapter 6

New Vaccination Strategies for the Improvement of DNA Vaccines Against Virus in Fish

Carolina Tafalla[1,*], Amparo Estepa[2] and Julio Coll[3]

[1] Centro de Investigación en Sanidad Animal (CISA-INIA). Carretera de Algete a El Casar km. 8,1. Valdeolmos 28130. Madrid, Spain

[2] Instituto de Biología Molecular y Celular (IBMC). Universidad Miguel Hernández. Elche. Alicante, Spain

[3] SGIT, INIA, Biotecnología, Carretera de La Coruña km. 7. 28040 Madrid, Spain

Abstract

Vaccination seems like the most adequate method to control viral diseases in aquacultured fish. Up to date, very few viral vaccines are commercialised, since it has been a difficult task to obtain effective vaccines in fish that are both safe and cost-effective. In the past years, DNA vaccination has been proven as a very effective method to control some rhabdoviral infections, such as those caused by infectious hematopoietic necrosis virus (IHNV) and viral hemorrhagic septicemia virus (VHSV), responsible for great losses in aquaculture production. These vaccines consisting in a plasmid in which the viral glycoprotein gene is expressed under the control of the citomegalovirus (CMV) promoter, are able to confer protection in fish after intramuscular injection. Even though the protection conferred is high and long-lasting, these vaccines are still not being used worldwide due to safety issues concerning the viral promoter, to difficulties in the administration of the vaccine at a large scale, and to high vaccination costs. Up to date, only an IHNV DNA vaccine has been licensed for use in Canada.

Moreover, the immune mechanism through which they confer protection is still unknown, and this is a major restriction for the optimisation of DNA vaccines for other viral pathogens also of great importance in fish, such as for example nodavirus, for which DNA vaccines as formulated for rhabdovirus are not effective.

* Correspomding author : Tel. +34-91-6202300 / Fax +34-91-6202247 Email: tafalla@inia.es

In this chapter, we discuss the trends in research dealing with DNA vaccination in fish, that will hopefully allow the improvement and finally the routine use of these vaccines in aquaculture. These will include the possible use of molecular adjuvants; the use of attenuated mutant glycoprotein genes and studies dealing with conformation versus lineal epitopes; the substitution of the CMV promoter for promoters of a non-viral origin; the use of induced plasmids or fish transposons, the study of new mass vaccination methods, or the inclusion of specific siRNA sequences within the plasmid. As well, we will review the most recent advances in studies related to the immune response towards these vaccines, some of them based on microarray technologies, that may help elucidate the immune mechanisms responsible for protection.

INTRODUCTION

One of the major problems for aquaculture is the impact that infectious diseases have on production. Among the different infectious agents, viruses are the major concern for the aquaculture industry since they usually produce very high mortalities upon infection and there are no effective treatments available such as antibiotics for bacterial infections.

Rhabdoviruses constitute one of the largest groups of viruses isolated from teleost fish, and include viruses such as infectious hematopoietic necrosis virus (IHNV) that affects different salmonid species like rainbow trout *Oncorhynchus mykiss*, Atlantic salmon *Salmo salar* or sockeye salmon *Oncorhynchus nerka* among others; and viral hemorrhagic septicemia virus (VHSV) that affects many salmonid (rainbow trout, Atlantic salmon, sockeye salmon, etc…), as well as many other non-salmonid and / or marine species (turbot *Scophthalmus maximus*, Atlantic cod *Gadus morhua*, etc…) [1]. Both IHNV and VHSV are responsible for great losses in aquaculture production, and even though IHNV was initially confined to America and VHSV was restricted to Europe, at the moment they are both spread worldwide. Moreover, rhabdoviruses not only affect fish at the early stages of development as most other viruses that infect fish, but can also produce a high percentage of mortality in adult fish of high economic value. Among other viruses affecting fish, infectious pancreatic necrosis virus (IPNV), a member of the *Birnaviridae* family, is also responsible for high mortalities, specially in salmonid fry [2], although it can also infect many other fresh water and marine species [3, 4]. Infectious salmon anaemia virus (ISAV), an orthomyxovirus, is a great problem for the salmon industry as well [5]. Regarding marine species, especially those held in moderate-warm water temperatures, nodavirus are nowadays a major concern [6-8]. Other virus belonging to the Togavirus, Aquareovirus, Iridovirus and Retrovirus families can also infect different fish species, but are a much lesser problem at a global scale [3].

There are no specific agents for the treatment of any of these diseases, consequently, the use of preventive measures such as vaccination seems as the most adequate method to control these viral agents in aquacultured fish [6]. Despite extensive research over many years, very few anti-viral vaccines are available. For instance, in China and Japan, an inactivated vaccine against spring viremia of the carp virus (SVCV), another fish rhabdovirus and a recombinant protein vaccine against red seabream iridovirus have been respectively licensed. In both Chile and Norway, different variants of a polyvalent vaccine against IPNV consisting in the recombinant VP2 viral protein are commercialised by different companies. One live attenuated VHSV vaccine is licensed in Germany, although its effectiveness is really limited.

Most recently, the first DNA vaccine has been approved for use in Canada against IHNV (APEX-IHN, Vical Inc. San Diego CA).

In fact, the most efficient vaccines developed against viral diseases in fish to date are DNA vaccines against rhabdoviruses [9]. These vaccines are based on a naked DNA plasmid coding for the rhabdoviral G glycoprotein gene under the control of the cytomegalovirus promoter (CMV). After intramuscular injection, plasmids are taken by muscle cells, and the viral G glycoprotein is expressed on the surface of muscle cells [9-12]. The gene encoding the G protein was chosen for rhabdovirus DNA vaccines because this is the protein against which protective neutralizing host antibodies are directed [13]. Accordingly, efficient vaccination upon intramuscular injection was only observed with the inclusion glycoprotein G gene and other rhabdoviral genes failed to induce protection [14]. Protection levels of more than 95 % have been obtained with these DNA vaccines [9, 11, 15-21], while the injection of trout with recombinant rhabdoviral G proteins produced in E*scherichia coli* [22, 23], yeast [23] or baculovirus [24] have not been able to confer good protection levels. Furthermore, DNA vaccines have been proved effective not only for VHSV and IHNV, but for all fish rhabdovirus tested: SVCV [25] and other more exotic rhabdovirus such as hirame rhabdovirus (HIRRV) [26] and snakehead rhabdovirus (SHRV) [25].

Although DNA vaccines have also been assayed for some other fish viruses, their effectiveness and the state of knowledge of their mechanism of action are actually very limited. For instance, for IPNV, five different plasmids coding for different regions of the segment A of the viral genome were tested [27], finding that the full segment A had to be included in the plasmid in order to obtain some protection. For ISAV, a DNA vaccine coding for the viral hemagglutinin-esterase gene only showed moderate protection [28]. Different open reading frames (ORFs) for early and late genes of channel catfish herpesvirus (IHV-1) were also tested with two of them inducing some protection [29]. For nodavirus, the recombinant capsid protein conferred protection while a plasmid coding for this same protein was not effective [30]. This last result is surprising, since even rhabdoviral DNA vaccines conferred protection against nodavirus at early times post-vaccination [8].

Even though, in the case of rhabdovirus, the protection conferred is high and relatively long-lasting, these vaccines are still not being used worldwide due to safety issues, mostly concerning the viral regulatory sequences present in plasmids, to difficulties in the administration of the vaccine at a large scale (mass vaccination methods) and to relatively high vaccination costs for fingerlings. In this chapter, we discuss the new trends in research dealing with DNA vaccination in fish, which might hopefully improve and finally allow their routine use in aquaculture. We have included possible improvements in the DNA vaccine formulations which might contribute to an increased effectiveness such as the potential use of molecular adjuvants, attenuated pathogen mutant genes, new non-viral promoters, inducible plasmid vectors, fish transposons or the inclusion of specific siRNA sequences as hairpin DNA, as well as trends in the possible search for new mass vaccination methods (Figure 1). Moreover, we have reviewed the most recent studies related to the early and late immune response towards these vaccines which may help elucidate the immune mechanisms responsible for protection and produce new ideas to improve DNA vaccination for rhabdovirus as well as for other fish viruses.

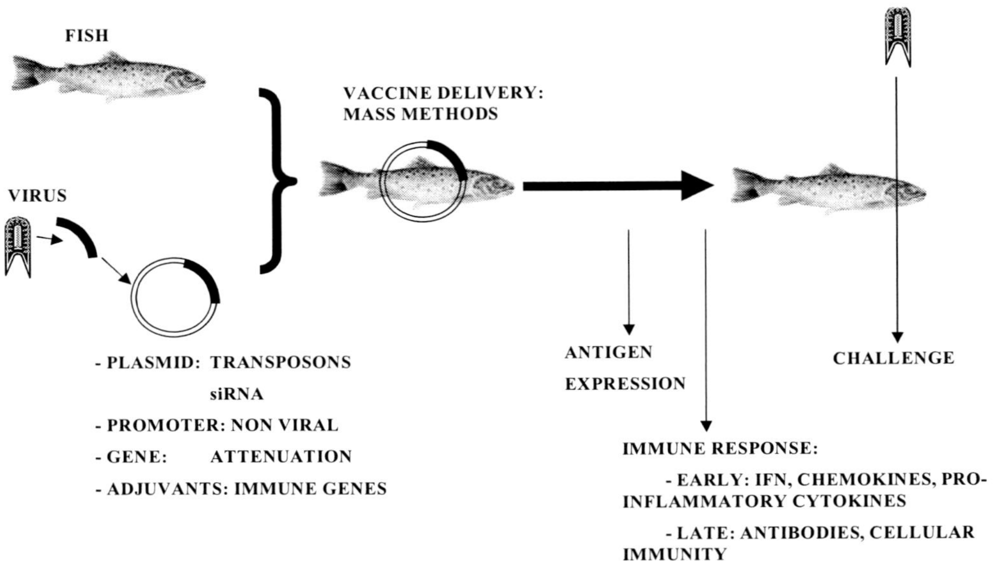

Figure 1. Research areas to be explored concerning fish DNA vaccination. The main elements to be taken into account for improvement of fish DNA vaccines include the vaccine plasmid (possible modifications in the type of promoter; use of adjuvants; attenuation of viral genes, possible use of transposons, siRNA etc.). Concerning the vaccine delivery, new mass delivery systems need to be optimised (oral or bath administration seem as the most suitable methods). From studies concerning the immune response to vaccination, new adjuvants might be elucidated.

IMMUNE RESPONSE TO DNA VACCINATION

At the moment, the immune mechanism(s) responsible for the protection conferred by injected DNA vaccines are still unknown. Moreover, DNA vaccines seem to be more effective in fish, that in other mammalian models [6], making difficult that studies performed in mammals may serve as a guide for what might be happening in fish. It seems however, that DNA vaccines are to some extent able to mimic the natural infection and thus produce high levels of protection [9, 11, 15-21].

Antigen-specific immune responses induced by DNA vaccines are elicited by the intramuscular inoculation of a plasmid containing foreign DNA expressed under the control of a viral promoter [31]. Because the inoculated genes (the antigens) are expressed intracellularly, epitope peptides are presented on MHC class I molecules, and are therefore capable of stimulating $CD8^+$ T cells. These gene products expressed by muscle cells are also phagocytosed by antigen presenting cells (APCs) and are either cross-presented on MHC class I molecules or presented by MHC class II molecules, inducing $CD4^+$ T cell responses [32]. DNA vaccines also induce the expression of co-stimulatory molecules on APCs, probably because the vaccine plasmids are recognised as bacterial DNA through the presence of specific "danger" motifs, leading to the initiation of a strong immune response [33].

Whether immunogenic plasmids directly transfect APCs or only muscle cells which then present the antigen to APCs has been the subject of considerable debate [32].

Non-Specific Early Stage of Protection

DNA vaccines against salmonid rhabdoviruses once intramuscularly injected induce an early stage of protection as soon as 4-7 days post-vaccination. During this period, protection is not only observed against the virus against which fish were vaccinated, but also against heterologous viruses [17], even to virus from families different than rhabdovirus, such as nodavirus [8] although not against bacterial pathogens [12]. No cross-protection was observed however, when fish were vaccinated with a plasmid coding for the G protein of rabies virus, a mammalian rhabdovirus [17]. All these results suggest that a non-specific but exclusively viral immune mechanism is responsible for protection at this early stage, because specific antibodies (Abs) and / or cellular responses are not developed in most fish species until at least two weeks post-immunization [34].

Type I interferon (IFN) seems like a solid candidate for being responsible for most if not all of the early protection induced by DNA vaccines. Since IFN genes were not identified in fish up until recently, most studies performed with DNA vaccines so far involved the analysis of the IFN-induced Mx gene expression. It has been demonstrated that Mx was induced in response to DNA vaccination [20, 35, 36] and correlation between the levels of Mx expression and the levels of trout early protection was observed by using a VHSV DNA vaccine [20]. However, in a further study Mx synthesis was not detectable until day 7 and its higher level of expression was reached 14 days post-vaccination [36]. In most species various isoforms of Mx proteins are present. In fish, three isoforms have been identified in rainbow trout [37] or Atlantic salmon [38], but studies concerning the differential regulation have not been approached for most of them, although differential antiviral activity has been demonstrated for human, murine and rat isoforms [39]. Moreover, some Mx isoforms do not seem to have antiviral activity, such as human MxB that exclusively plays a role in the control of the cell cycle [40]. In rainbow trout, we have recently demonstrated differential regulation of the three trout Mx isoforms depending on the cell or tissue cell type rather than to the stimuli used [41], however no antiviral activity against viruses has been demonstrated yet for any of the Mx isoforms [37]. Polymorphisms (variations of amino acid sequences) should also be taken into account when studying these immune early genes, and specially Mx, since differences in activity among polymorphisms have been observed in other species, with some polymorphisms having high and some low or even absent antiviral activity [42-44].

To date, it has been demonstrated that IFN is induced in response to DNA vaccines as expected from Mx studies [45], but as for Mx, different genes have identified for most fish species [46, 47], and studies concerning the role of each of the forms in protection have not been performed and should be carried out. In rainbow trout, for example, one of the IFN genes identified does not have an apparent antiviral activity [47]. Moreover, there are many other genes different than Mx with a direct antiviral role that are induced in response to IFN. In fact, more than 1000 genes are induced in a cell in response to IFN [48]. Therefore, in order to establish whether the protection conferred by DNA vaccination at this early stage is dependent of the IFN expression system at some point, much more work needs to be performed involving functional studies of Mx and IFN genes taking into account polymorphisms, as well as studies concerning other IFN-induced genes.

SPECIFIC LATE STAGE OF PROTECTION

After this stage of early non-specific immune protection, there is a later stage at which fish are specifically protected only against the virus they were DNA vaccinated for. The switching time from early (non-specific) to late (specific) immune response varies depending on the size of the trout, the dosage of challenge, the dosage of vaccine and the temperature [20, 49, 50]. It has been demonstrated that at this later stage, both humoral (Abs) [15] and cellular [51, 52] immune responses are induced. However, at this late stage again the mechanisms responsible for the protection are still unknown, since induced neutralizing Ab titers as estimated by *in vitro* assays did not always correlate with *in vivo* protection levels. For example, *in vitro* neutralizing Abs induced by injection of the G gene of VHSV could only be detected for up to 6 months after immunization, whereas protection lasted longer than 9 months [20]. Possible explanations for this lack of correlation could be that: i) the *in vitro* techniques used do not detect all induced neutralizing Abs with enough sensitivity; ii) some non-neutralizing Abs *in vitro* could be neutralizing *in vivo,* as it happened with some monoclonal Abs [53] or iii) there are other specific mechanisms mediating the protection such as cellular mechanisms. Not enough evidences are yet available in favour of any of these possible hypotheses.

Studies previously performed with VHSV revealed that it is not so simple to correlate neutralization titers *in vitro* with protection. Even though immunity to VHSV could be transferred by injection of the immunoglobulin M (IgM) fraction of the serum from survivors [54], contradictions still appear in the results. The *in vitro* neutralization assays of trout Abs are restricted to the detection of G conformational-dependent Abs [22, 55, 56], not detectable by immunoblotting [22, 57, 58]. Confirming these results, most G epitopes recognized by neutralizing monoclonal Abs are also conformation-dependent and some of them are even discontinuous in their sequence [59-61]. Attempts to map the B-cell epitopes on the VHSV G by the use of overlapping 15-mer synthetic peptides confirmed that none of the neutralizing monoclonal Abs could be mapped to lineal epitopes [62]. On the other hand, non-neutralizing Abs (those directed towards lineal epitopes) could also induce *in vivo* protection and persist longer than neutralizing Abs [53]. Furthermore, *in vitro* neutralizing Abs do not always correlate with its VHSV protection properties *in vivo* [53, 56, 63] and their *in vitro* neutralization was highly dependent on the VHSV isolate used [64]. Therefore, assays to detect trout Abs directed towards lineal epitopes on the G of VHSV (such as ELISA) or injection of fingerlings for *in vivo* protection assays should be also developed and conducted in vaccinated fish, as a complement to the *in vitro* neutralizing Ab assays.

Moreover, up until recently, teleost Igs were thought to be only of isotype IgM, whereas, we now know that IgD is also produced in fish [65, 66] and that a new exclusive isotype (IgT) is present at least in rainbow trout [67]. No studies have been yet performed on the possible significance of these new molecular species of fish Ig isotypes on the Ab response in vaccination or infection. To make these studies possible, however, more specific monoclonal Abs against the new fish Igs would be required.

Specific late protection to rhabdoviral infections can also be mediated by a delayed cellular memory consisting in both helper and/or cytotoxic activities (T-like cell responses). Up-regulation of MHC class II expression (a sign of T-cell activation) was observed in trout immunized with the G gene [51] and proliferation of kidney cells in response to the antigen encoded by the DNA vaccine was also demonstrated [50]. Recently, it has been shown that immunization of rainbow trout with both DNA encoding the glycoprotein G and the nucleocapsid protein N of VHSV provoked the activation of antiviral cytotoxic cells [68]. Since both VHSV-infected MHC class I mismatched (xenogeneic) and matched target cells were killed, cytotoxicity was probably executed by both cytotoxic cells of the adaptive (CTLs) and innate (NK cells) immune systems. As expected, the VHSV G protein was more potent in the triggering of cell-mediated immune responses than the VHSV N protein. Despite these studies, much more efforts on research in this area need to be made to elucidate the role of cellular immunity on antiviral protection.

With the use of microarrays, a technology is now available to study new genes involved in immune response induced by DNA vaccines. To design which fish sequences could be included in the microarrays, orthologous immune-related genes mapped in other species might be first selected and compared with fish EST (expressed sequence tags). This compared analysis will help identify the existence of possible candidate sequences not yet identified or mapped in the selected genome, but present in the genome maps of other fish or mammalian species. The cDNA obtained from tissue mRNA (transcriptomes) and / or oligo probes defined from EST fish sequences could be also used in microarrays. There are only a few examples where cDNA microarrays were used to compare transcriptomes obtained from DNA vaccinated and non-vaccinated samples [69-71]. As an example of the information that the use of microarrays gives concerning DNA vaccination [69, 70], by vaccinating Japanese flounder with a rhaddoviral G gene DNA vaccine and comparing it to fish vaccinated with the empty plasmid, using a cDNA chip with a total of 779 flounder sequences consisting in 228 known immune-related genes and 551 unknown genes, they observed that the greatest number of genes (16.6 %) with changed expression levels was observed at 3 days after injection. Of those 91.4 % were up-regulated (31 % known and 60.4 % unknown). Up-regulated genes include genes related to the non-specific immune response such as Kupffer cell receptor, TNF-α, MIP1α, IL-1 receptor, coagulation factor XIII, CC chemokine receptor, Mx, transcription factors such as IF-induced protein, TAP2 protein, caspase-10d, and even a few genes related to the late specific Ab response such as the CD20 receptor and B cell adhesion molecule. A great number of unknown genes were also up-regulated, in particular one gene was up-regulated 56-fold in vaccinated fish, pointing it as an interesting gene for future research. Even though microarray studies show a great amount of information, all these results will not be of relevance to future applications in vaccine technology if they are not taken as a first step to identify potential genes involved and if no functional studies are performed with the proteins that these genes encode. Moreover, up to date, all the studies performed to study genes regulated upon vaccination have only been performed at early stages post-vaccination. It would be also of great importance to perform these kind of studies after a viral challenge in vaccinated fish at later stages post-vaccination, in order to further studies genes that are involved in the specific immune response. The comparison of the gene response to a virus in vaccinated fish compared to fish susceptible to the disease would be of great importance in determining which genes are relevant in protection.

Possible New Adjuvants for DNA Vaccination

Adjuvants (from the Latin *adjuvare,* to help) are agents capable of increasing the immunogenicity of a vaccine, and thus increasing its effectiveness. Usually they act by increasing the presentation of the antigen to the immune system, or by inducing a slight inflammatory process that activates immune cells, specially antigen presenting cells. For many adjuvants, the mechanism of action remains unclear, and in some cases their discovery has been a matter of trial and error for many years, explaining why alum, an aluminium salt developed in the early 1920s, is still one of the most widely used adjuvant for traditional vaccines. Consequently, there is a major need for safer and more effective adjuvants suitable for human and animal use, and this constitutes a wide field of research that is being explored in different animal models. Moreover, an adjuvant that might be suitable for a vaccination model against one virus, does not necessarily have to be suitable for vaccination against another virus in the same host, since the pathogenesis of the virus has also to be taken into account. Therefore, each new adjuvant should be tested independently for each host, type of vaccine and virus. In this chapter, we review some of the trends in the current search for adjuvants that would be useful for DNA vaccination in fish.

Cytokines

In mammals, different cytokines have been proved to be successful adjuvants for DNA vaccination, while others expected to have positive effects according to their biological activity decreased the protection conferred by the DNA vaccine. For example, co-administration of interleukin 12 (IL-12) within a DNA vaccine resulted in an inhibition of the Ab response, whereas the co-injection of granulocyte-macrophage colony stimulating factor (GM-CSF) enhanced it [72]. Among cytokine adjuvants, chemokines (cytokines with chemoattractant capacity) are those more widely studied for vaccination against viruses in mammals [73-75], since they are not only capable of recruiting more specific cell types to the site of immunization, but they can also drive the immune system to a more favourable response.

In fish, almost no studies on the adjuvant capacity of cytokines have been performed to date, since the state of knowledge of cytokine and chemokine functionality is still scarce compared to mammals, and this is essential for the right choice of genes to be used as adjuvants. The co-administration of mouse GM-CSF had a marked effect on cytotoxic T-cell-like activity in fish by reducing the average number of myocites expressing the antigen [52]. In rainbow trout, our laboratory has studied the effects of the chemokine IL-8 as a potential adjuvant for DNA vaccination against VHSV. We have found that the co-administration of a plasmid coding for rainbow trout IL-8 with a DNA vaccine, increases the expression of pro-inflammatory cytokines such as IL-1β and TNF-α [76]. Moreover, an increase in the levels of expression of other chemokines was also observed [77], suggesting that IL-8 might be an efficient molecular adjuvant for DNA vaccination in fish. However, whether these effects on the immune response to the vaccine correlate with a higher protection still has not been demonstrated (ongoing experiments). A close attention must be given to the doses used, since the optimal dose for immunostimulation may not be the optimal dose for adjuvant activity, as an exaggerated inflammatory response can also be detrimental for the immunization, something widely demonstrated in mammals.

TLR Ligands

Toll-like receptors (TLRs) are a family of invariant pathogen-specific molecular pattern (PAMPS) recognition receptors, that initiate the immune response against a pathogen upon activation [78]. Examples of PAMPs that trigger TLR pathways are lipopolysaccharides (LPS; recognized by TLR4), lipopeptides (by cooperation of TLR2 with TLR1 or TLR6), viral single- or double-stranded RNA (by TLR7 with TLR8 and by TLR3, respectively), bacterial DNA containing CpG motifs (by TLR9) or flagellin (by TLR5). In addition, recent investigations have demonstrated that TLRs are the critical in the linkage between innate and adaptive immunity.

The involvement of TLRs in vaccine responses is explained by the control they exert over the activation of adaptive immunity. This control is mediated through several mechanisms, including maturation of DCs, induction of expression of cytokines and co-stimulatory proteins, and reversal of tolerance [79]. Although the importance of TLR signalling for vaccine response in humans has been demonstrated conclusively only for the OspA vaccine against [80], many promising experimental vaccine strategies for infectious diseases include the use of TLR ligands as adjuvants.

Among TLR ligands, those that have been more thoroughly studied in mammalian systems are the oligodeoxynucleotides containing CpG motifs. These CpG motifs enable the vertebrate immune system to recognise bacterial DNA via TLR 9 since while methylation of bacterial CpG sequences is rare, vertebrates usually methylate most CpG sequences at the 5′position of the cytosine [81]. It has been demonstrated that some of these CpG motifs in the plasmid backbone of DNA vaccines play an important role in the induction of immunity [82], and that the inclusion of additional CpG motifs increase the effectiveness of DNA vaccination [83]. In fish, activation of lymphocyte proliferation [84] and production of immunostimulatory [85, 86] or antiviral [87] factors have been demonstrated by CpG motifs, but few more studies have been performed to test their possible adjuvant activity, except for a study in which a CpG motif known to stimulate mice, increased the Ab response to an antigen in fish when included in the plasmid vector [88]. No other TLR ligands have been assayed in fish so far since fish TLR are poorly characterized to date. In mammalian systems, successful adjuvant activities have also been obtained by using other TLR ligands such as flagellin (as DNA encoding format) [89] as well as TLR adaptor molecules, particularly TRIF [90] and MyD88 [91]. Recently, there have been several studies on fish that have identified many fish orthologs of mammalian TLRs [78, 92-94], that appear to be functionally analogous, therefore, future studies should determine whether or not TLR signalling has a significant role in fish responses to DNA vaccines, and which TLR ligands are the most effective adjuvants.

POSSIBLE ATTENUATION OF THE VIRAL GENE(S)

One of the possibilities to reduce present concerns about the potential risks of introducing fish pathogen virulent genes into the wild through the use of DNA vaccines, could be the use of artificially attenuated mutant genes with low possibilities of reversion to wild type, such as multiple mutants. The desired mutant genes can be easily obtained by site directed mutagenesis once established that the mutations do not interfere with the induction of protective immune responses within the host. One of the few examples of this strategy is the collection of sixteen VHSV G gene mutants (two of them double mutants) that were recently obtained in our laboratory [95]. When susceptible cells are transfected with a plasmid coding for the wild type G protein, the transfected cells form syncitia equivalent to those formed upon infection. According to the sequences that affect fusion previously identified in the G protein [96, 97] and confirmed by the elucidation of the tridimensional structure of the G of other rhabdovirus [98, 99], these mutants were affected in fusion and conformation but nevertheless they were recognized by polyclonal as well as for some neutralizing mouse Abs. It remains to be demonstrated whether polyclonal trout Abs still recognize these mutants and if DNA vaccination using these mutant genes induces equivalent rates of protection.

SUBSTITUTION OF THE CMV PROMOTER BY NEW PROMOTERS

A main concern for the efficiency and safety of fish DNA vaccines is the promoter and regulatory sequences responsible for the transcription of the pathogen gene within the plasmid. To date, all DNA vaccines use the immediate early cytomegalovirus promoter (CMV). When used for gene expression, it acts as a very strong promoter capable of transcribing most genes assayed in a wide range of eukaryotic cells [100, 101], including fish cells [102]. For example, with this system and after intramuscular injection of 1 μg of a VHSV vaccine plasmid, the relative percent of survival (RPS) values were 98% [18].

Despite the effectiveness of this system, CMV is a viral promoter responsible for the regulation of eukaryotic transcription, and this is an impairment to the license and commercialisation of these vaccines worldwide, for example in Europe. Therefore, a search for alternative promoters seems necessary, although the efficiency and price of the vaccine should not be compromised. Various promoters from salmonid genes have already been tested to control gene expression in fish, such as the rainbow trout metallothionein-A promoter [103], the Atlantic salmon histone H3 promoter [104] and the sockeye salmon histone, metallothionein and protamine promoters [105] without very promising results. However the optimisation of β-actine promoters seems as a good alternative, since β-actine promoters of different species have proved to be almost as effective as CMV [106]. We have recently demonstrated that whereas the carp β-actine promoter induced levels of expression of the VHSV G protein lower than those of the CMV promoter at very early times, it then increased and reached levels higher than those induced by CMV at later times. This was assayed both *in vitro* after transfection of fish cell lines and *in vivo* after intramuscular injection, where we found an increase in the Ab titers induced when the β-actine carp promoter was used instead of that of CMV (unpublished data).

Much more research needs to be performed in the search for new fish promoters among those existing in fish or from synthetic and/or semi-synthetic sources, in order to optimise vaccination systems that are both safe and efficient.

Possible Use of Inducible Plasmids

Regulating the control of recombinant protein expression by inducible systems could some day help obtain vectors that induce stronger immune responses that are long-lasting. Further technical complications arise from the fact that the inducible constructs should be inserted into the genome of some of the somatic cells to obtain a long lasting capacity of response rather than a temporary response. In the inducible approach, the switching on / off of the target gene depends on a specific inducer acting over a control protein that binds the target promoter [107, 108]. The constitutive expression of genes is thus avoided and the most adequate time and place of the expression could be controlled, for instance upon a risk of infection. One prerequisite for a useful inducible system is that the inducer / controller must be innocuous and that the possible activation of other gene loci would be inconsequential. The ideal inducible system should also allow the switching of gene expression on and off, rapidly, reversibly, in a dose-dependent manner and only in the desired time and tissue [109]. Limitations that have arisen during the use of these systems in mammals include leaky expression, cellular toxicity, unstable transcripts and insensitivity to the inducer. In the targeted gene approach, the introduction of genes into cells is generally made by homologous recombination by incorporating long genomic inserts or specific recombination sequences. The combination of gene targeting with inducible gene systems constitutes the highest possible control of gene expression available today [109]. At the moment, no inducible systems have been yet tested as possible DNA vaccines, but the development of this technology could be useful in the future for fish vaccination uses.

Fish Transposon Technology Applied to DNA Vaccines

DNA-based active transposons are repetitive DNA sequences found in the genomes of animals and plants that move from one genome locus to another. Although active transposons have been difficult to detect in most vertebrates including fish [110], due to the fact that most of their existing copies are inactivated through deletions, frameshift mutations or premature termination codons in their transposase genes, the enzymes controlling transposon mobilization [111], expression systems based in transposons have been developed and widely studied for applications, such as the improvement of the efficiency of vaccine plasmids and / or the increase of the duration of protection in mammalian systems [112-115]. For example, a decade ago, an active transposase named *Sleeping Beauty* (SB) was artificially reconstructed from trout and salmon genome sequences [116] which recognized sequences flanked by inverted terminal repeats (ITR) of 225 bp. Dual transposition systems consisting in two independent plasmids were designed to transform mammalian cells. One plasmid contained an expression cassette flanked by left and right SB ITRs whereas the other plasmid contained the SB transposase (pSB) under the control of a constitutive promoter like CMV.

Concerning fish, complete SB transposase sequences suggesting their active transposition have been demonstrated in *Salvelinus namaycush* and *Pleuronectes platessa* [117, 118] and active transposition has been detected in zebrafish [119]. In addition, we have demonstrated that active transposases are present in fish cells such as RTG-2 from rainbow trout or EPC from carp and that the *in vitro* temporal expression of genes included in a transposon expression system is notably increased if compared to a plasmid in which the gene is not flanked by ITRs (unpublished data). A *priori* the integration in the genome would not be a problem for the commercialisation of transposon system-based vaccines, and the inclusion of transposon sequences within the plasmid would increase the efficiency and favour the integration with consequent effects of duration of protection. Thus, expectations now might be high for possible new applications of this technology to fish DNA vaccination.

Possible Use of Sirna Generating Sequences

During the last years, double stranded RNAs of 19-25-nucleotides (silencer RNA or siRNA) have been shown to inhibit specifically the translation of complementary mRNAs into proteins. This technology is being applied to specifically knockdown gene expression. Some examples of viral replication inhibition by using dsRNAs that are complementary in sequence to some viral proteins include respiratory synthitial virus [120], hepatitis virus [121, 122] and the mammalian rhabdovirus, vesicular stomatitis [123]. In most mammalian cells, transfected siRNAs that target different viral proteins can inhibit the synthesis of virus at nM concentrations without inducing IFN synthesis or other cellular genes [123]. However, in fish cells such as EPC cells, the transfection of siRNAs targeting the G glycoprotein was shown to induce IFN, being this IFN at least in part responsible for the inhibition of VHSV infection [124]. Many different possibilities of other siRNAs targeting other viral genes need to be explored to elucidate the mechanism of action of the interference in viral replication of these siRNAs in fish cells. In any case, a method including the injection of complexes of siRNAs with DOTAP has been described to delay VHSV *in vivo* infection during 4-5 days [125]. The siRNA can not only be delivered as synthetic RNAs but also through the use plasmids that include the corresponding hairpin DNA sequences (two complementary sequences of 21 nucleotides separated by a loop of 9 nucleotides) under the control of a strong promoter [126]. When transcribed, these sequences form a dsRNA forming an hairpin which behaves as a siRNA, and this could be the system through which this technology could be incorporated into DNA vaccines in fish.

In our laboratory, several potential targets to VHSV L polymerase sequences were designed, chemically synthesized and tested by Q-PCR and microneutralization. Some of them gave inhibitory results and were cloned into commercial plasmids codifying both siRNA and puromicin resistance genes (data not published). EPC transfected cell lines were derived for each of them to demonstrate that some clones acquired the capacity to inhibit VHSV replication. At the moment however, we still do not know whether the inhibition of replication is only due to IFN synthesis or to as in mammals to an interference with the translation of the targeted gene.

The use of siRNAs opens the possibility to interfere with one of the early steps of VHSV infection such as those affected by the polymerase L, thus offering possibilities to potentiate

DNA vaccines. For example, the delay seen in the effects of VHSV infection seen after the injection of complexes of siRNAs with DOTAP [125] might be useful to protect the fish from viral infection until the early immune response induced by the vaccine four days post-vaccination is established.

Moreover, in the future, as advances are available in the understanding of the immune mechanisms responsible for the protection conferred by DNA vaccines, the siRNA technology could be used to block the action of immune factors that interfere with the immune response. For example, the incorporation in a vaccine of siRNAs that target apoptotic proteins, prolonged the life of transfected dendritic cells and thus increased the T cell responses induced by a vaccine [127].

New Mass Vaccination Strategies

A suitable delivery of DNA vaccines allowing mass vaccination of fingerling fish should substitute the actual muscular injection, if DNA vaccines are to be of practical use in those most desirable sizes. Probably, the optimal route for the administration of a vaccine in aquaculture fish would be through bath administration, however, the only route that so far has proved effective for the administration of DNA vaccines has been the intramuscular injection, a very laborious methodology that almost makes impossible a mass vaccination method of small fish of low economic value. Through the use of low frequency ultrasounds [21] or hiperosmotic shock [128], the protection levels after bath DNA vaccination were significantly improved in comparison to the protection levels obtained after a regular bath vaccination, but were still far from the results obtained by intramuscular injection. The bombardment of fish muscle with gold particles covered with DNA [129] proved as an efficient delivery method and the antigen expression lasted for at least 60 days post-bombardment. This method, however, remains expensive and no further optimisation was yet reported. When different routes of administration of a plasmid DNA coding for luciferase coupled to liposomes or chitosan were tested, acceptable levels of luciferase expression were obtained after intraperitoneal, intramuscular or intravenous injection of these complexes, but no detectable luciferase levels were obtained after oral or bath administration [130]. In any case, there are many different possible formulations and improvements that could be tested for liposomes, that have been studied in different mammalian systems, therefore, the area of mass delivery still offers a wide room for further research to make fish DNA vaccines more suitable from a practical point of view.

Conclusion

Vaccination is the most adequate method to control viral diseases, specially in animals of relatively low economic value such as fish. With the efficacy of DNA vaccination against rhabdovirus seen in fish, we have a great opportunity to finally obtain efficient vaccines that are easy to produce, safe, cheap, and provide long-lasting immunity. We only need to continue our research efforts to solve some of the problems they still offer, which still impair their routine use worldwide. In this chapter, we have tried to outline some of the most important lines of research along which the improvement of these vaccines may be achieved.

It seems obvious that concerning vaccines, one of the most important fields of research is the study of immune mechanisms responsible for protection. Previous experience in mammals has demonstrated that from immunological studies, many new ideas can be obtained for the optimisation of vaccine systems to produce more successful and long-lasting fish DNA vaccines. Moreover, this will be a critical point for the development of efficient DNA vaccines against other viral pathogens in fish, that are still not as efficient as those against rhabdovirus.

Acknowledgements

This work was supported by INIA strategic project CPE03-016-C4 and Spanish MEC projects AGL2004-07404-CO2/ACU and AGL05 00339-ACU.

References

[1] Coll, J. M. (2001). Fish Rhabdoviruses. *Recent Research Developmental Virology, 3*, 229-241.

[2] Smail, D. A., Bruno, D. W., Dear, G., McFarlane, L. A. and Ross, K. (1992). Infectious pancreatic necrosis (IPN) virus sp serotype in farmed Atlantic salmon, *Salmo salar* L. post-smolt associated with mortality and clinical disease. *Journal of Fish Diseases, 15*, 77-83.

[3] Bernard, J. and Bremont, M. (1995). Molecular biology of fish viruses: a review. *Veterinary Research, 26*, 341-51.

[4] Macdonald, R. D. and Gower, D. A. (1981). Genomic and phenotypic divergence among three serotypes of aquatic birnaviruses (infectious pancreatic necrosis virus). *Virology, 114*, 187-95.

[5] Gregory, A. (2002). Detection of infectious salmon anaemia virus (ISAV) by *in situ* hybridisation. *Diseases of Aquatic Organisms, 50*, 105-110.

[6] Biering, E., Villoing, S., Sommerset, I. and Christie, K. E. (2005). Update on viral vaccines for fish. *Developmental Biology, 121*, 97-113.

[7] Frerichs, G. N., Rodger, H. D.and Peric, Z. (1996). Cell culture isolation of piscine neuropathy nodavirus from juvenile sea bass, *Dicentrarchus labrax*. *Journal of General Virology, 77*, 2067-2071.

[8] Sommerset, I., Lorenzen, E., Lorenzen, N., Bleie, H. and Nerland, A. H. (2003). A DNA vaccine directed against a rainbow trout rhabdovirus induces early protection against a nodavirus challenge in trout. *Vaccine, 21*, 4661-4667.

[9] Lorenzen, N. and LaPatra, S. E. (2005). DNA vaccines for aquacultured fish. *Reviews Science Technology Office International Epizooties, 24*, 201-213.

[10] Leong, J. A. and Munn, C. B. (1991). Potencial uses of recombinant DNA in development of fish vaccines. *Bulletin of the European Association Fish Pathologists, 11*, 30-40.

[11] Anderson, E. D., Mourich, D. V., Fahrenkrug, S. C., LaPatra, S. C., Shepherd, J. and Leong, J. C. (1996). Genetic immunization of rainbow trout (*Oncorhynchus mykiss*) against infectious hematopoietic necrosis virus. *Molecular Marine Biology and Biotechnology, 5*, 114-122.

[12] Lorenzen, N., Lorenzen, E.,Einer-Jensen, K.and LaPatra, S. E. (2002). Immunity induced shortly after DNA vaccination of rainbow trout against rhabdoviruses protects against heterologous virus but not against bacterial pathogens. *Developmental and Comparative Immunology, 26*, 173-179.

[13] Huang, C. J., Chien, M. S., Landolt, M.,Batts, W. and Winton, J. (1996). Mapping the neutralizing epitopes on the glycoprotein of infectious haematopoietic necrosis virus, a fish rhabdovirus. *Journal of General Virology, 77*, 3033-3040.

[14] Corbeil, S., LaPatra, S. E., Anderson, E. D., Jones, J., Vincent, B., Hsu, Y. L. and Kurath, G. (1999). Evaluation of the protective immunogenicity of the N, P, M, NV and G proteins of infectious hematopoietic necrosis virus in rainbow trout *Oncorhynchus mykiss* using DNA vaccines. *Diseases of Aquatic Organisms, 39*, 29-36.

[15] Anderson, E. D., Mourich, D. V. and Leong, J. C. (1996). Gene expression in rainbow trout (*Onchorynchus mykiss*) following intramuscular injection of DNA. *Molecular Marine Biology and Biotechnology, 5*, 105-113.

[16] Lorenzen, E., Einer-Jensen, K., Martinussen, T., LaPatra, S. E. and Lorenzen, N. (2000). DNA vaccination of rainbow trout against viral hemorrhagic septicemia virus: A dose-response and time-course study. *Journal of Aquatic Animal Health, 12*, 167-180.

[17] LaPatra, S. E., Corbeil, S., Jones, G. R., Shewmaker, W. D., Lorenzen, N., Anderson, E. D. and Kurath, G. (2001). Protection of rainbow trout against infectious hematopoietic necrosis virus four days after specific or semi-specific DNA vaccination. *Vaccine, 19*, 4011-4019.

[18] Lorenzen, N., Lorenzen, E. and Einer-Jensen, K. (2001). Immunity to viral haemorrhagic septicaemia (VHS) following DNA vaccination of rainbow trout at an early life-stage. *Fish and Shellfish Immunology, 11*, 585-591.

[19] Lorenzen, N., Lorenzen, E., Einer-Jensen, K., Heppell, J., Wu, T. and Davis, H. (1998). Protective immunity to VHS in rainbow trout (*Oncorhynchus mykiss*, Walbaum) following DNA vaccination. *Fish and Shellfish Immunology, 8*, 261-270.

[20] McLauchlan, P. E., Collet, B., Ingerslev, E., Secombes, C. J., Lorenzen, N. and Ellis, A. E. (2003). DNA vaccination against viral haemorrhagic septicaemia (VHS) in rainbow trout: size, dose, route of injection and duration of protection-early protection correlates with Mx expression. *Fish and Shellfish Immunology, 15*, 39-50.

[21] Fernandez-Alonso, M., Rocha, A. and Coll, J. M. (2001). DNA vaccination by immersion and ultrasound to trout viral haemorrhagic septicaemia virus. *Vaccine, 19*, 3067-3075.

[22] Lorenzen, N., Olesen, N. J. and Jorgensen, P. E. V. (1993). Antibody response in rainbow trout to VHS virus proteins. *Fish and Shellfish Immunology, 3*, 461-473.

[23] Estepa, A., Thiry, M. and Coll, J. M. (1994). Recombinant protein fragments from haemorrhagic septicaemia rhabdovirus stimulate trout leucocyte anamnestic *in vitro* responses. *Journal of General Virology, 75*, 1329-1338.

[24] Koener, J. F. and Leong, J. A. (1990). Expression of the glycoprotein gene from a fish rhabdovirus by using baculovirus vectors. *Journal of Virology, 64*, 428-30.

[25] Kim, C. H., Johnson, M. C., Drennan, J. D., Simon, B. E., Thomann, E. and Leong, J. A. (2000). DNA vaccines encoding viral glycoproteins induce nonspecific immunity and Mx protein synthesis in fish. *Journal of Virology, 74*, 7048-7054.

[26] Takano, T., Iwahori, A., Hirono, I. and Aoki, T. (2004). Development of a DNA vaccine against hirame rhabdovirus and analysis of the expression of immune-related genes after vaccination. *Fish and Shellfish Immunology, 17*, 367-374.

[27] Mikalsen, A. B., Torgersen, J., Alestrom, P., Hellemann, A. L., Koppang, E. O. and Rimstad, E. (2004). Protection of Atlantic salmon *Salmo salar* against infectious pancreatic necrosis after DNA vaccination. *Diseases of Aquatic Organisms, 60*, 11-20.

[28] Mikalsen, A. B., Sindre, H., Torgersen, J. and Rimstad, E. (2005). Protective effects of a DNA vaccine expressing the infectious salmon anemia virus hemagglutinin-esterase in Atlantic salmon. *Vaccine, 23*, 4895-905.

[29] Nusbaum, K. E., Smith, B. F., DeInnocentes, P. and Bird, R. C. (2002). Protective immunity induced by DNA vaccination of channel catfish with early and late transcripts of the channel catfish herpesvirus (IHV-1). *Veterinary Immunology and Immunopathology, 84*, 151-168.

[30] Sommerset, I., Skern, R., Biering, E., Bleie, H., Fiksdal, I. U., Grove, S. and Nerland, A. H. (2005). Protection against Atlantic halibut nodavirus in turbot is induced by recombinant capsid protein vaccination but not following DNA vaccination. *Fish and Shellfish Immunology, 18*, 13-29.

[31] Wolff, J. A., Malone, R. W., Williams, P., Chong, W., Acsadi, G., Jani, A. and Felgner, P. L. (1990). Direct gene transfer into mouse muscle *in vivo*. *Science, 247*, 1465-1468.

[32] Greenland, J. R. and Letvin, N. L. (2007). Chemical adjuvants for plasmid DNA vaccines. *Vaccine, 25*, 3731-41.

[33] Krieg, A. M. (1999). Direct immunologic activities of CpG DNA and implications for gene therapy. *Journal of General Medicine, 1*, 56-63.

[34] Chiller, J. M., Hodgins, H. O. and Weiser, R. S. (1969). Antibody response in rainbow trout (*Salmo gairdneri*). II. Studies of the kinetics of development of antibody-producing cells and on complement and natural haemolysin. *Immunology, 102*, 1202-1207.

[35] Purcell, M. K., Kurath, G., Garver, K. A., Herwig, R. P. and Winton, J. R. (2004). Quantitative expression profiling of imune response genes in rainbow trout following infectious haematopoietic necrosis virus (IHNV) infection or DNA vaccination. *Fish and Shellfish Immunology, 17*, 447-462.

[36] Acosta, F., Petrie, A., Lockhart, K., Lorenzen, N. and Ellis, A. E. (2005). Kinetics of Mx expression in rainbow trout (*Oncorhynchus mykiss*) and Atlantic salmon (*Salmo salar* L.) parr in response to VHS-DNA vaccination. *Fish and Shellfish Immunology, 18*, 81-89.

[37] Trobridge, G. D., Chiou, P. P. and Leong, J. C. (1997). Cloning of the rainbow trout (*Oncorhynchus mykiss*) Mx2 and Mx3 cDNAs and characterization of trout Mx protein expression in salmon cells. *Journal of Virology, 71*, 5304-5311.

[38] Robertsen, B., Trobridge, G. and Leong, J. A. (1997). Molecular cloning of double-stranded RNA inducible Mx genes from Atlantic salmon (*Salmo salar* L.). *Developmental and Comparative Immunology, 21*, 397-412.

[39] Staeheli, P., Pitossi, F. and Pavlovic, J. (1993). Mx proteins: GTPases with antiviral activity. *Trends Cell Biology, 3*, 268-72.

[40] King, M. C., Raposo, G. and Lemmon, M. A. (2004). Inhibition of nuclear import and cell-cycle progression by mutated forms of the dynamin-like GTPase MxB. *Proceedings of the National Academy of Sciences of the United States of America, 101*, 8957-62.

[41] Tafalla, C., Chico, V., Pérez, L., Coll, J. and Estepa, A. (2007). In vitro and in vivo differential expression of rainbow trout (*Oncorhynchus mykiss*) Mx isoforms in response to viral hemorrhagic septicemia virus (VHSV) G gene, Poly I:C and VHSV. *Fish and Shellfish Immunology, 23*, 210-21.

[42] Zurcher, T., Pavlovic, J. andStaeheli, P. (1992). Mechanism of human MxA protein action: variants with changed antiviral properties. *The EMBO Journal, 11*, 1657-61.

[43] Nakajima, E., Morozumi, T., Tsukamoto, K., Watanabe, T., Plastow, G. and Mitsuhashi, T. (2007). A naturally occurring variant of porcine Mx1 associated with increased susceptibility to influenza virus *in vitro*. *Biochemical Genetics, 45*, 11-24.

[44] Palm, M., Leroy, M., Thomas, A., Linden, A. and Desmecht, D. (2007). Differential anti-influenza activity among allelic variants at the *Sus scrofa* Mx1 locus. *Journal of Interferon and Cytokine Research, 27*, 147-55.

[45] Purcell, M. K., Nichols, K. M., Winton, J. R., Kurath, G., Thorgaard, G. H., Wheeler, P., Hansen, J. D., Herwig, R. P. and Park, L. K. (2006). Comprehensive gene expression profiling following DNA vaccination of rainbow trout against infectious hematopoietic necrosis virus. *Molecular Immunology, 43*, 2089-2106.

[46] Robertsen, B., Bergan, V., Rokenes, T., Larsen, R. and Albuquerque, A. (2003). Atlantic salmon interferon genes: cloning, sequence analysis, expression, and biological activity. *Journal of Interferon and Cytokine Research, 23*, 601-12.

[47] Zou, J., Tafalla, C., Truckle, J. and Secombes, C. (2007). Identification of a second group of type I interferons in fish sheds light on interferon evolution in vertebrates. *Journal of Immunology, In press*.

[48] Der, S. D., Zhou, A., Williams, B. R. and Silverman, R. H. (1998). Identification of genes differentially regulated by interferon alpha, beta, or gamma using oligonucleotide arrays. *Proceedings of the National Academy of Sciences of the United States of America, 95*, 15623-8.

[49] Traxler, G. S., Anderson, E., LaPatra, S. E., Richard, J., Shewmaker, B. and Kurath, G. (1999). Naked DNA vaccination of Atlantic salmon *Salmo salar* against IHNV. *Diseases of Aquatic Organisms, 38*, 183-190.

[50] Kanellos, T., Sylvester, I. D., Ambali, A. G., Howard, C. R. and Russell, P. H. (1999). The safety and longevity of DNA vaccines for fish. *Immunology, 96*, 307-13.

[51] Boudinot, P., Blanco, M., de Kinkelin, P. and Benmansour, A. (1998). Combined DNA immunization with the glycoprotein gene of viral hemorrhagic septicemia virus and infectious hematopoietic necrosis virus induces double-specific protective immunity and nonspecific response in rainbow trout. *Virology, 249*, 297-306.

[52] Kanellos, T., Sylvester, I. D., Howard, C. R. and Rusell, P. H. (1999). DNA is as effective as protein at inducing antibody in fish. *Vaccine, 17*, 965-972.

[53] Lorenzen, N., Olesen, N. J. and Vestergaard-Jorgensen, P. E. (1990). Neutralization of Egtved virus pathogenicity to cell cultures and fish by monoclonal antibodies to the viral G protein. *Journal of General Virology, 71*, 561-567.

[54] DeKinkelin, P., Geraed, J. P., Dorson, M. and Le Berre, M. (1977). Viral Haemorrhagic Septicemia: Demonstration of a protective immune response following natural infection. *Fish Health News, 6*, 43-45.

[55] Estepa, A., Fernandez-Alonso, M. and Coll, J. M. (1999). Structure, binding and neutralization of VHSV with synthetic peptides. *Virus Research, 63*, 27-34.

[56] Lorenzen, N. and LaPatra, S. E. (1999). Immunity to rhabdoviruses in rainbow trout: the antibody response. *Fish and Shellfish Immunology, 9*, 345-360.

[57] Ristow, S., Avila, J., La Patra, S. and Lauda, K. (1993). Detection and characterization of rainbow trout antibody against infectious hematopoietic necrosis virus. *Diseases of Aquatic Organism, 15*, 109-114.

[58] Lapatra, S. E., Fryer, J. L. and Rohovec, J. S. (1993). Virulence cmparison of different electrophero-types of infectious hematopoietic necrosis virus. *Diseases of Aquatic Organisms, 16*, 115-120.

[59] [59] Olesen, N. J., Lorenzen, N. and Jorgensen, P. E. V. (1993). Serological differences among isolates of viral haemorrhagic septicemia virus detected by neutralizing monoclonal and polyclonal antibodies. *Diseases of Aquatic Organisms, 16*, 163-170.

[60] Huang, C. J., Chien, M. S., Landolt, M. and Winton, J. (1994). Characterization of the infectious hematopoietic necrosis virus glycoprotein using neutralizing monoclonal-antibodies. *Diseases of Aquatic Organisms, 18*, 29-35.

[61] Bearzotti, M., Monnier, A. F., Vende, P., Grosclaude, J., DeKinkelin, P. and Benmansour, A. (1995). The glycoprotein of viral hemorrhagic septicemia virus (VHSV): antigenicity and role in virulence. *Veterinary Research, 26*, 413-422.

[62] Fernandez-Alonso, M., Lorenzo, G., Perez, L., Bullido, R., Estepa, A., Lorenzen, N. and Coll, J. M. (1998). Mapping of the lineal antibody epitopes of the glycoprotein of VHSV, a salmonid rhabdovirus. *Diseases of Aquatic Organisms, 34*, 167-176.

[63] Lefrancois, L. (1984). Protection against lethal viral infection by neutralizing and nonneutralizing monoclonal antibodies: distinct mechanisms of action *in vivo*. *Journal of Virology, 51*, 208-14.

[64] Fregeneda-Grandes, J. M. and Olesen, N. J. (2007). Detection of rainbow trout antibodies against viral haemorrhagic septicaemia virus (VHSV) by neutralisation test is highly dependent on the virus isolate used. *Diseases of Aquatic Organisms, 74*, 151-8.

[65] Wilson, M., Bengten, E., Miller, N. W., Clem, L. W., Du Pasquier, L. and Warr, G. W. (1997). A novel chimeric Ig heavy chain from a teleost fish shares similarities to IgD. *Proceedings of the National Academy of Sciences of the United States of America, 94*, 4593-7.

[66] Hordvik, I. (2002). Identification of a novel immunoglobulin delta transcript and comparative analysis of the genes encoding IgD in Atlantic salmon and Atlantic halibut. *Molecular Immunology, 39*, 85-91.

[67] Hansen, J. D., Landis, E. D. and Phillips, R. B. (2005). Discovery of a unique Ig heavy-chain isotype (IgT) in rainbow trout: Implications for a distinctive B cell developmental pathway in teleost fish. *Proceedings of the National Academy of Sciences of the United States of America, 102*, 6919-24.

[68] Utke, K., Kock, H., Schuetze, H., Bergmann, S. M., Lorenzen, N., Einer-Jensen, K., Kollner, B., Dalmo, R. A., Vesely, T., Ototake, M. and Fischer, U. (2007). Cell-

mediated immune responses in rainbow trout after DNA immunization against the viral hemorrhagic septicemia virus. *Developmental and Comparative Immunolo*gy. In press.

[69] Byon, J. Y., Ohira, T., Hirono, I. and Aoki, T. (2005). Use of a cDNA microarray to study immunity against viral hemorrhagic septicemia (VHS) in Japanese flounder (*Paralichthys olivaceus*) following DNA vaccination. *Fish and Shellfish Immunology, 18*, 135-147.

[70] Byon, J. Y., Ohira, T., Hirono, I. and Aoki, T. (2006). Comparative immune responses in Japanese flounder, *Paralichthys olivaceus* after vaccination with viral hemorrhagic septicemia virus (VHSV) recombinant glycoprotein and DNA vaccine using a microarray analysis. *Vaccine, 24*, 921-30.

[71] Yasuike, M., Kondo, H., Hirono, I. and Aoki, T. (2007). Difference in Japanese flounder, *Paralichthys olivaceus* gene expression profile following hirame rhabdovirus (HIRRV) G and N protein DNA vaccination. *Fish and Shellfish Immunology, 23*, 531-41.

[72] Kim, J. J., Bagarazzi, M. L., Trivedi, N., Hu, Y., Kazahaya, K., Wilson, D. M., Ciccarelli, R., Chattergoon, M. A., Dang, K., Mahalingam, S., Chalian, A. A., Agadjanyan, M. G., Boyer, J. D., Wang, B. and Weiner, D. B. (1997). Engineering of *in vivo* immune responses to DNA immunization via codelivery of costimulatory molecule genes. *Nature biotechnology, 15*, 641-6.

[73] Kim, J. J., Yang, J. S., Dentchev, T., Dang, K. and Weiner, D. B. (2000). Chemokine gene adjuvants can modulate immune responses induced by DNA vaccines. *Journal of Interferon and Cytokine Research, 20*, 487-498.

[74] Eo, S. K., Lee, S., Kumaraguru, U. and Rouse, B. T. (2001). Immunopotentiation of DNA vaccine against herpes simplex virus via co-delivery of plasmid DNA expressing CCR7 ligands. *Vaccine, 19*, 4685-4693.

[75] Lusso, P. (2002). HIV and chemokines: implications for therapy and vaccine. *Vaccine, 20*, 1964-1967.

[76] Jimenez, N., Coll, J., Salguero, F. J. and Tafalla, C. (2006). Co-injection of interleukin 8 with the glycoprotein gene from viral haemorrhagic septicemia virus (VHSV) modulates the cytokine response in rainbow trout (*Oncorhynchus mykiss*). *Vaccine, 24*, 5615-26.

[77] Sanchez, E., Coll, J. and Tafalla, C. (2007). Expression of inducible CC chemokines in rainbow trout (*Oncorhynchus mykiss*) in response to a viral haemorrhagic septicemia virus (VHSV) DNA vaccine and interleukin 8. *Developmental and Comparative Immunology, 31*, 916-26.

[78] Purcell, M. K., Smith, K. D., Hood, L., Winton, J. R. and Roach, J. C. (2006). Conservation of toll-like receptor signalling pathways in teleost fish. *Comparative Biochemistry and Physiology Part D: Genomics Proteomics, 1*, 77-88.

[79] van Duin, D., Medzhitov, R. and Shaw, A. C. (2006). Triggering TLR signalling in vaccination. *Trends in Immunology, 27*, 49-55.

[80] Steere, A. C., Sikand, V. K., Meurice, F., Parenti, D. L., Fikrig, E., Schoen, R. T., Nowakowski, J., Schmid, C. H., Laukamp, S., Buscarino, C. and Krause, D. S. (1998). Vaccination against Lyme disease with recombinant *Borrelia burgdorferi* outer-surface lipoprotein A with adjuvant. Lyme Disease Vaccine Study Group. *The New England journal of medicine, 339*, 209-15.

[81] Krieg, A. M. (2000). The role of CpG motifs in innate immunity. *Current Opinion in Immunology, 12*, 35-43.

[82] Klinman, D. M., Yamshchikov, G. and Ishigatsubo, Y. (1997). Contribution of CpG motifs to the immunogenicity of DNA vaccines. *Journal of Immunology, 158*, 3635-9.

[83] Sato, Y., Roman, M., Tighe, H., Lee, D., Corr, M., Nguyen, M. D., Silverman, G. J., Lotz, M., Carson, D. A. and Raz, E. (1996). Immunostimulatory DNA sequences necessary for effective intradermal gene immunization. *Science, 273*, 352-354.

[84] Carrington, A. C., Collet, B., Holland, J. W. and Secombes, C. J. (2004). CpG oligodeoxynucleotides stimulate immune cell proliferation but not specific antibody production in rainbow trout (*Oncorhynchus mykiss*). *Veterinary Immunology and Immunopathology, 101*, 211-22.

[85] Jorgensen, J. B., Zou, J., Johansen, A. and Secombes, C. J. (2001). Immunostimulatory CpG oligodeoxynucleotides stimulate expression of IL-1 beta and interferon-like cytokines in rainbow trout macrophages via a chloroquine-sensitive mechanism. *Fish and Shellfish Immunology, 11*, 673-682.

[86] Shen, K. Y., Hogstrand, C. and Davies, D. H. (2006). A newly identified CpG oligodeoxynucleotide motif that stimulates rainbow trout (*Oncorhynchus mykiss*) immune cells to produce immunomodulatory factors. *Developmental and Comparative Immunology, 30*, 311-24.

[87] Jorgensen, J. B., Johansen, A., Stenersen, B. and Sommer, A. I. (2001). CpG oligodeoxynucleotides and plasmid DNA stimulate Atlantic salmon (*Salmo salar* L.) leucocytes to produce supernatants with antiviral activity. *Developmental and Comparative Immunology, 25*, 313-321.

[88] Kanellos, T. S., Sylvester, I. D., Butler, V. L., Ambali, A. G., Partidos, C. D., Hamblin, A. S. and Russell, P. H. (1999). Mammalian granulocyte-macrophage colony-stimulating factor and some CpG motifs have an effect on the immunogenicity of DNA and subunit vaccines in fish. *Immunology, 96*, 507-10.

[89] Applequist, S. E., Rollman, E., Wareing, M. D., Liden, M., Rozell, B., Hinkula, J. and Ljunggren, H. G. (2005). Activation of innate immunity, inflammation, and potentiation of DNA vaccination through mammalian expression of the TLR5 agonist flagellin. *Journal of Immunolology, 175*, 3882-91.

[90] Takeshita, F., Tanaka, T., Matsuda, T., Tozuka, M., Kobiyama, K., Saha, S., Matsui, K., Ishii, K. J., Coban, C., Akira, S., Ishii, N., Suzuki, K., Klinman, D. M., Okuda, K. and Sasaki, S. (2006). Toll-like receptor adaptor molecules enhance DNA-raised adaptive immune responses against influenza and tumors through activation of innate immunity. *Journal of Virology, 80*, 6218-24.

[91] Zhu, J., Huang, X. and Yang, Y. (2007). Innate immune response to adenoviral vectors is mediated by both Toll-like receptor-dependent and -independent pathways. *Journal of Virology, 81*, 3170-80.

[92] Oshiumi, H., Tsujita, T., Shida, K., Matsumoto, M., Ikeo, K. and Seya, T. (2003). Prediction of the prototype of the human Toll-like receptor gene family from the pufferfish, *Fugu rubripes*, genome. *Immunogenetics, 54*, 791-800.

[93] Jault, C., Pichon, L. and Chluba, J. (2004). Toll-like receptor gene family and TIR-domain adapters in *Danio rerio*. *Molecular Immunology, 40*, 759-71.

[94] Tsujita, T., Tsukada, H., Nakao, M., Oshiumi, H., Matsumoto, M. and Seya, T. (2004). Sensing bacterial flagellin by membrane and soluble orthologs of Toll-like receptor 5 in

rainbow trout (*Onchorhynchus mykiss*). *Journal of Biological Chemistry, 279*, 48588-97.

[95] Rocha, A., Ruiz, S., Tafalla, C. and Coll, J. M. (2004). Conformation and fusion defective mutants in the hypothetical phospholipid-binding and fusion peptides of the protein G of viral haemorrhagic septicemia salmonid rhabdovirus. *Journal of Virology, 78*, 9115-9122.

[96] Nuñez, E., Fernandez, A. M., Estepa, A., Gonzalez-Ros, J. M., Gavilanes, F. and Coll, J. M. (1998). Phospholipid interactions of a peptide from the fusion-related domain of the glycoprotein of VHSV, a fish rhabdovirus. *Virology, 243*, 322-330.

[97] Estepa, A., Rocha, A., Pérez, L., Encinar, J. A., Nuñez, E., Fernandez, A., Gonzalez Ros, J. M., Gavilanes, F. and Coll, J. M. (2001). A protein fragment from the salmonid VHS rhabdovirus induces cell-to-cell fusion and membrane phosphatidylserine translocation at low pH. *Journal of Biological Chemistry, 276*, 46268-46275.

[98] Roche, S., Bressanelli, S., Rey, F. A. and Gaudin, Y. (2006). Crystal structure of the low-pH form of the vesicular stomatitis virus glycoprotein G. *Science, 313*, 187-91.

[99] Roche, S., Rey, F. A., Gaudin, Y. and Bressanelli, S. (2007). Structure of the prefusion form of the vesicular stomatitis virus glycoprotein G. *Science, 315*, 843-8.

[100] Ramesh, N., Shin, Y. K., Lau, S. and Osborne, W. R. (1995). High-level expression from a cytomegalovirus promoter in macrophage cells. *Human gene therapy, 6*, 1323-7.

[101] Artuc, M., Nurnberg, W., Czarnetzki, B. M. and Schadendorf, D. (1995). Characterization of gene regulatory elements for selective gene expression in human melanoma cells. *Biochemical and biophysical research communications, 213*, 699-705.

[102] Heppell, J., Lorenzen, N., Armstrong, N. K., Wu, T., Lorenzen, E., Einer-Jensen, K., Ahrens, P., Schorr, J. and Davis, H. L. (1998). Development of DNA vaccines for fish: vector design, intramuscular injection and antigen expression using viral haemorrhagic septicemia virus genes as model. *Fish and Shellfish Immunology, 8*, 271-287.

[103] Mayer, G. D., Leach, A., Kling, P., Olsson, P. E. and Hogstrand, C. (2003). Activation of the rainbow trout metallothionein-A promoter by silver and zinc. *Comparative Biochemistry and Physiology Part B: Biochemistry and Molecular Biology, 134*, 181-8.

[104] Hanley, S., Smith, T. J., Muller, F., Maclean, N., Uzbekova, S., Prunet, P. and Breton, B. (1998). Isolation and functional analysis of the histone H3 promoter from Atlantic salmon (*Salmo salar* L.). *Molecular marine biology and biotechnology, 7*, 165-72.

[105] Chan, W. K. and Devlin, R. H. (1993). Polymerase chain reaction amplification and functional characterization of sockeye salmon histone H3, metallothionein-B, and protamine promoters. *Molecular marine biology and biotechnology, 2*, 308-18.

[106] Hwang, G. L., Rahman, A., Razak, S. A., Sohm, F., Farahmand, H., Smith, A., Brooks, C. and Maclean, N. (2003). Isolation and characterisation of tilapia ß-actin promoter comparison of its activity with carp ß-actin promoter. *Biochimica and Biophysica Acta, 1625*, 11-18.

[107] Lewandoski, M. (2001). Conditional control of gene expression in the mouse. *Nature Review Genetics, 2*, 743-755.

[108] Ristevski, S. (2001). Transgenic studies in the mouse. Improving the technology towards a conditional temporal and spatial approach. *Methods in Molecular Biology, 158*, 319-334.

[109] Gao, X., Kemper, A. and Popko, B. (1999). Advanced transgenic and gene-targeting approaches. *Neurochemical Research, 24*, 1181-1188.

[110] Tafalla, C., Estepa, A. and Coll, J. M. (2006). Fish transposons and their potential use in aquaculture. *Journal of Biotechnology, 123*, 397-412.

[111] Ivics, Z., Kaufman, C. D., Zayed, H., Miskey, C., Wallsko, O. and Izsvak, Z. (2004). The sleeping beauty transposable element: evolution, regulation and genetic applications. *Current Issues in Molecular Biology, 6*, 43-56.

[112] Izsvak, Z.,Ivics, Z.andPlasterk, R. H. (2000). Sleeping beauty, a wide host-range transposon vector for genetic transformation in vertebrates. *Journal of Molecular Biology, 302*, 93-102.

[113] Dupuy, A. J., Clark, K., Carlson, C. M., Fritz, S., Davidson, A. E., Markley, K. M., Finley, K., Fletcher, C. F., Ekker, S. C., Hackett, P. B., Hom, S. and Largaespeda, D. A. (2002). Mammalian germ-line transgenesis by transposition. *Proceedings of the National Academy of Sciences of the United States of America, 99*, 4495-4499.

[114] Horie, K., Yusa, K., Yae, K., Odajima, J., Fisher, S. E., Keng, V. W., Hayakawa, T., Mizuno, A., Kondoh, G., Ijiri, T., Matsuda, Y., Plasterk, R. H. and Takeda, J. (2003). Characterization of sleeping beauty transposition and its application to genetic screening in mice. *Molecular Cell Biology, 23*, 9189-9207.

[115] Masuda, K.,Yamamoto, S.,Endoh, M.andKaneda, Y. (2004). Transposon-independent increase of transcription by the sleeping beauty transposase. *Biochemical and Biophysical Research Communications, 317*, 796-800.

[116] Ivics, Z., Izsvak, Z. and Hackett, P. B. (1997). Molecular reconstruction of sleeping beauty, a Tc1-like transposon system from fish and its transposition in human cells. *Cell, 91*, 501-510.

[117] Reed, K. M. (1999). Tc1-like transposable elements in the genome of lake trout (*Salvelinus namycush*). *Marine Biotechnology, 1*, 60-67.

[118] Leaver, M. J. (2001). A family of Tc1-like transposons from the genomes of fishes and frogs: evidence for horizontal transmission. *Gene, 271*, 203-214.

[119] Lam, W. L., Lee, T. S. and Gilbert, W. (1996). Active transposition in zebrafish. *Proceedings of the National Academy of Sciences of the United States of America, 93*, 10870-10875.

[120] Li, T., Zhang, Y., Fu, L., Yu, C., Li, X., Li, Y., Zhang, X., Rong, Z., Wang, Y., Ning, H., Liang, R., Chen, W., Babiuk, L. A. and Chang, Z. (2005). siRNA targeting the Leader sequence of SARS-CoV inhibits virus replication. *Gene Therapy, 12*, 751-761.

[121] Baltayiannis, G., Filis, S., Tsianos, E. and Karayiannis, P. (2005). Small interfering RNA(siRNA)-mediated inhibition of hepatitis C virus replication using bicistronic vector *in vivo* (cell line Huh-7) and *in vitro*. *Journal of Hepatology, 42*, 154-154.

[122] Korf, M., Jarczak, D., Beger, C., Manns, M. P. and Kruger, M. (2005). Inhibition of hepatitis C virus translation and subgenomic replication by siRNAs directed against highly conserved HCV sequence and cellular HCV cofactors. *Journal of Hepatology, 43*, 225-234.

[123] Barik, S. (2004). Control of nonsegmented negative-strand RNA virus replication by siRNA. *Virus Research, 102*, 27-35.

[124] Schyth, B. D., Lorenzen, N. and Pedersen, F. S. (2006). Antiviral activity of small interfering RNAs: specificity testing using heterologous virus reveals interferon-related effects overlooked by conventional mismatch controls. *Virology, 349*, 134-41.

[125] Schyth, B. D., Lorenzen, N. and Pedersen, F. S. (2007). A High throughput in vivo model for testing delivery and antiviral effects of siRNAs in vertebrates. *Molecular Ther*apy

[126] Wang, E., Sun, X., Qian, Y., Zhao, L., Tien, P. and Gao, G. F. (2003). Both heptad repeats of human respiratory syncytial virus fusion protein are potent inhibitors of viral fusion. *Biochemical and biophysical research communications, 302*, 469-75.

[127] Kang, T. H., Lee, J. H., Noh, K. H., Han, H. D., Shin, B. C., Choi, E. Y., Peng, S., Hung, C. F., Wu, T. C. and Kim, T. W. (2007). Enhancing dendritic cell vaccine potency by combining a BAK/BAX siRNA-mediated antiapoptotic strategy to prolong dendritic cell life with an intracellular strategy to target antigen to lysosomal compartments. *Internation Journal of Cancer, 120*, 1696-703.

[128] Huising, M. O., Guichelaar, T., Hoek, C., Verburg-vanKemenade, B. M., Flik, G., Savelkoul, H. F. J. and Rombout, J. H. W. M. (2003). Increased efficacy of immersion vaccination in fish with hyperosmotic pretreatment. *Vaccine, 21*, 4178-4193.

[129] Tucker, C., Endo, M., Hirono, I. and Aoki, T. (2000). Assessment of DNA vaccine potential for juvenile Japanese flounder *Paralichthys olivaceus*, through the introduction of reporter genes by particle bombardment and histopathology. *Vaccine, 19*, 801-9.

[130] Romoren, K., Thu, B. J. and Evensen, O. (2002). Immersion delivery of plasmid DNA II. A study of the potentials of a chitosan based delivery system in rainbow trout (*Oncorhynchus mykiss*) fry. *Journal of Controlled Release, 85*, 215-225.

In: Aquaculture Research Trends
Editor: Stephen H. Schwartz, pp. 237-261
ISBN 978-1-60456-217-0

Chapter 7

DIPLODUS PUNTAZZO, A NEW SPECIES IN AQUACULTURE: BIOLOGY AND CULTURE

María Dolores Hernández[(1)*]***, Francisco Javier Martínez***[(2)] ***and Benjamín García García***[(1)]

(1) IMIDA-Acuicultura. Consejería de Agricultura y Agua de la Región de Murcia. Apdo. 65. 30740. San Pedro del Pinatar. Murcia. Spain.
(2) Department of Physiology. Faculty of Biology. University of Murcia. Campus de Espinardo. 30100. Murcia. Spain.

ABSTRACT

Diversifying the supply of cultured fish would contribute to stabilizing market prices and making aquacultural facilities profitable, while offering consumers a larger variety of cultured fish at more competitive prices than those available from extractive fishing. As a result, the quest for new fish species suitable for farming is becoming a priority in aquacultural research.

Sharpsnout seabream is a sparid fish that has recently drawn the attention of researchers and farmers alike, as it has many qualities that would make it a promising new species for culture. Pilot pre-growout and growout studies have been conducted, under both intensive and extensive culture conditions, with quite encouraging results. These results are very similar to those obtained for gilthead seabream in terms of both growth and conversion rates. In addition, sharpsnout seabream is an omnivorous species that has yielded very positive results in pilot studies designed to test the possibility of replacing fish meal for soybean meal (with replacement levels of up to 60%) as a source of dietary protein. Therefore, one may expect sharpsnout seabream commercial diets to be more economical than those for gilthead seabream, European sea bass, or other carnivorous species.

A number of commercial issues must be addressed before considering the large-scale farming of sharpsnout seabream. Consumers are not very familiar with sharpsnout seabream, as the specimens caught by fishermen tend to be locally consumed in coastal Mediterranean towns.

* Corresponding author. Phone: 34-968-184518; Fax: 34-968-184518; E-mail: mdolores.hernandez6@carm.es.

INTRODUCTION

Over the last few years, research in the field of marine culture have been directed, to a large extent, towards diversifying the number of species that can be farmed, progressively refining the techniques used to raise them in captivity. Up until now, marine aquaculture in the Mediterranean and the Canary Islands has centered almost exclusively on the culture of two species, gilthead seabream and European sea bass, with turbot being raised in Galicia. Farming technology has been firmly consolidated for these species, having reached high production levels in barely a decade, particularly in the cases of gilthead seabream and turbot.

However, the spectacular growth experienced by gilthead seabream farming is causing a saturation of the market that would most likely lead to reduced production if current issues are not resolved. In fact, sales prices for gilthead seabream have gone down practically reaching their cost price (Gonzalez-Laxe, 2000). Faced by this situation, besides expanding the market and reducing production costs, one of the main strategies proposed has been to diversify the species being raised (Stephanis & Divanach, 1993), as this strategy would give farmers greater flexibility to respond to changes in market demand (Sweetman, 1992). Furthermore, diversification of the farmed fish offering could help stabilize the prices and make facilities more profitable, which is why research on new species has become a priority in aquaculture.

Although all the Mediterranean sparidae appear as potentially interesting species for culture, special consideration has been given to bream species, due to their satisfactory market price, good growth performance and easy adaptation to captivity. Among the breams, the sharpsnout seabream is of particular interest, as it has a number of qualities that would make it a promising species for farming (Abellán & García-Alcázar, 1995). Its culture on an experimental basis began in Cyprus in 1988. Larval rearing did not posed any major, unsurpassed difficulties. The growth rate at the pre-ongrowing and ongrowing stages was similar to that of gilthead seabream, and slightly better than that of seabass. The market potential was shown also very promising, a factor which play a decisive role in the adoption of its culture by the commercial fish farms (Georgiu & Stephanou, 1995).

Although this species is already farmed on a small scale in some Mediterranean countries (primarily in Greece, where 1500 Tm were produced in 2000) (Fernández & Villanueva, 2000), as typically occurs with many coastal species with populations that are not overly abundant, sharpsnout seabream is largely unknown in the big markets. It tends to be consumed locally, where it is caught by fishermen using fishing poles or other traditional techniques common in certain areas. Nonetheless, the lack of familiarity with this species in the most important markets is practically the same as it was only 15 years ago for gilthead seabream, which good marketing has turned into one of the species with the greatest demand throughout all the major Mediterranean markets. Furthermore, sharpsnout seabream is already highly valued in certain Mediterranean countries such as Italy, where it reaches high sales prices. Therefore, production for export should not be ruled out, and it may be a good option until such time as local markets accept this product on a larger scale.

Most species of currently farmed fish have carnivorous eating habits. This is the case of gilthead seabream, European sea bass, turbot, trout and tuna, which together represent nearly 96% of the total fish production in Spain. The culture of all these species is currently dependent on the use of fish meal or other fishing resources as the main source of protein for feed production. The overexploitation of marine resources, however, is making fish meal an increasingly scarce commodity. As a result, the availability and price of fish meal is beginning to be a limiting factor in the farming of carnivorous species. Therefore, besides finding alternative protein sources to replace fish meal, the culture of commercially attractive species with omnivorous eating habits, such as sharpsnout seabream, would also be of interest.

PHYLOGENETIC POSITION: MORPHOLOGIC CHARACTERISTICS

The Sparidae comprise a total of about 100 species, which are found predominantly in the coastal waters of tropical and temperate zones, with their center of diversity being in the Atlantic Ocean. The systematics of the Sparidae family still remains unclear. Their classification has so far been based on solely morphological criteria (Bauchot & Hureau, 1986; Smith & Smith, 1986). Fiedler (1991) distinguished three subfamilies: the Sparinae, comprising all Diplodus species, the Denticinae, and the Boopsinae, based mainly on their dentition and diet. Also according to their trophic specialization, Smith & Smith (1986) named the Pagellinae as a fourth subfamily, but Hanel & Sturmbauer (2000) pointed to the urgent need for a revision of the Sparidae by a careful re-evaluation of anatomical traits.

Diplodus puntazzo, Cetti 1777 (= *Puntazzo puntazzo,* Gmelin 1789) therefore belongs to the Sparid family. The nomenclature is often confusing in the literature. Tortonese (1973) refers to this species using several different names: *Sparus puntazzo, S. annularis, S. acutirostris, Charax puntazzo* and *Sargus puntazzo*. In view of the doubts expressed by Valenciennes with regards to assigning this sparid to a particular monotypic genus (*Puntazzo puntazzo* or *Charax puntazzo*), he refers to it as *Diplodus puntazzo*. Cadenat (1964) states that, aside from the great similarity of many other characters, the dental structure is typical of *Diplodus*, and is more reduced than that of *Puntazzo*. Bini (1968) also refers to the remarkable similarities with *Diplodus*, from which it may otherwise be distinguished by the pointed shape of its snout, which has a concave, rather than a convex, profile (Faranda et al., 1985).

Sharpsnout seabream most prominent characteristic is its long, pointed snout, a small mouth with eight very prominent incisors at the front part of the jaw, a single row of molars and no eye teeth whatsoever. Its head is compressed, with a concave profile, while its body is tall, oval-shaped and laterally compressed. Its maximum length is between 45-60 cm and it is silver-greyish in color, with 7-9 incomplete black transversal stripes, a black ring around the caudal peduncle and a blackish band on its forehead (Fig. 1).

Its habitat is found on the rocky, algae-covered bottoms and sometimes in *Posidonia oceanica* prairies. Adult specimens often appear isolated between banks of other sparids. *Diplodus puntazzo* is found throughout the Mediterranean Sea, the Black Sea and the Eastern

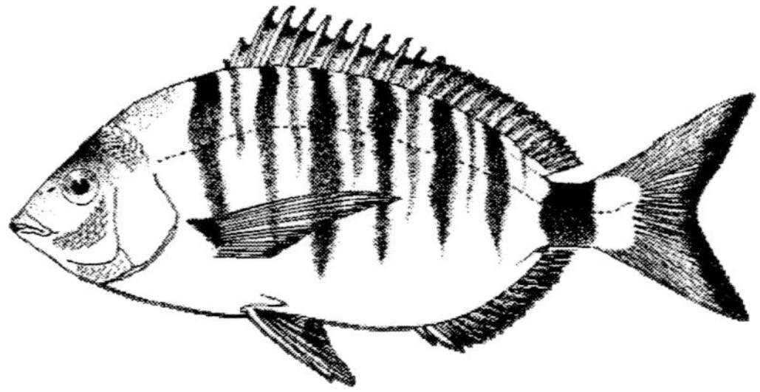

Fig. 1. *Diplodus puntazzo* (Whitehead et al., 1986).

coasts of the Atlantic Ocean, from the Bay of Biscay to Sierra Leone, the Canary Islands and Cape Verde (Kouttouki et al., 2006). In the northwestern Mediterranean waters, this genus is predominately found in shallow, rocky sublittoral habitats (Sala & Ballesteros 1997).

The presence of natural geographic barriers contributes to the emergence of differences between populations. Several authors (e.g., Borsa et al., 1997; Naciri et al., 1999) have reported a major phylogeographic break due to the Straits of Gibraltar, and the existence of significant differences between the Atlantic and Mediterranean populations of sharpsnout seabream has been confirmed (Palma & Andrade, 2002). In addition, some degree of differentiation, albeit with a lower level of significance, has also been found between Western and Eastern Mediterranean samples (Palma & Andrade, 2002).

Feeding Behavior

Omnivorous in its feeding habits, algae represent sharpsnout seabream most important prey type, followed by sponges and cnidarians. Its sharp teeth help it to cut algae and its molars to grind crustaceans, snails and molluscs. It exhibits morphological differences, having a longer intestine than either of the other species of *Diplodus*. It therefore appears to have adapted to feeding on epibenthic "low-digestible" organisms, such as algae and sponges (Sala & Ballesteros, 1997). Sharpsnout seabream takes advantage of a resource like sponges that apparently goes unused by other coastline fish species, in spite of the fact that these are important sources of prey in other environments, such as for certain reef-dwelling fish species (Mena Sellés & García García, 2002).

Rosecchi (1987) showed that these fish have very diverse diets throughout their lives, and although there are marked changes in the types of invertebrate prey targeted (amphipods, isopods and larvae for small fish; decapods, molluscs, echinoderms and polychaetes for larger fish), the fish may actually be selecting preys of relatively similar trophic levels. Furthermore, Sala & Ballesteros (1997) have demonstrated that even large *Diplodus* specimens tend to eat algal material. This may be one reason why we failed to find a trophic level increase for any of these wide-ranging omnivorous species (Deudero et al., 2004).

Reproduction in the Wild

Hermaphroditism in sharpsnout seabream was first reported by Syrski (1876). D'Ancona (1949) suggested a lack of sex inversion for this species, whereas according to Lissia-Frau & Pala (1968), protandrous individuals can be found, in addition to rudimentary hermaphrodites (either primary males or females) that never change their sex. All fish examined possessed bisexual gonads (ovotestes), consisting of a dorsomedial ovarian portion and a ventrolateral testicular portion, distinctly separated by a thin layer of connective cells and fibers. This indicates that *D. puntazzo* is an ambisexual species. The relative proportion of the two sexual territories within the ovotestis may provide a valuable indication as to the individual's functional sex, as one sexual area always predominates over the other throughout the annual cycle. Micale et al. (1996) studied two-year old specimens and found that 53% of the fish presented gonads with ovarian dominance and 47% with testicular dominance throughout the year. In this same study, where the ovarian and testicular cycles are described in detail, three distinct reproductive cycle phases were established for *D. puntazzo* according to the maturity of the gonad: a latent phase, which lasted from October to July for males, and from October to August for females; a development and pre-egg-laying stage, which occurred in August for males and in September for females; with a final egg-laying stage occurring between September and October. Some studies have shown early signs of differentiation in bisexual gonads, resulting in individuals that are predominantly male or female, and which do not subsequently change their sex. Sexual maturity is reached at two years of age (Georgiou & Stephanou, 1995; Hernández et al., 2003 (Fig.2)).

The natural spawning period in the Mediterranean Sea takes place from September to November (Micale et al., 1996). In the Paphos area (southwest of Cyprus), spawning starts when the water temperature reaches 23°C and stops when it falls below 20°C. In addition, it always occurs at the same time of the day: from 12 to 3 pm (Marangos, 1995).

Growth and Age

Very few studies have been conducted on the growth of wild sharpsnout seabream. Domínguez-Seoane (2006) studied their growth in the Canary Islands, and the Table 1 shows the parameters of individuals studied in this paper.

Lengths at age were described by specialised von Bertalanfy growth model, the Schnute growth model and the seasonalised von Bertalanfy growth model (Table 2).

No significant differences were found between the von Bertalanfy growth parameters of males and females, with males growing at slightly faster rate than females. The pattern exhibited by *D. puntazzo* is characterized by a high or moderate asymptotic length, and low or moderate growth coefficient. The seasonalised growth model reveals a reduction of somatic growth in winter period. Individuals grow fast during their immature first year, attaining approximately 45% of their maximum length. After the first year, their annual growth rate drops rapidly. This change in growth rate may be attributable to the use of available energy for reproduction instead of somatic growth, since in the studied area, the maturation process begins during the third year of life (Domínguez-Seoane et al., 2006). This is in agreement

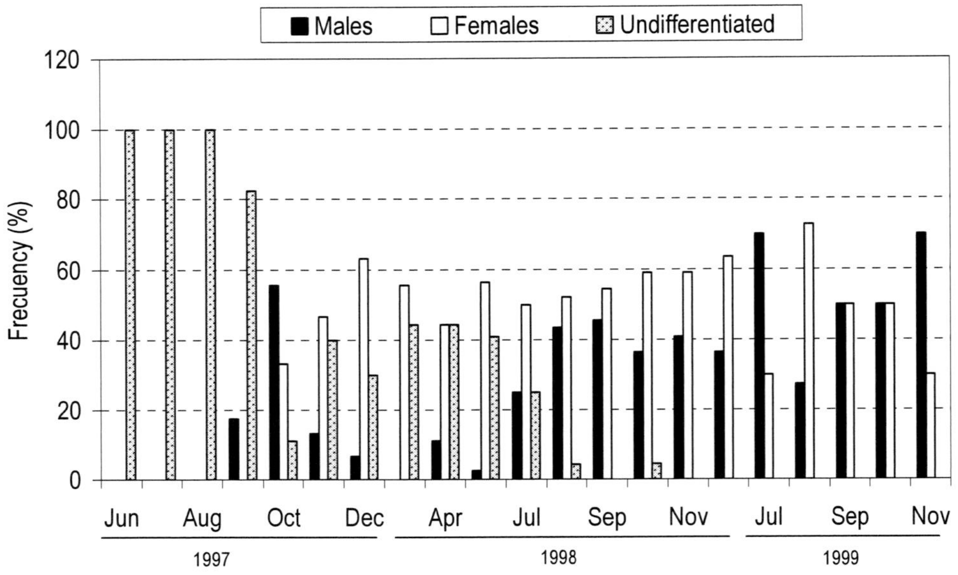

Fig. 2. Sex frequency distribution over three reproductive cycles.

Table 1. The number (n), minimum (min), maximum (max), and standard deviation (S.D.) values of the total length and the total weight for immatures, males, females and intersexual individuals of D. puntazzo from the Canary Islands sampled by Domínguez-Seoane et al. (2006).

	Total length (mm)				Total weight (g)				n
	Min.	Max.	Mean	S.D.	Min.	Max.	Mean	S.D.	
Immatures	140	365	204	47	60	858	178	128	155
Males	169	510	326	66	84	2057	616	356	168
Females	159	509	333	62	73	2187	662	350	348
Intersexuals	259	466	323	72	276	1542	591	450	23

Table 2. Parameters and standard errors for the specialised von Bertalanffy, Schnute, and the seasonalised von Bertalanffy growth models for all fish of D. puntazzo of the Canary Islands (Domínguez-Seoane et al., 2006).

Parameter	Estimate	Standard error
Specialised von Bertalanffy growth model ($r^2 = 0.947$)		
L_∞ (mm)	541	10.32
k ($year^{-1}$)	0.182	0.006
t_0 (year)	-2.531	0.093
Schnute growth model ($r^2 = 0.924$)		
y_1 (mm)	171	6.08
y_2 (mm)	509	
a	0.022	0.001
b	2.59	0.083
Seasonalised von Bertalanffy growth model ($r^2 = 0.932$)		
L_∞ (mm)	546	10.68
k ($year^{-1}$)	0.176	0.008
t_0 (year)	-2.21	0.097
C	0.712	0.031
t_s	0.798	0.046

with the observations made by García García et al. (2001) and Gatland (1995) of two growth-limiting circumstances: if water temperatures are below 15°C during the winter, the fish growth slows down or even stops; and during the reproductive season that runs from September to November and during the second year of ongrowing, when it also stops.

Broodstock

One of the advantages of culture sharpsnout seabream is that its spawning period in the Mediterranean occurs before that of gilthead seabream (January - March) and sea bass (February - March), which allows optimizing the use of fish hatcheries (Marangos, 1995).

Experiments on sharpsnout seabream reproduction in captivity conducted in different areas of the Mediterranean, have proved that temperature is the most important environmental factor for this species, and one that conditions both the maturation process as well as egg-laying (Marangos, 1995). Females lay eggs sequentially provided the temperature is kept within a narrow range around 20°C. If temperature variations occur, lasting for even just a few days, the egg-laying irreversibly stops. In addition, some overhydrated females may die if they are unable to reabsorb their eggs (Georgiou & Stephanou, 1995).

One of the most important factors in fish farming is that of controlling the spawning pattern of the cultured species in order to have a synchronized production of high-quality eggs to be able to provide ongrowing farms with a constant supply of fry. For this reason, although most experiments on sharpsnout seabream reproduction in captivity have been conducted using spontaneous egg-laying, there have also been different attempts of inducing egg-laying by injecting human chorionic gonadotropin (HCG) or gonadotropin releasing hormones. Such studies have obtained various unsatisfactory results, in most cases (Faranda et al., 1985; Greco et al., 1993; Georgiou & Stephanou, 1995). Since it was observed that temperature plays a predominant role in this species' spawning, temperature shocks have been used to induce *D. puntazzo* broodstock to spawn, however, these attempst have also failed to produce any positive results so far (Georgiou & Stephanou, 1995).

Larval Culture

Compared to other Sparidae, *D. puntazzo* is characterized by having a relatively small egg diameter and body size at hatching or feeding onset (Kouttouki et al., 2006). One gram of eggs contained 2250 eggs. Four hundred and seventy grams of eggs yielded 371000 larvae, with a hatching efficiency of 35% (Franicevic, 1989).

Immediately after release, the ripe eggs were spherical, white and transparent, with a very resistant chorion and a mean diameter of 845 μm. The eggs floated in sea water, due to the presence of an oil droplet with a diameter of between 180 and 200 μm (Table 3; Faranda et al., 1985). The length of the incubation period has been observed to vary with temperature, the best fertilization and hatching values being obtained at 22 °C. Upon emerging from the egg, the larvae measured around 1.69 mm. By the fourth day, the yolk sac was completely reabsorbed, although a residue from the lipid droplet was still visible. After the fifth or sixth day of life, the mouth functioned perfectly, the digestive tract was completely formed, and the larvae were able to feed (Faranda et al., 1985). Fin development has also been widely studied (Sfakianakis et al., 2005). Primary swim bladder inflation starts on day 5-6 and ends on day 8-9 (Marangos, 1995).

In farming operations, since the eggs are pelagic, they are collected in 500 μm nets placed on the surface overflow rims on each breeding tank. The fertilized egg (which floats on the surface) is incubated in cylindro-conical tanks at the temperature at which it was laid. The total lack of agonistic behavior makes it easy to attain high levels of larval survival, with no excessive complications either as the larvae are raised (Pastor et al., 2000).

Methods for determining biochemical egg quality have recently been published for *Diplodus puntazzo,* demonstrating that the percent of hatched larvae is correlated to carbohydrate metabolism parameters and lytic enzyme activities within the eggs during early developmental stages (Lahnsteiner & Patarnello, 2004). Neither the size or shape of the lipid

Table 3. Egg viability parameters and morphometric parameters for *Diplodus puntazzo* and those for other species (Faranda et al., 1985).

	Diplodus puntazzo	*Diplodus sargus*	*Sparus aurata*	*Dicentrarchus labrax*
Egg diameter (μm)	845	1115	931	1167
Oil droplet diameter (μm)	190	210	234	415
Eggs/day	1980	1390	1389	1053
Embryogenesis				
1st cleavage (h after fertilization)	1.05	1.50	1.40	1.30
Hours to hatching	24	64	51	110
Total length of larva (mm)	1.69	2.6	2.3	3.5
Yolk sac reabsorption (days)	3.5	5	5	5
Hatching temperature (°C)	22	16.8	21.0	15.0

vesicle, nor the mean egg diameter experienced any change during embryonic development. However, other changes did occur that have been attributed to two different processes: egg activation and embryonic development. The changes caused by egg activation occurred within the first 2h after fertilization, and included a decrease in mean yolk diameter, changes in the egg shape and an increase in the ratio of the mean diameter of the egg to the mean diameter of the yolk. The decrease in mean yolk diameter within the first 10 min after activation in sea water might indicate dehydration by sea water. Morphometric parameters, such as the ratio of the maximum to the minimum diameter of the lipid vesicle and the lipid vesicle shape coefficient, correlated with larval survival rates from viable eggs (Lahnsteiner & Patarnello, 2005).

The techniques used to cultivate sharpsnout seabream larvae are similar to those traditionally used for gilthead seabream, with some minor variations depending on the authors (Franicevic, 1989; Marangos, 1995). After hatching, *D. puntazzo* larvae were stocked in 1 to 3 m^3 rearing tanks, with a larval density of 50-65 l^{-1}. The water temperature ranged between 20-22°C, the salinity between 37-40‰ and the oxygen level between 70-100% of saturation. The culture water was renewed at a flow rate of 30% d^{-1}. It should also be observed that the reliability of the mesocosm technology for *D. puntazzo* larval rearing has been previously verified (Papandroulakis et al., 2004).

The best survival rates were found under dark conditions (Korkut et al., 2006). The larvae have a massive yolk sac containing the vitelline reserves and the lipid globule. This has been attributed to the transformation of a greater share of the total energy obtained from absorption into metabolic energy under illuminated conditions. It has been observed that illumination affects the speed of yolk sac consumption in sharpsnout seabream larvae in the lecithotrophic phase. It has also been suggested that illumination affects how the larvae use the energy they obtained from the absorption of the yolk sac.

Kouttouki et al. (2006) pointed out two main trends in shape ontogeny: (a) the transition from an elongated body to a deeper, more compressed profile, and (b) a negative allometric growth of the caudal peduncle area. Other important trends in shape ontogeny included the anterior transposition of the notochord posterior tip, the positive allometric growth of the caudal fin and abdominal area, and the anterior shift of the snout and eye.

At the beginning of external feeding, sharpsnout seabream larvae scan the surrounding environment in search of a signal for the presence of food using the olfactory ciliated and microvillate sensory cells. This does not exclude the possible role played by vision in sharpsnout seabream larval feeding behavior. In fact, larvae from other sparids are considered to be primarily visual feeders, with olfaction and the developing lateral line providing significant inputs to visually-oriented feeding. At the beginning of exogenous feeding, a high density of rapidly moving prey items should be offered to sharpsnout seabream larvae, as the frequency of vibratory movements and the density of prey is clearly important for attack and capture by the predators. The post-larval phase is characterized by a general improvement in mechanic and chemical responsiveness: the feeding behavior of post-larvae becomes polysensory as a consequence of such sensory implementation. During the post-larval phase of sharpsnout seabream, chemoreception (associated with food location) is more closely connected to olfactory neurons than to inner mechano-chemical receptors. The latter intervene in determining food palatability (Boglione et al., 2003).

The following larva feeding regimen was used (Marangos, 1995): the green water technique was applied during the first 30 days, using the algae *Chlorella sp.* at a concentration de 15×10^5 cells·ml^{-1}. Rotifers (*Brachionus plicatilis)* were provided from day 3 to day 40. The rotifers were cultured on bakers' yeast and algae. Before they were provided to the larvae, the rotifers were enriched with *Chlorella* or commercial enrichment diets that are highly rich in unsaturated fatty acids (C 20:5n-3 and C 22:6n-3). From day 3 to day 12, the larvae were fed small rotifers (80-150μm) and from day 12 to day 40, mixed rotifers (80-250μm). The concentration of rotifers in the rearing tanks was kept at 5-7 ind.·ml^{-1}. From day 20 to day 40, the larvae were fed freshly-hatched Artemia nauplii, and from day 30 to 60, they were fed enriched Artemia metanauplii. In addition to enriched rotifers, Artemia nauplii and enriched Artemia metanauplii, inert food was given on day 35. From day 40 to day 60, only dry food and enriched Artemia metanauplii were provided.

Larval survival rates of between 4% and 10% were achieved (Franicevic, 1989), with a rate of 18-22% (Marangos, 1995) at 60 days and 30-35% on day 70 (Pastor et al., 2000). The growth of the larvae is comparable to that obtained for gilthead seabream and European sea bass, and the feed appears to be adequate, with the fish reaching weights of 0.15 g at 60 days. A good acceptance of the artificial feed was also observed (Kenturi et al., 1992).

ONGROWING

The ongrowing phase for this species until it reaches its commercial size (350-400 g) has been carried out in both floating cages (Bermúdez et al., 1989; Gatland, 1995) and cultivation tanks (Kenturi et al., 1992).

The weight increases obtained in the various studies on the ongrowing of sharpsnout seabream are slightly lower than those for gilthead seabream, which reach their commercial size at 16 months. They are, however, very similar to that of sea bass (22 months) and snapper (20 months). Sharpsnout seabream gains 225 g in 10 months (between November and September), increasing its weight from 60 to 285 g (Faranda et al., 1983). Similar results were obtained by Divanach et al. (1993), with the fish reaching their commercial size in 20 months. However, in an experimental culture in cages on the Mar Menor coastal lagoon (Spain), Bermúdez et al. (1989) found a very similar growth rate for this species as that of gilthead seabream throughout the entire study period (12 months).

Sharpsnout seabream showed better growth performance in cages than they did in tanks, as is the case for *Sparus aurata* (Gatland, 1995). Growth performance was best when the fish were reared in a polyculture system with seabass and seabream in tanks (Sara et al., 1999), but not so much when the polyculture was done in cages. The main reason for this may be that sharpsnout seabream also feed on the bottom, and therefore can recuperate the food that is left uneaten by seabream on that part of the tank, whereas in cages this clearly cannot occur (Favaloro et al., 2002). The stocking ratio for the most effective duoculture rearing in a recirculated water system seems to be 40% *Diplodus puntazzo* – 60% *Sparus aurata* (Karakatsouli et al., 2006). This specific combination not only improved the growth rate of *D. puntazzo,* as compared to a monoculture system, but also reduced the fish aggressiveness and increased the biomass.

As can be seen in Figure 3, this species demonstrates rapid growth in the spring and summer months (May-September), and slow growth rates from October to April (Hernández et al., 2003). Growth rates, and as a result, the duration of the cultivation depends on the temperature range in the region where the ongrowing takes place.

Hernández et al. (2003) examined the seasonal condition and body composition changes for sharpsnout seabream raised in captivity. The condition factor, hepatosomatic index and intraperitoneal fat ratio all decreased (CF: 1.83-2.51, HSI: 0.80-2.47%, IPF ratio: 1.27-2.95%, minimum and maximum values, respectively) during the warmer months of the year. There were inverse correlations between the relative fat and moisture contents associated with growth periods and sexual maturation (fat: 13.5-18.5%, moisture: 59.9-64.6%), whereas mineral and protein proportions were more stable (ash: 3.8-4.8%, proteins: 17.5-21.4%).

Compensatory growth is defined as a rapid growth induced by a hyperphagic behavior after a period of starvation. Egea Nicolás et al. (2002) studied this phenomenon in sharpsnout seabream after starving the fish for 17 days. During the re-feeding period (18 days), the growth rate doubled that of the control groups, with feed intake also increasing, albeit the re-fed fish did not reach the final weight of the control fish.

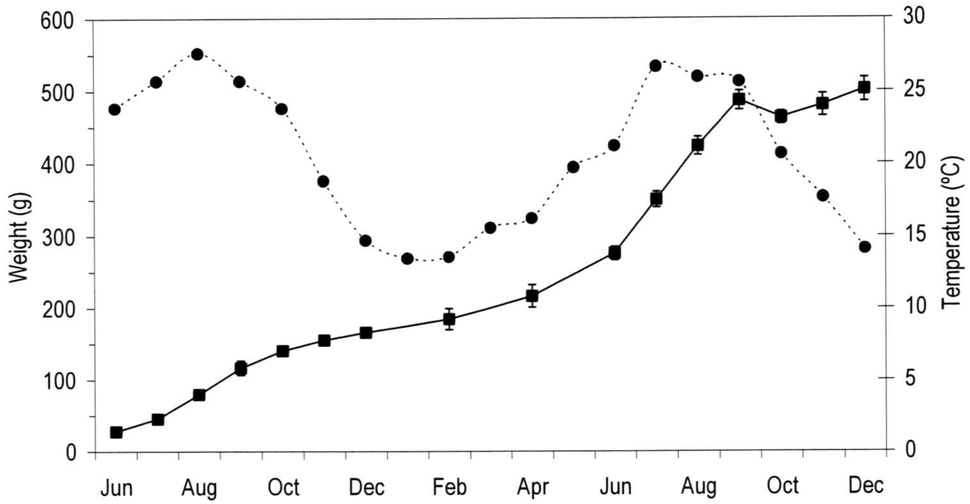

Fig. 3. Sharpsnout seabream body weight evolution throughout the entire ongrowing process.

Bioeconomic model

García García et al. (2001) studied the growth and conversion index of sharpsnout seabream under intensive ongrowing conditions in experimental tanks over 18 months. The fry were classified according to their weight into one of three sizes, with the large-sized group making up 24% of the culture population, the medium-sized group 61%, and the small-sized group 15%. For all three groups, the relationship between growth rate and ongrowing time could be best-fitted by a 6th order polynomial model ($p<0.001$; $R^2>0.997$), whereas the conversion index versus time relationship was well-fitted by a linear model of the form $Y=a+bLnX$ ($p<0.05$; $R^2>0.605$). The large specimens reached their commercial size (body weight of 300 g) at 11 months, with a GI of 1.85, the medium-sized specimens at 12 months, with a GI of 1.86, and the small specimens at 14 months, with a GI of 2.03. Based on these data, the authors adjusted the AGR (absolute growth rate, g/fish/day) and AFR (absolute feeding rate, g/fish/day) data to match the model: $LnY=Lna+b*LnW+c*T+d*T^2$, where Y is the AGR (or AFR), W is the body weight and T is the temperature. These models estimate the temperature at which the maximum AGR (27.5ºC) and AFR (28ºC), as well as the maximum feed efficiency (FE, 27ºC), values are reached (Fig. 4).

Furthermore, based on the previous data, García García et al. (2003) developed a mathematical viability/profitability model in order to estimate those economic parameters linked to production that had not previously been established at a commercial culture level for sharpsnout seabream. These included such factors as the final sales price (SP), and the cost of

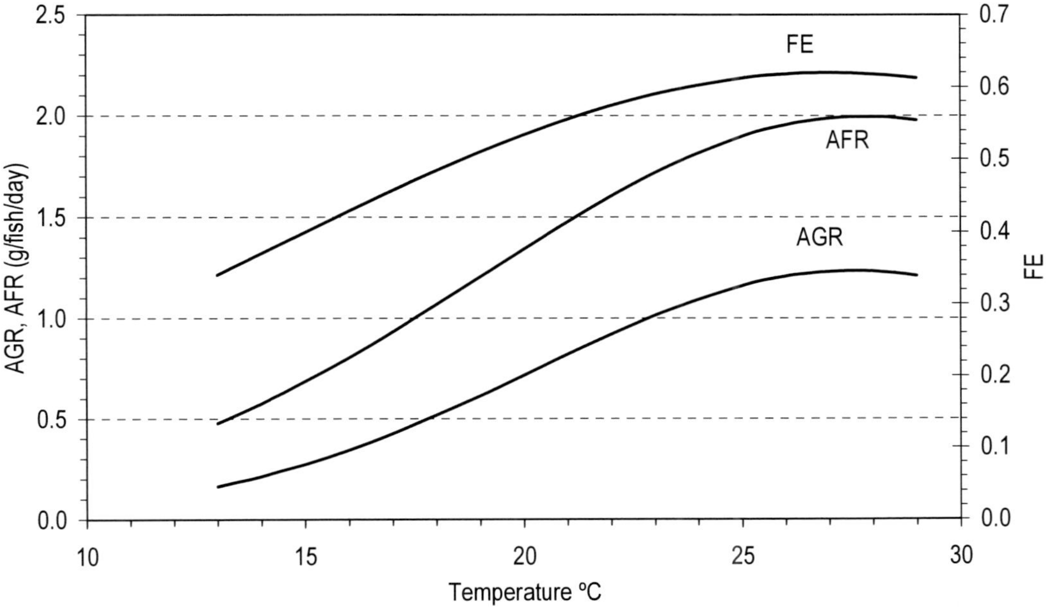

Fig. 4. Estimated AGR, AFR and FE changes as a function of temperature for a 100-g body weight sharpsnout seabream specimen.

juveniles (CJ) and feed (CF). The model is valid for a series of preset ranges for these parameters, which were defined based on the characteristics of this species. To accomplish this, they designed and evaluated the economic aspects of an 800 Tn ongrowing plant in open sea, based on the existing technology for the ongrowing of gilthead seabream (García García et al., 2005). The plant was equipped with 12 flexible, circular polyethylene cages with 10 meters of net height and an outside diameter of 22 m. The anchor and buoy were typical for this type of installation. The cultivation began with 2,285,714 specimens weighing 15 g each, which entered the installation in two groups (one in April, the other in August) and were marketed the following year, with a commercial size of 360-400g. The mortality rate was estimated to be 10%, with the maximum stock density being 17.5 Kg/m^3. Commercial gilthead seabream feed was used.

To calculate the investment, the following costs were considered: cage installation, a multi-purpose building, a dock, equipment and machinery and other miscellaneous items (project, concession, environmental impact study, etc.). Cash flow was established for a working life of 18 years, and calculated based on the corresponding ordinary and extraordinary in and out payments. The following variables were used to perform the sensitivity analysis: Investment (K), the interest rate (i), sales price in Euros/Kg (SP), the cost of juveniles in Euros (CJ) and the cost of feed in Euros/Kg (CF). The preset ranges for the variables are shown in Table 4. 243 alternatives were generated, using the corresponding

Table 4. Starting values and variable ranges used to develop the econometric model (García García et al., 2003).

	Starting value (gilthead seabream)	Range
K (Euros)	1,053,190 (1)	842,552-1,263,828
i (%)	6	4.5-7.5
SP (Euros/Kg)	3.75-4.20	3.30-4.20
CJ (Euros/indiv.)	0.36	0.31-0.41
CF (Euros/Kg)	0.66	0.53-0.79

(1) Before taxes.

spreadsheets. The data obtained were fit by multiple regression analysis to the following equation: $Y=a+bK+ci+dSP+e\ CJ+fCF$, where Y is the NPV (net present value) or the IRR (internal rate of return), and a, b, c, d, e and f are the coefficients that determine the multiple regression analysis.

The equations obtained were as follows:

$$NPV= -68\,649-1.60K-435\,304i+7\,534\,200SP-26\,949\,619CJ-17\,761\,915CF$$

$$(P<0.001;\ R^2_{adj.}=0.979)$$

$$IRR=0.035-7.4710\text{-}8K+0.203SP-0.725CJ-0.482CF$$

$$(P<0.001;\ R^2_{adj.}=0.959)$$

For the start-up alternative, updated prices for gilthead seabream ongrowing were used as variable values. Under these circumstances and for SP values of 3.75 and 4.20 Euros, the model estimated an NPV of 2,467,017 and 5,857,407 Euros and an IRR of 13.78 and 22.91%, respectively, values which are consistent with the current conditions for gilthead seabream production.

During the first phase of sharpsnout seabream culture development, it is likely that CF and CJ values will be similar to those for gilthead seabream. Under these conditions, the minimum market price with an interesting level of profitability for the investor (IRR=15%, which is the sum of the market interest rate plus an interest risk rate) is 3.81 Euro. During the second development phase, when a significant production level has been reached, it is anticipated that the feed price will drop, as this species makes an efficient use of protein and vegetable oils (Hernández et al., 2007; Piedecausa et al., 2007). Therefore, if we set a minimum SP (3.81 Euro), the CJ same as for gilthead seabream and the CF 20% lower, the IRR would be 21%, which is a very attractive value for investors. Obviously, if the SP is above the minimum, the profitability would increase. On the other hand, with feed costs 20% cheaper, in order to obtain a minimum profitability of 15% of IRR, the product could be offered on the market at a minimum SP of 3.5 Euros, resulting in a more competitive product.

In this manner, the model allows to analyze many different circumstances, which may also include variations in the investment and the market interest rate.

Nutrition

As a first step towards learning about the species' feeding demands, we studied *D. puntazzo* nutritional utilization of three commercial diets with different protein/energy ratios (Hernández et al., 2001a), concluding that sharpsnout seabream can make good use of dry feeds currently available on the market for gilthead seabream growout, yielding entirely acceptable growth rates and feed conversion ratios. It would appear that better growth rates can be obtained by lowering the protein/energy ratio below 22 g of protein/MJ, which would prompt the fish to use fat and carbohydrates as a non-protein energy source.

In order to determine the ability of this species to use carbohydrates effectively, Parpoura et al. (2002) prepared six semi-purified diets with two levels of crude protein (40 and 48%) and varying levels (19, 25, 29 and 35%) and types of corn starch (raw and gelatinized). Final mean weights were not different among the experimental groups. However, the groups that were fed high levels of gelatinized starch (29 and 35%) consumed the highest rations and exhibited the lowest feed efficiencies. In addition, the results indicate that raw corn starch is less effectively used than gelatinized corn starch by sharpsnout seabream.

With the objective of obtaining the ideal dietary amino acid profile, Saavedra et al. (2007) determined the bioavailability of individual amino acids. High relative bioavailabilities were found for isoleucine, leucine and valine, which means that these aminoacids are retained more efficiently. On the contrary, alanine, glutamate and aspartate presented a lower bioavailability, as compared to lysine.

In order to optimize the contents of sharpsnout seabream diet, we have tested five extruded diets with different lipid levels (4, 8, 12, 16 and 20%) and a constant level of protein (44%). We found no effects of dietary lipid levels on growth. Daily feed intake increased in those groups fed diets with lipid levels greater than 8%, which decreased the feeding performance. Protein retention reached maximum levels with a dietary lipid rate of 8%, and no savings effect was seen for higher lipid levels. Sharpsnout seabream fail to effectively use high levels of lipids, with optimal levels being above 4% and below 12%, while maintaining protein levels between 45% and 47%. Fat levels above these did not improve growth rates, nutritional utilization of the diet, or protein deposition, and furthermore produce a higher level of fat deposition on the fish body (Hernández et al., 2005).

These results agree with the findings of Torrejón et al. (2004), who used a geometric framework to investigate whether this species is able to create a balanced diet selecting from three diets with different macronutrient composition (fat and protein) that were offered to the fish in pairs. Their results show that macronutrient consumption appears to converge, suggesting a target intake (47% protein and 10% fat content). However, in a study using macronutrient self-selection, Vivas et al. (2006) demonstrated that sharpsnout seabream can select from incomplete diets to compose a balanced diet containing 63% protein, 19% fat and 18% carbohydrates, i.e., the fish selected a protein-rich diet similar to that of carnivores, even though they have omnivorous feeding habits.

We have studied the possible use of soybean meal in sharpsnout seabream diets by progressively increasing its inclusion level (0, 20, 40 and 60%) at the expense of fish meal in isonitrogenous (45%) and isoenergetic formulated (20 MJ/kg) diets (Hernández et al, 2007). The results show that sharpsnout seabream tolerate up to 60% soybean meal in their diets with no serious negative effects, aside from the need to increase feed consumption in order to attain similar growth values. From the customer's perspective, the use of soybean protein-rich diets means, in practice, lower prices, with no changes in the organoleptic properties of the fish. Given the current problems with fish meal, which will undoubtedly worsen over time, being able to produce feed with high levels of vegetable-origin contents may translate into an important reduction in production costs, and consequently the possibility of offering the product at prices below that for gilthead seabream, while still maintaining its profitability. Alternatively, if the price is higher, greater profits may be made (García García et al., 2003).

The possibility of replacing fish meal by increasing levels (7, 14 and 21%) of alfalfa protein concentrate (Chatzifotis et al., 2006) has also been evaluated, finding adverse effects on growth and feed utilization even with alfalfa levels as low as 7%, while higher levels of incorporation did not further worsen the parameters studied.

We have also studied the effects of dietary replacement of fish oil with vegetable oils on sharpsnout seabream (Piedecausa et al., 2007) finding that replacing fish oil with soybean or linseed oil in the fish diets does not affect growth or feed utilization after three months of feeding. The consumption of vegetable oils reduced the muscle content of ARA (arachidonic acid), EPA (eicosapentaenoic acid) and DHA (docosahexaenoic acid). In terms of economic performance, the SO diet was the least expensive one, and it also had the best economic conversion ratio.

PHYSIOLOGY

The respiratory behavior of sharpsnout seabream was studied by Cerezo & García García (2004). The mean value for critical and lethal oxygen saturation was 34% and 11%, respectively. The ventilatory frequency remained constant, reaching a mean value of 67% oxygen saturation, regardless of fish weight or temperature. These findings suggest an optimum oxygen saturation of above 70% for *D. puntazzo* culture.

Vera et al. (2006) investigated the circadian organization of sharpsnout seabream. This species displayed diurnal feeding and locomotor patterns. Daily plasma melatonin levels displayed a rhythmic profile, with low daytime values (111 pg/ml) and high nighttime concentrations (791 pg/ml).

Other parameters that have been determined by our working group but not yet published are those related to the hematology of this species. Table 5 shows the differences found between the two months in which measurements were taken, with an approximate temperature difference of 2 °C.

The kinetic profile of oxytetracycline has also been investigated in sharpsnout seabream, following intravascular and oral drug administration (Rigos et al., 2004). After a single intravascular injection (40 mg/kg), the distribution and elimination half-lifes were calculated to be 1.4 h and 35 h, respectively. The bioavailability of oxytetracycline following oral administration (75 mg/kg) revealed undetectable levels of the drug in plasma and muscle samples. Therefore, oral administration of oxytetracycline in farmed sharpsnout seabream should be discouraged in favor of other routes of administration or other antibiotic drugs. The residue depletion of oxolinic acid after a multiple (10-day) in-feed administration (30 mg/kg fish) has also been studied (Rigos et al., 2003). Oxolinic acid displayed a biphasic tissue elimination pattern, and reached "consumer safe levels" 24h after the treatment. The elimination half-life in sharpsnout seabream tissue was estimated to be 11-19h.

Table 5. Hematological parameters of sharpsnout sebream.

	November	December	Range
Hematocrit (%)	38.2±3.8	36.3±5.8	28-41
Hemoglobin (g/100 ml)	7.3±1.0	5.7±1.0	4.1-10.5
Red Blood Cell Count (x10^6)	2.4±0.4	2.5±0.4	1.73-3.22
Mean Corpuscular Volume (μm^3)	162.0±27.5	146.9±22.2	124.7-214.8
Mean Corpuscular Hemoglobin Concentration (g/dl)	19.2±1.8	15.7±1.8	12.6-21.3
Mean Corpuscular Hemoglobin (pg)	30.7±2.4	22.7±1.9	12.6-34.7

Enzymatic studies have also shown an enzymatic pattern that is well-suited to protein digestion, but that also has a high potential for digesting vegetable polysaccharides, in accordance with its omnivorous habits (Tramati et al., 2005).

The main problem with using plant seeds oils is the important differences in composition that exist between them and fish oil. Vegetal oils are rich in C_{18} polyunsaturated fatty acids (PUFA C_{18}), but they lack the n-3 highly unsaturated fatty acids (HUFAs) that are so abundant in fish oil. This means that fish feeding on vegetal oils would have to desaturate and elongate PUFAs, 18:3n-3 (linolenic acid, LNA) and 18:2n-6 (linoleic acid, LA) to their HUFA derivates, 20:5n-3 (eicosapentaenoic acid, EPA) and 22:6n-3 (docosahexaenoic acid, DHA), and 20:4n-6 (araquidonic acid, AA), respectively. For this reason, the fish's ability to compensate for oil composition differences in order to adapt to fish oil replacement is very important. Almaida-Pagán et al. (2007) showed that the inclusion of vegetable oils in sharpsnout seabream diet did not have any quantitative nutritional effects on the desaturation/elongation of [1-^{14}C] linolenic acid and [1-^{14}C] linoleic acid in isolated hepatocytes and total digestive tract enterocytes, clearly showing the lack of any significant desaturase/elongase activity in these cells. Only high levels of HUFA in fish tissues reveal any regulatory mechanisms, presumably based on HUFA bioaccumulation and PUFA C_{18} oxidation.

PATHOLOGY

Some studies conducted in Greece, Italy and Turkey have reported the presence of a number of pathogens in the farming of *D. puntazzo* that might be harmful to its development. These include *Candida* spp., *Ichthyophonus* sp., *Trichodina* sp., *Amyloodinium* sp. and monogeneans such as *Gyrodactylus* sp, *Atrispinum salpae*, *A. seminalis,* and *Sparicotyle chrysophrii* (Athanassopoulou et al., 1999; Di Cave et al., 1998; Rigos et al., 1998; Tampieri et al., 1999; Zaccone & La Ferla, 1991). These last parasites were apparently transmitted by wild sharpsnout seabream and other sparid specimens (Di Cave et al., 1998). Nonetheless, the parasite that appears to be the most virulent for this species is *Enteromyxum leei.* This myxosporidian is also a significant pathogen for gilthead seabream, causing acute enteritis, followed by anorexia, cachexia and death, in many cases (Padrós et al., 2001). This disease has been observed in facilities in several different Mediterranean countries, where it has led to high mortality rates (Caffara et al., 1999; Padrós et al., 2001). Mortality due to *E. Leei* has also been detected in *D. Puntazzo*, with this species apparently being even more susceptible than gilthead seabream (Caffara et al., 1999; Le Breton, 1999). On the Spanish coast (San Pedro del Pinatar and Gandía), it caused high mortality in specimens farmed in floating cages that came from Italy (Montero et al., 2006). In addition to the problem that this represents for the emerging farming of *D. puntazzo*, there is a very high probability that it will be transmitted and serve as a reservoir for gilthead seabream farms.

QUALITY AS FOOD

One of the factors to keep in mind when analyzing the viability of a species for industrial farming is that of consumer acceptance of the product, a variable that can be measured by sensory evaluation tests. Preliminary studies on the acceptance of this species on the part of the consumer conducted at the IMIDA (San Pedro del Pinatar, Murcia, Spain) showed that the people surveyed had a positive impression of sharpsnout seabream when they tried it. Participants gave it high scores in a hedonic scale designed to measure the level of consumer satisfaction, and 60% of them would consider buying it at a price similar to that of gilthead (Hernández et al., 2001b). Based on the results of the sensory test, even when sharpsnout seabream are fed fish meal substitutes in large proportions (40 or 60% soybean meal), the flesh quality is affected very little, although fish feeding on diets with partial replacement of fish meal tended to be somewhat softer (Hernández et al., 2007). The use of soybean meal in sharpsnout seabream diets produce fish with an acceptable fatty acid profile for human consumption. This is mainly due to an increased PUFA content, affecting especially linoleic acid. However, fish meal replacement causes a loss of n-3 fatty acids that has to be compensated with other sources rich in this type of fatty acids (Rondán et al., 2004a).

Another aspect that has been analyzed is the influence of the different culturing conditions on the nutritional quality of the fish (Orban et al., 2000). The proximate composition of fish reared separately in tanks and in cages did not differ significantly, except for the higher lipid content of the tank-reared specimens. Total lipid levels for cage-reared fish showed a significantly higher sum of saturated fatty acids than for tank-reared fish. Lipid fractionation revealed that neutral lipids from fish reared in tanks had a significantly higher content of total monounsaturated fatty acids compared to cage-reared fish.

The fatty acid composition and lipid content of white muscle, liver and mesenteric fat, in reared *vs.* wild sharpsnout seabream were compared (Rueda et al., 2001). The mesenteric fat index (MFI) and the lipid contents of both white muscle and liver proved consistently higher

Table 6. White muscle and liver lipid content and mesenteric fat index for wild and reared sharpsnout seabream (Rueda et al., 2001).

	Wild	Reared
White muscle lipid (g/kg)	38.7±5.1	75.9±13.1
Liver lipid (g/kg)	58.2±3.9	188.4±31.0
MFI (%)	1.7±0.2	2.7±0.4

Table 7. Fatty acid composition of white muscle, hepatic tissue and mesenteric fat for wild and reared sharpsnout seabream (g/100 g total fatty acids) (Rueda et al., 2001).

	White muscle		Hepatic tissue		Mesenteric fat	
	Wild	Reared	Wild	Reared	Wild	Reared
SFA	34.92±0.26	27.66±0.23	34.89±0.42	29.16±0.76	35.30±0.43	26.97±0.40
MUFA	32.80±1.58	36.95±1.34	24.22±1.41	47.38±2.34	38.63±1.24	40.49±0.53
PUFA	32.29±1.70	35.39±1.50	40.89±1.36	23.45±2.96	26.08±1.38	32.54±0.71

in farmed *vs.* wild sharpsnout seabream (Table 6). The higher values of linoleic, eicosapentaenoic, docosahexaenoic and n-3 series acids in reared fish muscle make reared sharpsnout more favorable for human consumption, as they also have higher atherogenic and thrombogenic indices (Table 7). Depending on the time of the year or the season, reared fish might be more suitable for human consumption than wild fish.

In sharpsnout seabream, restricted food intake increases the n-3/n-6 ratio by both decreasing the percentage of n-6 and increasing that of n-3 fatty acids. As is the case in other fish species, HUFA were the main contributors to the n-3 increase. Food intake restriction to one-third of satiety levels was enough to trigger the mechanisms that preserve essential fatty acids (Rondán et al., 2004b).

Corsi et al. (2004) have investigated how different facility locations and breeding conditions might affect fish metabolic capabilities and the accumulation of toxic chemicals in farmed sharpsnout seabream. Significantly higher activities of two metabolizing enzymes (EROD and B(a)PMO) were, in fact, recorded in land-based farmed fish than in sea farmed ones, these activities being also much higher in specimens from polyculture than from monoculture systems. By contrast, higher levels of DDE and PCBs were found in sea-farmed fish, except in the case of HCB. This may be the result of our limited ability to control chemicals in natural seawater.

REFERENCES

Abellán, E. & García Alcazar, A. (1995). Pre-growout and growout experiences with white seabream (*Diplodus sargus sargus*, Linnaeus, 1758) and sharpsnout seabream (Diplodus puntazzo, Cetti 1777). *Cah. Options Mediterr., 16*, 57-63.

Almaida-Pagán, P.F., Hernández, M.D., García García, B., Madrid, J.A., de Costa J., Mendiola, P. (2007). Effects of total replacement of fish oil by vegetable oils on n-3 and n-6 polyunsaturated fatty acid desaturation and elongation in sharpsnout seabream (*Diplodus puntazzo*) hepatocytes and enterocytes. *Aquaculture, 272*, 589-598.

Athanassopoulou, F., Prapas, T., Rodger, H. (1999) Diseases of *Puntazzo puntazzo* Cuvier in marine aquaculture systems in Greece. *J. Fish Dis. 22(3)*, 215-218.

Bauchot M.L.& Hureau, J.C. (1986). Sparidae. In: Whitehead, P.J.P., Bauchot, M.L., Hureau, J.C., Nielsen, J., Tortonese, E. (eds) Fishes of the northeastern Atlantic and the Mediterranean, Vol 2. Paris. United Nations Educational Scientific and Cultural Organization.

Bermúdez, L., García García, B., Gómez, O., Rosique, M. J., Faraco, F. (1989). First results of the ongrowing in cages of *Sparus aurata, Puntazzo puntazzo* and *Lithognatus mormirus* in the Mar Menor (Murcia, SE, Spain). *EAS Special Publication No. 10,* pp. 27-28.

Bini, G. (1968). Atlante dei pesci delle coste italiane. *Mondo Sommerso, 4,* 187-198.

Boglione., C., Giganti, M., Selmo, C., Cataudella, S. (2003). Morphoecology in larval fin-fish: a new candidate species for aquaculture, *Diplodus puntazzo* (Sparidae). *Aquacult. Int., 11*, 17-41.

Borsa, P., Blanquer, A., Berrebi, P., (1997). Genetic structure of flounders Platichthys flesus and P. stellatus at different geographic scales. *Mar. Biol., 129*, 233–246.

Cadenat, J., (1964). Les "Sars" des gensres *Puntazzo* et *Diplodus* des eaux tropicales ouest-africaines. *Bull. Inst. Fondam. Afr. Noire, 26,* 944-970.

Caffara, M., Delgado, M.L., Fioravanti, M.L., Quaglio, F., Restani, R. (1999). *Myxidium leei* (Myxozoa, Myxosporea) in sparidi (*Sparus aurata* e *Diplodus puntazzo*) allevati in Italia. *Boll. Soc. Ital. Patol. Ittica., 11(27)*, 10-17.

Cerezo, J. & García García, B. (2004). The effects of oxygen levels on oxygen consumption, survival and ventilatory frequency of sharpsnout seabream (*Diplodus puntazzo* Gmelin, 1789) at different conditions of temperature and fish weight. *J. Appl. Ichthyol., 20*, 488-492.

Corsi, I., Mazzola, A., Focardi, S. (2004). Mixed function oxidase activity and organochlorine levels in farmed sharpsnout seabream (*Diplodus puntazzo*) from two intensive aquaculture facilities. *Aquacult. Int., 12*, 357-375.

Chatzifotis, S., Esteban, A.G., Divanach, P. (2006). Fishmeal replacement by alfalfa protein concentrate in sharpsnout seabream *Diplodus puntazzo. Fisheries Science, 72*, 1313-1315.

D'Ancona, U., (1949). Il differenziiamento della gonade e l'inversione sessuale degli Sparidi. *Arch. Oceanogr. Limnol., VI,* 97-164.

Deudero, S., Pinnegar, J. K. , Polunin, N. V. C., Morey, G, B. Morales-Nin, (2004). Spatial variation and ontogenic shifts in the isotopic composition of Mediterranean littoral fishes. *Mar. Biol.*, 14, 971–981.

Di Cave, D., De Liberato, C., Berrilli, F., Orecchia, P. (1998) Parasitic infections of sharpsnout sea bream, *Diplodus puntazzo* and gilthead sea bream, *Sparus aurata* in cage-culture. *Parassitologia, 1,* 40- 45.

Divanach, P., Kentouri, M., Charalambakis, G., Pouget, F., Esterioti, A. (1993). Comparison of growth performance of six Mediterranean fish species reared under intensive farming conditions in Crete (Greece), in raceways with the use of self feeders. *E.A.S. Special Publication No 18*, pp: 285-297.

Domínguez-Seoane, R., Pajuelo, J.G., Lorenzo, J.M., Ramos, A.G. (2006). Age and growth of the sharpsnout seabream *Diplodus puntazzo*(Cetti, 1777) inhabiting the Canarian archipelago, estimated by reading otoliths and by backcalculation. *Fish. Res.*, *81*,142–148

Egea Nicolás, M.A., Rueda González, F., Martínez López, F.J., García García, B. (2002). Efecto de la realimentación un período de ayuno sobre el crecimiento en el sargo picudo *Diplodus puntazzo* (Cetti, 1777). *Bol. Inst. Esp. Oceanog. 18(1-4)*, 357-362.

Faranda, F., Cavaliere, A., Lo Paro, G., Manganaro, A., (1983). *Puntazzo puntazzo* and *Diplodus vulgaris* growth. Comparison of two different diets. *Mem. Biol. Mar. Ocean., XIII*, 37-53.

Faranda, F., Cavaliere, A., Lo Paro, G., Manganaro, A., Mazzola, A. (1985). Preliminary studies on reproduction of *Puntazzo puntazzo* (Gmelin 1789) (Pises, Sparidae) under controlled conditions. *Aquaculture, 49*, 111-123.

Favaloro, E., Lopiano, L., Mazzola, A. (2002). Rearing of sharpsnout seabream (*Diplodus puntazzo*, Cetti 1777) in a Mediterranean fish farm: monoculture versus polyculture. *Aquacul. Res., 33,* 137-140.

Fernández, B. & Rodriguez, X.L. (2000). Guía de la Piscicultura Europea. Xunta de Galicia, Consellería de Pesca e Asuntos Marítimos. Pontevedra. 68 pp.

Fiedler, K. (1991). Familie Sparidae. In: Starck, D. (ed) Lehrbuch der Speziellen Zoologie. Teil 2: Fische. Gustav Fischer Verlag, Jena, pp 354–355.

Franicevic, V. (1989). Preliminary results on the larviculture of *Puntazzo puntazzo* (Gmelin 1789) (Pisces, Sparidae). *EAS Special Publication No. 10,* pp. 139-141.

García García, B., Rueda, F.M., Hernández, M.D., Aguado, F., Egea, M.A., Faraco, F. (2001). Crecimiento e índice de conversión del sargo picudo (*Puntazzo puntazzo* Gmelin, 1789) en engorde intensivo en tanques. In: *Convergencia entre Investigación y Empresa: Un reto para el siglo XXI.* Fernández-Palacios, H y Izquierdo, M. (Eds.). Monografías del ICCM nº4.(pp: 385-390). Las Palmas de Gran Canaria.

García García, B., Hernández, M.D., García García, J., Martínez, F.J., Jover, M. (2003). Desarrollo de modelos de crecimiento y tasa de alimentación en función del peso, la temperatura y el contenido de harina de soja en la dieta del sargo picudo (*Diplodus puntazzo*). *IX Congreso Nacional de Acuicultura.* Cádiz, Spain.

García García, J., Rouco, A., García García, B. 2005. Economies of scale for ongrowing gilthead sea bream (*Sparus aurata*) on off shore cages in the Mediterranean Sea. *An. Vet. (Murcia) 21*: 69-76.

Gatland, P. (1995). Growth of *Puntazzo puntazzo* in cages in Selonda bay, Corinthos, Greece. *Cah. Options Mediterr., 16*, 51-55.

Georgiou, G. & Stephanou, D. (1995). Contribution to the study of maturation and spawning problems of the sharpsnout seabream *Puntazzo puntazzo. Cah. Options Mediterr., 16*, 47-55.

Gonzalez-Laxe, F .(2000). Los desafios estratégicos de la piscicultura marina en España.

Productos del Mar, 51: 1-26.

Greco, S., Lo Paro, G., Caridi, D., Perdichizzi, F., Cammaroto, S., Micale V., Genovese, L. (1993). Controlled spawning and larval development in the sharpsnout seabream (*Diplodus puntazzo*), Sparidae. *E.A.S. Special Publication No 18*, pp. 185-188.

Hanel, R. & Sturmbauer, Ch. (2000) Multiple recurrent evolution of trophic types in Northeastern Atlantic and Mediterranean Seabreams (Sparidae, Percoidei). *J. Mol. Evol.*, 50, 276-283.

Hernández, M.D., Egea, M.A., Rueda, F.M., Aguado, F., Martínez, F.J., García García, B. (2001a). Effects of commercial diets with different P/E ratios on sharpsnout seabream (Diplodus puntazzo) growth and nutrient utilization. *Aquaculture, 195*, 321-329.

Hernández, M.D., Martínez, F.J., García García, B. (2001b). Sensory evaluation of farmed sharpsnout seabream (*Diplodus puntazzo*). *Aquacult. Int., 9*, 519-529.

Hernández, M.D., Egea, M.A., Rueda, F.J., Martínez, F.J., García García, B. (2003). Seasonal condition and body composition changes in sharpsnout seabream (*Diplodus puntazzo*) raised in captivity. *Aquaculture, 220*, 569-580.

Hernández, M.D., García García, B., Martínez, F.J., Jover, M. (2005). Efecto del contenido en lípidos de la dieta sobre el crecimiento y la composición corporal en el sargo picudo (*Diplodus puntazzo*). *Actas del IX Congreso Nacional de Acuicultura*. Cádiz, Spain. pp. 107-110.

Hernández, M.D., Martínez, F.J., Jover, M., García García, B. (2007). Effects of partial replacement of fish meal by soybean meal in sharpsnout seabream (*Diplodus puntazzo*) diet. *Aquaculture, 263*, 159-167.

Karakatsouli, N., Papafotiou, P., Papoutsoglou, S.E. (2006). Mono- and duoculture of juvenile sharpsnout seabream *Diplodus puntazzo* (Cetti) and gilthead seabream *Sparus aurata* L. in a recirculated water system. *Aquacult. Res., 37*, 1654-1661.

Korkut, A.Y., Saka, S., Firat, K. (2006). The effects of different light intensities on early life development of sharpsnout seabream (Diplodus puntazzo, Cetti 1777) larvae. *Turk. J. Vet. Anim. Sci. 30*, 381-387.

Kouttouki, S., Georgakopoulou, E., Kaspiris, P., Divanach, P., Koumoundouros, G. (2006). Shape ontogeny and variation in the sharpsnout seabream, *Diplodus puntazzo* (Cetti 1777). *Aquacult. Res. 37*, 655-663.

Lahnsteiner, F. & Patarnello, P. (2004). Biochemical egg quality determination in the gilthead seabream, *Sparus aurata*: reproducibility of the method and its application for sharpsnout seabream, *Puntazzo puntazzo*. *Aquaculture, 237*, 433-442.

Lahnsteiner, F. & Patarnello, P. (2005). The shape of the lipid vesicle is a potential marker for egg quality determination in the gilthead seabream, *Sparus aurata*, and in the sharpsnout seabream, *Diplodus puntazzo*. *Aquaculture 246*, 423-435.

Le Breton, A.D. (1999) Mediterranean finfish pathologies: present status and new developments in prophylactic methods. *Bull. Eur. Assoc. Fish Pathol. 19(6)*, 250-253.

Lissia-Frau, A.M., Pala, M. (1968). Ricerche sull'ermafroditismo dei Saraghi: *D. sargus* (L.), *D. vulgaris* (Geofr.*)*, *D. annularis* (L.) e *P. puntazzo* (Cetti). *Studi Sassarei, 46*, 203-221.

Marangos, C. (1995). Larviculture of the sheepshead bream, *Puntazzo puntazzo* (Gmelin 1789) (Pisces, Sparidae). *Cah. Options* Mediterr.*, 16:* 41-45.

Mena Sellés, C. and García García, B. (2002). Importancia de la proteína vegetal en la dieta natural de poblaciones salvajes de sargo picudo, *Diplodus puntazzo* (Cetti, 1777): sus implicaciones en el cultivo intensivo. *AquaTic, 17*: 1-11.

Micale, V., Perdichizzi, F., Basciano, G. (1996). Aspects of the reproductive biology of the sharpsnout seabream *Diplodus puntazzo* (Cetti, 1777). Gametogenesis and gonadal cycle in captivity during the third year of life. *Aquaculture, 140*: 281-291.

Montero, F.E.; Cuadrado, M.; Holzer, A.S.; García García, B.; Jover, M.; Padrós, F.; Crespo, S., Raga, J.A. (2006). *Cryptocarion irritans* and *Enteromyxum leei* two potential threatens for the cultures of *Diplodus puntazzo* in the Mediterranean. 11th International Conference of Parasitology. Glasgow, Great Britain. 7-14 August 2006.

Naciri, M., Lemaire, C., Borsa, P., Bonhomme, F. (1999). Genetic study of the Atlantic/Mediterranean transition in the sea bass (*Dicentrarchus labrax*). *J. Hered., 90*, 591–596.

Orban, E., Di Lena, G., Ricelli, A., Paoletti, F., Casini, I., Gambelli, L., Caproni, R. (2000). Quality characteristics of sharpsnout seabream (Diplodus puntazzo). *Food Chemistry, 70*, 27-32.

Padrós, F., Palenzuela, O., Hispano, C., Tosas, O., Zarza, C., Crespo, S., Alvarez-Pellitero, P. (2001). *Myxidium leei* (Myxozoa) infections in aquarium-reared Mediterranean fish species. *Dis. Aquat. Org. 47*, 57-62.

Palma, P., Andrade, J.P. (2002). Morphological study of *Diplodus sargus, Diplodus puntazzo*, and *Lithognathus mormyrus* (Sparidae) in the Eastern Atlantic and Mediterranean Sea. *Fish. Res.*, 57, 1-8.

Papandroulakis, N., Kentouri, M., Maingot, E., Divanach, P. (2004). Mesocosm: a reliable technology for larval rearing of Diplodus puntazzo and Diplodus sargus sargus. 2004. *Aquacult. Int.,* 12, 345-355.

Parpoura, A.C., Alexis, M.N., van Ham, E.H. 2002. Investigation of the ability of juveniles of *Puntazzo puntazzo* to use carbohydrates as a dietary energy source. 10th International Symposium on Nutrition and Feeding in Fish (Feeding For Quality). Rhodes, Greece, 2nd-7th June, 2002. Abstract no. P13, p. 94.

Pastor, E., Grau, A., Riera, F., Pou, S., Massutí, E., Grau, A.M. (2000). Experiences in the culture of new species in the "Estación de Acuicultura" of the Balearic government (1980-1998). *Options Méditerranéennes Serie B. Etudes et recherches, 47*, 371-379.

Piedecausa, M.A., Mazón, M.J., García García, B., Hernández, M.D. (2007). Effects of total replacement of fish oil by vegetable oils in sharpsnout seabream (*Diplodus puntazzo*) diets. *Aquaculture, 263,* 211-219.

Rigos, G., Christophilogiannis, P., Giahnishi, M., Andriopoulou, A., Koutsodimoy, M., Nengas, I., Alexis, M. (1998) *Amyloodinium ocelatum* infestation on sharpsnout sea bream, *Puntazzo puntazzo* Cetti. *Bull. Eur. Assoc. Fish Pathol. 18(6)*, 198-200.

Rigos, G., Nengas, I., Alexis, M., Tyrpenou, A.E., Troisi, G.M. (2003). Tissue distribution and residue depletion of oxolinic acid in gilthead seabream (*Diplodus puntazzo*) following multiple in-feed dosing. *Aquaculture, 224*, 245-256.

Rigos, G., Tyrpenou, A.E., Alexis, M., Athanassopoulou, F., Troisi, G.M. (2004). Poor bioavailability of oxytetracycline in sharpsnout seabream *Diplodus puntazzo. Aquaculture, 235*, 489-497.

Rondán, M., Hernández, M.D., Egea, M.A., García, B., Jover, M., Rueda, F.M., Martínez, F.J. (2004a). Effects of fishmeal replacement with soybean meal as protein source, and protein replacement with carbohydrates as an alternative energy source on sharpsnout seabream, *Diplodus puntazzo*, fatty acid profile. *Aquacul. Res., 35*, 1220-1227.

Rondán, M., Hernández, M.D., Egea, M.A., García, B., Rueda, F.M., Martínez, F.J. (2004b). Effect of feeding rate on fatty acid composition of sharpsnout seabream (*Diplodus puntazzo*). *Aquaculture Nutrition, 10,* 301-307.

Rosecchi, E. (1987). The feeding of *Diplodus annularis, Diplodus sargus, Diplodus vulgaris* and *Sparus aurata* (Pisces, Sparidae) inthe Gulf of Lions and the coastal lagoons. *Rev. Trav. Inst. Pêches Marit.* 49, 125–141

Rueda, F.M., Hernández, M.D., Egea, M.A., Aguado, F., García García, B. (2001). Differences in tissue fatty acid composition between reared and wild sharpsnout sea bream, Diplodus puntazzo (Cetti, 1777). *Brit. J. Nutr., 86*, 617-622.

Saavedra, M., Beltran, M., Pousao-Ferreira, P., Dinis. M.T., Blasco, J., Conceicao, L.E.C. (2007). Evaluation of bioavailability of individual amino acids in *Diplodus puntazzo* larvae: Towards the ideal dietary amino acid profile. *Aquaculture, 263*, 192-198.

Sala, E. (1997). Fish predators and scavengers of the sea urchin *Paracentrotus lividus* in protected areas of the north-west Mediterranean Sea. *Mar. Biol.*, 129, 531-539.

Sala, E. & Ballesteros, E. (1997). Partitioning of space and food resources by three fish of the genus Diplodus (Sparidae) in a Mediterranean rocky infralitoral ecosystem. *Mar. Ecol. Prog. Series, 152*, 273-283.

Sara, M., Favaloro, E., Mazzola, A. (1999). Comparative morphometrics of sharpsnout seabream (*Diplodus puntazzo*) reared in different conditions. *Aquacult. Engin., 19*, 195-209.

Sfakianakis, D.G., Doxa, C.K., Kouttouki, S., Koumoundouros, G., Maingot, E., Divanach, P., Kentouri, M. (2005). Osteological development of the vertebral column and of the fins in Diplodus puntazzo (Cetti, 1777). *Aquaculture*, 250, 36– 46

Smith, J.L.B. & Smith, M.M. (1986). Family No. 183: Sparidae. In: Smith, M.M., Heemstra, P.C. (eds). *Smiths' sea fishes* Springer-Verlag, (pp 580–594). New York, Berlin, Heidelberg.

Stephanis J. & Divanach, P. (1993) Farming of Mediterranean finfish species. Present status and potentials. *European Aquaculture Society Special publications, 19*, 426.

Sweetman, J.W. (1992) Larviculture of mediterranean marine fish species: current status and future trends. *Journal of the World Aquaculture Society, 23,* 330-337.

Syrski, S. (1876). Ergebnise von untersuchungen der Geschlechtsorgane von Knochenfischen (De piscium osseorum organis genitalibus). Kosmos.Lemberg., 1, 418.

Tampieri, M.P., Delgado Montero, M.L., Galuppi, R. (1999) Prima segnalazione di Ittiofonosi in sarago pizzuto (*Diplodus puntazzo*) in Italia. *Boll. Soc. Ital. Patol. Ittica 11(27),* 18-26.

Tortonese, E. (1973). Sparidae. In: J.C. Hureau and T. Monod (eds.), Check-list of the fishes of the north-eastern Atlanctic and of the Mediterranean. UNESCO, Paris, Vol. 1, pp. 124-132.

Torrejón Atienza, M., Chatzifotis, S., Divanach, P. (2004). Macronutrient selection by sharpsnout seabream (*Diplodus puntazzo*). *Aquaculture, 232*, 481-491.

Tramati, C., Savona, B., Mazzola, A. (2005). A study of the pattern of digestive enzymes in *Diplodus puntazzo* (Cetti, 1777) (Osteichthyes, Sparidae): evidence for the definition of nutritional protocols. *Aquacult. Int., 13*, 89-95.

Vera, L.M., Madrid, F.J., Sánchez-Vázquez, F.J. (2006). Locomotor, feeding and melatonin daily rhythms in sharpsnout seabream (Diplodus puntazzo). *Physiology & Behaviour, 88*, 167-172.

Vivas, M., Rubio, V.C., Sánchez-Vázquez, F.J., Mena, C., García García, B., Madrid, J.A. (2006). Dietary self-selection in sharspnout seabream (*Diplodus puntazzo*) fed paired macronutrient feeds and challenged with protein dilution. *Aquaculture, 251*, 430-437.

Whitehead, P.J.P., Bauchot, M.L., Hureau, J.C., Nielsen, J., Tortonese, E. (1986). *Fishes of the North-eastern atlantic and the Mediterranean*. United Kingdom: UNESCO.

Zaccone, R., La Ferla, R. (1991) Lieviti associati ad animali marini in un impianto sperimentale di acquacoltura. *Mem. Biol. Mar. Oceanogr., 17*, 5-13.

In: Aquaculture Research Trends
Editor: Stephen H. Schwartz, pp. 263-277
ISBN 978-1-60456-217-0

Chapter 8

RECENT ADVANCES IN CHANNEL CATFISH NUTRITION AND FEEDING RESEARCH

Menghe H. Li and Edwin H. Robinson
Thad Cochran National Warmwater Aquaculture Center
Mississippi State University, P.O. Box 197, Stoneville,
Mississippi 38776 USA

ABSTRACT

Channel catfish *Ictalurus punctatus* is the leading aquaculture species in the United States with an annual production of approximately 275,000 metric tons. Extensive research has been conducted on the nutrition and feeding of this species and as a result, its nutrient requirements and feeding characteristics are well documented. Findings from laboratory and field studies conducted over the past few years have made a significant contribution to the existing channel catfish nutrition database, which is routinely used in the formulation of cost-effective feeds. Recent research provides a better understanding of the relationship between dietary protein and digestible energy and their effects on fish growth, processed yield, and body composition. Research has also shown that traditional feed ingredients can be replaced with less expensive alternative feedstuffs to reduce feed cost. For example, marine fish meal can be reduced or replaced with other animal protein sources, such as poultry by-product meal and meat and bone/blood meal. Soybean meal can be partially replaced with a combination of distillers dried grains with solubles, cottonseed meal, and supplemental lysine. Inorganic phosphorus supplements can be completely replaced with phytase enzymes without affecting fish growth and bone mineralization. While channel catfish are more tolerant to aflatoxins and deoxynivalenol, they appear to be as sensitive to T-2 toxin as rainbow trout. Channel catfish are sensitive to ochratoxin A in their diet. Various feeding strategies have been evaluated to optimize production efficiency under various culture conditions. Feeding channel catfish in food fish production ponds once daily to apparent satiation (without adversely affecting water quality) supports maximum growth. However, depending on water temperature and water quality and on the health of the fish, it may be prudent to restrict the daily feed allowance or to feed less frequently than daily.

Introduction

The channel catfish *Ictalurus puncatatus* (referred to as catfish throughout the chapter) is the primary aquaculture species grown in the United States. About 275,000 metric tons of catfish are processed annually, representing over 70% of the total U.S. aquaculture production and over 40% of total aquaculture value (National Marine Fisheries Service 2006). Since little was known about the nutrient requirements of catfish when the industry began, catfish were fed diets that were formulated based on information derived for other farmed animals and on general knowledge of fish behavior. These diets were considered as supplements to natural food sources found in pond waters and were not meant to provide all nutrients and energy needed by the fish. Modern catfish diets are based on a foundation that was developed through many years of extensive research during which their nutrient requirements and feeding habits were elaborated under culture conditions. Today catfish are typically fed a nutritionally complete diet that provides all known nutrients at required levels and the energy necessary for their utilization in a form that is water stable and readily digestible. It is essential to supply all nutrients in the diet because the contribution of microbially synthesized nutrients in the intestine of catfish is minimal. In addition, except perhaps for early life stages, the contribution of nutrients from natural food organisms in pond waters is relatively small in comparison to total nutrient requirements. Since feed cost represents about 50% of the variable operating costs in catfish farming, over the last 10 years or so the emphasis in nutrition and feeding research has been on reducing feed cost. In this regard, the trend has been to reduce dietary protein, use a wide variety of feedstuffs, use synthetic amino acids to improve the protein quality of inferior feedstuffs, reduce the amount of animal protein in the diet, and to feed in a manner that results in fast and efficient growth. An overview of catfish nutrition and feeding is presented herein with emphasis on current trends.

Optimum Dietary Protein Concentrations

Since feed cost is the major variable operating cost in catfish production, and protein is the most expensive nutrient in catfish feeds, considerable research efforts have been expended to determine dietary protein levels necessary to achieve optimum fish performance under various culture conditions. Optimum dietary protein levels vary depending on several factors including fish size, feeding rate, amount of non-protein energy in the diet, protein quality, natural food available, and management practices.

Catfish do not appear to require as much protein in the diet for maximum growth as has been typically assumed. We have examined a wide range of dietary protein levels (from 10% to 40%) in various studies and found no differences in weight gain and feed conversion in fish fed diets containing as low as 24% protein when fish were fed to apparent satiation (Robinson and Li 2007a). In addition, fish fed 16–20% dietary protein gain about 80–90% of that of fish fed a 32%-protein diet. However, as dietary protein decreases, the digestible energy/crude protein ratio increases beyond the optimum range (8.5–9.5 kcal digestible energy/g protein) resulting in an increase in body fat. This effect is quite dramatic in diets containing very low levels of dietary protein. Fish fed a 16%-protein diet contain about twice as much fillet fat as similar size fish raised on a 32%-protein diet.

The primary problem with increased fattiness is that too much fat may reduce processed yield. There are many factors other than diet that affect processed yield, but in general fish that are fed 26% or less dietary protein will have reduced processed yield (Robinson and Li 2007a). The data are not as consistent with fish fed a 28%-protein diet since some studies show a reduction in processed yield and others do not. Overall, there are marginal differences in processed yield of fish fed a 28%- or 32%-protein diet (about 0.3–0.4 percentage unit reduction in fish fed the 28%-protein diet), but the difference may be significant to catfish processors. Feeding diets containing 35–36% protein and above may reduce fattiness and improve processed yield compared with fish fed a 28%-protein diet. This approach to reduce fattiness and increase processed yield may not be economical for the catfish producer, but if the producer is paid a premium for low-fat fish products and higher processed yield it may be a more attractive proposition.

Low-protein diets (26–28%) are generally recommended for stockers and food fish, but not for small fingerlings. However, there were no differences in production of fingerling catfish raised from 10 to 45 g/fish and fed either a 28%-, 32%-, or 41%-protein diet (Robinson et al. 2002a). It would appear that high-protein diets offer no advantage for fingerlings from about 10 g and larger.

Generally, low-protein diets are effective for maximum growth when fish are fed to satiation, but higher-protein diets are required when feed is restricted. It appears that a 28%-protein diet provides for good growth at feeding rates of 90 kg/ha per day and above (Robinson and Li 1999; Robinson et al. 2004). Also, a 28%-protein diet can be fed throughout the growing season. We find no advantage to begin feeding in early spring with a higher-protein diet and then reducing dietary protein as feeding activity increases (Robinson and Li 2007a). Higher-protein diets (32–36%) may increase net production and weight gain when feeding every other day compared with daily feeding (Li et al. 2006b), but the value of this practice may or may not be economical. Regardless of diet, feeding every other day improves feed efficiency and reduces aeration time, but also reduces weight gain and processed yield compared with fish fed daily.

Although low protein diets do not affect fish growth, negative aspects are increased body fattiness and reduced processed yield. This makes it problematic to take advantage of the lower-cost diets. A possible solution that has been suggested would be to feed a low-protein diet for most of the growing season, and feed a high-protein diet for a set number of days before harvest (that is, use a "finishing" diet as used in other animal production). However, finishing catfish on high-protein diets to reduce body fat and improve processed yield does not appear to be an effective strategy (Robinson et al. 2006). There were no differences in body fat and processed yield of fish fed a 28% protein diet for most of the growing season and then feeding a high-protein diet (32%–36%) for 30–90 days before harvest compared with fish fed the 28%-protein diet for the entire growing season.

Our research has also demonstrated that stocking at high density does not increase the amount of dietary protein needed for maximum fish growth. There were no differences in fish growth and production using a 28%- or 32%-protein diet at various stocking densities (14,820, 29,640, or 44,460 per ha) (Li et al. 2003a). Also, disease resistance of catfish was not affected by different dietary protein levels (28%, 32%, or 36%) (Li et al. 2003b). Further, protein requirements did not differ among catfish strains (Mississippi normal, USDA102, NWAC103, and Norris) (Li et al. 1998; Jackson et al. 2003; Li et al. 2001). Thus it appears that dietary protein requirements for commonly used catfish strains are similar.

ALTERNATIVE FEED INGREDIENTS

Traditionally catfish feeds have been comprised primarily of soybean meal, corn, wheat middlings, and small amounts of marine fish meal (mainly menhaden). Research has demonstrated that these ingredients may be replaced by other less expensive alternative feedstuffs. For example, cottonseed meal is a locally available product that is generally priced competitively (on a protein basis) with soybean meal. It has 41% crude protein, 2–4% crude fat, and 11–13% crude fiber. However, it contains gossypol, a compound that can be toxic to catfish if fed at high levels. Barros et al. (2000; 2002) reevaluated the free gossypol toxicity to catfish and showed a reduction in weight gain and feed consumption of juvenile catfish fed a diet containing 671 mg free gossypol/kg from cottonseed meal. However, total cell count, red and white cell counts, hematocrit, and hemoglobin were not affected. Yildirim et al. (2003) reported a significant growth reduction in juvenile catfish fed chemically-defined diets containing free gossypol from gossypol-acetic acid as low as 300 mg/kg diet. Hematocrit, hemoglobin, and red blood cell count decreased at dietary free gossypol concentrations of 600, 900, and 1,200 mg/kg or higher, respectively.

Yildirim-Aksoy et al. (2004) compared the toxicity of natural free gossypol from cottonseed meal and that from gossypol-acetic acid using practical diets containing graded levels of glanded CSM (for natural free gossypol diets) and glandless CSM (for gossypol-acetic acid diets), respectively. They observed that neither sources of dietary free gossypol at levels up to 800 mg/kg affected weight gain, feed intake, feed efficiency, survival, red and white blood cell counts, hematocrit, and hemoglobin of juvenile catfish. Lim et al. (2005) reported that toxic concentrations of gossypol on catfish depend on type of diet. Levels of free gossypol up to 1,200 mg/kg in practical diet had no negative effect on weight gain, feed intake, feed and protein efficiencies, and survival. These variables, however, were significantly lower in the group fed purified diets supplemented with 1,200 mg/kg gossypol.

Although gossypol can be toxic to catfish at high levels, the main factor limiting its use in catfish feeds is lysine since gossypol binds to lysine making part of the total lysine inherent in the seed biologically unavailable. Research has shown that about 50% of soybean meal can be replaced with cottonseed meal (27% of the diet) plus supplemental lysine in catfish diets without negatively affecting fish performance (Li and Robinson 2006; Robinson and Li in press [a]).

Distillers dried gains with solubles (a by-product of producing ethanol from corn) is another feedstuff that can be used in catfish feeds. It contains about 27% protein, 9% fat, and 8.5% fiber. Ethanol production from corn has increased dramatically in the past few years because of increased emphasis on alternative energy sources. This has resulted in an increased supply of distillers dried gains by-products that are more competitively priced. Research with pond-raised catfish shows that distillers dried grains with solubles can be used at levels up to 30%–40% of the diet when supplemented with lysine to partially replace soybean meal (Robinson and Li in press [a]). Further, a combination of cottonseed meal, distillers dried grains with solubles, and supplemental lysine can be used to completely replace soybean meal in catfish diets. The use of distillers dried grains with solubles appears to improve feed efficiency, but it is not clear whether the increased feed efficiency is a result of the higher fat level or to a specific nutrient present in the product. The use of distillers dried grains with solubles at levels of 30–40% in the diet increases body fat because of their high fat content (about 9%) (Robinson and Li in press [a]). It is recommended that it be used at a level of ≤ 20% in catfish feeds. However, high levels may be used if the germ, which contains about 85% of the oil in corn, is removed before the fermentation process, or oil in the by-product is extracted. Such high-protein, low-oil products are presently commercially available.

Corn gluten feed is a co-product of wet milling of corn to produce starch and nutritive sweeteners that is a potential energy/protein source for catfish. This product typically contains about 21% protein, 2% crude fat, and 10% fiber, and it is usually competitively priced relative to corn and wheat middlings. Up to 50% of corn gluten feed can be used in catfish feeds without detrimental effects (Robinson et al. 2001a). Unlike high-protein corn gluten meal, corn gluten feed contains levels of yellow carotenoids (lutein and zeaxanthin) similar to that in corn grain, which is well below the carotenoid level that can cause an undesired yellow pigmentation in catfish flesh. The use of corn gluten feed to replace corn will reduce fattiness in catfish because it contains less energy.

Menhaden fish meal may be replaced by poultry by-product meal, catfish offal meal, and hydrolyzed feather meal plus lysine (Li et al. 2002). However, research on the replacement of fish meal with meat and bone/blood meal (60/40) in diets for catfish fingerlings is contradictory. One study demonstrated that meat and bone/blood meal could replace fish meal without affecting fish performance (Li et al. 2002), while another study showed a reduction in weight gain of fish fed meat and bone/blood meal compared with that of fish fed fish meal diets (Li et al. 2003b). The discrepancy may have resulted from the different quality of animal protein sources used between the two studies. More recently, a study conducted in ponds has shown that meat and bone/blood meal can replace menhaden fish meal in diets for food fish grow out (Robinson and Li 2007a).

Although these animal by-products are good sources of protein for catfish, their use depends on their quality, cost, and availability. Processing methods can have a significant effect on their quality; thus, only high quality products should be used. Further, since the perception of using beef products to feed catfish may be detrimental to marketing the fish products (because of its implication of bovine spongiform encephalopathy or "mad cow" disease), we do not recommend the use of beef by-products in catfish feeds unless the product is from animals that have been tested free of contaminant. Also, catfish offal meal is not recommended for use in catfish diets because of the concern of intra-species recycling unless it has been certified to be free of any contaminants.

ALL-PLANT DIETS

Feedstuffs of animal origin are generally considered to be of higher quality than those of plant origin. However, because animal protein is more expensive and its availability is often limited, efforts have been expended to reduce or eliminate animal proteins in diets for food fish grow out. Research has shown that the addition of up to 6% meat and bone/blood meal (60/40) does not improve weight gain and production over all-plant diets for food-sized catfish (Robinson and Li 1999, 2000). However, research results on comparing all-plant diets with diets containing fish meal or fish meal plus meat and bone/blood meal in both fingerlings and food fish are inconsistent. Some studies show a reduction in growth and feed efficiency (Robinson and Li 1998; Li et al. 2006a; Li et al. in press), while others do not

(Reigh 1999; Li et al. 2003b; Robinson and Li 2007a). The reason for different results is not clear, but it could be caused by differences in the quality of fish meal and other ingredients used or other unidentified factors.

If fingerling and food-sized catfish require fish meal in their diets, based on available information it appears that 4% fish meal in the diet is adequate for optimum growth and feed efficiency. In studies with positive effects of fish meal, fingerlings raised in aquaria and fed all-plant diets had about 12–15% reduction in weight gain (Li et al. 2006; Li et al. in press); whereas, food-sized catfish raised in ponds and fed all-plant diets had about 5% reduction in weight gain (Robinson and Li 1998; unpublished data).

Considering the inconsistent results from various studies and the small reduction of weight gain in food fish fed all-plant diets, the use of fish meal in diets for food fish grow out may not be justified. There are economic and marketing advantages of excluding fish meal and other animal feedstuffs from catfish food fish diets. Economic benefits depend on the cost of fish meal relative to soybean meal and other animal protein sources. Since consumers are particularly health oriented and are exposed to untold negative media reports (whether factual or not) on issues that influence food animal industries, it seems prudent to distance oneself from these issues if possible. The main issue is "mad cow disease" associated with use of beef by-products in animal feeds. We recognize that the health risk to humans that eat catfish fed these products is small, but the elimination of beef products from the diet would help dispel any negative perception that might arise. Also, there is a concern about fish meal use in aquaculture since fish meal production from the ocean may threaten wild fish stocks and also a concern about heavy metal contamination in the fish meal. The argument could also be made to remove all animal feedstuffs from the catfish diet because all-plant diets would certainly make farm-raised catfish unique among farmed fish in that it would be truly "grain" fed.

PHYTASE AS REPLACEMENT FOR INORGANIC PHOSPHORUS

Pond-raised catfish require about 0.3% available phosphorus for optimum growth and bone development (Robinson et al. 2001b). Because catfish feeds for grow out consist primarily of plant ingredients that contain low levels of biologically available phosphorus, commercial catfish feeds are typically supplemented with an inorganic phosphorus source to meet the requirement. A considerable portion of the phosphorus in plant ingredients is in the form of phytate, a bound form of phosphorus that has low availability to simple-stomach animals including catfish. Thus, a substantial amount of dietary phosphorus is unabsorbed and may accumulate in pond sediments, which essentially regulates phosphorus concentrations in pond water (Gross et al. 1998).

Phytase, an enzyme derived from the fungus *Aspergillus niger*, has been shown to be effective in improving bioavailability of phytate phosphorus in diets of simple-stomach animals. Results from our laboratory and field studies demonstrate that a phytase concentration of 250–500 phytase unit (FTU)/kg diet applied post-pelleting improves weight gain and bone phosphorus retention over a basal diet without a phosphorus supplement (Robinson et al. 2002b; Li et al. 2004a). This phytase level can effectively replace the dicalcium phosphate supplement in catfish diets without affecting growth, feed efficiency, or bone phosphorus deposition. Yan et al. (2002) also reported that phytase levels of 500 FTU/kg and above improved phosphorus retention in the bone of catfish. Since there is a chance that losses of enzyme activity may occur during and after phytase application, we recommend that a phytase level of 500 FTU/kg diet be used in the diet to ensure that an adequate amount of phosphorus is available to catfish. In formulating diets utilizing phytase, it is important that the diet be formulated to contain an adequate level of total phosphorus so that sufficient phosphorus can be released by phytase that will be available to the fish. Based on results from these studies, a total phosphorus level of 0.60% should be adequate. Among the most commonly used plant ingredients, cottonseed meal, wheat middlings, and distillers dried grains with solubles are good sources of phosphorus. Therefore, these and other ingredients high in phosphorus should be included to ensure sufficient total phosphorus levels in the feed.

One possible benefit of using phytase to replace inorganic phosphorus supplements in catfish feeds is to reduce the dietary phosphorus load in the pond. Phytase improves utilization of phytate phosphorus inherent in the feed ingredients and thus reduces fecal phosphorus concentrations. One would expect that minimizing the amount of phosphorus going into the pond water would lower the phosphorus level and thus reduce phytoplankton growth. However, the use of phytase in catfish feeds did not influence total phosphorus, soluble reactive phosphorus, or chlorophyl-a concentrations in pond water (Tucker et al. 2005). It should be noted that these pond studies were conducted for only one or two growing seasons. Since the uptake of phosphorus by pond sediments is a major factor controlling phosphorus levels in the pond water, feeding phytase diets over a long term may be beneficial in catfish ponds by reducing the P load in pond sediments. Based on current prices of phytase and inorganic phosphorus sources, it is economical to use phytase in catfish diets. Since phytase will not withstand the high temperature associated with manufacturing extruded catfish feed, it must be applied post extrusion.

Factors Affecting Feed Efficiency

Feed conversion ratio (amount of feed fed per unit of weight gain), a term commonly used in animal production, measures how efficiently the animal converts feed to body mass. The lower the value the better feed utilization. In commercial catfish production, feed conversion ratio dramatically influences profitability of the farm since production cost increases with increasing feed conversion ratio. For example, at a feed price of $250 per metric ton, it costs $0.50 to produce 1 kg of fish if feed conversion ratio is 2, while it will cost $0.75 (a 50% increase) at a feed conversion ratio of 3.

Feed conversion ratio of catfish is influenced by several factors including feed quality, fish size, feeding rate and frequency, stocking density, water temperature, and environmental conditions. Low quality diets reduce feed efficiency; however, this is not a problem since today's catfish feeds are made of high quality ingredients that meet or exceed all nutrient and energy requirements of the fish. Fish size appears to have a marked impact on feed efficiency. As fish grow larger, feed efficiency greatly diminishes (unpublished data from the National Warmwater Aquaculture Center, Mississippi State University). For example, in small experimental ponds, it took 1.8 kg of feed to produce 1 kg of fish when catfish were grown from 0.23 to 0.45 kg, while it took 2.9 kg of feed to produce 1 kg of fish when grown from 0.91 to 1.13 kg (Figure 1). Large fish use more energy for sexual maturation and reproduction than for growth, which decreases feed efficiency.

Theoretically, feed conversion ratio decreases as feeding rate increases, and minimum absolute or biological feed conversion ratio occurs when catfish are fed to satiation or near satiation. The relationship (Figure 2) between biological feed conversion and feeding rate has been proven by laboratory studies in which juvenile catfish were fed various ration levels up to satiation without feed waste (Li et al. 2004a). However, this is not the case for fish raised in production ponds where satiation feeding usually result in poorer feed efficiency compared with restricted feeding, mainly because of more wasted feed when fish are fed to satiation daily. Our data demonstrate that feeding fish to satiation daily during food fish production provides higher production and weight gain, but the feed conversion ratio increases compared with feeding no more than 90 kg/ha per day at stocking densities of 18,500–247,100 fish/ha (Robinson and Li 1999; Robinson et al. 2004). We also found that the total feed input, net production, weight gain, and feed conversion were about the same when fish were fed to apparent satiation or fed at a "cut-off" rate of ≤ 135 kg/ha per day under a single-batch cropping system. Generally, the higher the stocking density the lower the feed efficiency will be (Li et al. 2003a; Robinson and Li 2007b). However, this may be, in part, due to decreased fish survival at higher densities. Unfavorable environmental conditions, such as temperature extremes, may also negatively impact feed efficiency.

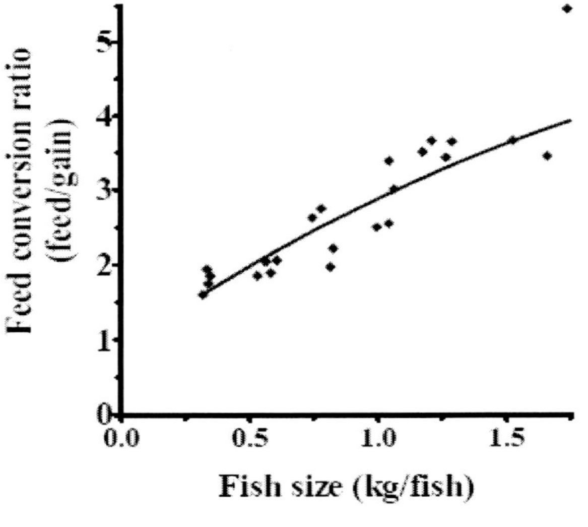

Figure 1. Feed conversion ratio (amount feed consumed per unit of weight gain) as a function of fish size for pond-raised, food-sized channel catfish.

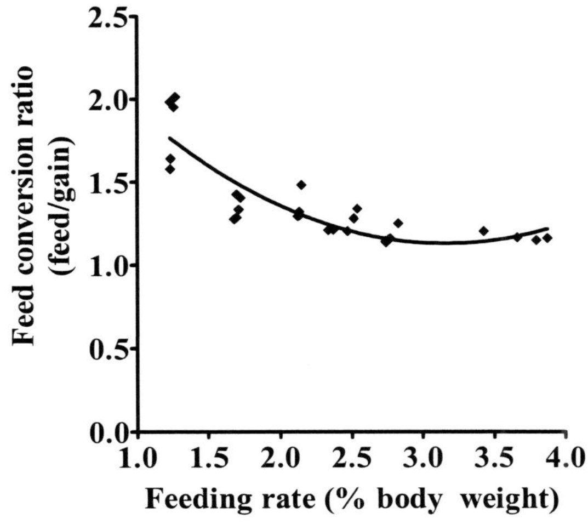

Figure 2. Weight gain and feed conversion ratio as a function of feeding rate for aquarium-raised, juvenile channel catfish.

MYCOTOXINS

Based on early work with purified aflatoxin B_1, it appears that catfish can tolerate high concentrations of aflatoxins (produced by molds *Aspergillus flavus or A. parasiticus*) (Lovell 1998). Recent studies conducted with catfish that were fed practical diets containing aflatoxins-contaminated corn (560 μg aflatoxin/kg corn) confirmed this assertion (Manning et al. 2005a). Aflatoxin levels of 220 and 88 μg/kg in aquarium and pond studies, respectively, had no effect on weight gain, feed conversion, hematocrit, and hepatosomatic index. No liver abnormalities were observed upon gross examination. Levels of aflatoxins were reduced by approximately 63% in the diets after the extrusion and drying processes during feed manufacturing, which involve high temperatures.

Catfish appear to be sensitive to ochratoxin A, primarily produced by molds *A. ochraceus and Penicilium verrucosum*. Results from an aquarium study showed that weight gain was depressed in catfish fed chemically-defined diets containing ≥ 1.0 mg ochratoxin A/kg supplied by contaminated corn meal, compared with fish fed either a diet containing 0.5 mg ochratoxin A/kg or the control diet (Manning et al. 2003a). Feed efficiency diminished in fish fed 4.0 and 8.0 mg ochratoxin A/kg diet. Survival and hematocrits were reduced in fish fed 8.0 mg ochratoxin A/kg of diet. Unlike other species studied, no necrosis of the renal tubules was observed. However, increased incidence and severity of multifocal melanomacrophage centers were noted in liver and kidney tissues of catfish fed ≥ 2.0 mg ochratoxin A/kg of diet. Exocrine pancreatic cells that normally surround the hepatic portal veins of catfish were reduced in number or absent in livers of fish fed ≥ 1.0 mg ochratoxin A/kg of diet. A subsequent study showed that dietary ochratoxin A levels of ≥ 4.0 mg/kg increased mortality of catfish experimentally infected with *Edwardsiella ictaluri*, a causative agent of enteric septicemia of catfish (Manning et al. 2005b).

Several mycotoxins produced by *Fusarium* species are toxic to catfish. Yildirim et al. (2000) reported that moniliformin reduced weight gain of juvenile catfish at dietary levels of ≥ 20 mg/kg. Levels below 20 mg/kg were not evaluated. Hematocrit values were reduced when catfish were fed ≥ 60 mg moniliformin/kg diet. When fed in combination, 20 or 40 mg moniliformin and 40 mg fumonisin B_1/kg diet had a negative synergistic effect on weight gain of juvenile catfish. Levels as low as 0.625 mg T-2 toxin/kg of diet depressed growth of juvenile catfish (Manning et al. 2003b). A dietary T-2 level of ≥ 1.25 mg/kg reduced hematocrit value, ≥ 2.5 mg/kg reduced survival, and ≥ 5.0 mg/kg reduced feed efficiency. Histopathological abnormalities of stomach, head, and trunk kidney were observed in fish fed diets containing ≥ 1.25 mg/kg. A subsequent study showed that dietary T-2 levels of ≥ 1.0 mg/kg increased mortality in catfish experimentally infected with *E. ictaluri* (Manning et al. 2005b). Dietary deoxynivalenol (purified) levels of up to 10 mg/kg did not affect feed consumption, weight gain, feed efficiency, hematocrit values, and liver weight in juvenile catfish (Manning 2005). A later study, in which juvenile catfish were fed practical diets containing deoxynivalenol-contaminated wheat, confirmed this observation (Manning 2005). However, fish fed 15 and 17.5 mg deoxynivalenol/kg of diet had reduced weight gain and feed efficiency. No feed refusal or vomiting, as that observed in other animals, was noted in catfish fed diets containing up to 17.5 mg deoxynivalenol/kg. At these levels, hematocrit values were not affected.

Feeding Strategies

Unlike other farmed animals, fish are not fed *ad libitum* (feed is available at all times and an animal can feed at will). Rather, the amount of feed offered to the fish is generally decided by the feeder. Therefore, feeding fish is a highly subjective process. There are no standard feeding practices across the catfish industry, mainly because many factors affect feeding and every pond of fish behave differently.

Generally, feeding once daily is satisfactory for food fish grow out. Research has shown that feeding food fish twice daily is not necessarily beneficial (Li et al. 2005a). Although fish fed twice daily were offered more feed than fish fed once daily, the increased feed input was not accounted for in weight gain. It is likely that feeding twice daily increases feed conversion because feed can be easily wasted by over-feeding.

Several factors dictate how much to feed catfish in a production pond. These include standing crop (number and weight of fish present in the pond), fish size, water temperature, water quality, and weather conditions. Generally catfish should be fed daily as much as they will eat (satiation) without wasting feed and without adversely affecting water quality. However, feeding to satiation is problematic since it is difficult to determine when the fish are satiated. Thus it is easy to overfeed, which results in wasted feed that may have deleterious effects on water quality. Our data demonstrate that feeding to satiation daily results in higher production and weight gain, but feed conversion ratio increases compared with feeding a restricted rate of ≤ 90 kg/ha per day (Robinson and Li 1999; Robinson et al. 2004).

Although we recommend that catfish grown for food be fed once daily, feeding less frequently than daily may be called for under certain circumstances, such as delayed harvest because of disease and off-flavor problems. Research has shown that fish fed every other day consume more feed on days fed compared with fish fed once daily to apparent satiation (Li et al. 2004b; Li et al. 2006b; Robinson and Li in press[b]). The increased feed consumption (on days fed) in fish fed less frequently than daily is primarily the result of compensatory growth (Lovell 1998; Gaylord and Gatlin 2001; Reigh et al. 2006), or at least a partial compensatory growth. Fish are capable of compensating for all or part of the weight loss during a short period of feed deprivation when full feeding is resumed. Although there are some advantages (reduced feed conversion, labor cost, and aeration) to feeding every other day, we do not recommend this type of feeding regimen for routine feeding since fish fed every other day cannot consume enough feed on days fed to compensate for the missed feed on days when they are not fed. In addition, feeding every other day appears to reduce fish processed yield and it extends the production cycle; therefore in the long term it may not be economical.

During the growing season, most catfish producers feed their fish seven days a week, but there are some producers who feed six days a week. Our data show that feeding six days a week (not feeding on Sundays) reduced net production by 3.3% and feeding five days a week (not feeding on both Saturdays and Sundays) reduced net production by 6.9%, compared with fish fed seven days per week for a growing season (Li et al. 2005b). Feed conversion ratio was reduced by 4.8% (1.80 vs. 1.89) and 7.9% (1.74 vs. 1.89), respectively for fish fed six days and five days a week, compared with fish fed seven days per week. Considering the slight decrease in net production, slight improvement in feed conversion, and possible reduction in other costs associated with feeding, feeding six days per week may reduce production cost for food-sized catfish. However, it should be noted that in our study, we used a single-batch cropping system and the fish were fed to satiation on days fed. If feed is restricted, one would expect a further reduction in net production by feeding six days per week compared with fish fed seven days per week. Also, if this strategy is used in a multi-batch cropping system, skipping feed days may have a more negative impact than in single-batch cropping systems because the smaller fish may gain less weight than was demonstrated in our study.

Maintenance feeding implies that all feed consumed by the fish is used to maintain the animal with no gain or loss of weight. This feeding regimen may be achieved by either feeding fish a maintenance ration daily or feeding to apparent satiation one or two times per week. Since fish of various sizes are typically present in a given pond at the same time, it is better to feed all they will eat on days fed than feeding a little every day. Feeding the fish all they eat on the days fed will allow the smaller, less aggressive fish to feed. The maintenance feeding regimen may be used to feed food-sized fish that cannot be harvested because of off-flavor or other issues. Based on our research results, it appears that feeding once a week to satiation can maintain the body weight of food-sized catfish under the single-batch cropping system (Li et al. 2005c). However, the condition factor, a nutritional status indicator that measures the relationship between fish body weight and length, is lower for fish fed once a week than fish fed more frequently per week.

References

Barros, M.M., Lim, C., Evans, J.J. and Klesius, P.H. (2000). Effect of iron supplementation to cottonseed meal diets on growth performance of channel catfish, *Ictalurus punctatus*. *Journal of Applied Aquaculture* 10 (1):65–86.

Barros, M.M., Lim, C. and Klesius, P.H. (2002). Effect of soybean meal replacement by cottonseed meal and iron supplementation on growth, immune response and resistance of channel catfish (*Ictalurus punctatus*) to *Edwardsiella ictaluri* challenge. *Aquaculture* 207:263–279.

Gaylord, T.G. and Gatlin, D.M. III. (2001). Dietary protein and energy modification to maximize compensatory growth of channel catfish (*Ictalurus punctatus*). *Aquaculture* 194:337–348.

Gross, A., Boyd, C.E., Lovell, R.T. and Eya, J.C. (1998). Phosphorus budgets for channel catfish ponds receiving diets with different phosphorus concentrations. *Journal of the World Aquaculture Society* 29:31–39.

Jackson, L.S., Robinson, E.H., Li, M.H., Wolters, W.R. and McKee, D.A. (2003). Restricted and satiate feeding of two genetically isolated strains of juvenile channel catfish *Ictalurus punctatus* reared on 28% and 32% protein diets. *Journal of the World Aquaculture Society* 34:478–486.

Li, M.H. and Robinson, E.H. (2006). Use of cottonseed meal in diets of aquatic animals: A review. *North American Journal of Aquaculture* 68:14–22.

Li, M.H., Robinson, E.H. and Wolters, W.R. (1998). Evaluation of three strains of channel catfish *Ictalurus punctatus* fed diets containing three concentrations of dietary protein. *Journal of the World Aquaculture Society* 29:156–161.

Li, M.H., Robinson, E.H., Manning, B.B., Bosworth, B.G. and Wolters, W.R. (2001). Comparison of growth, processing yield, and body composition of USDA103 and Mississippi "normal" strains of channel catfish *Ictalurus punctatus* fed diets containing three concentrations of protein. *Journal of the World Aquaculture Society* 32:402–408.

Li, M.H., Robinson, E.H. and Manning, B.B. (2002). Comparison of various animal protein sources for growth, feed efficiency, and body composition of juvenile channel catfish *Ictalurus punctatus*. *Journal of the World Aquaculture Society* 33:489–493.

Li, M.H., Manning, B.B., Robinson, E.H. and Bosworth, B.G. (2003a). Effect of dietary protein and stocking density on production characteristics of pond-raised channel catfish *Ictalurus punctatus*. *Journal of the World Aquaculture Society* 34:147–155.

Li, M.H., Wise, D.J., Manning, B.B. and Robinson, E.H. (2003b). Effect of dietary total protein and animal protein on growth and feed efficiency of channel catfish *Ictalurus punctatus* and their response to *Edwardsiella ictaluri* challenge. *Journal of the World Aquaculture Society* 34:223–228.

Li, M.H., Manning, B.B. and Robinson, E.H. (2004a). Summary of phytase studies for channel catfish. Research Report Vol. 23, No. 13. Mississippi State, Mississippi, USA: Mississippi Agricultural and Forestry Experiment Station.

Li, M.H., Manning, B.B. and Robinson, E.H. (2004b). Effect of daily feed intake on feed efficiency of juvenile channel catfish. *North American Journal of Aquaculture* 66:100–104.

Li, M.H., Robinson, E.H., Manning, B.B. and Bosworth, B.G. (2004c). Effect of dietary protein concentration on production characteristics of pond-raised channel catfish fed once daily or once every other day to satiation. *North American Journal of Aquaculture* 66:184–190.

Li, M.H., Manning, B.B. and Robi*nson, E.H. (2005a). Effect of feeding time and frequency on channel catfish production. Journal of Applied Aquaculture* 17 (3):51–60.

Li, M.H., Robinson, E.H. and Bosworth, B.G. (2005b). Effect of periodic feed deprivation on growth, feed efficiency, processing yield, and body composition of channel catfish *Ictalurus punctatus*. *Journal of the World Aquaculture Society* 36:444–453.

Li, M.H., Robinson, E.H., Manning, B.B. and Oberle, D.F. (2005c). Effects of maintenance feeding regimens on weight gain, feed efficiency, and body condition of pond-raised channel catfish. *North American Journal of Aquaculture* 67:129–132.

Li, M.H., Peterson, B.C., Janes, C.L. and Robinson, E.H. (2006a). Comparison of diets containing various fish meal levels on growth performance, body composition, and insulin-like growth factor-I of juvenile channel catfish *Ictalurus punctatus* of different strains. *Aquaculture* 253:628–635.

Li, M.H., Manning, B.B., Robinson, E.H., Oberle D.F. and Bosworth, B.G. (2006b). Effects of dietary protein concentration and feeding regimen on channel catfish *Ictalurus punctatus* production. *Journal of the World Aquaculture Society* 37:370–377.

Li, M.H., Peterson, B.C., Bates, T.D. and Robinson, E.H. In press. Growth and feed efficiency of juvenile channel catfish reared at different water temperatures and fed diets containing various levels of fish meal. *North American Journal of Aquaculture*.

Lim, C., Yildirim, M., Wan, P. and Klesius, P.H. (2005). Gossypol in purified and practical diets: Effect on growth performance, immune responses and resistance of channel catfish, *Ictalurus punctatus* to *Edwardsiella ictaluri* infection. Aquaculture America Conference; 2005 Jan 17–20 New Orleans, Louisiana; p. 242.

Lovell, T. (1998). Nutrition and feeding of fish, 2nd edition. Boston, Massachusetts: Kluwer Academic Publishers.

Manning, B.B. (2005). Mycotoxins in aquaculture. in D.E. Diaz (Ed.), The Mycotoxin Blue Book (pp. 139–156). Nottingham, United Kingdom: Nottingham University Press.

Manning, B.B., Ulloa, R.M., Li, M.H., Robinson, E.H. and Rottinghaus, G.E. (2003a). Ochratoxin A fed to channel catfish (*Ictalurus punctatus*) causes reduced growth and lesions of hepatopancreatic tissue. *Aquaculture* 219:739–750.

Manning, B.B., Li, M.H., Robinson, E.H., Gaunt, P.S., Camus, A.C. and Rottinghaus, G.E. (2003b). Response of channel catfish to diets containing T-2 toxin. *Journal of Aquatic Animal Health* 15:229–238.

Manning, B.B., Li, M.H. and Robinson, E.H. (2005a). Aflatoxins from moldy corn cause no reductions in channel catfish *Ictalurus punctatus* performance. *Journal of the World Aquaculture Society* 36:59–67.

Manning, B.B., Terhune, J.S., Li, M.H., Robinson, E.H. and Wise, D.J. (2005b). Exposure to feed-borne mycotoxins T-2 toxin or ochratoxin A causes increased mortality of channel catfish challenged with *Edwardsiella ictaluri*. *Journal of Aquatic Animal Health* 17:147–150.

National Marine Fisheries Service. (2006). Fisheries of the United States – 2006. Silver Spring, Maryland: National Marine Fisheries Service.

Reigh, R.C. (1999). Production characteristics of pond-raised channel catfish *Ictalurus punctatus* fed diets with and without animal protein for three growing seasons. *Journal of the World Aquaculture Society* 30:154–160.

Reigh R.C., Williams, M.B. and B.J. Jacob. (2006). Influence of repetitive periods of fasting and satiation feeding on growth and production characteristics of channel catfish *Ictalurus punctatus*. *Aquaculture* 254: 506–516.

Robinson, E.H. and Li, M.H. (1998). Comparison of practical diets with and without animal protein at various concentrations of dietary protein for performance of channel catfish *Ictalurus punctatus* raised in earthen ponds. *Journal of the World Aquaculture Society* 29:273–280.

Robinson, E.H. and Li, M.H. (1999). Effect of dietary protein concentration and feeding rate on weight gain, feed efficiency, and body composition of pond-raised channel catfish *Ictalurus punctatus*. *Journal of the World Aquaculture Society* 30:311–318.

Robinson, E.H. and Li, M.H. (2007a). Catfish protein nutrition, revised. Bulletin No. 1159. Mississippi State, Mississippi: Mississippi Agricultural and Forestry Experiment Station.

Robinson, E.H. and Li, M.H. (2007b). Effects of fish size and feeding frequency on catfish production. Research Report Vol. 24, No. 2. Mississippi State, Mississippi, USA: Mississippi Agricultural and Forestry Experiment Station.

Robinson, E.H. and Li, M.H. (In press [a]). Replacement of soybean meal in channel catfish, *Ictalurus punctatus*, diets with cottonseed meal and distillers dried grains with solubles. *Journal of the World Aquaculture Society*.

Robinson, E.H. and Li, M.H. (In press [b]). Effects of feeding diets with or without fish meal on production of channel catfish, *Ictalurus punctatus*, stocked at varying densities. *Journal of Applied Aquaculture*.

Robinson, E.H., Li, M.H. and Manning, B.B. (2000). Evaluation of various concentrations of dietary protein and animal protein for pond-raised channel catfish *Ictalurus punctatus* fed to satiation or at a restricted rate. *Journal of the World Aquaculture Society* 31:503–510.

Robinson, E.H., Li, M.H. and Manning, B.B. (2001a). Evaluation of corn gluten feed as a dietary ingredient for pond-*raised channel catfish Ictalurus punctatus. Journal of the World Aquaculture Society* 32:68–71.

Robinson, E.H., Li, M.H. and Manning, B.B. (2001b). A practical guide to nutrition, feeds, and feeding (second revision). Bulletin No. 1113. Mississippi State, Mississippi: Mississippi Agricultural and Forestry Experiment Station.

Robinson, E.H., Li, M.H. and Manning, B.B. (2002a). Evaluation of low-protein diets for production of fingerling channel catfish. *National Warmwater Aquaculture Center Newsletter* 5 (1):4–5.

Robinson, E.H., Li, M.H. and Manning, B.B. (2002b). Comparison of microbial phytase and dicalcium phosphate for growth and bone mineralization of pond-raised channel catfish *Ictalurus punctatus*. *Journal of Applied Aquaculture* 12 (3):81–89.

Robinson, E.H., Li, M.H., Manning, B.B., Mischke, C.C. and Bosworth, B.G. (2004). Effects of high protein diets on channel catfish *Ictalurus punctatus* production, composition of gain, processing yield, and water quality. *Journal of the World Aquaculture Society* 35:468–477.

Robinson, E.H., Li, M.H., Oberle, D.F. and Bosworth, B.G. (2006). Effects of high-protein finishing diets on channel catfish *Ictalurus punctatus* production. *Journal of Applied Aquaculture* 18 (3):37–52.

Tucker, C.S., Hargreaves, J.A. and Kingsbury, S.K. (2005). Dietary phosphorus modifications in practical feeds do not affect waterborne phosphorus concentrations and phytoplankton abundance in channel catfish ponds. *North American Journal of Aquaculture* 67:114–121.

Yan, W., Reigh, R.C. and Xu, Z. (2002). Effects of fungal phytase on utilization of dietary protein and minerals, and dephosphorylation of phytic acid in the alimentary tract of channel catfish *Ictalurus punctatus* fed an all-plant-protein diet. *Journal of the World Aquaculture Society* 33:10–22.

Yildirim, M., Manning, B.B., Lovell, R.T., Grizzle, J.M. and Rottinghaus, G.E. (2000). Toxicity of moniliformin and fumonisin B_1 fed singly and in combination in diets for young channel catfish *Ictalurus punctatus*. *Journal of the World Aquaculture Society* 31:599–608.

Yildirim, M., Lim, C., Wan, P.J. and Klesius, P.H. (2003). Growth performance and immune response of channel catfish (*Ictalurus punctatus*) fed diets containing graded levels of gossypol-acetic acid. *Aquaculture* 219:751–768.

Yildirim-Aksoy, M., Lim, C., Wan, P.J. and Klesius, P.K. (2004). Effect of natural free gossypol and gossypol-acetic acid on growth performance and resistance of channel catfish (*Ictalurus punctatus*) to *Edwardsiella ictaluri* challenge. *Aquaculture Nutrition* 10:153–165.

In: Aquaculture Research Trends
Editor: Stephen H. Schwartz, pp. 279-295

ISBN 978-1-60456-217-0

Chapter 9

THE UNDERWATER SILVICULTURE APPROACH FOR REEF RESTORATION: AN EMERGENT AQUACULTURE THEME

Shai Shafir and Baruch Rinkevich
Israel Oceanographic and Limnological Research, National Institute of Oceanography, Tel Shikmona, P.O. Box 8030, Haifa 31080, Israel

ABSTRACT

Coral reefs, often portrayed as "the rain forest of the sea", provide sustenance to hundreds of million of people worldwide. When severely degraded, reef restoration is expected to uphold the ecosystem biodiversity and retain, at least partially, the original biological and ecological portraits. Such active reef restoration practices are based on aquaculture of coral colonies (mariculture them in underwater nurseries) in as much as silviculture depends on rearing young trees in terrestrial nurseries. The management and cost efficiency of the nursery phase is crucial to the success of the whole restoration activities. However, contrast to terrestrial nurseries, coral nurseries are in the developing concepts and early operational regulation status. Here, we illustrate the practice of 'reef gardening' as a major reef restoration act, using common features, divergent approaches and ubiquitous rationale, as apply in terrestrial (forests) and marine (coral reefs) restoration principles. Even though basic protocols for reef gardening have not yet been thoroughly studied, several common agricultural aspects for silviculture and reef gardening are outlined. These include the use of reef restoration as mariculture theme, economic considerations and aquaculture perspectives such as dimension of nursery and its content, agricultural periods, growth rates, seedling spacing, maintenance issues, pest control, ecological goods and services and nursery construction. As restoration measures (terrestrial and marine alike) require significant amount of restoration material, simplifying coral seedling preparation and mariculturing them through low-tech technologies will ensure the success of this scientific discipline.

Introduction - What Underwater Silviculture Means?

Coral reefs boast exceptional species diversity but are poorly protected, highly degraded, and exposed to multiple persisting threats (Bruno and Selig 2007). Due to their economic importance and aesthetic attributes, there is increasing academic and public awareness towards the worldwide degradation of coral reefs and the associated decline in reef dwelling organisms (Bellwood et al. 2004; Feary et al. 2007). The decline in the abundance and diversity of reef fish through loss of structural reef heterogeneity is well studied (Jones et al. 2004; Wilson et al. 2006). Not only have approximately one fifth of the world's reefs been significantly destroyed by anthropogenic activities but it also shows ominous prospect of recovery (Bruno and Selig 2007; Wilkinson 2004). A further half of the worldwide reefs is under threats of collapse or, at least, under longer-term degradation. Following the predictions for global change, it has been suggested that some coral reefs will phase shift rather than disappear completely (Bruno and Selig 2007; Hughes et al. 2003), becoming different ecosystems, changing gradually into dissimilar ecological status and new consortia of reef associated vertebrates and invertebrates.

Coral reefs are often portrayed as "the rain forest of the sea". This assertion refers to their being one of the most productive biological ecosystems on earth, to their high level of biodiversity, and above all, to their function as the building blocks of the ecosystem framework (Connell 1978; Goreau et al. 1979; Epstein and Rinkevich 2001; Idjadi and Edmunds 2006). Coral colonies, as trees, share a common ecological role by virtue of their biological properties; both organisms constructed their ecosystems by engineering 3-D structures habitats. As the key organisms in their ecosystems, coral reefs and trees harbor numerous inhabitants and formulate local biological and environmental conditions (i.e., trees stabilize soil, enhance litter production and increase soil organic matter; corals enhance lithification processes and reef rock formation). As damaging anthropogenic activities are growing and intensifying alarmingly, coral reefs and rain forests are losing their resilience and self-rehabilitation are impaired. Clearing and logging forests and converting them to agricultural expanses intensify soil erosion and desertification. Similarly, marine recreational activities, pollution, over-fishing and the like, lead to progressive deterioration of coral coverage, collapse of reef communities, phase shift, and development of algal reefs (Lugo et al. 2000; Gunderson 2007; Hughes et al. 2007).

Forests cover 30% of the world land area, tending more than 50% of the terrestrial carbon and 60-90% of the animal and plant species. Noteworthy is the deforestation of tropical forests, which causes the lose of 14.2 million hectares a year (www1), sum up to nearly 50% of their area, which in turn, leads to an estimated 19% extinction of forest-associated endemic species (Wright and Muller-Landau 2006). Most forests (90%) in the temperate zone, especially those in the northern hemisphere, are the outcome of silviculture. 'Silviculture is the agriculture of trees; how to grow them, how to maximize growth and return, and how to manipulate tree species compositions to meet landowners' objectives' (Rinkevich 2006). Active rehabilitation of forests through silviculture has become the foremost mode of forest management (Charnley 2006), and planting practiced in 150 countries around the world account for 5% of the forests (www1). In contrast to forests, coral reefs cover less than 1% of the oceanic area (Hughes et al. 2003). However, this productive ecosystem supports almost one third of the world's marine fish species (Newton et al. 2007), yielding around 10% of the total fish consumed by humans (Smith 1978; Pauly et al. 2002), facilitating nursery sites for open sea fish fingerlings (Crépin 2003) and form the ground for tourism industry (Ahmed et al. 2007). More than 100 countries have coral reefs along their coasts, which sustain the livelihood of tens of millions of people (Salvat 1992). Unfortunately, coral reefs are declining much faster than forests, 1% annually before 1997 and 2% annually between 1997-2003 (Bruno and Selig 2007).

Silviculture and reef restoration (Epstein and Rinkevich 2001; Omori and Fujiwara 2004; Rinkevich 2005, 2006; Edwards and Gomez 2007; Skinner 2007) are complex biological and ecological procedures. Essentially, silviculture and reef restoration are expected to maintain the ecosystem biodiversity and retain, at least partially, the original biological and ecological portraits (Rinkevich 1995, 2000, 2006; Parviainen and Frank 2003). However, those disciplines differ in their targeted objectives. Many silvicultural practices favor regeneration of commercial species for the timber industry (Putz et al. 2001) whereas, silviculture endeavors to grow trees not only for timber but also for the environment, restoration of coral reef, at present, is designed solely for the environment. In contrast to silviculture practices and rationale, lack of state-of-the-art remediation protocols for reef restoration impedes the development of this ecological discipline. The following overview wishes to illustrate the practice of 'reef gardening' (Rinkevich 1995, 2000 2005, 2006; Shafir et al. 2006a,b) based on lessons learned from silvics and silviculture, their common features, divergent approaches and ubiquitous rationale as apply in terrestrial (forests) and marine (coral reefs) restoration principles.

The Gardening Concept In Reef Restoration - An Aquaculture Theme

The well-established discipline of silviculture led Rinkevich (1995) to suggest a new marine restoration concept that derives its rationale from silviculture, the two-steps protocol for the "gardening of the coral reef concept". The ground for this concept is restoration of declined areas by means of active rehabilitation to the stage where conservation acts are sufficient (Rinkevich 1995, 2000, 2005, 2006). The first step in this "gardening concept" entails rearing coral “seedlings”, in specially designed nurseries (Rinkevich 1995), to plantable size before applying the second step, rehabilitation (by active transplantation) of damaged area with the maricultured coral colonies (Rinkevich 1995).

Since target coral species need to be cultured in large numbers, the nursery step is pivotal to the success of the rehabilitation process. Careful maintenance ensures faster growth of healthy coral colonies. Issues like nursery structure and shape, temporary substrates for coral colony mariculture, realistic number of newly generated and maintained colonies, duration of nursery phase, growth rates and longevity of farmed colonies, and more (Shafir et al. 2006a,b), are some of the crucial issues associated with nursery care (Rinkevich 1995, 2000, 2005, 2006). The success of coral nursery depends of its initial set-up. The most successful nurseries studied up-to-date are those located in mid-water, at nutrient-enriched area, in isolated sites, away from the natural reef, protected from skin and scuba divers or predation by corallivorous organisms (Shafir et al. 2006a). A protected nursery phase provides cultured corals with a farming period that ensures better survivorship and growth to a size suitable for transplantation. A successful yield of the nursery depends also on management and protocols that produce healthy and plantable coral colonies at profitable cost. The nursery step in the "gardening concept" is the phase drawing most scientific investigation, and which conceptual and technical issues have already been addressed (Shafir et al. 2006a,b; Amar and Rinkevich 2007). The transplantation phase, which is in its preliminary steps, has already revealed interesting results, but its applications are still being developed (unpublished). Both phases are comparable to agricultural concepts and raise new themes in the discipline of farming of marine organisms.

Since the nursery stage in silviculture, in which trees are raised from seeds and seedlings, has been subjected to rigorous scientific testing for more than a century (Dumroese et al. 2005), its concepts and management tools are well established. The theories and practices shared by the "gardening concept" and silviculture led us to approach coral nursery with tree nursery concepts (Rinkevich 2006). Some of the common agricultural aspects are outlined below even though basic protocols have not yet been thoroughly studied.

Economic Considerations - Ecological Goods and Services

The economic value of an ecosystem is a practical tool for consistently evaluating different ecosystems (Naidoo and Ricketts 2006). Goods and services of each ecosystem are classified according to their direct and indirect contribution. For example, timber, fuel and tourism are considered as major direct goods and services derived from the forest ecosystem, whereas indirect benefits are protection of land and carbon storage (Pearce 2001). The total economic value of global forests exceeds US$ 4.7 trillion year^{-1} and the temperate/boreal forests economic value is evaluated at US$ 1719.2 ha^{-1}year^{-1} (Krieger 2001). There has yet to be developed a single, universal, and uniform model for economic evaluation that will serve as common ground for restoration economy.

However, this is not the case with coral reefs. As specified, coral reefs provide hundreds of million of people with goods and services. These include coastal protection, reef mining, food and proteins, agar and carrageenan, recreational activities, major source material for marine aquarium trade, ornamentals, new drugs for the pharmaceutical industry, raw material (coral skeletons) for medical purposes and cultural benefits (Moberg and Folke 1999; Ahmed et al. 2007). Coral reefs' services are vital for many less developed countries. For example, coral reef fisheries in the Philippines provide livelihood for more than a million small-scale fishermen who contribute almost US$ 1 billion annually to the country's economy (White et al. 2000). As reefs come in a great variety of forms, this is not simple to appraise their economic value or to assess the economic outcomes of reef degradation (Moberg and Folke 1999; Subade 2007). The benefits of each square kilometer of coral reef are often valued by parameters like the "willing to pay" factor (which we think is wrong as not based on robust economic considerations), the "total economic value" or the "biodiversity production values" (Subade 2007). For assessing the preferable extent of reef restoration or compensation, additional methodologies like 'habitat-equivalent analysis' (Milon et al. 2001, Spurgeon 2001; Precht et al. 2002) are used. The evaluations of coral reefs therefore vary between reef areas and between estimation methods. The economic value of Hawaii's reef area is estimated at US$ 878 $ha^{-1}year^{-1}$ (Cesar and van Beukering 2004) whereas the economic value of Tubbataha Reefs, the Philippine is estimated at less than US$ 19 $ha^{-1}year^{-1}$ (Subade 2007). Some methods estimated the economic value of the worldwide reefs to be between US$ 231 and US$ 2,700 $ha^{-1}year^{-1}$ (Bruno and Selig 2007).

ECONOMIC CONSIDERATIONS - ACTIVE RESTORATION ISSUES

When weighing the economic factors of coral reef restoration, one should favor the priorities of the underdeveloped countries by adopting "low-tech" methodologies that address the type of reef damages and the needs most common in local rural communities (Lindahl 1998; Bowden-Kerby 2001; Rinkevich 2005). Any economic estimation is bound to fluctuate because of the dynamic nature of the coral reef environment (i.e., lose/recruitment of coral colonies, reef invertebrates and fish, natural catastrophes). Anthropogenic loses associated with coral reefs (like fisheries, trade values of coral colonies and other exotic invertebrates or fish.) are not easy to quantify since, in most places, coral reefs are held in public trust by local, or governmental, agencies (Rinkevich 2005), that do not value the reefs by economic considerations. It is therefore important that greater awareness would be invested in better economic assessments (Spurgeon 2001), which will improve effectiveness of restoring coral reefs. Innovative economic templates guiding restoration initiatives are urgently needed (Rinkevich 2000).

Some of the economic considerations mirror those of silviculture. Active reef restoration is based on aquaculture of coral colonies (farming them in underwater nurseries) as silviculture depends on rearing young trees in terrestrial nurseries. However, while silviculture has been undergoing rigorous scientific testing for nearly two centuries on a global scale (Graham and Jain 1998), the budding discipline of active reef restoration is less than a decade old. Silviculture is the applied restoration act for rehabilitating terrestrial habitats, where natural self-regeneration of forests is not feasible (Putz et al. 2001). The same applies to the reef arena. Active reef restoration is drawing increasing attention because nearly all other practices for conserving degrading reefs have failed to yield significant results (Rinkevich, 1995; Risk, 1999; Epstein et al., 2001, 2003, 2005; Edwards and Gomez 2007). Such restoration acts carry economic significance. For example, conserving Southeast Asia reefs is estimated to be worth \$23,100– \$270,000 km^2/year (Burke et al. 2002).

Active restoration deals with scientific questions and with technical challenges. The scientific part focuses on issues like genetics, species combinations, landscape manipulations, biological engineering and key species (Palmer et al. 1997; Naveh 2007). The technical part focuses on cost, rates of success in each technical step, maintenance, number of new transplants and their survivorship after transplantation. This overview deals with the first phase of the 'gardening concept', the nursery phase. We outline below some aquaculture perspectives, comparing tree nursery (rearing trees for the timber industry and for reforestation) with the knowledge gained for mid-water coral nursery (rearing coral colonies for rehabilitation of degraded reef areas). Efficient nursery maintenance practices (Howell and Harrkngton 2004) are vital to the success of the entire restoration process.

Aquacultural Perspectives

Diverse types of tree nurseries for timber or reforestation exist. These include bare-root nursery and container nursery, each with its unique maintenance and taxon combinations (Riley and Steinfeld 2005) that influence nursery performance and cost-benefit values. Although still in its initial phases, the same applies to coral nurseries. Start up experiments have already designated several types of coral nurseries, including mid-water floating nursery, attach to the bottom nursery, leg-fixed nursery, *in-situ* vs. *ex-situ* nursery (Rinkevich 1995; 2000, Soong and Chen 2003; Shafir et al. 2006a), in addition to new types, such as rope nursery (unpubl.). Below is a comparison between a mid-water coral nursery (the most studied type of coral nursery) and a container tree nursery. Seedlings planted in containers have the advantage of a discrete medium that supplies the roots all their needs (water, air, nutrients). The containers also provide physical support to the seedling during the nursery period and during shipment (Landis et al. 1990).

A. Dimension of nursery and its content -

Both types of nurseries are designed to raise large numbers of propagates for a wide range of applications (restoration, education, special markets as timber or marine aquarium). These different applications may experience contradicting farming purposes, like the decision for culturing of biological material with high vs. low genetic variation or enhanced vs. reduced number of species, including numeral vs. few genotypes per specific species.

Mid-water coral nurseries farm primarily coral fragments and nubbins (Rinkevich 2005, 2006; Shafir et al. 2006a,b). Depending on its initial size, each single donor colony can yield hundreds to thousands of nubbins and fragments (Shafir et al. 2006a,b). Approximately 30 nubbins can be pruned from a thumb size coral branch with minimal damage to the donor colony (Shafir et al. 2006c). Therefore, a $10m^2$ size coral nursery (a size often used for experimental manipulations) could hold around 3,000 coral seedlings. In container tree nurseries, the source material is seeds, and a $10m^2$ nursery farms about 2,000-5,000 seedlings. However, since a profitable nursery area should exceed 200-300 acres, container tree nurseries usually hold millions of seedlings (Dumroese and Barnett 2004; www1). There are no plans yet to build large-scale mid-water coral nurseries because so far, restoration has been carried out with some success on scales of tens of square meters to several hectares (Edwards and Gomez 2007). Only large-scale mid-water nurseries would be able to cope with extensive reef degradation of several square kilometers and more.

B. Economic considerations -

Nursery costs are divided into fix costs, i.e., hardware, the initial cost of building the nursery; and variable costs, i.e., maintenance. Mid-water nursery fix costs are relatively small compared to the costs of terrestrial nurseries. Whereas the major cost in establishing a terrestrial nursery is the capital needed for acquisition the land, roads, irrigation, and machinery (Rantala et al. 2003), it is not relevant for any type of underwater coral nursery.

However, the cost of their outlay (which varies considerably depending type of construction) and their maintenance are comparable. The fix cost for a 36 m^2 coral nursery in Eilat, which include hardware as frame, anchors, buoys, ropes and cultivating plastic trays, is about US$ 1000 (Shafir et al. 2006a). The variable costs in coral and tree nurseries are mainly the cost of labor and culturing time, which could vary between counties (Rantala et al. 2003; Navarro et al. 2006; Shafir et al 2006a). The variable costs of material for coral nurseries are small, less than US$ 0.1 per coral seedling (Shafir et al. 2006a). The cumulative cost of mariculturing a single branching coral species in a mid-water coral nursery (from preparation to transplantation) is estimated at US$0.5-1 (Shafir et al. 2006a). It is comparable to the estimated cost of US$0.5-1 for a ready to plant tree seedling and the commercial price that may reach US$3.8-6.7 (Howell and Harrkngton 2004). Above figures support the estimation (Rinkevich 2005, 2006) that active reef restoration is economically feasible.

C. Agricultural periods -

Farming of coral colonies requires investing ample time for both, preparation and maintenance of cultured coral colonies. It seems somehow inappropriate to compare preparation of seedling in tree nursery were machines are sowing the seed into the containers (White 2004) with producing seedlings in a coral nursery. Coral seedlings are prepared individually and manually by pruning donor colonies into nubbin or fragment sizes, and then gluing the coral material onto different types of substrates adhering to transplantation protocols. The last step in this phase is attaching the seedlings onto the underwater nursery (Shafir et al. 2006a). We estimated that an untrained worker could prepare up to 100 coral seedlings per hour from an easily pruned colonial coral species (Shafir et al. 2006a). However, when pruning from massive corals, this figure is reduced considerably. While in some tree nurseries, (like bare-root nursery; Riley and Steinfeld 2005) seeding is performed by machines (making comparisons impossible), the container tree nursery does require some manual work (filling the containers with a soil mixture, seeding and watering; Landis et al. 1990). Calculation revealed that, in a container type nursery, a single worker can plant about 1500 seedlings per hour (Rantala et al. 2003), more than an order of magnitude faster than preparing coral seedlings in an underwater nursery.

The maintenance of cultured coral colonies is more comparable to terrestrial nursery maintenance. Regularly maintenance of the nursery includes relocation of seedlings, removing dead seedlings, spacing and general handling the whole nursery (irrigation system, roads, buildings vs. anchors, ropes, buoys) take in 36 m^2 coral nursery about 10 hours a month (Shafir et al. 2006a).

D. Growth rates –

The species-specific coral growth rates are easily influenced by an array of biological and environmental conditions (Veron 2000), including nutrient availability (Bongiorni et al. 2003). Installing a mid-water nursery in a nutrient-enriched environment could accelerate growth rates up to three times the rate in oligotrophic natural reef environment (Bongiorni et

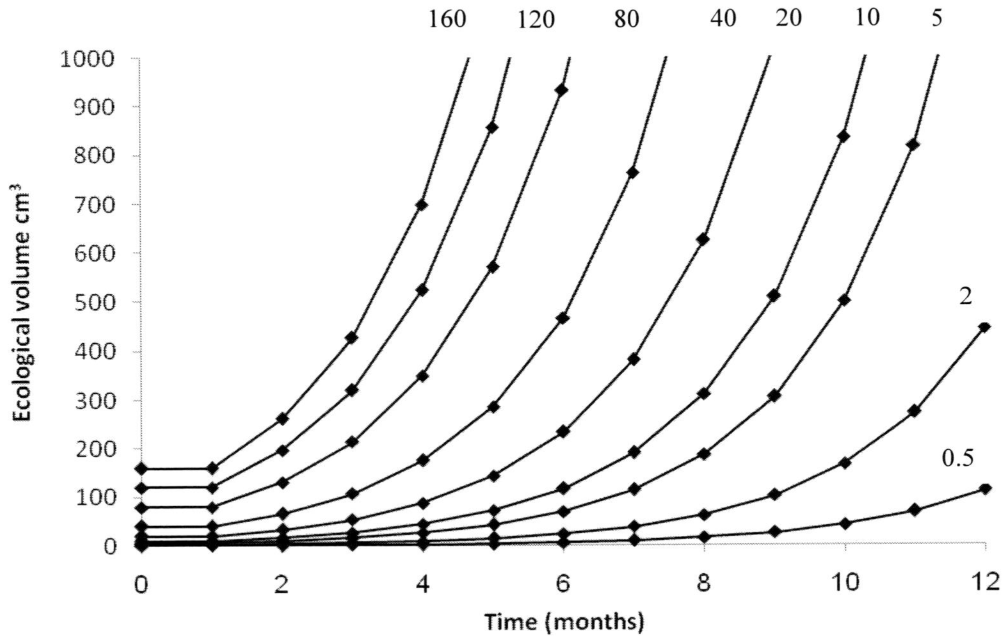

Fragment ecological volume (cm^3)	0.5	2	5	10	20	40	80	120	160
Fragment radius (cm)	0.65	1.02	1.39	1.75	2.21	2.78	3.50	4.01	4.41
Potential number of fragments	1400	350	140	70	35	18	9	6	4

Fig 1: Simulation for coral nursery periods (from start up to plantable ecological volume) as influenced by the initial fragment size (0.5 – 160 cm^3). An average constant growth rate k = 0.01637 was calculated for branching coral species residing in the nursery (Eilat's mid-water nursery; *Stylophora pistillata, Pocillopora damicornis, Acropora pharaonis, A. eurystoma* and *A. valida*; Shafir et al. 2006a). The simulation takes into consideration, regeneration period along the first month after coral fragmentation without added ecological volume. The table below details the size and number of fragments obtained from a 700 cm^3 size branching colony and the trade-off between number of fragments and their sizes.

al. 2003). As culture conditions influence growth rate constant (k=(ln Et/Eo)/t ;Shafir et al. 2006a) of fragments, this parameter may be employed to characterize overall nursery performance (Shafir et al. 2006a). Corals seedling initial size are of nubbins and fragments (Shafir et al. 2001; Shafir et al. 2006a,b,c), another parameter that influence growth rates. The correlations between k and fragment size affects not only growth rates, but also nursery time (Fig. 1). In a nutrient-enriched environment, most corals seedlings reach 'transplantation size' in less than one year (Shafir et al. 2006a) compared to 2-3 years under oligotrophic conditions (Bongiorni et al. 2003). Above rationale applies to seedling growth rates in tree nursery (Riley and Steinfeld 2005), since nutrient dependency influences nursery performance (Landis et al. 1990). However, since most tree nurseries start from seeds, initial size is not a decisive factor as in underwater coral nursery. Actually, they are comparable to coral nurseries that raise colonies from settled coral planula larvae, another promising coral source material (Rinkevich 1995, 2005). Tree seedlings are cultured from half a year to two years (Landis et al. 1990), analogous periods to mid-water coral nurseries (Shafir et al. 2006a,b).

E. Seedling spacing –

All seedlings require sufficient space for fast growth in order to avoid intraspecific or interspecific disturbances. Allogeneic and xenogeneic interactions can cause damage and growth abatement, they influence reproductive activities or can cause death (Rinkevich and Loya 1987; Rinkevich and Sakai 2001; Rinkevich 2004). To reduce intraspecific interactions, only nubbins from the same genotype were placed in close proximity (Shafir et al. 2006a), but too close isogeneic spacing resulted in coral tissue damage (outcome of the free movement of the floating coral nursery) and tissue fusions (Shafir et al. 2006b). Moreover, spacing coral "seedlings" on trays, offers herbivorous fish the opportunity to clean from fouling algae that develop between the farmed corals and on the culture nets (Shafir et al. 2006b). Spacing is also an important silviculture issue. Roots of tree seedlings need ample space for sufficient absorbing of water and nutrients from the soil (Riley and Steinfeld 2005). The size of the root system is the limiting factor for the developing tree seedlings and it determines the space between the seedlings or the size of their containers (South et al. 2005).

F. Maintenance issues –

Maintenance of mid-water coral nursery is an intricate business. Such a coral nursery, situated away from the reef and above sea bottom, in most cases, is also away from the shore, which necessitates frequent boat trips. This problem does not exist in terrestrial nurseries.

Coral reefs flourish typically in oligotrophic seas (Bythell 1990; Veron 2000). The successful symbiosis of hard corals with algae results in their becoming the main coral reef-builders (Little et al. 2004). Nevertheless, supplying corals with nutrients or food enhances their growth rates (Atkinson et al. 1995; Bongiorni et al. 2003; Houlbrèque et al. 2004). As aforementioned, corals maricultured in a nutrient-enriched mid-water nursery grew faster and reached sizes suitable for transplantation in six months (Bongiorni et al. 2003; Shafir et al. 2006a).

Developing seedlings in a tree nursery depends on a constant supply of water and nutrients. The initial soil treatment, adding organic material, exchange of gases and drainage is requires for successful development of seedlings (Riley and Steinfeld 2005). In subtropical and arid regions, water can be the major limiting factor and its supply incurs the major cost since it entails investment in irrigation and compensation for evaporation. In addition, nutrients are constantly added to soil during the nursery period to enhance growth rates (Landis et al. 1990, 2005; Oliet et al. 2005).

Of the major maintenance tasks in coral nursery, cleaning fouling organisms is the most significant and the most tedious. In nutrient enriched environments, algae can overgrow and shade the developing corals. In some places, as in the Eilat, Red Sea nursery, herbivores like schools of fish *Siganus rivulatus* and sea urchin *Diadema setosum* control the algal growth naturally by intensive grazing (Shafir et al. 2006a). When natural cleaning is not sufficient, manual scrapping of the substrates, net trays and other nursery components is required. In addition, prior to transplantation, much efforts and time is invested in removing all fouling organisms. To reduce the constant manual handling of fouling organisms, several methods have been suggested (Rinkevich 2005, 2006), of which the use of antifouling agents has recently been considered and successfully tested (unpublished). The time required for maintenance of coral nursery is an important factor. Since scuba diving time is limited and is influenced by depth (Slosman et al. 2004), some cleaning procedures were performed above water (Shafir et al. 2006b). The possibility of adjusting the nursery depth is another solution for reducing diving depth and increase performance (unpublished).

Similarly, tree nursery maintenance copes with weeds that plunder light, nutrients and water from the developing seedlings (Willoughby et al. 2006). One way for reducing weeds is keeping strict sanitation by growing weed-free seeds and watering them with filtered water (South and Enebak 2006). In large nurseries, the use of herbicides is the preferred management method (South and Carey 2005). Manual weeding is usually performed routinely only in small nurseries (Mangaoang et al. 2005). Tree nursery maintenance is not limited by working hours (days and nights) and does not require any special equipment.

G. Pest control

Under *in situ* conditions, coral nurseries can exercise only a limited control on pest infections. While placing the nursery away from the natural reef limits the presence of large corallivorous organisms as parrotfish and sea stars, it only delays the appearance of other pests, which settle from the plankton. One example experienced in the Red Sea nursery, is the major negative impacts of corallivorous gastropods such as *Drupella*, which is being kept at bay only by meticulous routine inspections and manual removal (Shafir et al. 2006a). Another common pest in Eilat nursery is aggregates of the sea anemone, *Boloceroides memurrichi* that cause stings to workers. . The sea anemone population sizes are controlled by fish predation (Shafir et al. 2006a).

In the tree nursery, pests appear inside the soil (i.e., fungi and nematodes) or on the seedlings and soil (insects), therefore maintenance is much more efficient than in marine nurseries. It is treated mainly by sterilizing the soil prior to seeding and by cleaning the tools (Landis et al. 1990; South and Enebak 2006). Combating insects is carried out by constant diurnal or nocturnal inspections, depending on the species. Pheromone traps help in detecting the presence of specific insects and selecting the suitable mode of pest control, usually vaporized chemicals (Landis et al. 1990; South and Enebak 2006).

H. Nursery construction

Constructing underwater facilities is complicated, especially the flexible three-dimension structure of floating mid-water coral nurseries. It involves various types of anchors for holding and stretching the floating nurseries (Fig. 2 a, b) to either sandy (fine or coarse sediment) or hard substrate. Both floating parts (with holding buoys) and anchors should be

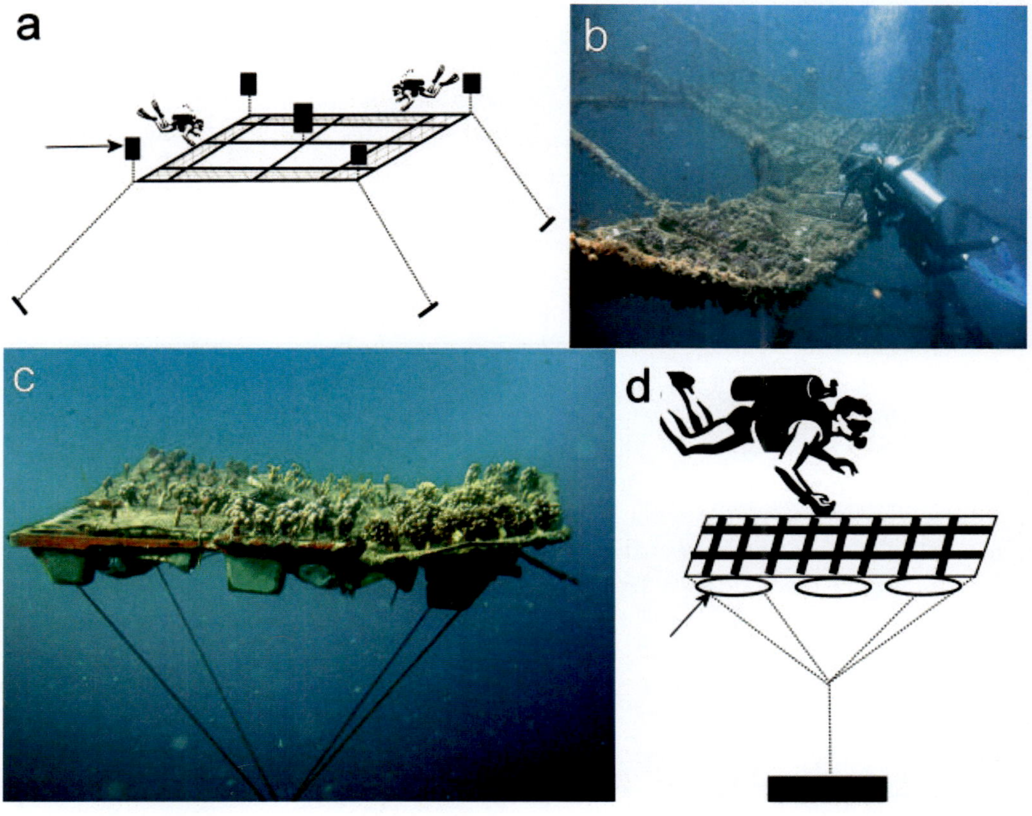

Fig 2: Nursery types: A scheme (a) and a photograph (b) of a large (10 X 10 m) prototype of peripheral mid-water nursery, that allows easier maintenance from both sides of the coral-carrying frames. A photograph (c) and a scheme (d) of small size (1 X 1 m) mid-water nursery that may also be part of a modular larger nursery. This type of nursery is accessible for maintenance from four sides, but holds limited number of farmed coral colonies. (a), (d), the diver's illustration depicts the relative scale for nursery size. Note unalike buoys locations (arrows) in both nursery types.

able to hold the nursery under adverse conditions, like storms (southern storm in Eilat, hurricanes in Jamaica), high tides and currents. Another option is constructing several smaller nurseries, each with smaller resistance to waves movement (unpublished; Fig 2 c, d).

In contrast to the underwater structure of nurseries, terrestrial nurseries are more two-dimensional, usually on larger scale (Landis et al. 1990). Constructed in large open-air regions, terrestrial nurseries require paving access roads and long-distance irrigation systems like pipes and sprinklers (Riley and Steinfeld 2005).

SYNOPSIS

Restoration acts (terrestrial and marine alike) require significant amount of restoration material. While this issue is self-evident for silviculture, it is not yet well rooted in the concepts for reef restoration (i.e., Edwards and Gomez 2007). In the USA alone, 45% of the more than 1.6 billion trees planted each year, are for the timber industry and 6% for national reforestation, totaling over 800 million trees each year for timber and restoration. Clearly, there is yet no major coral rehabilitation act that comes close to these figures, and all rehabilitation projects are still at experimental stages, not exceeding few hundred to few thousands corals per act.

The importance of coral reef ecosystems must spur us to improve our rehabilitation and active restoration methods. There is much to learn from silviculture and apply the knowledge to reef restoration. The basic act in both ecosystems active restoration is building nurseries for economical rearing of large numbers of 'seedlings'. For that purposes we need to simplify our coral seedling preparation and mariculture and use low-tech technology that will be available mainly to under-developed countries. This resembles tree nurseries that successfully are closing the gap between industrial agriculture in developed countries and the small communities in 3^{ed} world countries (Cedamon et al. 2005).

'Silviculture' of the rapidly declining coral reefs around the world is probably the most efficient way to perform reef restoration either on small or large-scales of reef area. It can be executed as a large operation, financed by international bodies such as the World Bank, or carried out as a small operation by local communities. The knowledge to construct the right type of coral nursery is building up rapidly because of comparative work of research groups in Israel, the Philippines, Singapore, Thailand, Jamaica and Tanzania. The same basic prototype of coral nursery, that was constructed for the work in Eilat, Red Sea (Shafir et al. 2006a,b), is examined and modified in these localities in accordance to local specific environmental conditions. Such approach may set conservation priority schemes towards more active restoration acts (Rinkevich, 1995, 2005) under frameworks that can guide the allocation of funds among alternate rehabilitation actions, each addressing different threats.

Therefore, restoration measures, as other aquaculture themes, should incorporate economic considerations and reef-landscape dynamics into a unified conceptual and practical scheme, providing a framework for decision-making authorities, where, when, and how much efforts should be invested for restoration measures in any specific reef site.

The literature details the plethora of methods (Rinkevich ,2005,2006) that have been applied in active restoration projects. However, much of the discussion in the literature does not take into account the use of already obtained results in an agricultural context, nor 'offer advice on technical know-how needed, chance of success, risks or likely costs' (Edwards and Gomez 2007). Adapting silviculture theories and practices into reef restoration concepts and acts (Epstein et al. 2003) may improve the development of this important scientific discipline, including its socioeconomic considerations.

ACKNOWLEDGEMENTS

Studies detailed here are supported by INCO-DEV project (REEFRES, no: 510657), by the World bank/GEF project (reef remediation/restoration working group), by a grant from the Israeli Ministry of Infrastructure and by the AID-CDR program (no. C23-004).

REFERENCES

Ahmed, M., Umali, G.M., Chong, C.K., Rull, M.F. & Garcia, M.C. (2007). Valuing recreational and conservation benefits of coral reefs - The case of Bolinao, Philippines. *Ocean & Coastal Management*, 50:103-118

Amar, K.O. & Rinkevich, B. (2007). A floating mid-water coral nursery as larval dispersion hub: testing an idea. *Marine Biology*, 151:713-718

Atkinson, M.J., Carlson, B., Crow, G.L. (1995) Coral growth in high-nutrient, low pH seawater: a case study of corals cultured at the Waikiki Aquarium, Honolulu, Hawaii. *Coral Reefs* 14:215-223

Bellwood, D.R., Hughes, T.P., Folke, C. & Nystrom, M. (2004). Confronting the coral reef crisis. *Nature*, 429:827-833

Bongiorni, L., Shafir, S., Angel, D. & Rinkevich, B. (2003). Survival, growth and gonadal development of two hermatypic corals subjected to in situ fish farm nutrient enrichment. *Marine Ecology Progress Series*, 253:137-144

Bowden-Kerby, A. (2001). Low-tech coral reef restoration methods modeled after natural transplantation processes. *Bulletin of Marine Science*, 69:915-931

Bruno, J.F. & Selig, E.R. (2007). Regional decline of coral cover in the Indo-Pacific: timing, extent, and subregional comparisons. *PLoS ONE*, 2:e711

Burke, L., Selig, E. & Spalding, M. (2002). Reefs at risk in southeast Asia. World Resources Institute, Washington DC, USA, 72

Bythell, J. C. 1990. Nutrient uptake in the reef-building coral *Acropora palmata* at natural environmental concentrations. *Marine Ecology Progress Series*, 68:65-69

Cesar, H.S.J. & van Beukering, P.J.H. (2004). Economic valuation of the coral reefs of Hawaii. *Pacific Science*, 58:231–242

Charnley, S. (2006). The northwest forest plan as a model for broad-scale ecosystem management: a social perspective. *Conservation Biology*, 20:330–340

Cedamon E.D., Mangaoang E.O., Gregorio N.O., Pasa A.E., Herbohn J.F. (2005) Nursery Management in Relation to Root Deformation, Sowing and Shading. Annals of Tropical Research 27: 1-10

Connell, J.H. (1978). Diversity in tropical rain forests and coral reefs. *Science*, 199:1302

Crépin, A.-S. (2003). Multiple species boreal forests – what Faustmann missed. *Environmental and Resource Economics*, 26:625–646

Dumroese, R.K. & Barnett, J.P. (2004). Container seedling handling and storage in the Southeastern States. *USDA Forest Service Proceedings RMRS-P-33*:22-25

Dumroese, R.K., Landis, T.D., Barnett, J.P. & Burch, F. (2005). Forest Service Nurseries: 100 years of ecosystem restoration. *Journal of Forestry* :241-247

Edwards, A.J. & Gomez, A.D. (2007). Reef restoration concepts and guidelines: Making sensible management choices in the face of uncertainty, Vol. Coral Reef Targeted Research & Capacity building for Management Programme, St. Lucia, Australia, 38pp.

Epstein, N., Bak, R.P.M. & Rinkevich, B. (2003). Applying forest restoration principles to coral reef rehabilitation. *Aquatic Conservation*, 13:387-395

Epstein, N. & Rinkevich, B. (2001). From isolated ramets to coral colonies: The significance of colony pattern formation in reef restoration practices. *Basic and Applied Ecology*, 2:219-222

Epstein, N., Vermeji, M.J.A., Bak, R.P.M. & Rinkevich, B. (2005). Alleviating impacts of anthropogenic activities by traditional conservation measures: can a small reef reserve be sustainedly managed ? *Biological Conservation*, 121:243-255

Feary, D.A., Almany, G.R., McCormick, M.I. & Jones, G.P. (2007). Habitat choice, recruitment and the response of coral reef fishes to coral degradation. *Oecologia*, 153:727–737

Goreau, T.F.S., Goreau, N.I. & Goreau, T.J. (1979). Corals and coral reefs. *Scientific American*, 241:110

Graham, R.T. & Jain, T.B. (1998). Silviculture's role in managing boreal forests. *Conservation Ecology*, 2:8

Gunderson, L. (2007). Ecology: a different route to recovery for coral reefs. *Current Biology*, 17:R27-R28

Houlbrèque F., Tambutté E., Allemand D. & Ferrier-Pagès C. (2004) Interactions between zooplankton feeding, photosynthesis and skeletal growth in the scleractinian coral *Stylophora pistillata .Journal of Experimental Biology*,1469-1461 207:

Howell, K.D. & Harrington, T.B. (2004). Nursery practices influence seedling morphology, field performance, and cost efficiency of containerized Cherrybark Oak. *Southern Journal of Applied Forestry*, 28:152-162

Hughes, T., Rodrigues, M., Bellwood, D., Ceccarelli, D., Hoegh-Guldberg, O., McCook, L., Moltschaniwskyj, N., Pratchett, M., Steneck, R. & Willis, B. (2007). Phase shifts, herbivory, and the resilience of coral reefs to climate change. *Current Biology*, 17:360-365

Hughes, T.P., Baird, H., Bellwood, D.R., Card, M., Connolly, S.R., Folke, C., Grosberg, R., Hoegh-Guldberg, O., Jackson, J.B.C., Kleypas, J., Lough, J.M., Marshall, P., Nyström, M., Palumbi, S.R., Pandolfi, J.M., B., R. & J., R. (2003). Climate change, human impacts, and the resilience of coral reefs. *Science*, 301:929-933

Idjadi, J.A. & Edmunds, P.J. (2006). Scleractinian corals as facilitators for other invertebrates on a Caribbean reef. *Marine Ecology and Progress Series*, 319:117–127

Jones, G.P., McCormick, M.I., Srinivasan, M. & Eagle, J.V. (2004). Coral decline threatens fish biodiversity in marine reserves. *Proceedings of the National Academy of Sciences of the USA*, 101:8251–8253

Krieger, D.J. (2001). Economic Value of Forest Ecosystem Services: A Review., The Wilderness Society, Washington, DC

Landis, T.D., Haase, D.L. & Dumroese, R.K. (2005). Plant nutrient testing and analysis in forest and conservation nurseries. *USDA Forest Service Proceedings* RMRS, P-35:76-83

Landis, T.D., Tinus, R.W., McDonald, S.E. & Barnett, J.P. (1990). The Container Tree Nursery Manual. Forest Service, U.S. Department of Agriculture, Washington D.C

Lindahl, U. (1998). Low-tech rehabilitation of degraded coral reefs through transplantation of staghorn corals. *Ambio*, 27:645-650

Little, A. F., M. J. H. van Oppen, and B. L. Willis. 2004. Flexibility in Algal Endosymbioses Shapes Growth in Reef Corals. *Science,*304:1492-1494

Lugo, E.L., Caroline, S.R. & Scott, W.N. (2000). Hurricanes, coral reefs and rainforests: resistance, ruin and recovery in the Caribbean. *Ambio*, 29:106

Mangaoang, E.O., Cedamon, E.D. & Pasa, A.E. (2005). Cost-effective systems for seedling production and tree farm establishment. *Annals of Tropical Research*, 27:11-18

Milon, J.W. & Dodge, R.E. (2001). Applying habitat equivalency analysis for coral reef damage assessment and restoration. *Bulletin of Marine Science*, 69:975-988

Moberg, F. & Folke, C. (1999). Ecological goods and services of coral reef ecosystems. *Ecological Economics*, 29:215-233

Naidoo, R. & Ricketts, T. (2006). Mapping the economic costs and benefits of conservation. *PLoS Biol*, 4:e360

Navarro, R.M., Retamosa, M.J., Lopez, J., del Campo, A., Ceaceros, C. & Salmoral, L. (2006) Nursery practices and field performance for the endangered Mediterranean species *Abies pinsapo* Boiss. *Ecological Engineering*, 27:93–99

Naveh, Z. (2007). The Transformation of Landscape Ecology and Restoration Ecology into Transdisciplinary Sciences of Holistic Landscape Study, Management and Planning, Conservation and Restoration. In: Transdisciplinary Challenges in Landscape Ecology and Restoration Ecology :353-401

Newton, K., Cote, I.M., Pilling, G.M., Jennings, S. & Dulvy, N.K. (2007). Current and future sustainability of island coral reef fisheries. *Current Biology*, 17:655–658

Oliet, J.A., Planelles, R., Artero, F. & Jacobs, D.F. (2005). Nursery fertilization and tree shelters affect long-term field response of *Acacia salicina* Lindl. planted in Mediterranean semiarid conditions. *Forest Ecology and Management*, 215:339–351

Omori, M. & Fujiwara, S. (eds) (2004). Manual for restoration and remediation of coral reefs, Vol. Nature Conservation Bureau, Ministry of Environment, Japan, 84 pp.

Palmer, M.A., Ambrose, R.F.N. & Poff, L. (1997). Ecological theory and community restoration ecology. *Restoration Ecology*, 5:291–300

Parviainen, J. & Frank, G. (2003). Protected forests in Europe approaches-harmonising the definitions for international comparison and forest policy making. *Journal of Environmental Management*, 67:27

Pauly, D., Christensen, V., Guenette, S., Pitcher, T.J., Sumaila, U.R., Walters, C.J., Watson, R. & Zeller, D. (2002). Towards sustainability in world fisheries. *Nature*, 418:689-695

Pearce, D.W. (2001). The economic value of forest ecosystems. *Ecosystem Health*, 7:284–296

Precht, W.F., Deis, D.R. & Gelber, A.R. (2002). Damage assessment protocol and restoration of coral reefs injured by vessel groundings. *Proceedings of the 9th International Coral Reef Symposium, Indonesia*, 2:963-968

Putz, F.E., Blate, G.M., Redford, K.H., Fimbel, R. & Robinson, J. (2001). Tropical forest management and conservation of biodiversity: an overview. *Conservation Biology*, 15:720

Rantala, J., Vaatainen, K., Kiljunen, N. & Harstela, P. (2003). Economic evaluation of container seedling packing and disinfection machinery. *Silva Fennica*, 37:121–127

Riley, L.E. & Steinfeld, D. (2005). Effects of bareroot nursery practices on tree seedling root development: an evolution of cultural practices at J. Herbert Stone nursery. *New Forests*, 30:107–126

Rinkevich, B. (1995). Restoration strategies for coral reefs damaged by recreational activities: The use of sexual and asexual recruits. *Restoration Ecology*, 3:241-251

Rinkevich, B. (2000). Steps towards the evaluation of coral reef restoration by using small branch fragments. *Marine Biology*, 136:807-812

Rinkevich, B. (2004). Allorecognition and xenorecognition in reef corals: A decade of interactions. *Hydrobiologia*, 430/531:443-450

Rinkevich, B. (2005). Conservation of coral reefs through active restoration measures: recent approaches and last decade progress. *Environtal Science and Technology*, 39:4333-4342

Rinkevich, B. (2006). The coral gardening concept and the use of underwater nurseries; lesson learned from silvics and silviculture. In: W.F. Precht (ed.) *Coral Reef Restoration Handbook*. Boca Raton, Florida, CRC Press, pp 291-301

Rinkevich, B. & Loya, Y. (1987). Variability in the patterns of sexual reproduction of the coral *Stylophora pistillata* at Eilat, Red Sea: A long-term study. *The Biological Bulletin*, 173:335-344

Rinkevich, B. & Sakai, K. (2001). Interspecific interactions among species of the coral genus *Porites* from Okinawa. *Japanese Zoology*, 104:91-97

Risk, M.G. (1999). Paradise lost: how marine science failed the world's coral reefs. *Marine & Freshwater Research*, 50:831-837

Salvat, B. (1992). Coral reefs-a challenging ecosystem for human societies. *Global Environal Change*, 2:12-18

Shafir, S., Van Rijn, J. & Rinkevich, B. (2006a). Steps in the construction of underwater coral nursery, an essential component in reef restoration acts. *Marine Biology*, 149:679-687

Shafir, S., van Rijn, J. & Rinkevich, B. (2006b). A mid-water coral nursery. *Proceedings of the 10th International Coral Reef Symposium,* pp. 1674-1679

Shafir, S., Van Rijn, J. & Rinkevich, B. (2006c). Coral nubbins as a source material for coral biological research: a prospectus. *Aquaculture*, 259:444-448

Skinner, C.N. (2007). Silviculture and forest management under a rapidly changing climate. *USDA Forest Service Gen. Tech. Rep*, PSW-GTR-203:21-32

Slosman, D.O., de Ribaupierre, S., Chicherio, C., Ludwig, C., Montandon, M.-L., Allaoua, M., Genton, L., Pichard, C., Grousset, A., Mayer, E., Annoni, J.-M. & de Ribaupierre, A. (2004). Negative neurofunctional effects of frequency, depth and environment in recreational scuba diving: the Geneva "memory dive" study. *British Journal of Sports Medicine*, 38:108–114

Smith, S.V. (1978). Coral-reef area and the contribution of reefs to processes and resources of the world's oceans. *Nature*, 273:225-226

Soong, K. & Chen, T. (2003). Coral transplantation: regeneration and growth of Acropora fragments in a nursery. *Restoration Ecology*, 11:62-71

South, D.B. & Carey, W.A. (2005). Weed Control in Bareroot Hardwood nurseries. *USDA Forest Service Proceedings* RMRS-P-35:34-38

South, D.B. & Enebak, S.A. (2006). Integrated pest management practices in southern pine nurseries. *New Forests*, 31:253–271

South, D.B., Harris, S.W., Barnett, J.P., Hainds, M.J. & Gjerstad, D.H. (2005). Effect of container type and seedling size on survival and early height growth of *Pinus palustris* seedlings in Alabama, U.S.A. *Forest Ecology and Management*, 204:385–398

Spurgeon, J.P.G. (2001). Improving the economy effectiveness of coral assessment and restoration. *Bulletin of Marine Science*, 69:1031-1045

Subade, R.F. (2007). Mechanisms to capture economic values of marine biodiversity: The case of Tubbataha Reefs NESCO World Heritage Site, Philippines. *Marine Policy*, 31:135–142

Veron, J.E.N. (2000). Corals of the World. Australian Institute of Marine Science, Townsville, Australia.

White, A.T., Vogt, H.P. & Arin, T. (2000). Philippine coral reefs under threat: The economic losses caused by reef destruction. *Marine Pollution Bulletin*, 40:598-605

White, B. (2004). Container Handling and Storage in Eastern Canada. *USDA Forest Service Proceedings 3*:10-14

Wilkinson C. (2004). Status of coral reefs of the world: 2004, vol. 1. Townsville, Australia: Australian Institute of Marine Science. 301pp.

Willoughby, I., Jinks, R.L. & Stokes, V. (2006). The tolerance of newly emerged broadleaved tree seedlings to the herbicides clopyralid, cycloxydim and metazachlor. *Forestry*, 79:599-608

Wilson, S.K., Graham, N.A.J., Pratchett, M.S., Jones, G.P. & Polunin, N.V.C. (2006). Multiple disturbances and the global degradation of coral reefs: are reef fishes at risk or resilient? *Global Change Biology*, 12:2220–2234

Wright, S.J. & Muller-Landau, H.C. (2006). The Future of Tropical Forest Species. *BIOTROPICA*, 38:287–301

WWW1 - http://www.safnet.org/aboutforestry/world.cfm - Foresters, S.o.A. (2007). Facts on forests around the world. Society of American Foresters

In: Aquaculture Research Trends
Editor: Stephen H. Schwartz, pp. 297-320
ISBN 978-1-60456-217-0

Chapter 10

EMERGING TRENDS OF AQUACULTURE RESEARCH IN THE ARABIAN GULF REGION

S. M. Almatar [*] ***and C.M. James*** [#]
Mariculture and Fisheries Department, Kuwait Institute for Scientific Research, P.O. Box 1638, Salmiya 22017, Kuwait

ABSTRACT

The Arabian Gulf Region is bordered by eight countries including Bahrain, Iran, Iraq, Kuwait, Oman, Qatar, Saudi Arabia and United Arab Emirates (UAE). All these countries have a combined coastline of about 23,000 km in the Arabian Gulf (also known as Persian Gulf) and in the Gulf of Oman, in addition to 500 km of Caspian Sea coastline bordering Iran and about 1840 km of Red Sea coast bordering Saudi Arabia. Except Oman, all other Gulf countries consume more seafood than their availability from local waters. Furthermore, wild capture fisheries in Kuwait, Iran and UAE are declining and it is anticipated that the gap between the fresh fish supply and demand will widen due to the declining natural fish stocks. To fill up the gap, several research organizations in Arabian Gulf countries are involved in intensive research programs to develop aquaculture.

Although the major focus of research is on developing marine species for aquaculture, Iran, Iraq and Saudi Arabia have also inland fishery research programs. Among Gulf countries, Kuwait was the first country to achieve captive spawning and hatchery seed production for local grouper and sobaity seabream as well as acclimation of Tilapia in seawater for commercial production. Species selection for aquaculture depends on the local demand of each of these countries and based on this research is focused on both endemic as well as exotic species. The research outcome has enabled the establishment of commercial marine fish farming using sea-cages, marine shrimp farming using earthen ponds and commercial tilapia farming as well as integrated tilapia culture along with agricultural crops.

As a research outcome, Bahrain is producing commercial quantities of fish fingerlings in their hatchery for sea-cage farming. Research efforts have enabled Iranian total aquaculture production to increase from 4,935 t in 1978 to 124,000 t in 2004.

[*] E-mail address: smattar@mfd.kisr.edu.kw
[#] E-mail address: charlesm.james@hotmail.com

Although aquaculture production in Iraq declined from 3,622 t in 1980 to 1,600 t in 1990, it is steadily increasing and in 2005 the total production reached 12,870 t. Kuwait was the first country to establish commercial sea-cage fish farming in 1985 and its production reached up to 400 t during 2000. In Oman the total production through aquaculture was 514 t during 2004.Aquaculture is not well established in Qatar and produced 11 t during 2005. Marine shrimp farming is the major aquaculture activity in Saudi Arabia and has a targeted annual production of 13,500 t during Phase-I and 17,500 t in Phase-II. Marine fish production in UAE has increased from 570 t in 2004 to 900 t in 2006/07 period.

Each country has its own constraints and priority issues for developing aquaculture. One of the major challenges is the non-availability of fish fingerlings from hatchery for sea-cage farming, especially with reference to local prime species. Furthermore, the low domestic market price is compelling to search for export oriented production involving exotic species. In some region, shrimp farming is restricted by adverse environmental conditions. Aquaculture is expected to have a significant growth in the Arabian Gulf countries with an overall trend in species diversification as well as development of intensive culture systems.

INTRODUCTION

The Arabian Gulf Region, also known as Persian Gulf, that will be referred as the 'Gulf' in this report is situated in the semi-arid region of the Middle-East just north of Tropic of Cancer and bordered by eight countries, Bahrain, Iran, Iraq, Kuwait, Oman, Qatar, Saudi Arabia and United Arab Emirates (UAE). All these countries have about 23,000 km of coastline in the Gulf, in addition to 990 km of Caspian Sea coastline bordering Iran and about 1840 km of Red Sea coast bordering Saudi Arabia. It has a shallow marginal sea with an average depth of 35m and has a continental shelf area of about 608,000km^2 [Feidi, 1998]. The Gulf region is with extremes of climatic conditions with long summer air temperature of about 48°C and having mean winter temperature generally below 20°C and may drop to 0°C in the north western part of the Gulf. Seawater temperature in the north varies between 35°C in summer to 16 ^{o}C in winter, while in the south it may decrease to 19 ^{o}C. Seawater salinity ranges from 43 ppt in the north to 39 ppt in the south. Average annual rainfall does not exceed 15 cm and there is virtually no river runoff flow except in the northern tip of the Gulf. Under arid and extreme climatic conditions, aquaculture was not a traditional venture in this region. Development and expansion of modern aquaculture has taken place recently, since the 1980's [Poynton, 2006].

The total fish landing in the Gulf is about 250,000 t in recent years that include finfish, crustaceans and mollusks. Except Oman, all other Gulf countries consume more seafood than their availability from local fisheries. Furthermore, the wild capture fisheries in Kuwait, Iran and UAE are showing a declining trend. In the northern part of Iran, in the south coast of Caspian Sea, the fish landings declined from 110,000 t during 1999 to 44,887 t during 2005, at the same time period in the south at the Iranian side of the Gulf the fish catch has increased from 234,200 t to 343,492 t. However, technical efficiency analysis for the Iranian fishery in the Gulf shows that despite increasing fishing effort, the per capita catch has declined in recent years [Esmaeili, 2006]. During the same period, Kuwait fishery declined from 7,811 t to 4,600 t and in UAE it declined from 117,608 t to 86,734 t. Furthermore, about 90% of world's sturgeon catch is from Caspian Sea and the catches rapidly declined from 3000 t in 2000 to 220 t in 2004 in Iranian Caspian Sea coast. It is anticipated that the gap between fresh fish supply and demand will widen in the future due to the declining of natural fish stocks and increasing human population as well as increase in per capita fish consumption. To fill up the gap, several research organizations are focusing their efforts in developing aquaculture technologies to enhance the primary food fish production.

RESEARCH ESTABLISHMENTS

The aquaculture research and development in the Gulf countries are governed by national agencies under the ministry of agriculture, fisheries or science and technology, which plans and coordinates the national aquaculture research agenda and programs. The research sector supports national aquaculture development with improved and applicable aquaculture policies, technologies and management methods and tools for environmentally sound and socially responsible aquaculture. The responsible institutions working closely with the government agencies in charge of overall development of national aquaculture determine research priorities and formulate national aquaculture research programs [FAO and NACA, 1997]. Various international agencies as well as consultants are also involved in the implementation of research programs for the development of sustainable aquaculture technologies in this region. Among the factors considered in setting aquaculture research priorities are national development plans and priorities, field problems and farmers' needs, availability of funds, land, water, expertise and manpower, and local and provincial development needs.

Bahrain

Bahrain is a small island country which has limited land resources for aquaculture development. In 1979, FAO carried out an initial assessment on the possibility of aquaculture development in Bahrain. Based on this the National Mariculture Centre (NMC) located at Ras Hayan on the South-Eastern coast of Bahrain was established in 1993 under the Directorate of Fisheries and Marine Resources (DFMR), Ministry of Works and Agriculture. The facility initiated with a small fish hatchery aimed at applied research activities that include studies in the areas of nutrition, reproduction, hatchery techniques, nursery and in developing growout production systems for some of the commercially important marine fish species. Over the years the hatchery facilities have been enhanced and the current research is focused on the development and refinement of appropriately efficient technologies and the production of marine fin fish juveniles both for sale to commercial sea-cage farmers as well as for restocking programs. Its major objectives are to create the scientific and technical base for the development of mariculture operations in Bahrain; to carry out appropriate scientific research projects and developmental programs on the mariculture of living marine resources and to mass produce seed of suitable and commercially important species for the fish farming sector [National Aquaculture Sector Overview – Bahrain, www.fao.org/fi/website/ FIRetrieve Action.do?dom=countrysectorandxml=naso_bahrain.xml 7/15/2007]. The Bahrain Center for Scientific Research and the Biology Department of the University of Bahrain also conduct related academic oriented research in this field.

One of the aquaculture research centre – South of Iran Aquaculture Research Center

Iran

The first aquaculture attempts were initiated in Iran as early as 1959 with the culture of rainbow trout near Tehran and the first warm water farm was established in 1971. Following this Cold-Water Fish Research Center in Tonekabon and Sturgeon Fish Research Center/International Sturgeon Research Institute in Rasht are involved in research projects relevant to rainbow trout, sturgeon and other cold water fish species native to Iran. The Khouzestan Fisheries Research Center (KFRC) initiated in 1988 focused its research on aquaculture development in southern Iran. With the expansion of research activities KFRC was renamed as South of Iran Aquaculture Research Center (SIARC) with seven hectare research farm facility. A mariculture station located in Bandar Imam Khomaini Port adjacent to the Gulf is also affiliated to SIARC. Iranian Fisheries Research and Training Organization (IFRTO) is responsible for planning and implementation of national aquaculture research programs and dealing with the aquaculture research priorities and training of the personnel. It has six affliated centres situated along the Caspian Sea coast, Gulf and in Oman Sea coasts [FAO]. Iran Shrimp Research Center (ISRC) in Bushehr is involved in shrimp culture research activities. Apart from these, Artemia and Aquatic Research Center (AARC) affiliated to Urmia University is Collaborating with University of Ghent, Belgium in developing Artemia production for aquaculture. By 1997, sixteen research organizations, University centers and stations were involved dealing with all aspects of aquaculture, emphasizing sturgeon, marine finfish, freshwater fishes, coldwater fishes, shrimp, oyster and seaweed [FAO and NACA, 1997].

Iraq

In Iraq, fisheries research is carried out by a number of institutions such as the Fish Research Center at Baghdad, the Marine Science Centre in Basrah, the Agriculture Research Center, the Central Hatchery at Swairah, Colleges of Agriculture and Colleges of Science in Iraqi Universities and by several Technical Institutes. The Fish Research Center (Zeafaraniyah, Baghdad) has five sections: fish culture, nutrition, diseases, water environment and artificial propagation. The center provides some facilities for post-graduate studies, and has two fish farms and a hatchery for research purposes. The State Fisheries Department is carrying out research and field studies on aquaculture. It has carried out research on fish feeds and fish diseases and has undertaken feasibility studies for fish farms based on non-indigenous fish. Iraq is a member of the Regional Commission for Fisheries (RECOFI belonging to FAO) which addresses regional fisheries research and management issues. The Fish Research Center provides some facilities for post-graduate studies, and has two fish farms and a hatchery for research purposes. The Marine Science Centre is carrying out research on aquaculture and food nutrition and provides supervision and facilities to post graduate students. The Agriculture Research Center has an experimental fish farm and hatchery to artificially propagate local fish species. The Central Hatchery at Swairah produces various species of carp fingerlings for research experiments as well as for commercial farming [www.fao.org/fi/website/FIRetrieveAction.do/dom=countrysectorandxml=FI-CP_IQ.xmlandlang=en 9/2/2007].

Kuwait

With the establishment of Kuwait Institute for Scientific Research (KISR) as early as 1968, the Mariculture and Fisheries Department (MFD) was founded considering the importance of fishery resources and aquaculture. MFD is located in Ras Al-Ardh in Salmiyah and has easy access to Kuwait Bay and to Kuwait's southern waters. MFD's facilities cover an area of 3.8 ha, including a boat marina. The major thrust of the aquaculture component is to conduct research to develop commercially viable technology for the private sector. This involves applied research into the life cycle, reproduction, nutrition and production of those local species most likely to support viable commercial aquaculture in Kuwait. Because of its early success of hatchery breeding of grouper under captive culture conditions as early as 1975, MFD quickly established itself as the region's premier aquaculture research facility [Hussain *et al.* 1975]. Towards supporting the fisheries industry in Kuwait, the MFD is dealing with intensive research and development projects relevant to fisheries, aquaculture and oceanography and advise the government agencies in fisheries regulations. In 1983, the government established the Public Authority for Agricultural Affairs and Fish Resources (PAAF) which is responsible for agriculture, livestock and fisheries in Kuwait.

The Mariculture & Fisheries Dept., KISR, Kuwait

Oman

Research program in aquaculture was initiated in 1992 through the establishment of the Marine Science and Fisheries Center (MSFC) by the Oman Ministry of Agriculture and Fisheries. With successful research output and involvement of private sector lead the Ministry to establish an Aquaculture Committee in 1996 to evaluate and manage the private investors. During the period 2000, a Department of Aquaculture Development (DAD) was established to collaborate with private sector and to administer aquaculture development in Oman. The Ministry of Agriculture and Fisheries is planning to build a new aquaculture centre to meet the needs of the sector which is fast expanding [FAO].

Qatar

Aquaculture is not a traditional activity in Qatar and the aquaculture research is limited to the establishment of some experimental facilities during 1988 at the University of Qatar. The experimental facility has a farm and there is no commercial production. The Ministry of Agriculture is responsible for collection of landing and market data for the Fisheries Database.

Kingdom of Saudi Arabia

Aquaculture developments in Saudi Arabia was initiated in 1980 through the establishment of the Fish Culture Project at the Saudi Arabian National Center for Science and Technology, at present known as King Abdulaziz City for Science and Technology (KACST). As part of the project, an aquaculture technical team from Taiwan Fisheries Research Institute was collaborating in aquaculture development programs. The KACST deals with the research on freshwater fish culture. The Saudi Arabia Ministry of Agriculture and Water together with the cooperation of FAO/UN, established the Fish Farming Center (FFC) in 1982 for developing marine fish and shrimp culture in the Kingdom. From the very beginning of its inception, FAO has helped the Fish Farming Centre with technical expertise and has trained the National staff working in this project. The FFC plays a major role in developing culture technology for the local marine shrimp, fish and also in developing local feed formulations to meet with the nutritional requirement of cultured fish and shrimps in the Kingdom [James, 1998]. The establishment of FFC is the result of one of 15 agricultural projects undertaken at various locations in Saudi Arabia with FAO assistance since 1964 towards enhancing the agricultural and fisheries output [www.fao.org/English /newsroom /news/2002/9346-en.html].

UAE

Towards developing aquaculture in UAE, the Marine Resources Research Center (MRRC) in Umm Al Qaiwain was established in 1984. The MRRC initiated its activity with the technical co-operation from Japan International Co-operation Agency (JICA). The major role of the centre is to conserve and to replenish the marine resources in the territorial waters of UAE. Aquaculture is one of its major activities undertaken by MRRC since its inception. As part of the conservative program, MRRC has undertaken sea-ranching by releasing hatchery produced fish fingerlings at various locations in the coastal waters of UAE. The Ministry of Agriculture and Fisheries is promoting aquaculture through offering fingerlings and diverting technical know-how to private investors. There are two main sections in the Ministry, the Living Marine Resources Department and the Fisheries Department. The MRRC comes under the former and contains Fishery Resources Development Section, the Fisheries Extension and Marine Research Section and an Aquarium.

Aquacultured Species

Research in Bahrain focused on developing the hatchery and grow-out production technology for the orange spotted grouper (*Epinephelus coioides*), yellow-finned seabream (*Acanthopagrus latus*), mangrove red snapper (*Lutjanus argentimaculatus*), streaked rabbit fish (*Siganus javus*) and the green tiger shrimp (*Penaeus semisulcatus*). Hatchery technology for the rabbit fish (*Siganus canaliculatus*), sobaity bream (*Sparidentex hasta*), gilthead seabream (*Sparus aurata*) have been achieved for commercial scale production.

In Iran, research is mainly focused on the culture of grass carp (*Ctenopharyngodon idella*), silver carp (*Hypophthalmichthys molitrix*), beghead carp (*Aristichthys nobilis*), commom carp (*Cyprinus carpio*), freshwater bream (*Abramis brama*), pike-perch (*Stizostedion lucioperca*), beluga (*Huso huso*), rainbow trout (*Oncorhynchus mykiss*), milkfish (*Chanos chanos*), marine shrimp (*Penaeus indicus*, *P. semisulcatus* and *P.*

merguiensis) and the freshwater prawn (*Macrobrachium rosenbergii*). Cage and pen culture systems have been introduced for feasibility studies on the culture of Caspian salmon, rainbow trout, sturgeon, carps and fourteen species of marine finfish [Matinfar and Nikouyan]

In Iraq, aquaculture is limited to the fresh water fish such as common carp (*C. carpio*), silver carp (*H. molitrix*) and grass carp (*C. idella*). The main species cultured is common carp and to a lesser extent grass and silver carp. Barbus sp. is also cultured in small quantities.

Kuwait, research focused on prime species such as silver black seabream (*S. hasta*), orange spotted grouper (*E. coioides*), yellow-finned seabream (*A. latus*), and more recently the silver pomfret (*Pampus argenteus*). MFD has also produced hybrids between *S. hasta* and *A. latus*. Research on shrimp culture initiated in 1974 toward enhancing the natural stock by releasing the hatchery produced juveniles of *P. semisulcatus*. From 1980's onwards, this research was switched over to developing culture technology for this indigenous shrimp species. Research on Tilapia culture in Kuwait introduced *Oreochromis spp*. in brackishwater in 1982 and further research fully adopted them to seawater. Current research is focused on developing genetically improved strains of tilapia (*O. niloticus*).

Oman research and development on marine fish is focused on the local black seabream *S. hasta*, organge spotted grouper *E. coioides*, yellowfin seabream *A. latus*, thin lip grey mullet *Liza ramada* and fattening of yellowfin tuna *Thunnus albacares*. Commercial cage farming is focused on using the exotic gilthead seabream *S. aurata* and European seabass *Dicentrarchus labrax*. In addition to the local shrimp *P. semisulcatus*, much attention is focused on farming the tiger shrimp *P. monodon* and *P. indicus*. The Nile tilapia *O. niloticus* is also considered for farming.

Candidate species for aquaculture in Qatar is limited to the research on the rabbit fish (*S. canaliculatus*) and grouper (*Epinephelus* spp.).

In the Kingdom of Saudi Arabia, the FFC is dealing with developing technology for the local groupers (*E. fuscoguttatus* and *E. polyphekadion*), marine shrimp (*P. indicus* and *P. monodon*), rabbitfish (*Siganus rivulatus*). During the late 90's, the Asian seabass (*L. calcarifer*) was introduced by FFC/FAO project through importing fertilized eggs from Singapore. Tilapia (*O. spilurus*) was also introduced to culture them using brackish as well as in full strength seawater.

Aquaculture activities in the UAE initiated with the culture trials of the rabbitfish *S. canaliculatus* and the orange spotted grouper *E. coioides* during 1985-1996 culture period. From 1997-2000 MRRC studied on the culture of seabreams Rhabdosargus sarba and *A. latus*. Fingerling production of sobaity seabream *S. hasta* and large-scale mullet *L. macrolepis* intiated in 2001. Postlarvae and adults of the endemic shrimp species *P. semisulcatus* and *P. indicus* were produced since 1985 and the shrimp culture was halted in 2001. The exotic species such as the gilthead seabream *S. aurata* and the European seabass *D. labrax* are used for commercial farming in sea-cages.

Table 1. Candidate species considered for aquaculture in the Arabian Gulf

Country	Species	Status
Bahrain	Rabbit fish, *S. canaliculatus*	Technology developed
	S. javus	Ongoing research
	Sobaity seabream, *S. hasta*	Commercial hatchery prod
	Mangrove red snapper, *L. argentimaculatus*	
		Ongoing research
	Brown-spotted grouper, *E. coioides*	
	Gilthead seabream, *S. aurata*	Ongoing research
	Green tiger shrimp, *P. semisulcatus*	Commercial hatchery prod.
		Ongoing research
Iran	Rainbow trout, *O. mykiss*	Commercial farming
	Caspian trout, *Salmo trutta caspius*	Ongoing research
	Sturgeon, *A. persicus*	Ongoing research
	A.guldenstaedi, A. stellatus,	
	A. nudiventries, H. huso	
	Mahi sephid, *R. frisii kutum*	Ongoing research
	Bream, *A. brama*	Ongoing research
	Pike-perch, *S. lucioperca*	Ongoing research
	L. lucioperca	
	Silver carp, *H. molitrix*	Commercial farming
	Big head, *A. nobilis*	Commercial farming
	Chinese carps: silver carp, grass carp	Commercial farming
	Grouper, *E. coioides*	Ongoing research
	B. sharpeyi, B. grypus,	Technology developed
	B. xanthopterus	
	White shrimp *P. indicus*	Adopted technology
	Endemic shrimp *P. semisulcatus*	Ongoing research
Iraq	Common carp, *C. carpio*	Research and commercial
	Silver carp, *H. molitrix*	Research and commercial
	Grass carp, *C. idellus*	Research and commercial
	Barbus sp.	Research and commercial
Kuwait	Silver black seabream, S. hasta	Technology developed plus ongoing research
	Yellow-finned seabream, A. Latus	Ongoing research
	Hybrids of S. hasta x A. latus	Ongoing research
	Grouper, E. coioides	Technology developed plus ongoing research
	Tialpia strains, O. niloticus and O. spilurus	Technology developed plus ongoing research
	Endemic shrimp P. semisulcatus	Ongoing research
	Silver pomfret, P. argenteus	Ongoing research

Table 1. Candidate species considered for aquaculture in the Arabian Gulf (continued)

Country	Species	Status
Oman	Silver black seabream, *S. hasta*	Ongoing research
	Yellow-finned seabream, *A. latus*	Ongoing researchOngoing research
	Grouper, *E. coioides*	Ongoing research and farming
	Thin lip grey mullet *L. ramada*	Fattening unit
	Yellowfin tuna *T. albacares*	Seacage trial
	Gilthead seabream *S. aurata*	Seacage trial
	European seabass *D. labrax*	Research and farming
	Tialpia, *O. niloticus*	Ongoing research
	Endemic shrimp *P. semisulcatus*	Research and farming
	White shrimp *P. indicus*	Research and farming
	Tiger shrimp *P. monodon*	
Qatar	Rabbit fish, *S. canaliculatus*	Ongoing research
	Grouper, *Epinephelus* spp.	Ongoing research
Saudi Arabia	Tiger grouper, *E. fuscoguttatus*	Ongoing research
	Camouflage grouper, *E. polyphekadion*	Ongoing research
	Rabbit fish, *S. rivulatus*	Ongoing research
	Asian seabass, *L. calcarifer*	Techology developed plus ongoing research
	Tialpia strains, *O. spilurus*	Technology developed plus ongoing research
	Tiger shrimp, *P. monodon*	Technology developed plus ongoing Research
	White shrimp, *P. indicus*	Technology developed plus ongoing research
UAE	Rabbit fish, *S. canaliculatus*	Technology developed plus Ongoing research
	Sobaity seabream, *S. hasta*	Commercial hatchery prod.
	Local seabream, *R. sarba*	Ongoing research
	Yellow-finned seabream, *A. latus*	Ongoing research
	Large-scale mullet, *L. macrolepis*	
	Brown-spotted grouper, *E. coioides*	Ongoing research
	Gilthead seabream, *S. aurata*	
	European seabass, *D. labrax*	Ongoing research
	Green tiger shrimp, *P. semisulcatus*	Commercial production
	White shrimp, *P. indicus*	Commercial farming
		Ongoing research
		Farming trials

RESEARCH ACCOMPLISHMENTS AND PRODUCTION STATUS

Bahrain

The NMC has successfully achieved mass seed production for the commercially important species such as the local sobaity bream *S. hasta* and has adopted technology for the seed production of gilthead seabream *S.aurata*. At present these species are produced in commercial quantities in their hatchery facility and sold to the private companies involved in sea-cage farming. Hatchery technology has also been developed for other commercially important local species such as the brown-spotted grouper, rabbitfish and mangrove snapper. The success has been such that the NMC is supplying marine finfish seed to other member countries of the Regional Commission for Fisheries (RECOFI) and the Gulf Cooperation Council (GCC) [Shams and Al-Ansari, 2004]. During 2006/2007 culture period, the hatchery produced more than 400,000 marine fish seeds for stocking in sea-cages. Being a small island nation, extensive land-based aquaculture developments are limited and accordingly land-based shrimp culture from a practical point of view is not feasible. However, the country is promoting private sector for investing in tank-based mariculture activities and open water cage culture.

Bahrain hatchery for local species

Bahrain hatchery for exotic species

Iran

Fish farming in Iran initiated with fish stock enhancement in the Caspian Sea and then continued for the semi-intensive aquaculture, utilizing Chinese carp species, rainbow trout and recently the Indian white shrimp. Aquaculture developments in Iran had a rapid growth since 1980. It has expanded to culture of food fish in raceways (trout), ponds (cyprinids) and farming of penaeid shrimp in the 90's [Rana and Bartley, 1998]. Production from aquaculture source and aquaculture based fisheries has been increased from 4,935 t in 1978 to 124,000 t in 2004 [National Aquaculture Sector Overview –Iran, www.fao.org/fi/website /FIRetrieve Action.do?dom=countrysectorandxml=naso_iran.xml 7/17/2007]. Table 2, shows the

Hatchery produced sturgeon fish for releasing in Caspian Sea

Caviar produced from the sturgeon *Acipenser persicus* Borodin, 1897.

Seacage for marine finfish production at Qeshm, Iran [Regunathan & Kitto, 2005]

categorical production from different production units and facilities during 2004 culture period. The rapid growth has placed Iran among the top ten aquaculture producers in the world, in terms of growth, between 2002-2004 periods [FAO, 2007]. This will increase in the next five year plan period (Table 3) to more than 300,000 t through aquaculture production by the year 2009 [FAO]. Furthermore, due to its stocking of Persian sturgeon in the Caspian sea, UN has exempted Iran from banning the export of caviar. Site surveys have been carried out and SHILAT, the Iranian Fisheries Organization, has studied the potential for marine fish farming in the Northern coast of Gulf and Oman Sea. Iranian government assessment of prospects for sea cage aquaculture development along the Iranian coastline has targeted development of an 8,000 t industry in the Caspian Sea, and a staggering 25,000 t in the Gulf and Sea of Oman. According to SHILAT, five companies have received licenses for to pursue cage culture in the Gulf. Iran is actively pursuing a holistic approach for aquaculture development, building on one of its key strengths, technology for mass seed production from the hatcheries and infrastructure for restocking inland and coastal waters. In support of the hatchery seed production, Iran has initiated production of *Artemia* cysts from the hypersaline (100-150 ppt) waters of Uromia Lake through its technical collaboration with University of Ghent, Belgium.

Shrimp culture, based on the Indian white shrimp *P. indicus,* started in 1995 and expanded rapidly to more than 2000 ha in four southern provinces in 1999. In 2003, shrimp production exceeded 7,500 t from 3590 ha of coastal ponds along the Iranian coast of the Gulf and the Gulf of Oman. In order to provide post-larvae for shrimp culture, about 32 hatcheries have been established and were in operation in 2003 [FAO]. Shrimp culture activities are on rapid development, based on earthen ponds, and Iran has a potential shrimp/fish pond culture area of about 26,239 ha in the southern coastal area. Iranian government has established eight hatcheries to produce fin fish and sturgeon fingerlings for releasing in the Caspian Sea as well as for private fish farming. Iran's International Sturgeon Research Institute has been chosen as the international center for the development of aquaculture and sturgeon in Asia-Pacific region by NACA (Network of Aquaculture Centers in Asia-Pacific) and introduced Iran as an ideal model for the aquaculture research during the year 2006 [www.payvand. com/ news /06/dec/1303.html]. Furthermore, two of Iran's aquaculture research centers in the northern cities of Rasht and Tonkabon have been selected as ultimate patterns for other NACA members. Mazandaran province stands first in aquaculture production by providing 25 percent of the national requirement. Iran Fisheries Organization is issuing permits for breeding cold-water fish species, releasing red-spotted trout fingerlings in the rivers and establishing facilities for breeding fish in Behshahr. Culture of rainbow trout is restricted to the cooler northern areas of the country. Seed production and intensive ongrowing of rainbow trout is well developed using standard methods of breeding, larval rearing and farming in aerated grow-out raceways [Rana and Bartley, 1998]. The government supports the private sector by giving low interest rate loan for aquaculture and suitable land for low price. Iranian Fisheries Organization has invested a considerable amount of money to establish research centers in different parts of the country for supporting the aquaculture industry.

Table 2. Aquaculture production from different facilities in Iran during 2004 (source FAO, National Aquaculture Sector Overview – Iran)

Culture Category	Production Unit/Facility	Production (t)
Warm water fish	21,700 ha ponds	65,000
Cold water fish	Concrete raceways	30,000
Marine Shrimp	4,253 ha coastal ponds	09,000
Culture-based fisheries	514,000 ha lakes and reservoirs	20,000
	Total	*124,000*

Table 3. Aquaculture production projections in the 4th Five-Year Plan for Fisheries, 2005-2009 in Iran [FAO]

Category	2003	2005	2006	2007	2008	2009
Shrimp	7,492	14,110	23,800	32,300	40,500	47,200
farming	23,138	35,000	40,000	46,000	51,000	59,000
Cold-water						
fish farming	79,545	106,527	129,158	153,806	173,314	208,206
Warm-water						
fish farming	-	800	1,500	2,900	4,400	7,000
Mariculture						
Total	110,175	156,437	194,458	235,006	269,214	321,406

Iraq

The political turmoil in Iraq since the 1960's does not provide an opportunity to enhance aquaculture production. In some parts of the country, from the Southwestern Bedouins to the Northeastern Kurds, traditional self-reliant food systems have been maintained. Aquaculture venture in Iraq is restricted to freshwater fish species. Though the aquaculture production declined from 3,622 t in 1980 to 1,600 in 1990, it increased to 7500 t in 2000. Recently in 2005 the total production through aquaculture was 12,870 t [FAO, 2007].

Kuwait

The research on most of the species focused on hatchery production, feed and nutritional requirement, husbandry, breeding under captivity and mass grow-out using tanks, ponds and sea-cages. The MFD/KISR, Kuwait pioneered in Mariculture research for the seabream *S. hasta,* grouper *E. coioides* and silver pomfret *P. argenteus* among the Gulf Countries. Kuwait was one of the pioneers in the successful breeding and larval rearing of domesticated grouper [Hussain and Higuchi, 1980]. Furthermore, Kuwait was the first country in the world to culture silver pomfret under domesticated culture conditions from eggs collected from the wild spawners [Almatar *et al.,* 2000; Al-Abdul-Elah *et al.,* 2002]. Recently, MFD/KISR has also achieved a breakthrough in successful spawning and larval rearing of silver pomfret under domesticated culture conditions, thus offering this highly priced fish as a potential candidate species for aquaculture around the world [Almatar and James, 2007]. The research accomplishments enabled Kuwait as the first country among the Gulf Cooperation Council (GCC) states to initiate commercial marine fish farming using sea-cages through the establishment of Bubiyan Fisheries Co., in 1985. Actual fish production was initiated in 1992, after the Gulf war. Initial production was based on the European gilthead seabream *S. aurata.* The sea-cage fish production was up to 400 t during 2000. Towards meeting the demand for marine fish seeds for commercial farming in Kuwait and for the grow-out in UAE and Iran, the Gulf International Aquaculture Co., (GIAC) was incepted in the year 2000 for producing marine fish fingerlings availing part of the hatchery facilities within MFD/KISR.

Cultured grouper *E. coioide*

Cultured juvenile silver pomfret *P. argenteus*

Silver black seabream *S. hasta*

Tilapia strain *O. niloticus*

Marine fish hatchery at MFD/KISR

Integrated tilapia farm with agriculture

Tilapia culture introduced in Kuwait by MFD/KISR during 1982, paved the way for the establishment of commercial tilapia farming in Kuwait from 1986 onwards. During the year 2000, the number of farms increased to 55 and tilapia production reached 50 t. The increase in tilapia production resulted from a fish feed subsidy provided by PAAF. During 2001, a total of 100 t of *O. spilurus* and *O. niloticus* were produced and marketed in Kuwait. Integrated aquaculture established in Kuwait use the ground water first for tilapia production and then the effluent water is used for the production of vegetable crops. Currently 58 farms produce tilapia utilizing brackish ground water. The farmers are supported by subsidies from government agencies. Current research programs aims at developing commercially applicable production technologies or improvements in the available culture systems.

Based on the research outcome, several economic feasibility studies were carried out by MFD/KISR. Several species were opted for commercial farming using different culture systems. Table 4 summarize the targeted production level and their economics.

Table 4. Options of species for grow-out production and their techno-economic assessment for aquaculture development in Kuwait

Fish species and Production systems	Prod. Level (t/year)	Prod. Cost (KD/kg)	Economic indicators		
			IRR (%)*	Net profit (KD/t)	Payback Period (yrs)
Marine Fish					
1. Sobaity (*S. hasta*)					
Sea-cages	200	1.846	2.23	154	6.06
	400	1.670	11.30	330	2.91
	600	1.590	15.47	410	2.27
2. Sobaity production in flow-through water tanks	400	1.906	-.065	94	11.78
3. Sobaity production in PVC-lined ponds	400	1.720	5.88	280	3.93
4. Shaem (*A. latus*)					
Sea-cages	400	1.795	3.90	205	3.67
5. Integration of Sobaity + Tilapia in sea-cages	Sobaity + Tilapia				
	200 + 100	1.535	10.52	302	3.02
	200 + 200	1.382	16.46	372	2.16
	400 + 100	1.529	14.72	374	2.36
	400 + 200	1.420	18.00	418	1.96
6. Integration of Shaem	Sobaity + Shaem				
(*A. latus*) + Sobaity or	400 + 100	1.618	13.65	382	2.39
Hamoor (*E. coioides*) + Sobaity or	400 + 100	1.583	15.44	417	2.22
Gilthead Sea Bream (*S. aurata*) + Sobaity in sea-cages	400 + 100	1.687	7.57	265	3.29
Tilapia					
1. Recirculating brackish	200	1.477	-2.43	23	16.61
groundwater culture system	400	1.323	04.75	177	07.10
using tanks	800	1.243	08.61	257	05.24
2. Integration with Agriculture	1.0 t/yr/well	1.279	10.96	221	06.72
at a water flow rate of 500 l/min	2.5 t/yr/well	1.016	27.17	484	01.73

* IRR = Internal Rate of Return; Source – KISR Interim Report No. 6 and Technical Appendix 13 (1988); 1 KD = 3.3 US

Oman

Prior to the initiation of aquaculture research in 1992, during 1986 a commercial tiger shrimp *P. monodon* farm was developed in Sur region, but the facility ceased its operation two years later due to technical and administrative issues. Since site evaluation is crucial for aquaculture development, the Aquaculture Laboratory has accomplished two projects on site surveys along the Omani coast (Al-Yahyai, et al., 2004). The Laboratory has also carried out experimental farming of the Indian white shrimp *P. indicus* and studies on seacage farming of exotic gilthead seabream *S. aurata* and the local black seabream *S. hasta*. Based on the research outcome two privately operated commercial cage farms owned by Quriyat Aquaculture Company have been established (Al-Qasmi et al., 1998). The company has two cage sites located in the Muscat area. One of the cage site is 2,400 m^2 and another is 3,500 m^2. Commercial seacage farms produce mainly the gilthead seabream *S. aurata* and small quantities of European seabass *D. labrax* and yellowfin seabream *A. latus*. The gilthead seabream production increased from 331 t in 2003 to 460 t in 2004 and the European seabass production increased from 13 t to 27 t during the same period. In 2004 yellowfin seabream production reached 13 t. Further seacage production is concentrating on fattening of the yellowfin tuna *T. albacares*. Oman is the first country in the Middle East to initiate tuna fattening project. This is carried out through the establishment of a large 47,000 m^3 capacity seacage in the Bandar Khayran area in 2004 and produced 14 t during that period. Small scale tilapia farming is also a widespread commercial activity in Oman. The total production through aquaculture reached 514 t during 2004 [FAO].

Qatar

Aquaculture is not well established in Qatar. In 2005, the total production amounted to 11 t from experimental facilities. There is no major commercial aquaculture production in Qatar [FAO].

Saudi Arabia

The research outcome at FFC/FAO has paved the way for the rapid development of aquaculture industry in Saudi Arabia, especially relevant to marine shrimp farming and tilapia production by private investors. The distinct research accomplishments are that FFC succeeded in the spawning and hatchery larval rearing of one of the expensive food fish, the camouflage grouper *E. polyphekadion* for the first time [James et al., 1997]. The centre was able to demonstrate the growout production of the brown-marbled grouper *E. fuscoguttatus* and the camouflage grouper *E. polyphekadion* under hatchery and growout culture conditions [James et al., 1998]. Furthermore, the research outcome has demonstrated the potential of grouper hybrid (*E. fuscoguttatus* x *E. polyphekadion*) for aquaculture [James et al., 1999]. Considering shrimp farming as the most successful aquaculture sector, FFC carried out several studies on the farming of the tiger shrimp *P.* monodon and the Indian white shrimp *P. indicus* and formulated shrimp growout feeds with local supplies. The outcome of the

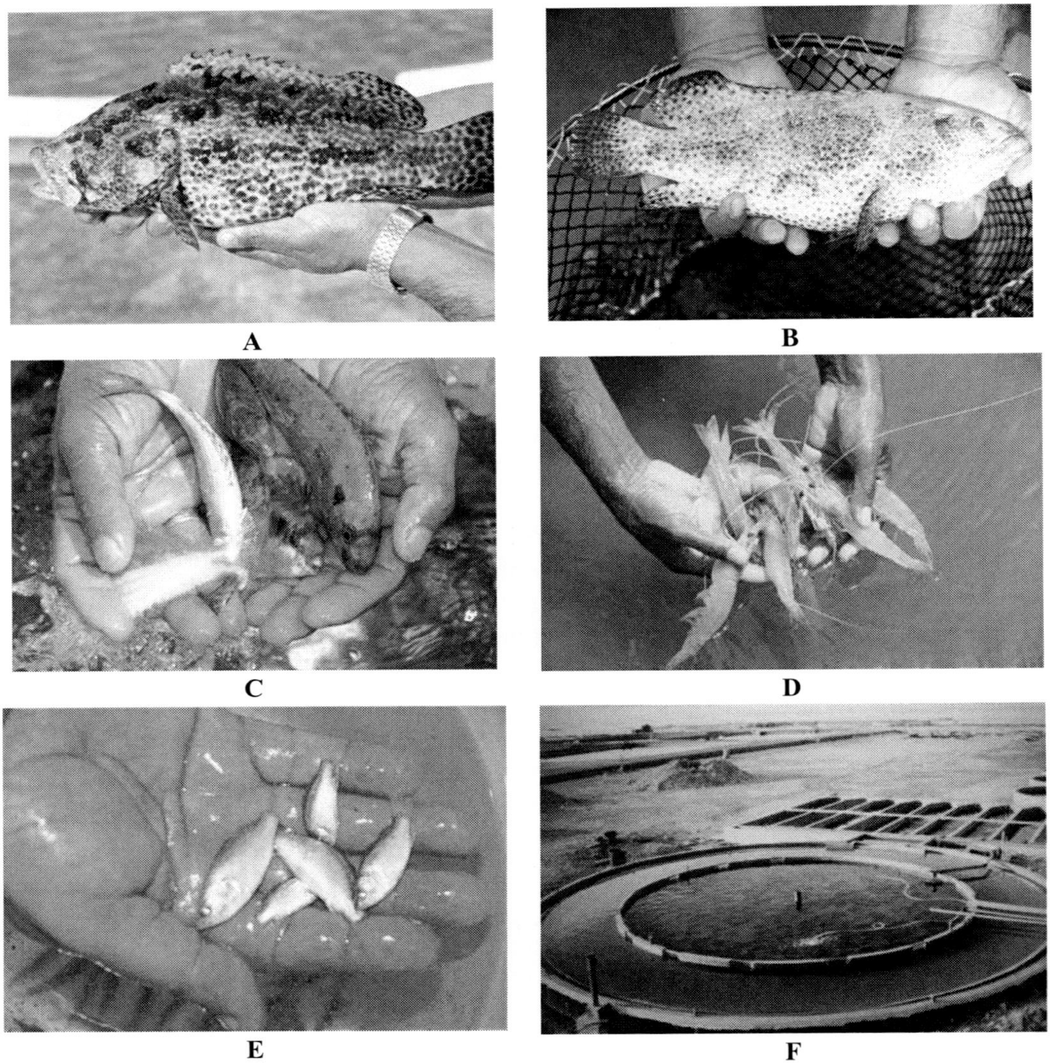

Aquaculture activities in the Kingdom of Saudi Arabia: A) Cultured camouflage grouper *Epinephelus polyphekadion* at FFC; B) Grouper hybrid cultured at FFC; C) Cultured rabbitfish *Siganus rivulatus* at FFC; D) Cultured white shrimp *Penaeus indicus* in Gizan; E) First hatchery produced Asian seabass *Lates calcarifer* at FFC; F) Baobab tilapia production experimental and demonstration unit at FFC.

technology by FFC has paved the way for successful farming of shrimps by private sector in the hypersaline waters of Saudi Arabia [Al-Thobaity and James, 1998]. The 10 km^2 Red Sea Shrimp Farm produces over 1,500 t of shrimp annually. The research programs have paved the way for establishing the world's largest, fully integrated desert coastal aquaculture project located on the Red Sea coast of Saudi Arabia by National Prawn Company. The National Prawn Company proposed to have 2,800 ha shrimp farm with a projected production of 13,500 t of shrimps per year in Phase-I and will have 3,500 ha of more shrimp ponds capable of producing 17,500 t of shrimp per year in Phase-II. In general, the aquaculture production has a rapid increase from 1,988 t in 1990 to 14,375 t in 2005. Tilapia farming is also intensively carried out in Saudi Arabia. Currently there are 125 licensed aquaculture farms of which 56 are fully operational. In addition to tilapia, the Asian seabass introduced in the late 90's by FFC/FAO has transformed into commercial ventures in recent years.

UAE

The country has achieved food security in the availability of fresh fish. The government feels that the annual restocking/releasing of fish fingerling program into the sea is making a contribution to the wild fish stock of this region, which indirectly helps the economy [Environmental Research and Wildlife Development Agency, 2003]. The MRRC is also providing necessary assistance for establishing aquaculture facilities by the private investors in UAE. The MRRC targeted fish fingerling production in its hatchery during 2005 was 250,000 comprising sobaity seabream (*S. hasta*), rabbit fish (*S. canaliculatus*) and the large scale mullet (*L. macrolepis*). Commercial aquaculture has been established in UAE through the induction of the International Fish Farming Company (ASMAK) in 1999 with the support of the Abu Dhabi government with a paid up capital of US$ 82 million. ASMAK has about 50 circular sea-cages (19m dia. x 12m deep) off the coast of Dibba. An estimated 570 t of fish comprising of the gilthead seabream, European seabass (*S. aurata, D. labrax*) and the silvery black seabream (*S. hasta*) produced during 2004 [FAO, 2006]. Recently, the production has been increased to 900 t per annum during 2006/2007 culture period.

Constraints and Priority Issues

Each country has its own constraints and priority issues for developing aquaculture. Some of these issues are relevant to farming systems, developing technologies for local species, economy of marketing and processing, policies, legal framework, institutions and investment constraints. Limited availability of suitable sites for new aquaculture activities is a commonly encountered problem in the region due to the shortage of land, insufficient freshwater, insufficient tidal amplitude for land-based culture and restrictions in the availability of suitable marine sites. In some cases, suitable sites may exist but access to them is restricted due to human settlement, tourism, conservation, military defense and local government restrictions. Economic sanctions imposed on Iraq in 1991 and the aftermath of war and political turmoil has restricted the aquaculture development in that country. Business model for aquaculture in Iraq has been proposed [USAID, 2006]. Weather is a challenging issue for aquaculture in several regions in the Gulf. Extremely hot and arid climatic conditions during summer and very cold weather during winter restricts the economic productivity of land-based aquaculture systems. Restrictions on suitable soil and land availability combined with harsh weather conditions are some of the constraints facing the aquaculture industry.

Aquaculture activities in UAE: A) Hatchery facility at MRRC; B) Out door culture tanks; C) Broodstock holding facility; D) Live food Artemia production facility at MRRC; E) Fish sorting for stocking; F) Commercial sea-cage farming by International Fish Farming Company (ASMAK) at the eastern coast of UAE'

One of the major challenge for the development of commercial aquaculture, especially for marine fish farming, is the non-availability of adequate fish fingerling supply, especially with reference to local prime species (such as grouper and sobaity seabream) from the hatchery. In addition to inadequate local marine finfish seed supply there is also a need to depend on imported feed supply in most of these countries. Because of the non-availability of endemic marine fish fingerlings, the commercial sea-cage culture systems depends on imported exotic species with a possibility for unknown environmental impacts and constraints in market profitability due to the hesitant acceptability of foreign species by the local population. Market profitability is a challenging issue not only for exotic fish species but also for some of the endemic fish species for which the culture technology is an expensive issue compared to its market value. Research and technology transfer cooperation between the Gulf countries relevant to adoption of local prime marine fish species for aquaculture could resolve some of these issues.

FUTURE TREND

Aquaculture is expected to have a significant growth in the Gulf countries. The major trend in aquaculture is contained in the overall increase in the culture of marine species (fish and shrimps), intensification of the culture system and more integrated agriculture-aquaculture developments. Within marine species, both diversification and intensification are anticipated, driven by such forces as successful research by government organizations providing technical knowledge and seed stock, availability of private investment funds and potential export markets. Intensification approach is mainly to make use of the suitable site and water to achieve maximum sustainable production since as mentioned above availability of suitable land is restricted in some countries for example in Kuwait and Bahrain. Another trend is the increasing interests in marine fish culture using sea-cages. Due to the limitations in marine fish hatchery production, integrated aquaculture activities are emerging such as the commercial fish hatchery located in Kuwait while the grow-out sea-cages located in UAE and in Iran. Interests in the integrated farming is also due to the availability of deeper and warmer waters in the south for locating grow out sea-cages at the same time locating the hatcheries in the north. Furthermore Oman and Saudi Arabia have high potential for shrimp farming because of the availability of suitable farming sites. National Prawn Company in Saudi Arabia is considered to be the world's largest fully integrated desert coastal aquaculture project with hatchery, farm and processing facilities and pose to increase its future shrimp production to 45,000 t per annum and bid to tap into the booming global seafood market.

Growth of aquaculture sector demands for adoption of new species, some of which are endemic to a particular country and developing technology for farming in that country, while others are non-endemic to a particular country and are imported for aquaculture purposes. A number of these newly introduced species are already contributing significantly to aquaculture production in some Gulf countries such as the gilthead seabream (*S. aurata*), the European seabass (*D. labrax*), the Nile tilapia (*O. niloticus*) and the rainbow trout (*O. mykiss*). Future trend demands for the development of technology and adoption of highly priced new endemic species such as the silver pomfret (*P. argenteus*) that is fast declining in the local waters. To enhance the seafood export and marketability, improvement in seafood processing and adoption of new technical know how in seafood processing is also required to compete with other countries.

Fish is not a popular part of the peoples' diet in the Gulf, especially in the interior central cities. In general, the per capita fish consumption has been increased from about 1kg/year in 1980 to over 6 kg in recent years. The Iranian Fisheries Organization has aimed at enhancing the per capita fish consumption to about 10 kg by 2009.

The major driving force behind the growth of aquaculture industry in this region is the demand for local seafood production combined with the declining trend of some of the most commercially important food fish stocks in the wild. Other factors include the interests in export market due to its high price and compulsion of socio-economic development via provision of employment and career opportunities. Within the aquaculture sector, progress in technical organizations, improvements in infrastructure, availability of government and private funding agencies for research and development as well as for commercial ventures, organized legal framework including effective certification and licensing, are important driving forces to support the aquaculture development in Gulf Countries.

ACKNOWLEDGEMENTS

Sincere thanks are due to Dr. Jasim Meramazi, Head South of Iran Aquaculture Research Centre; Mr. Anwar Al-Hariri and Mr. Adli Al-Ansari of the National Mariculture Centre of Bahrain; Mr. Abdul Razaq Abdulla and Mr. Mustafa Al-Shaer of the Marine Resources Research Centre in Umm Alqaiwain, UAE; Mr. Abdul Rehuman A. Yousef of the mariculture and Fisheries Department of KISR, Kuwait; for providing the photographs included in this report.

REFERENCES

Al-Abdul-Elah, K.M., Almatar, S., Abu-Rezq, T., James, C.M. and S. El-Dakour. (2002).Development of hatchery techniques for the silver pomfret *Pampus argenteus* (Euphrasen). Asian Fisheries Science, 15: 107-121.

Almatar, S., Al-Abdul-Elah, K. and T. Abu-Rezq. (2000). Larval development stages of laboratory reared silver pomfret, *Pampus argenteus*. Ichthyological Research, 47: 137-141.

Almatar, S. and C.M. James. (2007). Prospects of silver pomfret *Pampus argenteus* Euphrasen as a new candidate species for aquaculture. Paper presented during 8th Asian Fisheries Forum, 20-23 November, 2007, Kochi, India. Published abstract.

Al-Qasmi, A., Al-Farsi, I., Gindy, A., Al-Busaidi, Y. and A. Al-Mazroai. (1998). Finfish cage culture demonstration project: Final report. Oman Ministry of Agriculture and Fisheries. 20 pp.

Al-Thobaiti, S.A. and James, C.M. (1998). Saudi Arabian shrimp success in hypersaline waters. *Fish Farmer International File* July/August, '98: 20-21.

Al-Yahyai, D.S., Mevel, J.Y., Al-Farsi, I., Al-Farsi, E., and Y. Al-Ruqishi. (2004). General introduction to suitable sites for aquaculture in Oman. Agriculture and Fisheries Research Bulletin, 1 (5): 5-10.

Environmental Research and Wildlife Development Agency. (2003). Fish Resource Assessment Survey Project of Abu-Dhabi and UAE Waters – Project Report. Environmental Research and Wildlife Development Agency, United Arab Emirates.

Esmaeili, A. (2006). Technical efficiency analysis for the Iranian fishery in the Persian Gulf. *ICES Journal of Marine Science,* 63: 1759-1764.

FAO. Fishery and Aquaculture country profile – Islamic Republic of Iran. www.fao. org/fi/ website/FIRetrieveAction.do/dom=countrysectorandxml=FI-CP_IR. xmlandlang=en 8/12/2007.

FAO. National Aquaculture Sector Overview – Oman.www.fao.org/fi/website/FIRetrieve Action.do?xml=naso_oman.xmlanddom=countrysector 7/15/2007.

FAO. Fishery Country Profile – State of Qatar. www.fao.org/fi/fcp/en/QAT/profile.htm 9/27/2007.

FAO and NACA (1997). *Survey and Analysis of Aquaculture Development Research Priorities and Capacities in Asia.* FAO Fisheries Circular No. 930 FIRI/C930, Rome, FAO. 1997. pp.263.

FAO, (2006). Aquaculture production, 2004. Year book of Fishery Statistics, 96/2. Food and Agriculture Organization of the United Nations, Rome, Italy.

FAO, (2007). *The State of World Fisheries and Aquaculture 2006.* Food and Agriculture Organization of the United Nations, Fisheries and Aquaculture Department, Rome, 2007. pp. 180.

Feidi, I.H. (1998). Fisheries Development in the Arab World: Section IV. Marine Environments. In: J. Coppock, and J.A. Miller (Eds), *Transformations of Middle Eastern Natural Environments: Legacies and Lessons.* Bulletin No. 103, pp. 492. Yale University, New Haven, Connecticut.

Hussain, N.A., Saif, M. and Ukawa, M. (1975). On the culture of *Epinephelus tauvina* (Forskal). Kuwait Institute for Scientific Research, Kuwait. 12 pp.

Hussain, N.A. and Higuchi, M. (1980). Larval rearing and development of the brown-spotted grouper, *Epinephelus tauvina* (Forsskal). *Aquaculture* 19: 339-350.

James, C.M. (1998). Fish Farming Centre Project –UTFN/SAU/010/SAU, Kingdom of Saudi Arabia, Terminal Report. FAO, Rome: pp.1-31.

James, C.M., Al-Thobaity, S.A., Rasem, B.M. and Carlos, M.H. (1997). Breeding and larval rearing of the camouflage grouper *Epinephelus polyphekadion* (Bleeker) in the hypersaline waters of the Red Sea coast of Saudi Arabia. *Aquaculture Research*, 28: 671-681.

James, C.M., Al-Thobaity, S.A., Rasem, B.M. and Carlos, M.H. (1998). Comparative growth of brown marbled grouper *Epinephelus fuscoguttatus* (Forsskal) and camouflage grouper *E. polyphekadion* (Bleeker) under hatchery and growout culture conditions. *Asian Fisheries Science*, 11: 133-147.

James, C.M., Al-Thobaity, S.A., Rasem, B.M. and Carlos, M.H. (1999). Potential of grouper hybrid (*Epinephelus fuscoguttatus* x *E. polyphekadion*) for aquaculture. *NAGA, The ICLARM Quarterly*, 22(1): 19-23.

Matinfar, A. and A. Nikouyan. FAO report on a regional study and workshop. www.fao.org/docrep.field/003/AC279E/AC279E14.htm.

Poynton, S.L. (2006). *Regional Review on Aquaculture Development. 2. Near East and North Africa – 2005.* FAO Fisheries Circular No.1017/2. Rome, FAO. 79 pp.

Rana, K. and D.M. Bartley. (1998). *Iran Promotes Aquaculture Development.* FAO Aquaculture Newsletter, No. 19, August 1998, pp. 26-30.

Regunathan, C. and M.R. Kitto. (2005). Persian Gulf fish culture in Iran – pointers for success. *Aquaculture Asia,* 10 (2): 40-42.

Shams, A.J. and A.A. Al-Ansari. (2004). Update of National Mariculture Sector Review for the Kingdom of Bahrain. Presented in the First Meeting of RECOFI-Working Group on

Aquaculture, WGA/NASA/04-04/ at Manama, Bahrain, pp.5.USAID. (2006). Iraq private sector growth and employment generation: Business Models for Aquaculture in Iraq. United States Agency for International Development, Contract # 267-C-00-04-00435-00, pp.1-38.

INDEX

A

A.D.A., 180
abalone, x, 22, 43, 44, 105, 108, 110, 149, 153, 154, 174, 176, 177, 179, 183
abatement, 287
abdomen, 21, 23, 24, 27, 43, 44
Abu Dhabi, 315
Abundance, 9
AC, 94
access, 34, 105, 108, 171, 289, 301, 315
accounting, 154
accuracy, 63, 65, 68, 78, 81, 97, 98, 100, 188, 190
acetic acid, 121, 266, 277
acid, 32, 35, 57, 117, 118, 119, 155, 156, 158, 160, 178, 181, 190, 199, 200, 204, 207, 211, 212, 253, 255, 256, 259, 260, 266, 277
acrosome, 150
actin, 233
activation, 181, 219, 221, 223, 232, 245
AD, 94
adaptability, 137
adaptation, 138, 238
additives, 118, 119
adenosine, 151
adenosine triphosphate, 151
adenovirus, 171
adhesion, 20, 30, 32, 55, 219
adhesions, 20, 47
adjustment, 72
administration, xi, 32, 35, 37, 55, 57, 213, 215, 216, 220, 225, 252
adults, 123, 124, 304
aerobic, 157, 173, 176, 181
aerobic bacteria, 176, 181
aflatoxins, xii, 263, 271
agar, 131, 158, 160, 282
age, 35, 118, 119, 138, 143, 203, 241
ageing, 151, 183
agent, 20, 37, 39, 41, 175, 176, 272
agglutination, 26, 47
agglutination test, 26, 47
aggregates, 288
aggression, 142
aggressive behavior, 136
aggressiveness, 247
agonist, 232
agricultural crop, xiii, 297
agriculture, 280, 290, 299, 301, 311, 317
Alabama, 294
alanine, 251
alfalfa, 252, 256
algae, vii, 25, 46, 114, 120, 123, 124, 127, 128, 132, 239, 240, 246, 287
algal paste, ix, 113
algorithm, 67, 70, 93, 94
alimentation, 140
alkaline, 119, 158, 160
alkaline phosphatase, 119
allele, ix, 60, 62, 63, 65, 66, 67, 71, 73, 74, 76, 80, 81, 83, 85, 86, 87, 88, 89, 95, 104
alleles, 62, 63, 64, 65, 66, 67, 71, 72, 73, 74, 75, 76, 77, 81, 83, 84, 85, 86, 87, 90, 93, 94, 96, 103, 104, 105
ALT, 30
alternative, xii, 2, 65, 67, 68, 74, 76, 92, 95, 120, 126, 137, 222, 239, 250, 259, 263, 266
alternative energy, 259, 266
alternatives, 249
alters, 147
aluminium, 220
amino, 2, 117, 118, 129, 142, 144, 217, 251, 260, 264
amino acid, 2, 117, 118, 129, 142, 144, 217, 251, 260, 264

amino acids, 117, 118, 129, 144, 251, 260, 264
ammonia, 123, 132
amplitude, 315
Amsterdam, 111, 179, 180, 182, 207
amylase, 119
anaemia, 214, 226
anaerobic, 154
anemia, 228
animal diseases, 15, 38
animal models, 220
animal sperm, ix, 149, 152, 169, 171, 172
animals, vii, ix, 1, 7, 8, 10, 14, 15, 16, 32, 35, 106, 116, 128, 149, 150, 151, 152, 153, 154, 155, 156, 158, 159, 163, 164, 170, 172, 180, 193, 223, 225, 264, 267, 268, 272, 274
annealing, 65, 94
anorexia, 254
ANOVA, 191
anthropogenic, 280, 292
antibacterial, 30, 55, 158, 182
antibiotic, 35, 164, 166, 168, 169, 173, 176, 178, 252
antibiotic resistance, 176
antibiotics, 14, 36, 152, 163, 164, 166, 169, 173, 214
antibody, 20, 31, 47, 53, 54, 228, 229, 230, 232
antigen, 34, 216, 219, 220, 221, 225, 233, 235
antigenicity, 230
antiviral, 15, 32, 38, 56, 217, 219, 221, 228, 229, 232, 235
antiviral drugs, 32
anxiety, 37
apoptotic, 225
applied research, 299, 301
Arabian Gulf, vi, xiii, xiv, 297, 298, 305, 306
arachidonic acid, 117, 130, 134, 138, 139, 140, 143, 147, 252
argument, 268
arithmetic, 71, 87
ascites, 24, 45
ascorbic acid, 130, 138, 145, 146
aseptic, 172
ash, 247
Asia, 1, 140, 291, 309, 319
Asian countries, 3
aspartate, 251
Aspergillus niger, 268
assessment, 2, 4, 8, 66, 104, 110, 168, 171, 174, 293, 294, 299, 308, 312
assignment, 66, 73, 74, 75, 77, 79, 80, 82, 83, 103, 105, 106, 108, 110, 111
assimilation, 114, 119, 143, 204
assumptions, ix, 60, 70, 89, 90, 94
asymptotic, 70, 241
Atlantic cod, 107, 112, 151, 152, 175, 214
Atlantic Ocean, 239, 240
atmosphere, 152
ATP, 151, 183
atrophy, 20, 29
attention, xii, 2, 3, 129, 150, 172, 220, 237, 283, 304
attractiveness, 2, 131
AU, 56
Australia, vii, 1, 2, 105, 113, 142, 174, 292, 295
availability, viii, x, xiii, xiv, 2, 59, 60, 75, 78, 81, 123, 134, 135, 144, 185, 187, 199, 201, 202, 239, 267, 268, 285, 297, 298, 299, 315, 316, 317, 318
averaging, 85
awareness, 283

B

B cell, 219, 230
Bacillus, 25, 155, 157, 158, 159, 160, 161, 162, 165, 166, 168, 169, 170, 176
bacteria, vii, x, 7, 8, 9, 10, 11, 14, 20, 21, 31, 32, 37, 43, 44, 45, 46, 47, 53, 122, 136, 141, 149, 153, 154, 155, 156, 157, 158, 159, 160, 162, 163, 164, 166, 167, 168, 169, 170, 171, 172, 173, 174, 175, 176, 177, 178, 180, 181
bacterial, vii, x, 3, 7, 8, 9, 10, 11, 16, 21, 22, 25, 29, 31, 32, 39, 42, 45, 46, 50, 52, 134, 149, 152, 153, 154, 155, 157, 158, 159, 160, 162, 163, 164, 165, 166, 167, 168, 169, 171, 172, 173, 174, 178, 179, 214, 216, 217, 221, 227, 232
bacterial contamination, x, 149, 164, 171, 172
bacterial infection, 3, 172, 214
bacterial strains, vii, 7, 10
bacterium, 20, 43, 154, 174, 178, 183
Baghdad, 301
Bahrain, xiii, 297, 298, 299, 303, 305, 307, 317, 318, 319
banking, 174
banks, 152, 171, 239
barriers, 240
base pair, 62
battery, 100
Bayesian, 68, 70, 71, 74, 79, 103, 106, 109
B-cell, 218
BD, 94
beef, 267, 268
beer, 127
behavior, 115, 141, 244, 246, 247, 252, 264
Beijing, 37, 38, 49, 57
Belgium, 145, 300, 309
bending, 136
benefits, 3, 123, 172, 268, 282, 291, 293
bias, 8, 73, 97, 103

binding, 32, 141, 230, 233
bioaccumulation, 253
bioavailability, 251, 252, 259, 260, 268
biochemistry, 208
biodiversity, xii, 279, 280, 281, 282, 292, 293, 295
biogeography, 145
biological activity, 220, 229
biological control, 32, 35
biomass, 123, 127, 128, 134, 247
biosecurity, x, 149, 173
biota, viii, 7, 8, 141
biotechnology, 151, 231, 233
birds, 150
Bizert lagoon, x, 185, 187, 188, 191, 192, 194, 195, 197, 198, 199, 200, 201, 202, 203, 204, 205, 206, 207
Black Sea, 239
bladder, 139, 142, 244
blocks, 98
blood, xii, 29, 30, 46, 52, 60, 263, 267
blood group, 60
body composition, xii, 247, 258, 263, 274, 275, 276
body fat, 264, 265, 266
body size, 137, 203, 244
body temperature, 21
body weight, x, 136, 185, 248, 249, 273
bone, xii, 171, 263, 267, 268, 276
bone marrow, 171
boreal forest, 282, 291, 292
bovine, 119, 169, 178, 191, 267
bovine spongiform encephalopathy, 267
BP, 7
brain, 29, 141
brain development, 141
branching, 26, 285, 286
Brazil, 2
breeding, 60, 61, 62, 78, 82, 98, 100, 102, 105, 106, 152, 168, 208, 244, 255, 301, 309, 310
broodstock, viii, x, 56, 59, 60, 61, 74, 75, 78, 92, 100, 101, 102, 110, 149, 159, 163, 172, 207, 208, 210, 244
brothers, 94
budding, 283
buffer, 159, 160, 162, 163
building blocks, 280
Burkholderia, 157, 168, 170
by-products, 266, 267, 268

C

Ca^{2+}, 180
cachexia, 254
calcium, 151, 202
calcium carbonate, 202
calculus, 66
Canada, xi, 180, 210, 213, 215, 295
Candida, 25, 26, 47, 253
candidates, vii, viii, 1, 59, 71, 72, 73, 74, 75, 77, 78, 79, 122
Capacity, 292
capillary, 190
capital intensive, 1
capsule, 18, 19, 20, 21, 22, 23, 24, 25, 131
carbohydrate, 210, 244
carbohydrate metabolism, 210, 244
carbohydrates, 251, 259
carbon, 121, 145, 191, 204, 280, 282
carbon atoms, 191, 204
Caribbean, v, 1, 2, 3, 4, 5, 292, 293
carotenoids, 267
carrier, 130
case study, 291
caspase, 219
Caspian, xiii, 297, 298, 300, 304, 305, 307, 308, 309
Caspian Sea, xiii, 297, 298, 300, 307, 308, 309
catalase, 175
catastrophes, 283
catfish, xii, 49, 61, 75, 110, 111, 140, 153, 155, 156, 174, 175, 177, 179, 183, 215, 228, 263, 264, 265, 266, 267, 268, 269, 270, 271, 272, 273, 274, 275, 276, 277
cattle, 151, 155
CD8+, 216
cDNA, 15, 30, 31, 38, 53, 56, 219, 231
cell, 19, 24, 25, 26, 29, 30, 32, 34, 36, 40, 41, 51, 53, 55, 117, 158, 162, 169, 171, 172, 176, 204, 216, 217, 219, 220, 222, 224, 229, 232, 233, 234, 235, 266
cell culture, 34, 229
cell cycle, 217
cell fusion, 233
cell line, 29, 51, 222, 224, 234
cell lines, 29, 51, 222, 224
cell membranes, 117, 204
cellular immunity, 219
certificate, 34
certification, 318
channel blocker, 178
chemical, 2, 137, 151, 208, 210, 246
chemical composition, 208, 210
chemokine receptor, 219
chemokines, 220, 231
Chile, 214
China, v, viii, 13, 14, 15, 16, 17, 20, 21, 25, 26, 27, 28, 29, 30, 31, 32, 33, 34, 35, 36, 37, 38, 39, 40,

41, 42, 43, 44, 45, 46, 47, 48, 49, 50, 51, 52, 53, 54, 55, 56, 57, 125, 142, 186, 214
Chinese, 14, 15, 18, 31, 32, 34, 35, 36, 37, 38, 39, 40, 41, 42, 45, 46, 47, 49, 50, 51, 52, 54, 55, 56, 57, 105, 142, 305, 307
chitosan, 225, 235
chloride, 160
chloroform, 190
chlorophyll, x, 186, 188, 189, 199, 201, 202, 203, 204, 206
chloroquine, 232
chorion, 244
chromosome, 60, 99
chymotrypsin, 118
ciliate, 48, 116
circulation, 30
citomegalovirus (CMV), xi, 213
cladocerans, ix, 113, 125, 126
classes, 71
classification, 61, 75, 91, 92, 97, 100, 239
cleaning, 287, 288
cleavage, 245
climate change, 292
clone, 41, 76, 77
clones, 77, 224
cloning, 9, 56, 228, 229
clustering, 95
CMV, xi, 213, 214, 215, 222, 223
cnidaria, ix, 149, 153
CO2, 226
coagulation, 219
coagulation factor, 219
coding, 215, 217, 220, 222, 225
codominant, 63, 65, 86, 95, 96, 99
cofactors, 234
cohesiveness, 117
cohort, 107
collaboration, 309
collateral, 61
colonization, 8, 158
Columbia University, 103
commercial, x, xii, xiii, 1, 33, 61, 74, 75, 78, 105, 107, 114, 117, 123, 128, 132, 137, 142, 149, 193, 224, 237, 238, 246, 247, 248, 249, 251, 258, 268, 269, 281, 285, 297, 299, 301, 302, 303, 304, 305, 307, 310, 311, 312, 313, 315, 316, 317, 318
commodity, 239
communication, 180
community, 146, 293
compatibility, 93, 97
compensation, 282, 287
competitiveness, 37
complement, 218, 228
complex interactions, 187
complexity, viii, 7, 60, 100, 118, 132
complications, 223, 244
components, xi, 8, 33, 36, 61, 95, 130, 131, 139, 171, 176, 186, 190, 287
composition, x, 8, 10, 137, 138, 139, 140, 142, 143, 145, 146, 147, 152, 153, 157, 185, 187, 191, 197, 200, 206, 207, 208, 209, 210, 211, 212, 251, 253, 254, 255, 256, 260, 276
compositions, 145, 164, 199, 280
compounds, 37, 55, 130, 159, 179
compulsion, 318
computation, 100
concentrates, 129
concentration, 36, 122, 130, 140, 150, 152, 166, 169, 183, 189, 201, 202, 204, 206, 210, 246, 268, 275, 276
concordance, 87
Concrete, 309
condensation, 172
conditioning, 179, 208, 210, 211
conduction, 36
confidence, 66, 67, 68, 70, 71, 73, 77, 80, 108, 110
confidence interval, 68, 70, 71
confidence intervals, 68, 71
configuration, 93, 95, 97, 133
confusion, 63
congruence, 75
Connecticut, 319
connective tissue, 189, 190, 191, 192, 201
Connective tissue, 190, 192
conservation, ix, 63, 109, 149, 151, 281, 290, 291, 292, 293, 315
constraints, xiv, 36, 145, 212, 298, 315, 317
construction, xiii, 63, 70, 86, 87, 279, 285, 288, 294
consultants, 299
consumer satisfaction, 254
consumers, xii, 237, 268
consumption, 127, 131, 181, 187, 246, 251, 252, 254, 255, 266, 272, 273, 299, 317
contaminant, 169, 267
contaminants, x, 149, 267
contamination, ix, 149, 151, 154, 159, 163, 164, 167, 168, 171, 172, 174, 175, 176, 177, 178, 179, 268
control, viii, xi, 13, 14, 16, 31, 32, 34, 35, 37, 46, 48, 55, 61, 119, 121, 123, 124, 130, 152, 164, 166, 168, 169, 172, 180, 210, 213, 214, 215, 216, 217, 221, 222, 223, 224, 225, 233, 247, 255, 271, 287, 288
control group, 130, 247
conversion, xii, 237, 248, 251, 252, 264, 269, 270, 271, 272, 273

conversion rate, xii, 237
convex, 239
cooling, 152, 175
copepods, ix, 113, 114, 122, 123, 124, 126, 137, 146, 147
copper, 32
coral nurseries, xiii, 279, 284, 285, 286, 288
coral reefs, xiii, 279, 280, 281, 282, 283, 290, 291, 292, 293, 294, 295
corn, 30, 251, 266, 267, 271, 275, 276
coronavirus, 18
correlation, 47, 85, 90, 94, 95, 97, 191, 202, 203, 205, 217, 218
correlation coefficient, 191, 205
correlations, xi, 186, 203, 206, 247, 286
cortisol, 143
cost effectiveness, 166
cost saving, 131
costs, xi, 33, 121, 123, 134, 213, 215, 249, 250, 264, 269, 273, 284, 290, 293
couples, 75, 78, 82
coverage, 280
covering, 60, 114
crack, 20, 41
Crassostrea gigas, x, 83, 103, 111, 175, 185, 187, 193, 195, 197, 198, 199, 200, 202, 204, 206, 207, 208, 209, 210, 211, 212
CRC, 179, 294
creatinine, 30
Crete, 257
critical period, 135
critical value, 74
Croatia, 181
crops, 311
crustaceans, ix, 60, 81, 149, 153, 181, 182, 240, 298
cryogenic, 171
cryopreservation, ix, 149, 150, 152, 153, 167, 168, 169, 170, 171, 172, 175, 176, 177, 178, 179, 182, 183, 184
cryotherapy, 177
crystalline, 144
CSF, 220
Cuba, 1, 2
cultivation, vii, 123, 124, 147, 164, 247, 249
cultural practices, 293
culture, vii, viii, ix, xii, xiii, xiv, 1, 2, 3, 4, 7, 8, 11, 14, 16, 37, 39, 42, 43, 45, 55, 56, 113, 114, 120, 121, 122, 123, 126, 127, 128, 132, 134, 137, 138, 140, 141, 142, 143, 144, 146, 148, 159, 162, 179, 183, 188, 204, 209, 211, 212, 226, 237, 238, 243, 245, 247, 248, 250, 252, 257, 259, 263, 264, 286, 287, 297, 298, 300, 301, 303, 304, 307, 308, 309, 310, 311, 312, 313, 315, 316, 317, 318, 319
culture conditions, xii, 3, 137, 237, 263, 264, 286, 301, 310, 313, 319
culture media, 159, 162
cultured pearls, vii
current prices, 269
cycles, 70, 202, 241, 242
Cyprus, 238, 241
cyst, 128
cysts, 117, 120, 128, 134, 137, 144, 147, 309
cytokine response, 231
cytokines, 220, 221, 232
cytology, 178
cytomegalovirus, 215, 222, 233
cytopathology, 39, 41, 51
cytoplasm, 192
cytosine, 221
cytotoxic, 219, 220
cytotoxicity, 219

D

danger, 216
data set, 68
database, xii, 108, 263
death, 3, 23, 44, 254, 287
decontamination, 147
defense, 315
deficiency, 104, 107, 134, 142
deficit, 76, 77, 81, 108
definition, 70, 85, 87, 90, 260
deforestation, 280
deformability, 30
degenerate, 164
degradation, 280, 282, 284, 292, 295
dehydration, 169, 245
delivery, 130, 142, 216, 225, 231, 235
demand, vii, ix, xiii, 1, 114, 120, 134, 149, 152, 187, 238, 297, 299, 310, 318
dendritic cell, 225, 235
density, 68, 120, 121, 127, 128, 129, 132, 133, 134, 137, 138, 148, 245, 246, 249, 265, 269, 270, 274
Department of Agriculture, 292
dephosphorylation, 277
deposition, 251, 269
depression, 61
deprivation, 134, 273, 275
dermatosis, 45
destruction, 295
detection, viii, 13, 17, 26, 31, 32, 33, 35, 36, 38, 40, 41, 47, 54, 56, 71, 76, 77, 97, 106, 108, 142, 218
deterministic, 8
developed countries, viii, 13, 33, 282, 290
developing countries, 3

developmental change, 118
deviation, 96
DHA, 117, 121, 128, 129, 130, 134, 137, 141, 145, 252, 253
diet, ix, xii, 2, 4, 113, 117, 118, 119, 123, 131, 133, 135, 137, 138, 139, 140, 141, 145, 148, 181, 204, 209, 239, 251, 252, 253, 258, 263, 264, 265, 266, 268, 271, 272, 277, 317
dietary intake, 119
diets, xii, 5, 114, 117, 118, 119, 120, 130, 131, 132, 135, 136, 138, 139, 142, 143, 144, 146, 147, 204, 209, 211, 212, 237, 240, 246, 251, 252, 254, 257, 258, 259, 264, 265, 266, 267, 268, 269, 271, 272, 274, 275, 276, 277
differentiation, 9, 99, 116, 119, 240, 241
digestibility, 2, 131, 147
digestion, vii, ix, 7, 113, 114, 116, 118, 119, 131, 137, 140, 147, 253
digestive enzymes, ix, 113, 114, 116, 118, 119, 135, 139, 140, 141, 143, 146, 148, 260
digestive process, vii, 7
digestive tract, vii, 7, 8, 30, 52, 114, 115, 118, 119, 135, 138, 148, 182, 244, 253
diluent, 175
diploid, 30, 52, 175, 202, 211
discipline, xiii, 279, 281, 282, 283, 290
discrimination, 9
disequilibrium, 71
disinfection, 174, 293
dispersion, 97, 291
dissolved oxygen, 127, 129
distribution, 2, 16, 30, 38, 47, 51, 67, 68, 69, 70, 71, 105, 123, 143, 192, 193, 210, 252, 259
divergence, 103, 150, 226
diversification, xiv, 9, 238, 298, 317
diversity, viii, 8, 9, 10, 11, 34, 53, 59, 61, 67, 82, 104, 239, 280
DNA, v, xi, 5, 8, 15, 18, 19, 26, 32, 34, 38, 49, 55, 56, 57, 86, 103, 104, 105, 106, 107, 108, 109, 110, 111, 213, 214, 215, 216, 217, 218, 219, 220, 221, 222, 223, 224, 225, 226, 227, 228, 229, 231, 232, 233, 235
docosahexaenoic acid, 117, 128, 134, 137, 139, 141, 142, 145, 252, 253
dominance, 8, 136, 241
dominant allele, 66, 96
dosage, 35, 37, 218
dosing, 259
double bonds, 191, 204
drainage, 287
drinking water, 173
Drosophila, 60, 109
drought, 8
drug metabolism, 35
drug resistance, 32, 35, 37, 42
drug use, 33
drugs, viii, 13, 31, 34, 35, 36, 37, 55, 57, 252, 282
dry ice, 177
dry matter, 123
drying, 169, 180, 271
dsRNA, 224
duration, 150, 223, 224, 227, 247, 281
dynamin, 229

E

E. coli, 157, 170
E.coli, 31, 41
earth, 280
eating, 238
echinoderms, x, 149, 153, 240
ecology, 2, 60, 74, 145, 209, 293
economic development, 318
economic losses, 295
economic performance, 252
economics, 312
ecosystem, xii, 8, 260, 279, 280, 281, 282, 291, 294
ecosystem restoration, 291
edema, 23
education, 284
Education, 13
effluent, 311
egg, 31, 53, 122, 130, 132, 143, 150, 174, 181, 183, 241, 243, 244, 245, 258
eggs, 55, 117, 121, 122, 123, 132, 134, 137, 138, 141, 177, 205, 208, 244, 245, 304, 310
eicosapentaenoic acid, 117, 126, 128, 129, 134, 137, 252, 253
electrochemical, 151
electrolyte, 169
electrolyte imbalance, 169
electron, 29, 50
electron microscopy, 29, 50
electrophoresis, 3, 10, 11
ELISA, 20, 26, 31, 32, 46, 49, 54, 218
ELISA method, 46
elongation, 253, 256
embryo, 31, 53, 169
embryogenesis, 204
embryonic, 31, 169, 245
embryonic development, 31, 169, 245
emission, x, 186, 209
employment, 318, 320
emulsification, 26
emulsifying, 129
emulsions, 128, 137, 144, 145

encapsulation, 144
encoding, 31, 215, 219, 221, 228, 230
endocrine, 30, 52, 210
endogenous, ix, 114, 116, 119, 132, 135
endurance, 8
energy, x, xii, 116, 117, 128, 132, 185, 187, 191, 201, 203, 205, 206, 207, 208, 211, 212, 241, 245, 251, 259, 263, 264, 267, 269, 274
England, 60, 108
enlargement, 18
enteritis, 21, 25, 42, 254
environment, 4, 8, 35, 117, 118, 121, 122, 123, 150, 154, 159, 171, 181, 210, 246, 281, 283, 285, 286, 294, 301
environmental change, 8, 137, 211
environmental conditions, xiv, 3, 8, 142, 187, 201, 207, 269, 270, 280, 285, 290, 298
environmental factors, 3, 156, 201, 202
environmental impact, 249, 317
environmental influences, 210
enzyme, 26, 49, 118, 119, 138, 139, 143, 145, 244, 268
enzyme-linked immunosorbent assay, 26, 49
enzymes, vii, ix, xii, 2, 7, 30, 113, 114, 118, 119, 143, 144, 223, 255, 263
eosinophils, 30
EPA, ix, 113, 117, 128, 129, 130, 134, 137, 145, 252, 253
EPC, 29, 224
epidemic, 16, 26, 47, 177
epidemiology, viii, 13, 14, 16, 39
epithelial cells, 55, 116, 118
epithelium, 29
epitopes, xi, 214, 218, 227, 230
equating, 68
equilibrium, 66, 73, 77, 90, 93, 94, 96
equipment, 152, 249, 288
Escherichia coli, 25, 45, 154, 157, 169, 215
essential fatty acids, 129, 147, 255
EST, 219
esterase, 215, 228
esters, 190
estimating, ix, 59, 60, 63, 66, 88, 101, 104, 108, 111
ethanol, 189, 266
Ethanol, 266
etiology, viii, 13
eukaryotic, 222
eukaryotic cell, 222
Euro, 250
Europe, 146, 214, 222, 293
European, xii, 35, 43, 54, 61, 67, 153, 155, 173, 176, 179, 183, 208, 212, 226, 237, 238, 246, 260, 304, 306, 310, 313, 315, 317
evacuation, 135, 139
evaporation, 188, 287
evidence, x, 2, 5, 16, 56, 107, 185, 202, 209, 234, 260
evolution, 103, 105, 109, 141, 179, 186, 195, 202, 207, 229, 234, 248, 258, 293
exchange rate, 128, 133
exchange rates, 133
exclusion, viii, 59, 66, 67, 71, 72, 76, 77, 78, 79, 80, 81, 82, 110, 111
excretion, 123
exercise, 77, 288
exogenous, ix, 114, 116, 119, 135, 140, 142, 143, 246
experimental condition, 8, 9, 181
expertise, 299, 303
exploitation, 206
exposure, 157, 171, 175, 202
Exposure, 275
expressed sequence tag, 219
extinction, 280
extracellular, 181
extraction, 111, 190, 191, 207
extrusion, 269, 271

F

FAA, 117
facilitators, 292
failure, 152, 159
false negative, 41
family, 49, 61, 62, 65, 67, 72, 75, 76, 77, 81, 93, 94, 95, 96, 97, 98, 100, 104, 110, 214, 221, 232, 234, 239
family relationships, 110
family structure, 77, 100
FAO, 114, 140, 143, 187, 208, 299, 300, 301, 302, 303, 304, 308, 309, 310, 313, 315, 319
farm, 61, 74, 174, 188, 257, 268, 269, 284, 291, 293, 300, 301, 302, 311, 313, 314, 317
farmers, xii, 237, 238, 299, 311
farms, 61, 238, 244, 254, 284, 301, 311, 313, 315
fasting, 276
fat, 131, 138, 247, 251, 254, 255, 264, 265, 266, 267
fatty acid, x, 117, 129, 134, 138, 139, 140, 142, 143, 144, 145, 146, 186, 191, 198, 199, 200, 203, 204, 205, 207, 208, 209, 210, 211, 246, 253, 254, 255, 259, 260
fatty acids, x, 117, 134, 138, 139, 143, 186, 191, 198, 199, 203, 204, 205, 207, 209, 210, 211, 246, 253, 254, 255
fecal, 11, 154, 180, 269
feces, 151

females, 75, 122, 123, 150, 151, 189, 192, 241, 242, 244
fermentation, vii, 7, 267
fertility, 179
fertilization, 134, 146, 150, 152, 163, 169, 174, 178, 183, 244, 245, 293
fiber, 266, 267
fibers, 241
filament, 28
film, 190
filters, 122, 123
filtration, 132
finfish, ix, 113, 114, 120, 126, 128, 130, 131, 132, 135, 137, 138, 144, 146, 147, 175, 258, 260, 298, 300, 304, 307, 308, 316
fingerprinting, 82, 86
fish, vii, viii, ix, xi, xii, xiii, xiv, 2, 7, 8, 9, 10, 13, 14, 15, 16, 17, 20, 24, 26, 29, 30, 31, 32, 33, 34, 35, 36, 37, 38, 39, 46, 48, 49, 51, 52, 55, 56, 60, 61, 63, 75, 81, 100, 105, 113, 114, 115, 116, 117, 118, 119, 120, 121, 122, 123, 125, 126, 127, 128, 129, 130, 131, 132, 133, 134, 135, 136, 137, 138, 139, 140, 142, 143, 144, 145, 146, 147, 148, 149, 150, 151, 153, 154, 155, 156, 157, 159, 172, 174, 175, 176, 177, 178, 179, 181, 182, 183, 184, 204, 211, 213, 214, 215, 216, 217, 218, 219, 220, 221, 222, 223, 224, 225, 226, 227, 228, 229, 230, 231, 232, 233, 234, 235, 237, 238, 240, 241, 243, 244, 246, 247, 248, 251, 252, 253, 254, 255, 256, 257, 258, 259, 260, 263, 264, 265, 266, 267, 268, 269, 270, 271, 272, 273, 275, 276, 280, 283, 287, 288, 291, 292, 297, 298, 299, 300, 301, 303, 304, 305, 306, 307, 308, 309, 310, 311, 313, 315, 316, 317, 318, 319
fish disease, viii, 13, 14, 16, 29, 32, 33, 35, 36, 37
fish meal, xii, 2, 118, 237, 238, 252, 254, 263, 267, 268
fish oil, 252, 253, 256, 259
fisheries, vii, xiii, 1, 4, 37, 39, 114, 282, 283, 293, 297, 298, 299, 301, 303, 307, 309
fishing, vii, xii, 237, 238, 280, 298
fitness, 97
flagellum, 21, 22, 23, 24, 25
flame, 190
flavor, 273
flexibility, 72, 238
floating, 247, 254, 284, 287, 288, 291
flora, 9, 10, 153, 154, 157, 167, 173, 178, 179, 180, 181, 183
fluctuations, 119
fluid, 42, 45
fluorescence, 142, 143
focusing, 299
follicle, 189, 192, 201
follicles, x, 185, 189, 191, 192, 201
food, ix, x, xii, 2, 14, 60, 113, 114, 116, 117, 118, 119, 120, 121, 122, 123, 125, 126, 127, 128, 129, 130, 131, 134, 135, 137, 138, 139, 140, 141, 142, 143, 144, 145, 147, 163, 176, 181, 185, 187, 201, 202, 204, 206, 207, 211, 212, 246, 247, 254, 255, 260, 263, 264, 265, 267, 268, 270, 272, 273, 282, 287, 299, 301, 307, 310, 313, 315, 316, 318
food intake, 255
food production, 121, 130, 137, 141
forest ecosystem, 282, 293
forest management, 280, 293, 294
forest restoration, 292
Forest Service, 291, 292, 294, 295
forests, xiii, 279, 280, 281, 282, 283, 293, 295
formaldehyde, 32
fouling, 287
fractionation, 254
fragility, 187
fragmentation, 286
frameshift mutation, 223
France, 7, 10, 103, 139, 187, 201, 207
freezing, 153, 168, 169, 170, 171, 172, 174, 179, 183
frequency distribution, 242
freshwater, 14, 16, 19, 24, 26, 32, 41, 45, 49, 120, 121, 125, 126, 133, 140, 141, 142, 150, 151, 154, 155, 156, 157, 159, 163, 173, 174, 176, 178, 180, 181, 183, 300, 303, 310, 315
fuel, 206, 282
functional analysis, 233
funding, 318
funds, 290, 299, 317
fungal, 3, 277
fungi, 171, 288
fungus, 14, 25, 26, 32, 268
furunculosis, 20
fusion, 31, 222, 233, 235

G

G protein, 215, 217, 219, 222, 227, 229
Gadus morhua, 82, 107, 112, 151, 152, 175, 211, 214
gall bladder, 27
gamete, 151, 182, 201, 211
gametes, x, 175, 179, 181, 185, 187, 190, 192, 201
gametogenesis, x, 185, 187, 192, 196, 201, 202, 203, 205, 206, 208, 209, 211
gametogenic cycle, x, 185, 187, 194, 196, 203, 204, 205, 206, 207, 210
Gamma, 158, 160

gas chromatograph, 210
gases, 287
gastric, 115, 116, 119
gastric glands, 115, 116, 119
gastrin, 30
gastroenteritis, 178, 179
gastrointestinal tract, 10, 118, 157, 180
GC, 32
GCC, 307, 310
gel, 10, 11
gelation, 148
gender, 35
gene, xi, 3, 8, 10, 11, 17, 34, 36, 47, 56, 60, 213, 215, 216, 217, 218, 219, 222, 223, 224, 227, 228, 229, 231, 232, 233
gene expression, 11, 17, 217, 222, 223, 224, 229, 231, 233
gene targeting, 223
gene therapy, 228, 233
gene transfer, 228
genealogy, viii, ix, 59, 60, 85, 93, 97, 100, 108
general knowledge, 264
generation, viii, 59, 62, 93, 94, 100, 103, 126, 320
genes, viii, xi, 8, 9, 13, 15, 34, 36, 42, 47, 56, 214, 215, 216, 217, 219, 220, 222, 223, 224, 228, 229, 230, 231, 233, 235
genetic diversity, 60, 61, 75, 82, 83, 98, 100, 102, 108
genetic drift, 90
genetic marker, viii, 3, 59, 63, 65, 66, 75, 104, 105, 106, 111
genetic screening, 234
genetics, 4, 60, 66, 80, 83, 105, 108, 111, 141, 283
Geneva, 294
genome, 15, 42, 65, 108, 215, 219, 223, 224, 232, 234
genome sequences, 223
genotype, 20, 65, 74, 81, 86, 87, 92, 93, 94, 96, 103, 287
germ cells, 201
germ line, 109
Germany, 214
Gibbs, 70
gill, 15, 21, 24, 27, 28, 29, 30, 44, 45, 50, 57, 146, 154, 157, 202
gilthead seabream, xii, 119, 131, 135, 143, 146, 147, 148, 175, 237, 238, 243, 245, 246, 247, 249, 250, 251, 252, 254, 258, 259, 303, 304, 307, 310, 313, 315, 317
gland, x, 185, 189, 190, 191, 199, 200, 205, 206
glass, 189
global demand, 128
glucose, 121, 158, 160, 191
glutamate, 251
glycine, 177
glycogen, x, 114, 185, 187, 190, 199, 200, 205, 206, 208
glycoprotein, xi, 34, 56, 213, 214, 215, 219, 224, 227, 229, 230, 231, 233
glycoproteins, 228
goals, 134
goblet cells, 30
gold, 225
gonad-digestive gland, 190
gonadotropin, 244
gonads, x, 185, 189, 191, 192, 201, 211, 241
goods and services, xiii, 279, 282, 293
gossypol, 266, 277
government, iv, 259, 299, 301, 308, 309, 311, 315, 317, 318
GPO, 113
gracilis, 208
grading, 136
graduate students, 301
grains, xii, 263, 266, 269, 276
Gram-negative, 153, 154, 157, 159, 182
Gram-positive, 154, 155, 157, 158, 159, 162, 170
gram-positive bacteria, 183
graph, 103
grass, 14, 17, 18, 21, 22, 29, 32, 34, 38, 39, 42, 51, 303, 304, 305
gravity, 128, 133
grazing, 287
Great Britain, 259
Greece, 238, 253, 256, 257, 259
Greenland, 228
ground water, 311
groundwater, 312
grouper, xiii, 17, 44, 50, 52, 297, 301, 303, 304, 305, 306, 307, 310, 311, 313, 314, 316, 319
groups, viii, ix, 20, 59, 70, 71, 72, 74, 81, 85, 87, 92, 93, 94, 95, 97, 98, 99, 103, 105, 123, 168, 214, 248, 249, 251, 290
growth, x, xii, xiii, xiv, 1, 2, 55, 61, 75, 82, 107, 108, 111, 114, 116, 117, 118, 119, 121, 127, 129, 130, 131, 135, 136, 139, 141, 142, 143, 144, 145, 146, 152, 163, 164, 166, 169, 185, 186, 187, 191, 194, 196, 202, 203, 204, 205, 206, 207, 208, 209, 211, 212, 237, 238, 241, 243, 246, 247, 248, 251, 252, 257, 258, 263, 264, 265, 266, 267, 268, 269, 270, 272, 273, 274, 275, 276, 277, 279, 280, 281, 285, 286, 287, 291, 292, 294, 298, 307, 308, 317, 318, 319, 320
growth factor, 275

growth rate, xiii, 1, 2, 107, 114, 121, 131, 202, 204, 208, 238, 241, 247, 248, 251, 279, 281, 285, 286, 287
Guangdong, 37, 40, 49
guidelines, 291
Gulf of Oman, xiii, 297
gut, vii, ix, 7, 8, 9, 10, 11, 113, 114, 118, 135, 153, 157, 179, 180, 183

H

habitat, 10, 207, 239, 282, 293
Haifa, 279
half-life, 252
halibut, 11, 141, 146, 152, 173, 208, 228, 230
haplotypes, 74
hard clams, x, 149
harm, 14
harmful effects, viii, 59, 61
harvesting, 127
Hawaii, 122, 140, 143, 282, 291
head, 15, 22, 27, 30, 43, 45, 52, 150, 239, 272, 305
health, x, xii, 3, 8, 117, 149, 163, 168, 170, 172, 193, 263, 268
heart, 29
heavy metal, 37, 268
hedonic, 254
height, 249, 294
height growth, 294
hemagglutinins, 56
hematocrit, 266, 271, 272
hematology, 252
hemisphere, 124, 280
hemoglobin, 266
Hemoglobin, 253
hemorrhage, 17, 18, 20, 34, 39, 41
hepatic portal vein, 271
hepatitis, 171, 177, 224, 234
Hepatitis B, 177, 183
hepatocytes, 253, 256
herbal medicine, 35, 36, 37
herbicides, 288, 295
herbivores, 287
herbivorous fishes, vii, 7
herbivory, 292
heritability, 61, 75, 110, 111
herpes, 20, 171, 231
herpes simplex, 171, 231
heterogeneity, ix, 10, 114, 280
heterotrophic, 157, 167, 168, 169, 173
heterozygosity, 63, 66, 67
heterozygote, 62, 65, 76, 77, 81, 104, 107, 108
heterozygotes, 63, 76, 77, 81
high fat, 266
high scores, 254
higher quality, 267
high-performance liquid chromatography, 142
Hippocampus, 107
Hiroshima, 5
histochemistry, 15, 146, 179
histone, 222, 233
histopathology, 15, 29, 41, 46, 48, 49, 50, 51, 52, 180, 235
HIV, 231
homeotherms, vii, 7
homozygote, 65, 76
hormone, 31, 53
hormones, 179, 244
host, vii, 7, 10, 11, 16, 33, 36, 38, 215, 220, 222, 234
HPLC, 32
hub, 291
human chorionic gonadotropin, 244
humoral immunity, 30
hunting, 129
hurricanes, 289
hybrid, 9, 55, 111, 182, 313, 314, 319
hybridization, 15, 17, 36, 41
hydrodynamics, 136
hydrolysates, 118
hydrolyzed, 267
hygiene, 171
hypertrophy, 29, 51
hypothesis, 3, 8, 68, 208

I

IB, 208
IBD, 62, 63, 68, 83, 85, 87, 96
identification, viii, 10, 39, 40, 42, 43, 44, 45, 46, 47, 52, 54, 59, 60, 61, 73, 75, 78, 110, 173, 176, 182
identity, 79, 83, 89, 90, 96, 158
Identity By Descent, ix, 59
Identity By State, viii, 59
IFN, 217, 224
IHNV, xi, 34, 213, 214, 215, 228, 229
IL-8, 220
illumination, 128, 245
immersion, 5, 171, 172, 227, 235
immune function, 53
immune response, xi, 32, 34, 38, 56, 214, 215, 216, 218, 219, 220, 221, 222, 223, 225, 230, 231, 232, 274, 275, 277
immune system, 3, 30, 31, 52, 219, 220, 221
immunity, 3, 37, 56, 218, 221, 225, 227, 228, 229, 231

immunization, 30, 34, 56, 217, 218, 219, 220, 221, 227, 229, 231, 232
immunocompromised, 155
immunogenicity, 220, 227, 232
immunoglobulin, 31, 53, 218, 230
immunoglobulins, 53
immunohistochemistry, 52
immunological, 33, 36, 226
immunomodulatory, 232
immunostimulatory, 221
implementation, 82, 246, 299, 300
in situ, 10, 95, 183, 226, 288, 291
in vitro, 56, 174, 178, 218, 222, 224, 227, 229, 234
in vivo, 218, 222, 224, 228, 229, 230, 231, 234, 235
inbreeding, viii, 59, 61, 76, 85, 96, 100, 105, 108, 109, 137
inbreeding coefficient, 85, 96, 105, 108, 109
incentives, 120
incidence, x, 131, 149, 155, 271
inclusion, xi, 19, 20, 31, 71, 214, 215, 221, 224, 251, 253
inclusion bodies, 31
income, 3
incomes, 3
incompatibility, 65
incubation, 128, 132, 244
incubation period, 244
India, 318
Indian, 154, 177, 307, 309, 313
indication, 141, 241
indicators, 36, 312
indices, x, 8, 9, 87, 95, 186, 202, 255
indigenous, 15, 301, 304
Indonesia, 293
inducer, 223
induction, 122, 143, 177, 221, 222, 315
industrial application, 34
industry, vii, viii, 1, 2, 3, 4, 13, 14, 33, 120, 123, 128, 186, 187, 214, 264, 272, 281, 282, 283, 290, 301, 309, 313, 315, 318
infection, 3, 22, 41, 43, 50, 51, 52, 159, 171, 175, 176, 178, 181, 182, 214, 216, 218, 222, 223, 224, 228, 230, 275
infectious disease, 40, 214, 221
infectious diseases, 40, 214, 221
infectious hematopoietic necrosis virus (IHNV), xi
inferences, 96
inflammation, 232
inflammatory, 220, 221
inflammatory response, 221
inflation, 244
influenza, 229, 232
influenza a, 229, 232
infrastructure, 114, 309, 318
ingest, 116, 126, 147
ingestion, 137, 143, 147, 202
inhibition, 32, 55, 220, 224, 234
inhibitors, 235
inhibitory, 36, 224
initiation, 150, 180, 201, 216, 313
injury, iv
innate immunity, 9, 232
innovation, 128
inoculation, 174, 216
insecticide, 14, 33
insecticides, 32
insects, 288
inspections, 288
institutions, 299, 301, 315
insulin, 275
integration, 35, 70, 224
integrity, 131
intensity, 61, 132, 136, 150
interactions, 10, 11, 36, 61, 117, 233, 287, 294
interference, 33, 224
interferon, 217, 229, 232, 234
interferon (IFN), 217
interferons, 229
interleukin, 56, 220, 231
internal mechanisms, 119
internal rate of return, 250
international trade, 173
interpretation, 68, 74, 90
interval, 203
intervention, 182
intestinal tract, 11, 155
intestine, viii, 7, 23, 27, 28, 53, 116, 118, 157, 174, 181, 182, 240, 264
intramuscular injection, xi, 213, 215, 222, 225, 227, 233
inversion, 241
invertebrates, ix, 145, 149, 151, 153, 167, 204, 208, 211, 280, 283, 292
investment, 249, 251, 287, 315
investors, 250, 302, 303, 313, 315
ionization, 190
ions, 173
IQ, 301
IR, 319
Iran, xiii, 297, 298, 300, 303, 305, 307, 308, 309, 310, 317, 318, 319
Iraq, xiii, 297, 298, 301, 304, 305, 310, 315, 320
Ireland, 211
iron, 158, 160, 274
IRR, 250, 312
irrigation, 284, 285, 287, 289

Islamic, 319
isoforms, 217, 229
isolation, 10, 41, 77, 106, 112, 208, 226
isoleucine, 251
isopods, 240
isozyme, 3
Israel, 279, 290
Italy, 176, 180, 210, 238, 253, 319
iteration, 77

J

Jamaica, 2, 289, 290
Japan, vii, 1, 5, 120, 121, 141, 147, 148, 177, 178, 187, 214, 293, 303
Jung, 178
juveniles, 4, 9, 11, 46, 114, 115, 118, 119, 134, 143, 147, 187, 249, 259, 299, 304

K

K^+, 30, 150
kelp, vii
kidney, 17, 18, 27, 28, 30, 52, 53, 219, 271, 272
killing, 33, 55, 164
kin selection, 107
kinetics, 228
King, 146, 229, 303
kinship analysis, viii, 59, 63, 66, 71, 103, 108
Kurds, 310
Kuwait, xiii, 297, 298, 301, 302, 304, 305, 310, 311, 312, 317, 318, 319

L

labor, 14, 123, 134, 273, 285
lactic acid, 155, 156, 157, 174, 176
lactic acid bacteria, 155, 156, 174, 176
Lactobacillus, 155, 156, 175, 182
lakes, 125, 154, 309
land, vii, 1, 7, 255, 280, 282, 284, 299, 307, 309, 315, 317
larva, vii, 1, 49, 138, 142, 245, 246
larvae, ix, 1, 2, 3, 11, 29, 40, 44, 46, 51, 60, 61, 107, 108, 113, 114, 115, 116, 117, 118, 119, 120, 121, 122, 125, 126, 128, 129, 130, 131, 132, 133, 134, 135, 136, 137, 138, 139, 140, 141, 142, 143, 144, 145, 146, 147, 148, 157, 172, 178, 204, 205, 208, 210, 212, 240, 244, 245, 246, 258, 260, 286, 309
larval, vii, ix, 1, 3, 4, 5, 114, 115, 116, 117, 118, 119, 120, 122, 123, 126, 128, 129, 130, 131, 132, 133, 134, 135, 137, 138, 139, 140, 141, 142, 143, 144, 145, 146, 147, 148, 155, 157, 174, 179, 181, 204, 206, 208, 209, 211, 244, 245, 246, 256, 258, 259, 291, 309, 310, 313, 319
larviculture, ix, 113, 114, 126, 129, 132, 135, 137, 142, 144, 147, 257
latex, 26
laws, 62
leaching, 117, 131, 138, 144
lead, viii, 3, 4, 14, 59, 63, 72, 78, 91, 92, 115, 136, 138, 157, 238, 280, 302
learning, 251
lecithin, 118
lesions, 275
leucine, 251
leucocyte, 227
Leuconostoc, 155
licenses, 309
life cycle, 301
life span, 212
life style, 14
ligands, 221, 231
light conditions, 135
likelihood, viii, 59, 68, 69, 70, 71, 73, 74, 78, 89, 90, 93, 95, 96, 97, 99, 106, 107, 108, 109
linear model, 248
linkage, 77, 81, 93, 99, 221
linoleic acid, 253, 254
linolenic acid, 207, 253
lipase, 119
lipid, 116, 117, 138, 139, 143, 145, 146, 148, 187, 190, 191, 197, 199, 207, 208, 211, 212, 244, 245, 251, 254, 255, 258
Lipid, x, 117, 146, 186, 208, 254
lipids, x, 114, 117, 185, 190, 202, 203, 204, 205, 208, 209, 210, 211, 251
lipopolysaccharides, 221
lipoprotein, 231
liposome, 38
liposomes, 130, 142, 225
liquid nitrogen, 153, 167, 168, 169, 170, 171, 174, 180, 181, 182
liquid phase, 171
literature, 7, 14, 93, 239, 290
liver, 115, 208, 254, 255, 271, 272
livestock, 60, 301
lobsters, vii, 1, 2, 3, 4
local government, 315
localization, 30, 52
location, 3, 8, 27, 137, 246
locus, 62, 63, 66, 71, 73, 74, 76, 80, 83, 85, 86, 87, 88, 89, 90, 93, 94, 97, 103, 223, 229
logging, 280
London, 111, 140, 147, 174, 181, 182
longevity, 229, 281

lordosis, 136, 139
Louisiana, 212, 275
low temperatures, 163
lower prices, 252
LPS, 221
luciferase, 225
lumen, 189, 192, 201
lutein, 267
lymphatic, 31
lymphocyte, 221
lymphocytes, 30
lymphoid, 19
lysine, xii, 251, 263, 266, 267

M

machinery, 249, 284, 293
macromolecules, 116, 147
macrophages, 30, 232
mad cow disease, 268
Maintenance, 273, 287
maintenance tasks, 287
Malaysia, 157, 173
males, 74, 75, 150, 151, 201, 241, 242
mammal, 171
mammalian cells, 223, 224
management, ix, xiii, 2, 60, 61, 75, 103, 104, 107, 110, 114, 115, 134, 137, 149, 151, 152, 159, 172, 174, 181, 210, 264, 279, 282, 288, 291, 292, 299, 301
management practices, 159, 264
manipulation, 122, 136
manpower, 299
mantle, 27, 29, 51, 190, 202
manufacturing, 121, 269, 271
mapping, 73
marginal distribution, 70
marine environment, vii, 177, 181
market, vii, xii, xiv, 1, 34, 61, 114, 128, 237, 238, 250, 251, 298, 302, 317, 318
market prices, xii, 237
market value, 317
marketability, 317
marketing, 238, 267, 268, 315
markets, 238, 284, 317
Markov chain, 70, 111
Maryland, 275
MAS, 154
Massachusetts, 275
matrix, 94, 95, 97, 131
maturation, x, 116, 145, 148, 186, 190, 192, 201, 203, 204, 205, 209, 210, 221, 241, 243, 247, 257, 270
maturation process, 241, 243
Maximum Likelihood, ix, 60, 67, 73, 89
meals, 141
measurement, 209
measures, viii, xiii, 10, 13, 32, 214, 269, 273, 279, 290, 292, 294
meat, xii, 114, 210, 263, 267
media, 268
medicine, 33, 34, 45, 57, 231
Mediterranean, xii, 140, 203, 210, 237, 238, 239, 240, 241, 243, 254, 256, 257, 258, 259, 260, 261, 293
Mediterranean countries, 238, 254
melanoma, 233
melatonin, 252, 260
melting, 128
membranes, 169, 204
memory, 219, 294
meningoencephalitis, 175, 176
metabolism, 5, 9, 32, 33, 202, 207, 208, 209, 211
metabolites, 35, 201
metabolizing, 255
metamorphosis, 116, 120, 205, 206, 209, 210, 211
methanol, 190
Method of Moments Estimators, ix, 60
methyl group, 191
methylation, 221
methylene, 209
Mexico, 2, 208
MFI, 254, 255
MHC, 216, 219
MHC class II molecules, 216
mice, 10, 221, 234
microarray, xii, 15, 214, 219, 231
microbial, 8, 10, 11, 152, 168, 169, 170, 172, 174, 176, 182, 276
Microbial, x, 9, 150, 172, 174, 176, 180, 183
microbial community, 10
microbiota, vii, 7, 8, 9, 10, 11
Microbiota, v, 7
micronutrients, vii, 7
microorganisms, 9, 153, 156, 168, 173, 176, 180
microparticles, 148
microsatellites, 63, 65, 66, 72, 75, 76, 77, 81, 92, 100, 101, 103, 104, 105, 106, 108, 109
microscope, 40, 189
Middle East, 313, 319
migration, 62, 192
military, 315
mineralization, xii, 263, 276
minerals, 277
Ming dynasty, 14
mining, 282

Mississippi, 263, 265, 269, 274, 276
mitochondria, 177
mitochondrial, 3, 182
mitochondrial DNA, 3
MLE, ix, 60, 89, 90, 93
MME, ix, 60, 87, 88, 89, 90, 91, 96, 97
models, viii, x, 59, 74, 83, 186, 216, 243, 248
moisture, 190, 247
moisture content, 247
molecular biology, 26, 35, 38, 56
molecules, 114, 216, 221, 232
mollusks, 48, 187, 203, 298
money, 309
monoclonal antibodies, 31, 53, 54, 229, 230
Monoclonal antibodies, 54
monoclonal antibody, 31, 54
monocytes, 30
monounsaturated fat, 204, 254
monounsaturated fatty acids, 204, 254
Monte Carlo, 70, 93, 111
morphogenesis, 41, 51
morphology, 40, 52, 114, 115, 151, 158, 159, 160, 164, 166, 183, 292
morphometric, 191, 245
mortality, ix, 2, 23, 44, 114, 130, 131, 134, 135, 144, 176, 178, 179, 181, 208, 214, 226, 249, 254, 272, 275
mortality rate, 2, 249, 254
mountains, 49, 128
movement, 4, 61, 75, 98, 126, 150, 287, 289
mRNA, 219
mucormycosis, 26, 47
mucosa, 116
mucus, 53
multiple regression, 250
multiple regression analysis, 250
multiplexing, 111
multiplication, 189, 191, 201
muscle cells, 29, 215, 216
muscles, 20
mutagenesis, 222
mutant, xi, 214, 215, 222
mutation, 65, 76, 80, 81
mutation rate, 65, 76
mutations, 65, 67, 76, 105, 107, 109, 222
mycelium, 26
Mycobacterium, 154

N

Na^+, 30
NaCl, 158, 159, 160, 163
naming, 8
NASA, 320
nation, 307
National Research Council, 210
natural environment, 291
natural food, 264
natural resources, 123
natural selection, 208
neck, 20, 41
necrosis, 16, 17, 18, 29, 33, 34, 39, 50, 56, 57, 154, 213, 214, 226, 227, 228, 229, 230, 271
nematode, 49
nematodes, 288
Netherlands, 157, 187
neural development, 117
neurobiological, 2
neurons, 246
neuropathy, 226
neutral lipids, 254
New England, 231
New Orleans, 275
New York, iii, 9, 39, 103, 145, 180, 206, 207, 208, 209, 211, 260
New Zealand, vii, 1, 207, 210
Newton, 280, 293
Nile, 55, 106, 180, 304, 317
nitrogen, 30, 132, 167, 168, 169, 170, 171, 176
nitrogen gas, 171
NK cells, 219
nodes, 94, 95
norfloxacin, 32, 35, 57
normal curve, 91
normal development, 136
North Africa, 319
North America, 207, 274, 275, 277
North Atlantic, 109
North Carolina, 2
Norway, 214
notochord, 246
nucleic acid, 36, 39
nucleolus, 192
nucleotide sequence, 171, 177
nucleotides, 224
nucleus, 192
nutrients, vii, 7, 116, 117, 130, 134, 138, 142, 264, 284, 287, 288
nutrition, vii, ix, xii, 3, 7, 9, 113, 116, 130, 135, 136, 138, 146, 263, 264, 276, 299, 301

O

observations, 29, 48, 68, 70, 175, 203, 243
obstruction, 154
oceans, 294

oil, 116, 144, 159, 160, 162, 163, 164, 165, 166, 167, 168, 179, 244, 252, 253, 267
oils, 128, 138, 252, 253
olfaction, 246
olfactory, 246
oligodeoxynucleotides, 221, 232
oligonucleotide arrays, 229
Oman, xiii, 297, 298, 300, 302, 304, 306, 308, 309, 313, 317, 318, 319
ontogenesis, 115
oocyte, x, 186, 189, 192, 204
oocytes, 189, 192, 205
oogenesis, 204
operator, 133
optimization, 152, 175
organ, 31
organic matter, xi, 186, 194, 201, 203, 280
organism, 2, 47, 56, 187
organization, 150, 252
organizations, xiii, 297, 299, 300, 317, 318
organoleptic, 252
osmolality, 150, 152, 173, 180
osmosis, 169, 178
osmotic, 150, 181
Ottawa, 210
oxidation, 138, 253
oxygen, 120, 127, 128, 129, 151, 152, 207, 245, 252, 256
oxygen consumption, 127, 207, 256
oxygen consumption rate, 127, 207
oyster, 56, 103, 111, 151, 153, 181, 187, 188, 189, 190, 192, 201, 202, 204, 205, 206, 207, 208, 209, 210, 211, 212, 300
oysters, x, 149, 175, 185, 187, 188, 189, 190, 192, 201, 202, 203, 205, 207, 208, 209, 210, 211
ozone, 55, 133

P

Pacific, v, 103, 108, 111, 119, 143, 156, 175, 185, 187, 202, 206, 207, 208, 209, 210, 211, 291, 309
pancreas, 52, 115
pancreatic, 18, 33, 39, 119, 214, 226, 228, 271
Paracentrotus lividus, 260
paradox, 72
parameter, ix, 9, 59, 68, 69, 70, 85, 93, 150, 193, 212, 286
parameter estimation, 69
parasite, 14, 16, 29, 32, 254
parasites, 14, 26, 27, 28, 33, 37, 52, 254
Parasites, 26
parentage, viii, 59, 60, 61, 62, 63, 65, 66, 71, 74, 77, 80, 81, 103, 104, 105, 106, 107, 108, 109, 110, 111
parents, viii, 59, 62, 66, 67, 71, 72, 73, 74, 75, 76, 77, 78, 79, 81, 85, 94, 100, 105, 106, 107, 108
Paris, 10, 108, 256, 260
particles, 116, 117, 131, 138, 177, 225
partition, 103, 110
passive, 30
paternity, viii, 59, 60, 61, 62, 63, 65, 66, 67, 71, 72, 73, 74, 75, 77, 78, 81, 82, 100, 105, 108, 109, 110, 112
pathogenesis, 14, 20, 56, 220
pathogenic, viii, 5, 13, 15, 20, 21, 25, 26, 27, 28, 32, 33, 35, 36, 37, 39, 40, 43, 44, 45, 46, 47, 56, 136, 154, 155, 164, 168, 172, 174, 178, 183
pathogenic agents, 20, 36
pathogenic mechanism, viii, 13, 33, 36
pathogens, viii, x, xi, 13, 16, 31, 33, 34, 35, 36, 37, 42, 44, 45, 46, 52, 133, 149, 154, 155, 159, 164, 168, 182, 183, 213, 217, 226, 227, 253
Pathologists, 173, 183, 226
pathology, viii, 13, 14, 29, 36, 40, 42, 51, 56, 57, 61, 178
pathways, 221, 231, 232
PCBs, 255
PCR, 8, 9, 10, 15, 17, 20, 26, 31, 36, 38, 41, 47, 50, 56, 60, 63, 224
pedigree, 60, 93, 94, 95, 96, 99, 100, 103, 104, 105, 106, 108, 109, 110, 112
Pediococcus, 155
penicillin, 164, 165, 166, 167, 168
pepsin, 118
peptidase, 118
peptide, 233
peptides, 118, 216, 218, 230, 233
perception, 267, 268
performance, ix, 55, 78, 87, 90, 113, 132, 135, 187, 238, 247, 251, 257, 264, 266, 267, 274, 275, 276, 277, 284, 286, 288, 292, 293
peripheral blood, 30, 53
permeability, 153
permit, 61, 81
Persian Gulf, xiii, 297, 298, 318, 319
personal, 4
pest control, xiii, 279, 288
Pest control, 288
pest management, 294
pesticides, 34
pests, 288
pH, 21, 47, 118, 121, 128, 132, 134, 144, 156, 180, 181, 233, 291
phage, 54

phagocytosis, 30
pharmaceutical, 282
pharmacodynamics, viii, 13, 36, 37
pharmacokinetics, viii, 13, 31, 35, 54
phenol, 209
phenotype, 63, 96
phenotypes, 96
Philippines, 178, 282, 290, 291, 295
phosphatases, 119
phosphate, 159, 160, 162, 268, 276
Phosphate, 163
phosphatidylcholine, 144
phosphatidylserine, 233
phospholipids, 117, 134, 141, 211
phosphorus, xii, 37, 263, 268, 269, 274, 277
phosphorylation, 151
photographs, 318
photosynthesis, 292
phylogenetic tree, 95
physiology, 15, 114, 115, 174, 182, 208, 211
phytoplankton, 120, 121, 137, 201, 204, 269, 277
pigs, 111
plague, 19, 41
plankton, 133, 145, 146, 202, 288
planning, 300, 302
plants, 55, 223
plasma, 32, 150, 177, 180, 252
plasma membrane, 177
plasmid, xi, 213, 214, 215, 216, 217, 219, 220, 221, 222, 223, 224, 225, 228, 231, 232, 235
Plasmids, 223
plasticity, 105
policy making, 293
pollutants, 201
pollution, 2, 280
polycheates, ix, 149, 153
polyethylene, 249
polymerase, 41, 63, 224
polymerase chain reaction, 41
polymorphism, viii, 59, 63, 65, 66, 98, 106, 110
polymorphisms, 60, 217, 218
polyploid, 110
polypropylene, 124
polysaccharide, 32, 55
polysaccharides, 131, 253
polyunsaturated fat, 146, 199, 203, 204, 212, 253, 256
polyunsaturated fatty acids, 146, 199, 203, 204, 253
pools, 105
poor, 3, 116, 128, 131, 137
population, vii, ix, 3, 4, 5, 7, 16, 38, 60, 63, 66, 67, 68, 70, 71, 72, 73, 74, 76, 77, 81, 83, 85, 86, 87, 88, 89, 90, 91, 92, 93, 94, 96, 97, 98, 101, 103, 104, 106, 107, 108, 111, 126, 127, 135, 142, 163, 248, 288, 299, 317
population density, vii, 7, 127
population growth, 126, 142
population size, 60, 90, 104, 107, 288
positive correlation, 21
potassium, 180
potential energy, 267
poultry, xii, 175, 263, 267
poverty, 4
poverty alleviation, 4
power, 63, 66, 67, 71, 77, 97, 100, 103
preference, 5, 133
present value, 250
preservative, 164
pressure, 150
prevention, 14, 32, 35, 36, 37, 38, 43, 47
prices, xii, 134, 237, 238, 250, 252
private investment, 317
private sector, 301, 302, 307, 309, 313, 320
probability, 61, 66, 67, 68, 69, 70, 71, 73, 74, 77, 79, 81, 83, 85, 87, 88, 89, 90, 91, 93, 96, 254
proband, 87
probe, 56
probiotic, 180
Probiotics, 37
producers, 273, 308
production, vii, ix, xi, xii, xiii, xiv, 1, 4, 60, 113, 114, 120, 122, 123, 124, 127, 128, 129, 130, 131, 132, 133, 134, 136, 137, 140, 141, 142, 143, 144, 145, 150, 151, 152, 159, 173, 174, 183, 186, 187, 204, 208, 211, 213, 214, 221, 232, 238, 244, 248, 250, 252, 263, 264, 265, 266, 267, 268, 269, 270, 272, 273, 274, 275, 276, 280, 282, 293, 297, 298, 299, 300, 301, 302, 303, 304, 306, 307, 308, 309, 310, 311, 312, 313, 314, 315, 316, 317, 318, 319
production costs, 134, 238, 252
production technology, ix, 113, 122, 303
productivity, 133, 315
profit, 312
profitability, 248, 250, 252, 269, 317
profits, 252
program, 77, 78, 81, 82, 83, 101, 102, 104, 105, 106, 107, 108, 110, 191, 291, 302, 303, 315
pro-inflammatory, 220
proliferation, 9, 123, 201, 219, 221, 232
promote, viii, 9, 13, 122, 131
promoter, xi, 213, 214, 215, 216, 222, 223, 224, 233
propagation, 1, 147, 151, 159, 160, 163, 301
prophylactic, 258
proposition, 265
proteases, 5, 119
protected areas, 260

protein, x, xii, 2, 3, 5, 9, 17, 30, 31, 32, 36, 41, 42, 116, 117, 118, 131, 138, 139, 147, 151, 186, 187, 190, 199, 205, 206, 214, 215, 219, 222, 223, 227, 228, 229, 231, 233, 235, 237, 238, 247, 250, 251, 252, 253, 256, 259, 261, 263, 264, 265, 266, 267, 268, 274, 275, 276, 277
protein function, 17
protein hydrolysates, 2, 118, 139
protein synthesis, 228
proteins, viii, 2, 13, 31, 42, 53, 114, 119, 131, 144, 201, 205, 206, 215, 217, 219, 221, 224, 225, 227, 228, 247, 267, 282
Proteins, 36, 205
proteobacteria, 8
proteolytic enzyme, 140
protocol, 120, 132, 151, 159, 172, 173, 184, 281, 293
protocols, ix, xiii, 113, 133, 137, 152, 172, 260, 279, 281, 282, 285
prototype, 232, 289, 290
protozoa, 145
PRT, 99
pruning, 285
Pseudomonas, 22, 43, 154, 157, 158, 159, 160, 162, 163, 167, 168, 169, 170, 177, 180
Pseudomonas aeruginosa, 158, 159, 160, 162, 167, 177
public awareness, 280
Puerto Rico, 2
purification, 36, 40, 42, 53, 207, 208
PVC, 124, 312

Q

Qatar, xiii, 297, 298, 302, 304, 306, 313, 319
quality improvement, 140

R

rabies virus, 217
radius, 286
rain, xii, 279, 280
rain forest, xii, 279, 280
rainfall, 298
random mating, 74
range, viii, 2, 7, 60, 65, 68, 70, 76, 86, 90, 91, 117, 127, 132, 136, 153, 222, 234, 244, 247, 264, 284
reading, 69, 209, 215, 257
reasoning, 96
receptors, 47, 221, 246
recessive allele, 96
recognition, 5, 221
recombinant DNA, 31, 60, 226
recombination, 107, 223
reconstruction, ix, 60, 65, 71, 74, 75, 78, 79, 93, 94, 96, 98, 99, 100, 103, 104, 107, 108, 109, 110, 111, 234
recovery, x, 182, 186, 202, 204, 280, 292, 293
recruiting, 220
recycling, 267
red blood cell, 266
red blood cell count, 266
reduction, 3, 121, 163, 168, 241, 252, 265, 266, 267, 268, 273
reef gardening, xiii, 279, 281
reef restoration, xii, 279, 281, 282, 283, 285, 290, 291, 292, 294
refining, 238
regeneration, 281, 283, 286, 294
regional, 37, 116, 301, 319
regression, 87, 191, 195, 196, 250
regression analysis, 250
regression equation, 196
regulation, xiii, 36, 122, 143, 217, 219, 222, 234, 279
regulations, 301
rehabilitation, 280, 281, 283, 290, 292
relationship, viii, x, xii, 59, 60, 62, 65, 67, 71, 83, 85, 88, 89, 90, 91, 93, 97, 105, 106, 107, 108, 111, 144, 151, 174, 183, 185, 195, 196, 197, 202, 203, 204, 208, 248, 263, 270, 273
relationships, viii, 59, 60, 61, 62, 63, 66, 67, 71, 85, 90, 91, 93, 94, 97, 98, 100, 103, 107, 108, 111, 187, 191, 196, 206, 207
relatives, 78, 95, 98
relevance, 8, 60, 193, 219
reliability, 63, 97, 245
remediation, 281, 291, 293
renal, 271
replication, 224, 234
reproduction, x, 34, 137, 141, 142, 171, 185, 187, 203, 206, 207, 209, 211, 241, 243, 244, 257, 270, 299, 301
reproductive activity, 201, 202, 210
reserves, x, 116, 128, 132, 185, 187, 191, 202, 206, 208, 211, 212, 245, 292
residues, 31, 35
resilience, 280, 292
resistance, 32, 35, 36, 55, 129, 130, 138, 224, 265, 274, 275, 277, 289, 293
resolution, 104
resources, 65, 81, 95, 114, 238, 260, 294, 299, 301, 303
respiratory, 30, 224, 235, 252
respiratory syncytial virus, 235
responsiveness, 246

restriction fragment length polymorphis, 10
retail, 154, 176
retention, 191, 251, 268
retirement, 3
RFLP, 8
rhabdoviral, xi, 213, 215, 219
rhythms, 260
ribosomal RNA, 8
rice, 53
rice field, 53
rings, 172
ripeness, x, 186, 192, 196, 205
risk, x, 70, 97, 149, 168, 169, 171, 172, 175, 179, 223, 250, 268, 291, 295
risk assessment, x, 149, 173
rivers, 309
RNA, 17, 18, 19, 221, 224, 228, 234
Roads, 209
robustness, 104
rods, 154, 155
Rome, 140, 319
rotifer, ix, 113, 120, 121, 122, 125, 126, 133, 137, 138, 140, 141, 142, 143, 144, 146, 147, 148
Rotifers, ix, 113, 120, 121, 133, 144, 246
Royal Society, 107, 111
runoff, 298
rural communities, 283

S

SA, 113
safety, xi, 33, 60, 210, 213, 215, 222, 229
sales, 238, 248, 249
saline, ix, 113, 125, 126, 137, 142, 144
salinity, 35, 120, 125, 127, 128, 130, 136, 143, 156, 181, 188, 189, 202, 206, 245, 298
salmon, 8, 10, 11, 57, 109, 118, 136, 138, 144, 146, 155, 156, 164, 173, 175, 177, 178, 181, 182, 214, 217, 222, 223, 226, 228, 229, 230, 232, 233, 304
Salmonella, 154, 181, 184
salt, 37, 141, 220
salts, 141
sample, 67, 70, 72, 77, 81, 83, 102, 150, 171, 173
sampling, viii, 8, 59, 66, 70, 75, 85, 87, 160, 161, 162, 163, 165, 166, 167, 168, 171, 188, 190
sanctions, 315
sanitation, 172, 288
SARS, 234
SARS-CoV, 234
saturated fat, 204, 254
saturated fatty acids, 204, 254
saturation, 132, 238, 245, 252
Saudi Arabia, xiii, 9, 297, 298, 302, 304, 306, 313, 314, 315, 317, 318, 319
savings, 131, 251
scaling, 4
Scanning electron, 48, 49
scanning electron microscopy, 49
scarcity, 34, 37
Schmid, 231
school, 287
science, 294, 299
scoliosis, 136
scores, 73, 78
sea cucumber, x, 45, 149, 153
sea urchin, x, 149, 153, 173, 260, 287
sea-cage farming, xiii, xiv, 297, 298, 307, 316
seafood, xiii, 114, 297, 298, 317, 318
seals, 107
search, xiv, 74, 93, 94, 114, 215, 220, 222, 223, 246, 298
searching, 70, 73, 74, 78, 82
seasonal variations, x, 185, 187, 210
seasonality, 151
seawater, xiii, 16, 32, 55, 123, 129, 142, 204, 255, 291, 297, 304
seaweed, vii, 141, 300
secrete, 118
secretion, 118
security, 14, 172, 315
sediment, 171, 180, 181, 288
sedimentation, 152
sediments, 268, 269
seed, xiii, 1, 2, 3, 114, 127, 132, 163, 266, 285, 297, 299, 307, 309, 316, 317
seeding, 285, 288
seedlings, 281, 282, 284, 285, 286, 287, 288, 290, 294, 295
seeds, ix, 149, 164, 253, 282, 284, 286, 288, 307, 310
segregation, 26, 66, 77, 92, 93
selecting, 240, 251, 288
selectivity, 144
semen, 151, 152, 169, 173, 174, 175, 176, 178, 179, 180, 181, 182
Senegal, 61, 72, 104, 109, 141
sensitivity, 36, 44, 141, 153, 168, 218, 249
separation, 40, 71
septum, 26
sequencing, 8, 9
series, 40, 71, 120, 189, 203, 208, 249, 255
serine, 47
serum, 30, 31, 53, 191, 218
serum albumin, 191
severity, 271

sewage, 154
sex, 74, 79, 107, 189, 191, 241
sexual reproduction, 122, 141, 294
shade, 287
shape, 17, 18, 19, 20, 21, 22, 23, 24, 25, 26, 159, 162, 190, 192, 239, 244, 245, 246, 258, 281
shares, 62, 71, 74, 230
sharing, 73, 84, 85, 87, 95
Sharpsnout seabream, xii, 237, 239, 240, 247, 248, 251
Shell, 208, 212
shellfish, vii, ix, 140, 149, 153, 154, 175, 181, 184
Shigella, 154
shock, 201, 225
shortage, 117, 315
shrimp, vii, ix, xiii, xiv, 2, 15, 16, 29, 31, 32, 38, 50, 51, 53, 55, 105, 107, 108, 113, 116, 122, 126, 129, 134, 137, 147, 149, 150, 151, 152, 153, 154, 157, 158, 159, 160, 161, 162, 163, 164, 165, 166, 167, 168, 169, 172, 175, 177, 179, 180, 182, 183, 184, 297, 298, 300, 303, 304, 305, 306, 307, 309, 313, 314, 317, 318
sibling, 52, 103, 142
side effects, 37
Sierra Leone, 240
sign, 219
signal transduction, 181
signalling, 221, 231
signals, 9
signs, 241
silver, 15, 43, 143, 155, 233, 239, 303, 304, 305, 310, 311, 317, 318
similarity, viii, 59, 62, 63, 86, 87, 88, 95, 96, 239
simulation, 67, 71, 73, 87, 97, 286
Singapore, 290, 304
single-nucleotide polymorphism, 103
siRNA, xi, 214, 215, 216, 224, 225, 234, 235
sites, 31, 203, 210, 281, 313, 315, 317, 318
skeleton, 141
skills, 134
skin, 14, 20, 21, 22, 23, 24, 25, 30, 37, 42, 43, 44, 46, 157, 281
Sleeping Beauty, 223
small intestine, 56
SNP, 109
sobaity seabream, xiii, 297, 304, 315, 316
society, 4, 212
sodium, 141, 160
software, viii, ix, 59, 60, 78, 80, 99, 100, 103, 106, 111, 191
soil, 158, 159, 280, 285, 287, 288, 315
soil erosion, 280
somatic cell, 110, 223
somatostatin, 30, 52
sorting, 316
South Africa, 105, 174
Southeast Asia, 283
soybean, xii, 118, 237, 251, 252, 254, 258, 259, 266, 268, 274, 276
Spain, 59, 183, 187, 211, 213, 237, 238, 247, 254, 256, 257, 258
spawning, x, xiii, 16, 38, 61, 74, 75, 81, 151, 180, 181, 186, 192, 193, 194, 196, 199, 201, 202, 205, 241, 243, 244, 257, 258, 297, 310, 313
specialization, 239
species, vii, viii, ix, x, xii, xiii, xiv, 1, 2, 5, 7, 8, 14, 16, 20, 26, 29, 32, 35, 38, 48, 49, 52, 53, 60, 61, 63, 65, 72, 74, 75, 77, 81, 95, 104, 106, 107, 109, 113, 114, 116, 117, 118, 119, 120, 121, 122, 123, 125, 126, 129, 130, 133, 135, 136, 137, 138, 150, 151, 152, 153, 154, 155, 156, 157, 159, 163, 167, 168, 172, 175, 176, 177, 178, 183, 184, 187, 202, 205, 214, 217, 218, 219, 222, 237, 238, 239, 240, 241, 243, 244, 245, 247, 249, 250, 251, 252, 254, 255, 256, 257, 259, 260, 263, 264, 267, 271, 272, 280, 281, 283, 284, 285, 286, 288, 291, 293, 294, 297, 298, 299, 300, 301, 304, 305, 306, 307, 309, 310, 312, 315, 316, 317, 318
specificity, 60, 234
spectrum, 164, 177, 190, 192
speed, 202, 246
sperm, ix, 106, 149, 150, 151, 152, 158, 159, 163, 164, 166, 169, 170, 171, 172, 174, 175, 176, 177, 178, 179, 180, 181, 182, 183
sperm preservation, x, 149, 172
spermatids, 189, 192
spermatocytes, 189, 192
spermatogonia, 189, 192
spermatozoa, 150, 151, 152, 153, 173, 174, 175, 178, 179, 180, 181, 182, 183, 189, 192, 201
spine, 140
spleen, 17, 18, 30, 31, 52
sporangium, 26
spore, 21, 22, 23, 24, 25, 154, 157, 158, 162
spreadsheets, 250
stability, 34, 153
stabilizers, 174
stages, ix, 11, 113, 120, 123, 128, 130, 133, 134, 135, 136, 141, 147, 155, 183, 190, 192, 193, 202, 206, 214, 219, 238, 244, 264, 290, 318
standard deviation, 242
standard error, 243
Standard error, 243
standards, 35, 36
Staphylococcus, 155, 158, 159, 160, 161, 162, 166, 168, 169, 170

starch, 251, 267
stars, 288
starvation, 116, 247
statistics, 67, 68, 208
stem cells, 171
sterile, 164, 170, 172
stochastic, 8, 70
stock, 3, 5, 98, 107, 109, 110, 171, 208, 249, 304, 307, 315, 317
stomach, 24, 28, 45, 115, 116, 118, 131, 135, 147, 157, 268, 272
stomach ulcer, 24
stomatitis, 171, 224, 233
storage, ix, x, 34, 121, 122, 129, 149, 150, 151, 152, 153, 158, 159, 160, 161, 162, 163, 164, 165, 166, 167, 168, 169, 170, 171, 172, 173, 174, 175, 177, 180, 182, 183, 184, 185, 202, 206, 282, 291
storms, 289
strain, 47, 48, 110, 121, 122, 132, 133, 143, 183, 311
strategies, xii, 2, 3, 85, 106, 115, 221, 238, 263, 294
strength, 304
streptococci, 155, 183
stress, 129, 130, 137, 138, 142, 147, 208
stretching, 288
subgroups, 75, 79
subsidy, 311
subsistence, 3
subsistence farming, 3
substitutes, 34, 254
substitution, xi, 214
substrates, 187, 281, 285, 287
suffering, 41, 51
sugar, 158, 160
sugars, 159, 207
sulfate, 32, 55
summer, x, 138, 139, 186, 188, 192, 193, 199, 200, 201, 204, 207, 247, 298, 315
Sun, 22, 26, 39, 43, 45, 47, 51, 53, 55, 99, 108, 235
supervision, 301
supply, xii, xiii, 2, 3, 114, 120, 125, 127, 134, 136, 201, 205, 237, 244, 264, 266, 287, 297, 299, 316
suppression, 15, 169
surface area, 146
survival, vii, 1, 3, 107, 114, 116, 117, 121, 127, 129, 130, 131, 134, 135, 136, 139, 140, 141, 143, 144, 145, 146, 202, 204, 205, 208, 209, 222, 244, 245, 246, 256, 266, 270, 272, 294
survival rate, vii, 1, 135, 245, 246
survivors, 218
susceptibility, 182, 229
sustainability, 2, 3, 4, 293
sustainable development, 14
switching, 201, 218, 223
symbiosis, 287
symmetry, 17, 18, 19, 91
syndrome, 3, 15, 16, 17, 19, 20, 21, 22, 24, 25, 29, 31, 36, 39, 40, 41, 42, 45, 46, 50, 51, 52, 54, 55, 142, 143, 146, 177
synergistic effect, 272
synthesis, 217, 224
systems, ix, xiv, 1, 105, 106, 108, 113, 114, 122, 123, 126, 132, 133, 140, 141, 144, 154, 174, 183, 216, 221, 223, 225, 226, 255, 256, 273, 289, 293, 298, 299, 304, 310, 311, 312, 315, 316

T

T cell, 216, 225
T cells, 216
tactics, 210
Taiwan, 40, 42, 51, 52, 54, 303
tanks, 60, 61, 75, 98, 121, 122, 123, 127, 128, 132, 133, 136, 138, 141, 171, 244, 245, 246, 247, 248, 254, 310, 312, 316
Tanzania, 290
targets, 224
taxonomic, 8, 26
taxonomy, 15, 26, 45
T-cell, 219, 220
technical efficiency, 298
technology, 1, 2, 35, 41, 54, 60, 115, 120, 126, 137, 139, 148, 151, 171, 219, 223, 224, 225, 233, 238, 245, 249, 259, 290, 299, 301, 303, 304, 305, 307, 309, 313, 317
technology transfer, 317
teeth, 239, 240
temperature, x, xii, 1, 10, 17, 35, 47, 121, 122, 124, 127, 128, 132, 134, 136, 142, 151, 152, 156, 159, 160, 174, 177, 185, 188, 189, 201, 202, 203, 204, 206, 207, 208, 209, 211, 218, 241, 243, 244, 245, 247, 248, 249, 252, 256, 263, 269, 270, 272, 298
temperature gradient, 10
temporal, 10, 74, 224, 233
termination codon, 223
Thailand, 110, 149, 155, 290
theoretical assumptions, viii, 59, 77, 104
theory, 31, 94, 208, 293
therapeutic, 182
therapy, 231
threat, 114, 164, 295
threats, 14, 280, 290
threshold, vii, 7, 127
threshold level, 127
thresholds, 71, 73, 91, 100
tides, 289

Tilapia, xiii, 25, 46, 156, 181, 182, 297, 304, 311, 312, 314
timber, 281, 282, 283, 284, 290
time, vii, ix, 5, 9, 15, 16, 17, 30, 32, 33, 59, 63, 85, 92, 121, 123, 124, 126, 127, 128, 132, 134, 135, 141, 153, 155, 168, 175, 187, 190, 205, 218, 223, 227, 238, 241, 248, 252, 255, 265, 273, 275, 285, 286, 287, 298, 313, 317
timing, 119, 133, 187, 291
TIR, 232
tissue, x, 186, 189, 190, 201, 202, 203, 204, 209, 210, 217, 219, 223, 253, 255, 260, 275, 287
TLR, 221, 231
TLR2, 221
TLR3, 221
TLR4, 221
TLR9, 221
TNF, 219, 220
Togo, 208
Tokyo, 177, 178, 180, 183
total energy, 245
total plasma, 30
total product, xiii, 187, 298, 310, 313
tourism, 281, 282, 315
toxicity, 37, 54, 223, 266
toxin, xii, 263, 272, 275
toxins, 61
trade, 187, 282, 283, 286
trade-off, 187, 286
training, 152, 300
traits, 26, 111, 239
transcription, 219, 222, 234
transcription factors, 219
transcripts, 223, 228
transfection, 222, 224
transformation, 76, 85, 234, 245
transformations, 211
transgenic, 233
transition, ix, 114, 116, 119, 135, 246, 259
transition period, 119
transitions, 133
translation, 224, 234
translocation, 233
transmission, 62, 71, 108, 164, 168, 172, 174, 182, 183, 234
transplantation, 281, 282, 283, 285, 286, 287, 291, 292, 294
transport, 137, 151, 171
transportation, 174
transposases, 224
transposons, xi
trees, xiii, 104, 279, 280, 281, 282, 283, 290
trend, xiv, 14, 264, 298, 317, 318
trial, 136, 188, 220, 306
trial and error, 220
triglycerides, 138
triploid, 30, 52, 136, 146, 211
tropical forest, 280
tropical forests, 280
tropical rain forests, 291
trust, 283
trypsin, 118, 119
tumors, 232
Tunisia, v, x, 140, 185, 187, 207
turbulence, 171
Turkey, 175, 253
turtle, 20, 41, 43
type II error, 91

U

UAE, xiii, 297, 298, 303, 304, 306, 310, 315, 316, 317, 318
UK, 4, 208, 211
ulcer, 22, 23, 24, 25, 26, 43, 44, 45, 46
ulceration, 21, 42, 44
ultrasound, 227
ultrastructure, 15, 29, 40, 51, 180
ultraviolet, 3
ultraviolet irradiation, 3
UN, 303, 308
uncertainty, 8, 73, 100, 292
UNESCO, 260, 261
uniform, 70, 282
United Arab Emirates, xiii, 297, 298, 318
United Kingdom, 146, 206, 209, 261, 275
United Nations, 256, 319
United States, xii, 144, 145, 229, 230, 234, 263, 264, 275, 320
urea, 30, 158, 160
ureter, 151
urine, 151, 175
USAID, 315, 320
USDA, 291, 292, 294, 295
users, 81
Utah, 132, 180
UV, 106, 132, 133

V

vaccination, xi, 32, 34, 55, 56, 176, 213, 214, 215, 216, 217, 218, 219, 220, 221, 222, 223, 224, 225, 227, 228, 229, 231, 232, 235
Vaccination, v, xi, 213, 216, 220, 225, 231